A TEXTBOOK OF
MODERN TOXICOLOGY

A TEXTBOOK OF MODERN TOXICOLOGY

FOURTH EDITION

Edited by

Ernest Hodgson
North Carolina State University
Raleigh, North Carolina

A JOHN WILEY & SONS, INC., PUBLICATION

Published by John Wiley & Sons, Inc., Hoboken, New Jersey.
Published simultaneously in Canada.

For general information on our other products and services or for technical support, please contact our Customer Care Department within the United States at (800) 762-2974, outside the United States at (317) 572-3993 or fax (317) 572-4002.

Wiley also publishes its books in a variety of electronic formats. Some content that appears in print may not be available in electronic formats. For more information about Wiley products, visit our web site at www.wiley.com.

Library of Congress Cataloging-in-Publication Data:

A textbook of modern toxicology / edited by Ernest Hodgson. —4th ed.
   p. cm.
 ISBN 978-0-470-46206-5 (cloth)
 1. Toxicology.   I. Hodgson, Ernest, 1932–
  RA1211.H62 2010
  615.9—dc22

                                                          2009045883

Printed in the United States of America

10 9 8 7 6 5 4 3 2 1

CONTENTS

There are some excellent general reference works in toxicology, including *Casarett and Doull's Toxicology, 6th edition,* edited by Curt Klaassen, and the 13-volume *Comprehensive Toxicology,* the second edition currently being edited by Charlene McQueen, as well as many specialized monographs on particular topics. However, the scarcity of textbooks designed for teacher and student to use in the classroom setting that impelled us to produce editions 1 through 3 of this work is still apparent and the choice continues to be limited. The authors are, or have been, involved in teaching general toxicology at North Carolina State University and thus have insights into the actual teaching process and in the broader scope of toxicology as well as the subject matter of their areas of specialization.

Rapid advances are occurring in toxicology, particularly in the molecular and integrative aspects, and we hope these are reflected in this textbook. As an aid to students and teaching faculty, we have added sample questions to each chapter. Answering these questions not only indicates that the material presented has been understood but is, in itself, a learning experience.

At North Carolina State University, we continue to teach a course in general toxicology (TOX801) that is open to graduate students and undergraduate upper-classmen. Our experience leads us to believe that this textbook is suitable, in the junior or senior year, for undergraduate students with some background in chemistry, biochemistry, and animal physiology. For graduate students, it is intended to lay the foundation for subsequent specialized courses in toxicology, such as those in biochemical and molecular toxicology, environmental toxicology, chemical carcinogenesis, risk assessment, and so forth.

We share the view that an introductory text must present all of the necessary fundamental information to fulfill this purpose, but in as uncomplicated a manner as possible. To enhance readability, references have been omitted from the text, although Suggested Reading or Bibliography is recommended at the end of each chapter.

As with previous editions, the amount of material and the detail with which some of it is presented, is more than is needed for the average general toxicology course. This, however, will permit each instructor to select and emphasize those areas they feel need particular emphasis. The obvious biochemical and molecular bias of some chapters is not accidental; rather, it is based on the philosophy that progress in toxicology continues to depend on further understanding of the fundamental basis of toxic action at the cellular and molecular levels. The depth of coverage of each topic represents that chapter author's judgment of the amount of material appropriate to the beginning level as compared to that appropriate to a more advanced course or text such as Smart and Hodgson, *Molecular and Biochemical Toxicology*, 4th edition (John Wiley and Sons, 2008).

Thanks to all of the authors and to the students and faculty of the Department of Environmental and Molecular Toxicology at North Carolina State University. Particular thanks to Jonathan Rose of John Wiley and Sons, who facilitated the project by his hard work, his goodwill and, not least, for his patience.

Raleigh, North Carolina   ERNEST HODGSON
March 2010

CONTRIBUTORS

Jill A. Barnes, Molecular Biomedical Sciences, College of Veterinary Medicine, North Carolina State University, Raleigh, North Carolina

Ronald E. Baynes, Center for Chemical Toxicology Research and Pharmacokinetics, North Carolina State University, Raleigh, North Carolina

Bonita L. Blake, Department of Pharmacology and Neuroscience Center, University of North Carolina at Chapel Hill, Chapel Hill, North Carolina

James C. Bonner, Department of Environmental and Molecular Toxicology, North Carolina State University, Raleigh, North Carolina

David B. Buchwalter, Department of Environmental and Molecular Toxicology, North Carolina State University, Raleigh, North Carolina

W. Gregory Cope, Department of Environmental and Molecular Toxicology, North Carolina State University, Raleigh, North Carolina

Helen Cunny, National Institute for Environmental Health Sciences, Research Triangle Park, North Carolina

Ernest Hodgson, Department of Environmental and Molecular Toxicology, North Carolina State University, Raleigh, North Carolina

Chris Hofelt, Department of Environmental and Molecular Toxicology, North Carolina State University, Raleigh, North Carolina

Seth W. Kullman, Department of Environmental and Molecular Toxicology, North Carolina State University, Raleigh, North Carolina

Gerald A. LeBlanc, Department of Environmental and Molecular Toxicology, North Carolina State University, Raleigh, North Carolina

Carolyn J. Mattingly, Mount Desert Island Biological Laboratory, Salisbury Cove, Maine

Joel N. Meyer, Nicholas School of the Environment, Duke University, Durham, North Carolina

Sharon A. Meyer, Department of Toxicology, University of Louisiana, Monroe, Louisiana

Heather Patisaul, Department of Biology, North Carolina State University, Raleigh, North Carolina

Randy L. Rose, (deceased), Department of Environmental and Molecular Toxicology, North Carolina State University, Raleigh, North Carolina

MaryJane Selgrade, National Health and Environmental Effects Research Laboratory, United States Environmental Protection Agency, Research Triangle Park, North Carolina

Damian Shea, Departments of Biology and Environmental and Molecular Toxicology, North Carolina State University, Raleigh, North Carolina

Robert C. Smart, Department of Environmental and Molecular Toxicology, North Carolina State University, Raleigh, North Carolina

Joan Tarloff, Department of Pharmaceutical Sciences, Philadelphia College of Pharmacy, University of the Sciences in Philadelphia, Philadelphia, Pennsylvania

Andrew D. Wallace, Department of Environmental and Molecular Toxicology, North Carolina State University, Raleigh, North Carolina

Ida M. Washington, Department of Comparative Medicine, University of Washington School of Medicine, Seattle, Washington

Andrew Whitehead, Department of Biological Sciences, Louisiana State University, Baton Rouge, Louisiana

PART I

INTRODUCTION

Introduction to Toxicology

ERNEST HODGSON

Since the publication of the 3rd edition of this textbook (2004) major changes have been initiated in toxicology as the tools of molecular biology, genomics, proteomics, metabolomics, bioinformatics, and systems biology are increasingly brought to bear on the critical areas of mode of action, toxicity testing, and risk analysis. Chapter 2 provides information on new methodology and Part VIII—New Approaches in Toxicology is composed of two chapters of commentary on the current and expected impact of these new methods. While the traditional aspects and subdisciplines of toxicology, as outlined below, are still active and viable, during the next few years all are likely to be impacted and their development accelerated by these new approaches.

1.1 DEFINITION AND SCOPE

Toxicology can be defined as that branch of science that deals with poisons, and a poison can be defined as any substance that causes a harmful effect when administered, either by accident or by design, to a living organism. By convention, toxicology also includes the study of harmful effects caused by physical phenomena, such as radiation of various kinds, noise, and so on. In practice, however, many complications exist beyond these simple definitions, both in bringing more precise definition to the meaning of poison and to the measurement of toxic effects. Broader definitions of toxicology, such as "the study of the detection, occurrence, properties, effects, and regulation of toxic substances," although more descriptive, do not resolve the difficulties. Toxicity itself can rarely, if ever, be defined as a single molecular event, but is, rather, a cascade of events starting with exposure, proceeding through distribution and metabolism, and ending with interaction with cellular macromolecules (usually DNA or protein) and the expression of a toxic end point (Figure 1.1). This sequence may be mitigated by excretion and repair. It is to the complications, and to the science behind them and their resolution, that this textbook is dedicated, particular to the *how* and *why* certain substances cause disruptions in biologic systems that result in toxic effects. Taken together, these

A Textbook of Modern Toxicology, Fourth Edition. Edited by Ernest Hodgson
Copyright © 2010 John Wiley & Sons, Inc.

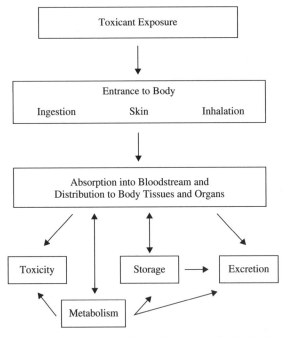

Figure 1.1   Fate and effect of toxicants in the body.

difficulties and their resolution circumscribe the perimeter of the science of toxicology.

The study of toxicology serves society in many ways, not only to protect humans and the environment from the deleterious effects of toxicants, but also to facilitate the development of more selective toxicants such as anticancer and other clinical drugs, pesticides, and so forth.

Poison is a quantitative concept, almost any substance being harmful at some doses but, at the same time, being without harmful effect at some lower dose. Between these two limits, there is a range of possible effects, from subtle long-term chronic toxicity to immediate lethality. Vinyl chloride may be taken as an example. It is a potent hepatotoxicant at high doses, a carcinogen with a long latent period at lower doses, and apparently without effect at very low doses. Clinical drugs are even more poignant examples because, although therapeutic and highly beneficial at some doses, they are not without deleterious side effects and may be lethal at higher doses. Aspirin (acetylsalicylic acid), for example, is a relatively safe drug at recommended doses and is taken by millions of people worldwide. At the same time, chronic use can cause deleterious effects on the gastric mucosa, and it is fatal at a dose of about 0.2–0.5 g/kg. Approximately 15% of reported accidental deaths from poisoning in children result from ingestion of salicylates, particularly aspirin.

The importance of dose is well illustrated by metals that are essential in the diet but are toxic at higher doses. Thus, iron, copper, magnesium, cobalt, manganese, and zinc can be present in the diet at too low a level (deficiency), at an appropriate level (maintenance), or at too high a level (toxic). The question of dose–response relationships is fundamental to toxicology (see Section 1.4).

The definition of a poison, or toxicant, also involves a qualitative biological aspect because a compound, toxic to one species or genetic strain, may be relatively harmless to another. For example, carbon tetrachloride, a potent hepatotoxicant in many species, is relatively harmless to the chicken. Certain strains of rabbit can eat *Belladonna* with impunity while others cannot. Compounds may be toxic under some circumstances but not others or, perhaps, toxic in combination with another compound but nontoxic alone. The methylenedioxyphenyl insecticide synergists, such as piperonyl butoxide, are of low toxicity to both insects and mammals when administered alone, but are, by virtue of their ability to inhibit xenobiotic-metabolizing enzymes, capable of causing dramatic increases in the toxicity of other compounds.

The measurement of toxicity is also complex. Toxicity may be acute or chronic, and may vary from one organ to another as well as with age, genetics, gender, diet, physiological condition, or the health status of the organism. As opposed to experimental animals, which are highly inbred, genetic variation is a most important factor in human toxicity since the human population is highly outbred and shows extensive genetic variation. Even the simplest measure of toxicity, the LD_{50} (lethal dose; the dose required to kill 50% of a population under stated conditions) is highly dependent on the extent to which the above variables are controlled. LD_{50} values, as a result, vary markedly from one laboratory to another.

Exposure of humans and other organisms to toxicants may result from many activities: intentional ingestion, occupational exposure, environmental exposure, as well as accidental and intentional (suicidal or homicidal) poisoning. The toxicity of a particular compound may vary with the portal of entry into the body, whether through the alimentary canal, the lungs, or the skin. Experimental methods of administration such as injection may also give highly variable results; thus, the toxicity from intravenous (IV), intraperitoneal (IP), intramuscular (IM), or subcutaneous (SC) injection of a given compound may be quite different. Thus, toxicity may vary as much as 10-fold with the route of administration. Following exposure, there are multiple possible routes of metabolism, both detoxifying and activating, and multiple possible toxic end points (Figure 1.1).

Attempts to define the scope of toxicology, including that which follows, must take into account that the various subdisciplines are not mutually exclusive and are frequently interdependent. Due to overlapping of mechanisms as well as use and chemical classes of toxicants, clear division into subjects of equal extent or importance is not possible.

Many specialized terms are used in the various subdisciplines of toxicology as illustrated in the *Dictionary of Toxicology*, 2nd edition (Hodgson et al., 1998). However, some terms are of particular importance to toxicology in general; these and some more recent terms are defined in the glossary to be found at the end of this volume.

Although B through F (following) include subdivisions that encompass essentially all of the many aspects of toxicology, there are two new approaches (A, following) that serve to integrate the discipline as a whole.

A. Integrative Approaches

    1. *Bioinformatics.* In the narrow and original meaning, bioinformatics was the application of information technology to molecular biology. While this is still

the most important aspect of bioinformatics, it is increasingly applied to other fields of biology, including molecular and other aspects of toxicology. It is characterized by computationally intensive methodology and includes the design of large databases and the development of techniques for their manipulation, including data mining.

2. *Systems Biology.* Although systems biology has been defined in a number of ways, some involving quite simple approaches to limited problems, in the currently most commonly accepted sense, it is an integrative approach to biological structure and function that will be of increasing importance to biology in general and toxicology in particular. In large part, biology has been reductionist throughout its history, studying organs as components of organisms, cells as components of organs, enzymes, nucleic acids, and so on, as components of cells, with the goal of describing function at the molecular level. Systems biology, on the other hand, is holistic and has the objective of discerning interactions between components of biological systems and describing these interactions in rigorous mathematical models. Furthermore, the proponents of systems biology aim to integrate these models at higher and higher levels or organization in order to develop an integrated model of the entire organism.

Clearly, systems biology is in its infancy; however, the ultimate value of having an integrative model that could clarify all of the effects, from the most proximate to the ultimate, of a toxicant on a living organism, will provide enormous benefits not only for fundamental studies but in such applied areas as human health risk assessment.

B. Modes of Toxic Action. This includes the consideration, at the fundamental level of organ, cell, and molecular function, of all events leading to toxicity *in vivo*: uptake, distribution, metabolism, mode of action, and excretion. The term mechanism of toxic action is now more generally used to describe an important molecular event in the cascade of events leading from exposure to toxicity, such as the inhibition of acetylcholinesterase in the toxicity of organophosphorus and carbamate insecticides. Important aspects include the following:

1. *Biochemical and molecular toxicology* consider events at the biochemical and molecular levels, including enzymes that metabolize xenobiotics, generation of reactive intermediates, interaction of xenobiotics or their metabolites with macromolecules, gene expression in metabolism and modes of action, signaling pathways in toxic action, and so on.

2. *Behavioral toxicology* deals with the effects of toxicants on animal and human behavior, which is the final integrated expression of nervous function in the intact animal. This involves both the peripheral and central nervous systems, as well as effects mediated by other organ systems, such as the endocrine glands.

3. *Nutritional toxicology* deals with the effects of diet on the expression of toxicity and with the mechanisms of these effects.

4. *Carcinogenesis* includes the chemical, biochemical, and molecular events that lead to the large number of effects on cell growth collectively known as cancer.

5. *Teratogenesis* includes the chemical, biochemical, and molecular events that lead to deleterious effects on development.

6. *Mutagenesis* is concerned with toxic effects on the genetic material and the inheritance of these effects.

7. *Organ toxicity* considers effects at the level of organ function (e.g., neurotoxicity, hepatotoxicity, and nephrotoxicity).

C. Measurement of Toxicants and Toxicity. These important aspects deal primarily with analytical chemistry, bioassay, and applied mathematics, and are designed to provide the methodology to answer certain critically important questions. Is the substance likely to be toxic? What is its chemical identity? How much of it is present? How can we assay its toxic effect, and what is the minimum level at which this toxic effect can be detected? A number of important fields are included:

1. *Analytical toxicology* is a branch of analytical chemistry concerned with the identification and assay of toxic chemicals and their metabolites in biological and environmental materials.

2. *Genomics.* The sometimes stated distinction that genomics deals with genomes while molecular biology deals with single genes is unrealistic and unnecessary; it is more appropriate to regard genomics as an aspect of molecular biology that deals not only with genomes and gene expression but also such important aspects as genetic polymorphisms, particularly single nucleotide polymorphisms (SNPs). Techniques, such as microarrays, are now available to examine simultaneously the expression of very large numbers of genes.

3. *Proteomics* deals with the protein complement of organisms, the entire complement being known as the proteome. Thus, while genomics is concerned with gene expression, proteomics examines the products of the expressed genes.

4. *Metabolomics* is the next step in the sequence from genomics through proteomics and is concerned with the profile of small molecules produced by the metabolic processes of an organism. Changes in the profile in response to chemical stress are of importance to both fundamental and applied toxicology.

5. *Toxicity testing* involves the use of living systems to estimate toxic effects. It covers the gamut from short-term tests for genotoxicity such as the Ames test and cell culture techniques to the use of intact animals for a variety of tests from acute toxicity to lifetime chronic toxicity. Although the term "bioassay" is used properly only to describe the use of a living organism to quantitate the amount of a particular toxicant present, it is frequently used to describe any *in vivo* toxicity test.

6. *Toxicologic pathology* is that branch of pathology that deals with the effects of toxic agents manifested as changes in subcellular, cellular, tissue, or organ morphology.

7. *Structure-activity* studies are concerned with the relationship between the chemical and physical properties of a chemical and toxicity and, particularly, the use of such relationships as predictors of toxicity.

8. *Biomathematics and statistics* relate to many areas of toxicology. They deal with data analysis, the determination of significance, and the formulation of risk estimates and predictive models.

9. *Epidemiology*, as it applies to toxicology, is of great importance as it deals with the relationship between chemical exposure and human disease in actual populations, rather than in experimental settings.

D. Applied Toxicology. This includes the various aspects of toxicology as they apply in the field or the development of new methodology or new selective toxicants for early application in the field setting.

1. *Clinical toxicology* is the diagnosis and treatment of human poisoning.

2. *Veterinary toxicology* is the diagnosis and treatment of poisoning in animals other than humans, particularly livestock and companion animals, but not excluding feral species. Other important concerns of veterinary toxicology are the possible transmission of toxins to the human population in meat, fish, milk, and other foodstuffs, and the care and ethical treatment of experimental animals.

3. *Forensic toxicology* concerns the medicolegal aspects, including detection of poisons in clinical and other samples.

4. *Environmental toxicology* is concerned with the movement of toxicants and their metabolites and degradation products in the environment and in food chains, and with the effect of such contaminants on individuals and, especially, populations. Because of the large number of industrial chemicals and possibilities for exposure, as well as the mosaic of overlapping laws that govern such exposure, this area of applied toxicology is well developed.

5. *Industrial toxicology* is a specific area of environmental toxicology that deals with the work environment and constitutes a significant part of *industrial hygiene*.

E. Chemical Use Classes. This includes the toxicology aspects of the development of new chemicals for commercial use. In some of these use classes, toxicity, at least to some organisms, is a desirable trait; in others, it is an undesirable side effect. Use classes are not composed entirely of synthetic chemicals; many natural products are isolated and are used for commercial and other purposes and must be subjected to the same toxicity testing as that required for synthetic chemicals. Examples of such natural products include the insecticide, pyrethrin, the clinical drug, digitalis, and the drug of abuse, cocaine.

1. *Agricultural chemicals* include many compounds, such as insecticides, herbicides, fungicides, and rodenticides, in which toxicity to the target organism is a desired quality whereas toxicity to "nontarget species" is to be avoided. Development of such selectively toxic chemicals is one of the applied roles of comparative toxicology.

2. *Clinical drugs* are properly the province of pharmaceutical chemistry and pharmacology. However, toxic side effects and testing for them clearly fall within the science of toxicology.

3. *Drugs of abuse* are chemicals taken for psychological or other effects and may cause dependence and toxicity. Many of these are illegal but some are of clinical significance when used correctly.

4. *Food additives* are of concern to toxicologists only when they are toxic or being tested for possible toxicity.

5. *Industrial chemicals* are so numerous that testing them for toxicity or controlling exposure to those known to be toxic is a large area of toxicological activity.

6. *Naturally occurring substances* include many phytotoxins, mycotoxins, minerals, and so on, all occurring in the environment. The recently expanded and now extensive use of herbal "remedies" and dietary supplements has become a cause of concern for toxicologists and regulators. Not only is their efficacy frequently dubious, but their potential toxicity is also largely unknown.

7. *Combustion products* are not properly a use class but are a large and important class of toxicants, generated primarily from fuels and other industrial chemicals.

F. Regulatory Toxicology. These aspects, concerned with the formulation of laws, and regulations authorized by laws, are intended to minimize the effect of toxic chemicals on human health and the environment.

1. *Legal aspects* are the formulation of laws and regulations and their enforcement. In the United States, enforcement falls under such government agencies as the Environmental Protection Agency (EPA), the Food and Drug Administration (FDA) and the Occupational Safety and Health Administration (OSHA). Similar government agencies exist in many other countries.

2. *Risk assessment* is the definition of risks, potential risks, and the risk–benefit equations necessary for the regulation of toxic substances. Risk assessment is logically followed by *risk communication* and *risk management*. Risk assessment, risk communication, and risk management are frequently referred to as *risk analysis*.

1.2 RELATIONSHIP TO OTHER SCIENCES

Toxicology is a highly eclectic science and human activity drawing from, and contributing to, a broad spectrum of other sciences and human activities. At one end of the spectrum are those sciences that contribute their methods and philosophical concepts to serve the needs of toxicologists, either in research or in the application of toxicology to human affairs. At the other end of the spectrum are those sciences to which toxicology contributes.

In the first group, chemistry, biochemistry, pathology, physiology, epidemiology, immunology, ecology, and biomathematics have long been important while molecular biology has, in the last two or three decades, contributed to dramatic advances in toxicology.

In the group of sciences to which toxicology contributes significantly are such aspects of medicine as forensic medicine, clinical toxicology, pharmacy, and pharmacology, public health, and industrial hygiene. Toxicology also contributes in an important way to veterinary medicine, and to such aspects of agriculture as the development and safe use of agricultural chemicals. The contributions of toxicology to environmental studies have become increasingly important in recent years.

Clearly, toxicology is preeminently an applied science, dedicated to the enhancement of the quality of life and the protection of the environment. It is also much more. Frequently, the perturbation of normal life processes by toxic chemicals enables us to learn more about the life processes themselves. The use of dinitrophenol and other uncoupling agents to study oxidative phosphorylation and the use of α-amanitin to study RNA polymerases are but two of many examples. The field of toxicology has expanded enormously in recent decades, both in numbers of toxicologists and in accumulated knowledge. This expansion has brought a change from a primarily descriptive science to one which utilizes an extensive range of methodology to study the mechanisms involved in toxic events.

1.3   A BRIEF HISTORY OF TOXICOLOGY

Much of the early history of toxicology has been lost, and in much that has survived, toxicology is of almost incidental importance in manuscripts dealing primarily with medicine. Some, however, deal more specifically with toxic action or with the use of poisons for judicial execution, suicide, or political assassination. Regardless of the paucity of the early record, and given the need for people to avoid toxic animals and plants, toxicology must be one of the oldest practical sciences.

The Egyptian papyrus, *Ebers*, dating from about 1500 BC, must rank as the earliest surviving pharmacopeia, and the surviving medical works of Hippocrates, Aristotle, and Theophrastus, published during the period 400–250 BC, all include some mention of poisons. The early Greek poet Nicander treats, in two poetic works, animal toxins (Therica) and antidotes to plant and animal toxins (Alexipharmica). The earliest surviving attempt to classify plants according to their toxic and therapeutic effects is that of Dioscorides, a Greek employed by the Roman emperor Nero about 50 AD.

There appear to have been few advances in either medicine or toxicology between the time of Galen (131–200 AD) and that of Paracelsus (1493–1541). It was the latter who, despite frequent confusion between fact and mysticism, laid the groundwork for the later development of modern toxicology by recognizing the importance of the dose–response relationship. His famous statement "All substances are poisons; there is none that is not a poison. The right dose differentiates a poison and a remedy" succinctly summarizes that concept. His belief in the value of experimentation was also a break with earlier tradition.

There were some important developments during the eighteenth century. Probably the best known is the publication of Ramazini's *Diseases of Workers* in 1700 which led to his recognition as the father of occupational medicine. The correlation between the occupation of chimney sweeps and scrotal cancer by Percival Pott in 1775 is almost as well-known although it was foreshadowed by Hill's correlation of nasal cancer and snuff use in 1761.

Orfila, a Spaniard working at the University of Paris in the early nineteenth century, is generally regarded as the father of modern toxicology. He clearly identified toxicology as a separate science and, in 1815, published the first book devoted exclusively to toxicology. An English translation in 1817 was entitled *A General System of Toxicology or, A Treatise on Poisons, Found in the Mineral, Vegetable and Animal Kingdoms, Considered in Their Relations with Physiology, Pathology and Medical Jurisprudence.* Workers of the late nineteenth century who produced treatises on toxicology include Christian, Kobert, and Lewin. The recognition of the site of action of curare by Claude Bernard (1813–1878) began the modern study of the mechanisms of toxic action. Since then, advances have been numerous—too numerous to list in detail. They have increased our knowledge of the chemistry of poisons, the treatment of poisoning, the analysis of toxicants and toxicity, as well as modes of toxic action and detoxication processes, and specific molecular events in the poisoning process.

With the publication of her controversial book, *The Silent Spring*, in 1962, Rachel Carson became an important influence in initiating the modern era of environmental toxicology. Her book emphasized stopping the widespread, indiscriminate use of pesticides and other chemicals and advocated use patterns based on sound ecology. Although sometimes inaccurate and with arguments often based on frankly anecdotal evidence, her book is often credited as the catalyst leading to the establishment of the U.S. EPA and she is regarded by many as the mother of the environmental movement.

It is clear, however, that since the 1960s, toxicology has entered a phase of rapid development and has changed from a science that was largely descriptive to one in which the importance of mechanisms of toxic action is generally recognized. Since the 1970s, with increased emphasis on the use of the techniques of molecular biology, the pace of change has increased even further, and significant advances have been made in many areas, including chemical carcinogenesis and xenobiotic metabolism, among many others.

1.4   DOSE–RESPONSE RELATIONSHIPS

As mentioned previously, toxicity is a relative event that depends not only on the toxic properties of the chemical and the dose administered but also on individual and interspecific variation in the metabolic processing of the chemical. The first recognition of the relationship between the dose of a compound and the response elicited has been attributed to Paracelsus (see Section 1.3). It is noteworthy that his statement includes not only that all substances can be toxic at some dose, but that "the right dose differentiates a poison from a remedy," a concept that is the basis for pharmaceutical therapy.

A typical dose–response curve is shown in Figure 1.2, in which the percentage of organisms or systems responding to a chemical is plotted against the dose. For many chemicals and effects, there will be a dose below where no effect or response is observed. This is known as the *threshold dose*. This concept is of significance because it implies that a *no observed effect level* (NOEL) can be determined and that this value can be used to determine the safe intake for food additives and contaminants such as pesticides. Although this is generally accepted for most types of chemicals and toxic effects, for chemical carcinogens acting by a genotoxic mechanism, the shape of the curve is controversial, and for regulatory purposes, their effect is assumed to be a no-threshold phenomenon. Dose–response relationships are discussed in more detail in Chapter 10—Acute Toxicity and Chapter 20—Toxicity Testing.

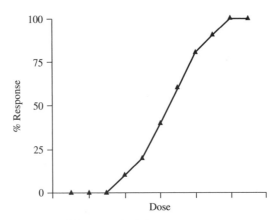

Figure 1.2   A typical dose–response curve.

1.5   SOURCES OF TOXIC COMPOUNDS

Given the enormous number of toxicants, it is difficult to classify them, either chemically, by function, or by mode of action since many of them would fall into several classes. Some are natural products, many are synthetic organic chemicals of use to society, while some are byproducts of industrial processes and waste disposal. It is useful, however, to categorize them according to the expected routes of exposure or according to their uses.

A. Exposure Classes. Exposure classes include toxicants in food, air, water, and soil as well as toxicants characteristic of domestic and occupational settings. Toxicant use classes are described in detail in Chapter 3.
B. Use Classes. Use classes include drugs of abuse, therapeutic drugs, agricultural chemicals, food additives and contaminants, metals, solvents, combustion products, cosmetics, and toxins. Some of these, such as combustion products, are the products of use processes rather than being use classes. All of these groups of chemicals are discussed in detail in Chapter 4.

1.6   MOVEMENT OF TOXICANTS IN THE ENVIRONMENT

Chemicals released into the environment rarely remain in the form, or at the location, of release. For example, agricultural chemicals used as sprays may drift from the point of application as air contaminants or enter run-off water as water contaminants. Many of these chemicals are susceptible to fungal or bacterial degradation and are rapidly detoxified, frequently being broken down to products that can enter the carbon, nitrogen, and oxygen cycles. Other agricultural chemicals, particularly halogenated organic compounds, are recalcitrant to a greater or lesser degree to metabolism by microorganisms and persist in soil and water as contaminants; they may enter biologic food chains and move to higher trophic levels or persist in processed crops as food contaminants. This same scenario is applicable to

Figure 1.3 Examples of simplified food chains.

any toxicant released into the environment either for a specific use or as a result of industrial processes, combustion, and so on. Chemicals released into the environment are also susceptible to chemical degradation, a process often stimulated by ultraviolet light.

Although most transport between inanimate phases of the environment results in wider dissemination, but, at the same time, dilution of the toxicant in question, transfer between living creatures may result in increased concentration or bioaccumulation. Lipid-soluble toxicants are readily taken up by organisms following exposure in air, water, or soil. Unless rapidly metabolized, they persist in the tissues long enough to be transferred to the next trophic level. At each level, the lipophilic toxicant tends to be retained while the bulk of the food is digested, utilized, and excreted, thus increasing the toxicant concentration. At some point in the chain, the toxicant can become deleterious, particularly if the organism at that level is more susceptible than those at the level preceding it. Thus, the eggshell thinning in certain raptorial birds was almost certainly due to the uptake of DDT (1,1,1-trichloro-2,2-bis(4-chlorophenyl) ethane) and DDE (1,1-dichloro-2,2-bis(4-chlorophenyl) ethane) and their particular susceptibility to this type of toxicity. Simplified food chains are shown in Figure 1.3.

It is clear that such transport can occur through both aquatic and terrestrial food chains, although in the former, higher members of the chains, such as fish, can accumulate large amounts of toxicants directly from the medium. This accumulation occurs because of the large area of gill filaments, their intimate contact with the water, and the high flow rate of water over them. Given these characteristics and a toxicant with a high partition coefficient between lipid membranes and water, considerable uptake is inevitable.

These and all other environmental aspects of toxicology are discussed in Part VII.

BIBLIOGRAPHY AND SUGGESTED READING

Hodgson, E., R. B. Mailman, and J. E. Chambers, eds. *Dictionary of Toxicology*, 2nd ed. London: Macmillan, 1998, 504 pp.

Klaassen, C. D., ed. *Casarett and Doull's Toxicology: The Basic Science of Poisons*, 6th ed. New York: McGraw-Hill, 2001, 1236 pp.

Smart, R. C. and E. Hodgson, eds. *Molecular and Biochemical Toxicology*, 4th ed. Hoboken, NJ: Wiley, 2008, 901 pp.

Wexler, P., ed. *Encyclopedia of Toxicology*, 2nd ed. Oxford, UK: Elsevier, 2005, 4 volumes.

SAMPLE QUESTIONS

1. Briefly define the following terms:
   a. Toxicology
   b. Poison
   c. Genomics
   d. Proteomics
   e. Metabolomics

2. Toxicity has been described as a cascade of events initiated by exposure to a harmful chemical. Name the principal steps in this cascade.

3. Name and define three important chemical use classes.

Introduction to Biochemical and Molecular Methods in Toxicology

ERNEST HODGSON, GERALD A. LEBLANC, SHARON A. MEYER, and ROBERT C. SMART

2.1 INTRODUCTION

This chapter is not designed to summarize biochemical methods long used in toxicology such as colorimetric and radiometric methods for the investigation of xenobiotic metabolism, either *in vivo* or *in vitro*, but rather to give a brief summary of the methods of molecular, biochemical, and cellular biology that have become, more recently, of critical importance in toxicological research. Areas of methodology that have achieved prominence since the publication of the 3rd edition of this textbook (Hodgson, 2004) include proteomics (Section 2.5, below), metabolomics (Section 2.6, below), and bioinformatics (Section 2.7, below). The current chapter owes much to Chapters 2 through 8 of *Molecular and Biochemical Toxicology* (Smart and Hodgson, 4th edition, 2008; see Bibliography), and the reader is referred to these chapters for additional information.

2.2 CELL CULTURE TECHNIQUES

While scientists have had the ability to culture many unicellular organisms for some time, recent advances in the culture of cells from multicellular organisms have played a pivotal role in recent advances in toxicology. Cells can be isolated and either maintained in a viable state for enough time to conduct informative experiments or, in some cases, can be propagated in culture. The advantages of cultured cells are that they can provide living systems for the investigation of toxicity that are simplified relative to the intact organism or they can be used as replacements for whole animal toxicity testing if the toxic end point can be validated. Human cells play an important role in the extrapolation of toxic effects, discovered in experimental animals, to humans. Cultured cells, either from humans or other mammals, are utilized in many of the molecular methods mentioned below. There

A Textbook of Modern Toxicology, Fourth Edition. Edited by Ernest Hodgson
Copyright © 2010 John Wiley & Sons, Inc.

are, however, limitations in the use of cellular methods. It has not been possible to culture many cell types, and of those that have been cultured, the loss of differentiated cell function is a common problem. Extrapolation of findings to the intact animal is often problematical, and the use of undefined media constituents such as serum, often essential for cell viability, may have unwanted or undefined effects on cell function and toxicant bioavailability.

Studies have been carried out on cells isolated from tissues and maintained in suspension culture or on cells that have formed monolayers.

2.2.1 Suspension Cell Culture

Circulating blood cells or cells easily obtained by lavage such as peritoneal and alveolar macrophages can normally survive in suspension culture when provided with a suitable nutrient medium. Cells from organized solid organs or tissues must be separated from the tissue and, if possible, separated into cell types, before being suspended in such a medium.

Cell association within organs depends on protein complex formation, which in turn is Ca^{2+}-dependent. Consequently, dissociation media generally contain a proteolytic enzyme and the Ca^{2+} chelator EDTA (ethylenediaminetetraacetic acid). There are a number of methods available to separate cell types from the mixture of dispersed cells, the commonest being centrifugation without a density gradient, wherein cells are separated by size, or centrifugation through a density gradient wherein cells are separated on the basis of their buoyant density.

Cells in suspension may be maintained for a limited period of time in defined media or for longer periods in nutrient, but less well-defined, media. In either case, these cultures are often used for studies of xenobiotic metabolism.

2.2.2 Monolayer Cell Culture

Proliferation of most cells in culture requires attachment to a substrate and occurs until limited by cell-to-cell contact, resulting in the formation of a cellular monolayer. The substrate provided for attachment is usually polystyrene modified to carry a charge. The medium for continued maintenance and growth contains salts and glucose, usually with a bicarbonate buffer. Because of the bicarbonate buffering system, these cultures are maintained in a 5–10% CO_2 atmosphere in a temperature and humidity controlled incubator. Many cells require serum for optimal growth, inducing considerable variability into the experimental system. Since the factors provided by serum are numerous and complex, defined serum substitutes are not always successful. The factors provided by serum include proteins such as growth factors, insulin, and transferrin (to provide available iron), small organic molecules such as ethanolamine, and pyruvate and inorganic ions, such as selenium.

2.2.3 Indicators of Toxicity in Cultured Cells

Routine observation of cultured cells is usually carried out by phase contrast microscopy, utilizing the inverted phase contrast microscope. More recently, more detailed observations have become possible utilizing fluorescent tags and inverted fluorescent microscopes. Fluorescent tags currently in use permit the assessment of oxidant

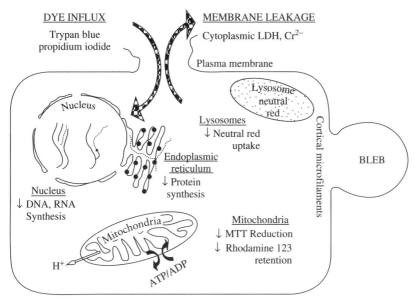

Figure 2.1   Idealized diagram of a cell to illustrate parameters often used to measure cyto-toxicity and the corresponding affected subcellular organelle. From *A Textbook of Modern Toxicology*, 3rd ed., ed. E. Hodgson. New York: Wiley, 2004.

status and mitochondrial function as well as the intracellular concentration of sulf-hydryl groups, Ca^{2+}, H^+, Na^+, and K^+.

Toxicity to cultured cells may be the result either of inadequacies in the culture or the toxicity effects of the chemical being investigated. Short-term toxicity is usually evaluated by examination of end points that indicate effects on cellular organelles such as leakage of cell constituents into the medium, uptake of dyes into the cell, and the formation of surface "blebs." This is illustrated in Figure 2.1.

Longer-term assessments of cell toxicity are highly dependent upon the relevant toxic end point. They may include measurement of growth competence, apoptosis, and/or necrosis, incorporation of radioactive precursors into essential cellular con-stituents such as RNA, DNA, and protein and specialized cellular functions. Some examples of the use of cultured cell lines in the study of toxicity effects are shown in Table 2.1.

2.2.4 Use of Stem Cells

Although the use of human stem cells is still controversial, their use in biomedical research is increasing. At the same time, stem cells from surrogate animals have long been used in biomedical, including toxicological, research. Probably the best example is the use of cultured mouse embryonic stem cells for generation of "knock-out" mice. Such mice have been widely used in the study of xenobiotic-metabolizing enzymes (XMEs), nuclear receptors, and the like.

The value of cultured stem cells in toxicology lies in their ability to provide a continuous source of cells that can be manipulated to provide a desired mature cell

TABLE 2.1 Application of Human Cell Lines Retaining Differentiated Properties for the Study of Toxic Effects and a Comparison with Primary Human Hepatocytes

Cell Type	Source	Differentiated Cell Type	Toxicant	Measured End Point
Cell lines				
SK-N-SH	Human neuroblastoma	Neuron	Anesthetic N_2O	Depressed cholinergic Ca^{2+} signaling
HepG2	Human hepatoblastoma	Hepatocyte	Cyclophosphamide (antineoplastic)	Cytochrome P450 (CYP)-dependent genotoxicity
			Rifampicin (PXR ligand)	Inhibition of bile acid synthesis
Caco-2	Human colon adenocarcinoma	Intestinal epithelial cell	Arsenic	Transepithelial leakiness
Primary hepatocyte			Fipronil, fipronil sulfoxide	Adenylate kinase release, Induction of caspase 3/7 Induction of CYP isoforms
			Deltamethrin (pyrethroid)	Adenylate kinase release, Induction of caspase 3/7 Induction of CYP isoforms
			Permethrin (pyrethroid)	Adenylate kinase release, Induction of caspase 3/7 Induction of CYP isoforms
			DEET (N,N-diethyl-meta-toluamide) (repellent)	Adenylate kinase release, Induction of caspase 3/7 Induction of CYP isoforms
			Chlorpyrifos (OP)	Adenylate kinase release, Induction of caspase 3/7 Induction of CYP isoforms

PXR, pregnane X receptor; OP, organophosphorus.

type. This could alleviate the use of surrogate animals in toxicity testing and provide metabolically competent human cell types.

2.2.5 Cell Culture Models as "Alternative" Toxicity Tests

Due primarily to the fact that the cell represents an excellent intermediate level of biological organization between the intact organism and the cellular organelle or enzyme/receptor levels, the primary use of cell culture models to date has been in mechanistic studies of chemical toxicity. At present, however, much effort is being placed on development of cell culture models as replacements for surrogate animals in toxicity. This arises not only from ethical concerns over animal use but also for economy of time and expense. Further, the use of human-derived cell lines may be an advantage in studies related to human health assessment.

Although difficulties are often encountered, particularly in agreement between the cell culture method and *in vivo* results as well as quantitative relationships between toxicants of related chemical structure or mode of toxic action, it appears that cell culture methods will be useful as early screens in tiered protocols for product safety testing.

Another emerging application of cell culture toxicity testing techniques is the development of cell lines engineered for a particular function, often for high-throughput screening protocols. An excellent example is in area of testing for endocrine disruptors and the recent mandate that chemicals in commerce be tested for endocrine disrupting activity. This involves the development of cell lines engineered to contain a vector with a reporter gene whose expression is responsive to activation of a cotransfected steroid hormone receptor. A similar approach is being adopted for the detection of dioxin-like compounds through their interaction with the aryl hydrocarbon receptor (Ah receptor).

2.3 MOLECULAR TECHNIQUES

Recombinant DNA techniques, including molecular cloning, have provided recent dramatic advances in many areas of both fundamental and applied biology, toxicology not excepted. Responses to toxicants often involve changes in gene expression and the microarray techniques enable the simultaneous examination of the global level of expression of thousands of genes in a single experiment. The completion of the human genome project now permits toxic effects in humans to be investigated and will facilitate extrapolation from experimental animals. The human genome will also provide the essential genetic background information for studies of polymorphisms in xenobiotic-metabolizing and other enzymes. Such polymorphisms have already been shown to be very important in individual sensitivity to clinical drugs and in the definition of populations and/or individuals at increased risk from particular toxicants. Identification of carcinogen-induced mutations, particularly in oncogenes and tumor-suppressor genes, are important in chemical carcinogenesis. The ability to develop "knock-out" and "knock-in" animals that lack a particular gene or express an altered gene in place of the wild-type gene, respectively, as well as knockdown of specific genes in cell culture are proving important in toxicological studies. Polymerase chain reaction (PCR) is an extremely versatile technique that

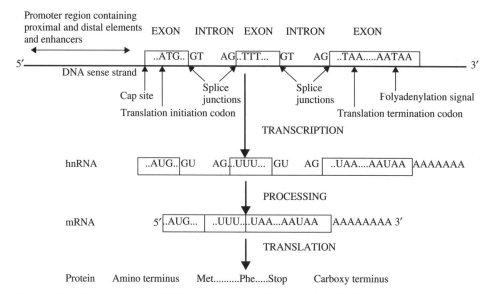

Figure 2.2 Transcription, mRNA processing, and translation. DNA sense strand is designated by bold lines, hnRNA and mRNA by thinner lines. Exons are shown as rectangles and introns as the intervening spaces between exons. From *A Textbook of Modern Toxicology*, 3rd ed., ed. E. Hodgson. New York: Wiley, 2004.

can be used for many applications including gene cloning, gene mutagenesis, and quantitative gene expression analysis.

Gene structure and any of the processes involved in DNA expression, including transcription, mRNA processing, and translation and protein synthesis (Figure 2.2), can all be examined by molecular techniques. In toxicology, this may include toxic effects on these processes or the role of the processes in the mechanism of toxic action.

2.3.1 Molecular Cloning

The basic principle of molecular cloning is the insertion of a DNA segment into a suitable vector. The vector is an autonomously replicating DNA molecule, and the inserted DNA segment may be as large as a gene or as small as a few nucleotides. The vector containing the DNA is inserted into a cell such as a bacterium, where it can be replicated many times and either the DNA or the expressed protein subsequently isolated (Figure 2.3).

2.3.2 cDNA and Genomic Libraries

cDNA or genomic libraries are collections of DNA fragments incorporated into a recombinant vector and transformed into an appropriate host cell. In the case of cDNA libraries, the cDNAs complementary to all of the mRNAs in the tissue or cell sample are synthesized in a procedure using reverse transcriptase, before incorporation into the vector. With genomic DNA libraries the genomic DNA is digested,

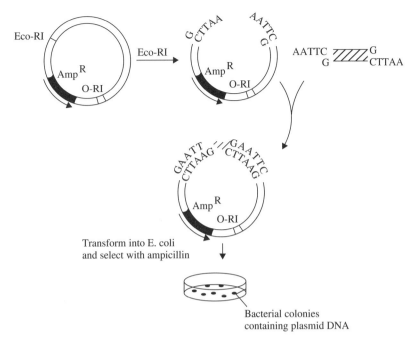

Figure 2.3 Molecular cloning using a plasmid vector. From *A Textbook of Modern Toxicology*, 3rd ed., ed. E. Hodgson. New York: Wiley, 2004.

before cloning into the vector, with a restriction enzyme to produce an overlapping set of DNA fragments of some 12–20 kb.

These libraries have been used in many screening procedures, including gene identification and gene regulation. Today, with availability of genomic information/ annotation for numerous species including mouse, rat, and human, direct bioinformatic analysis of such information allows for PCR approaches for the cloning of genes, promoter regions, and mRNA (cDNA). In fact, most applications that used cDNA and genomic libraries have been superceded by other methods, particularly those based on PCR.

2.3.3   Northern and Southern Blot Analysis

Northern analysis is usually used to identify and quantitate specific mRNAs in a sample. Southern analysis is used to determine whether or not a gene of interest is present as well as its copy number. Other uses for Southern analysis include identifying restriction fragment length polymorphisms and changes in heterozygosity.

In both Southern and Northern analyses, restriction digested DNA fragments or RNA, respectively, are separated by size when electrophoresed on agarose gel. The separated molecules are transferred, by electroblotting or capillary blotting, onto a nylon or nitrocellulose membrane. The immobilized RNA or DNA is reacted with a radiolabeled, chemiluminescent, or fluorescent probe that is complementary to the DNA/RNA of interest, unbound probe is washed off and the membrane exposed, in the case of radioactive probes, to radioautographic film to visualize the sample of interest.

2.3.4 PCR

PCR is a powerful technique that can, starting with amounts of DNA as small as those found in single cells, amplify the DNA until large amounts are available for many different kinds of research. Twenty to forty cycles of PCR can provide up to 10^5 times the original DNA sample.

It is necessary to know the flanking sequence of the DNA of interest in order to construct appropriate primers. These primers are complementary to the sequence at each end of the DNA sequence to be amplified. The DNA is incubated in a thermal cycler with thermostable DNA polymerase, all four deoxyribonucleotide triphosphates (dNTP), and the primers. The incubation temperature is raised to separate the DNA strands, lowered to permit annealing of the primers to the complementary regions of the DNA, and then raised to permit the polymerase to synthesize DNA. This cycle is then repeated up to 40 times. The PCR technique has been used for many types of toxicological investigation including uncovering polymorphisms in XMEs, cloning genes for functional studies as well as promoter regions of genes for gene regulation studies.

2.3.5 Evaluation of Gene Expression, Regulation, and Function

The methods used for the evaluation of regulation of gene expression are too numerous to be described in detail here. They include Northern analysis to determine levels of a particular mRNA, nuclear run-on to determine whether an increase in mRNA is due to an increase in the rate of transcription, and promoter deletion analysis to identify specific elements in the promoter region responsible for the control of expression, and the electrophoretic mobility shift assay (EMSA) is used to measure binding of a transcription factor to its specific DNA consensus sequence. High-throughput reporter gene assays are currently used to examine molecular pathways altered by toxicants. These assays employ specific regulatory promoter elements that respond to specific types of stressors/inputs; for example, estrogenic agents, reactive oxygen stress, and dioxin-related agents are engineered upstream of a reporter gene (i.e., luciferase), and cell lines containing these constructs can be treated with the toxicant of interest and reporter output quantified.

Of much current interest is the use of microarrays that permit the study of the expression of hundreds to thousands of genes at the same time. Microarrays are based on the principle that any gene being expressed at any point in time is giving rise to a specific, corresponding mRNA. The microarray itself consists of spots of DNA (c. 200μ) bound to a suitable matrix. The mRNAs in the biological sample in question bind to the corresponding DNA and can be visualized by techniques involving dyes. Given the complexity of the data obtained (often thousands of genes are evaluated on a single microarray), special techniques have been developed for array scanning, data extraction, and statistical analysis. A typical microarray experiment is illustrated in Figure 2.4.

Real-time reverse transcriptase–polymerase chain reaction (RT-PCR) is commonly used to amplify and quantitate mRNAs of interest. In fact, this technique is

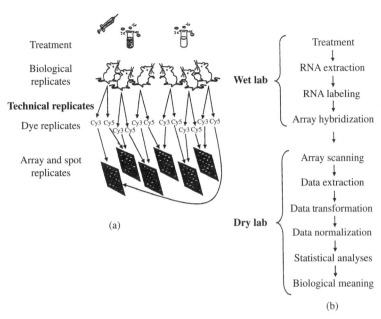

Figure 2.4   Schematic of a microarray experiment. (a) Experimental design incorporating both biological and technical replication. There are three treated mice and three control mice, providing biological replication. RNA from each mouse is labeled with each dye and hybridized more than once, providing technical replication. (b) Outline of the wet and dry laboratory steps involved in a microarray experiment. From *Molecular and Biochemical Toxicology*, 4th ed., eds. R.C. Smart and E. Hodgson. Hoboken, NJ: Wiley, 2008.

replacing the Northern technique described above as the preferred technique to measure changes in gene expression and the mRNA level. Gene function in cultured cells can be investigated by the forced expression of the gene product in a suitable expression system or through the use of small interfering RNAs (siRNAs), where the expression of the gene of interest can be knocked down in cultured cells. Gene function can also be studied *in vivo* through the creation of transgenic mice which overexpress the gene of interest or knock-out mice in which the gene in question has been functionally deleted or knock-in mice where an altered gene (i.e., serine is replaced by alanine to study the role of posttranslational modifications involving phosphorylation) is expressed in place of the wild-type gene.

A general, but more detailed and specific, account of these methods may be found in Smart (2008) and Oleksiak (2008) (see Bibliography).

2.4   IMMUNOCHEMICAL TECHNIQUES

Most of the recently developed methods for the detection, characterization, and quantitation of proteins (Leblanc, 2008) are immunoassays based on the fact that proteins are antigens, compounds that can be recognized by an antibody. It is also true that, by combining small molecules (haptens) with a larger carrier molecule such as a protein, these methods can be extended to small molecules of interest

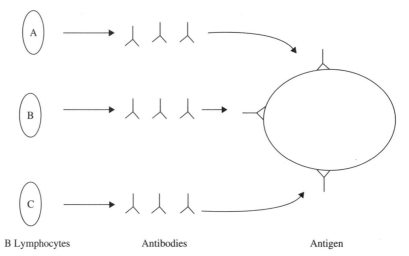

B Lymphocytes   Antibodies   Antigen

Figure 2.5 The generation of antibodies of several clonal origins (polyclonal antibodies) with antibodies from each clonal origin (monoclonal antibodies A, B, and C) recognizing a distinct epitope on the antigen. From *A Textbook of Modern Toxicology*, 3rd ed., ed. E. Hodgson. New York: Wiley, 2004.

since antibodies can be produced that recognize epitopes (specific sites on the antigen recognized by the antibody) that include the hapten.

The antibodies used may be polyclonal or monoclonal, each with characteristics fitting them for use in particular immunochemical methods. Injection of a mammal with a foreign protein (immunogen) gives rise to an immune reaction that includes the generation of antibodies from B lymphocytes. Each B lymphocyte gives rise to only a single antibody type that recognizes a single epitope on the antigen. However, since these antibodies are derived from many different B lymphocytes, the mixture of antibodies can recognize and bind to many different epitopes on the antigen. This mixture of antibodies can be isolated from the serum of the treated animal and is known, collectively, as *polyclonal antibodies*. However, if individual B lymphocytes from a treated animal can be isolated and cultured, because they are of a single clonal origin, they will produce a specific *monoclonal antibody* that recognizes only a single epitope on the antigen (Figure 2.5). Because of the multiple sites for binding, polyclonal antibodies are highly reactive. They are also relatively easy to produce. Monoclonal antibodies, although more difficult to produce are, on the other hand, more specific. The advantages and disadvantages of each must be considered to determine which is the antibody of choice for a particular application. The most important immunochemical methods include the following:

Immunolocalization is a technique for identifying the presence of a protein within the cell, its relative abundance, and its subcellular localization. After suitable preparation of the cells, they are treated with an antibody (the primary antibody) which binds to the protein of interest. An antibody that binds to the primary antibody (the secondary antibody) is then allowed to bind and form an antigen–primary antibody–secondary antibody complex. The detection system generally consists of

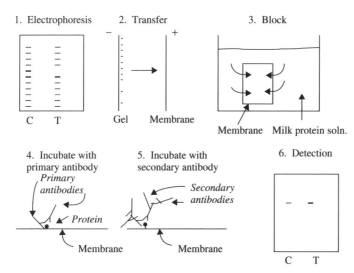

1. Electrophoresis   2. Transfer   3. Block

C   T   Gel   Membrane

Membrane   Milk protein soln.

4. Incubate with
primary antibody   5. Incubate with
secondary antibody   6. Detection

Primary
antibodies

Protein

Membrane

Secondary
antibodies

Membrane

C   T

Figure 2.6 Diagrammatic representation of the use of immunoblotting to assess relative levels of a P450 protein following treatment of rats with a polychlorinated biphenyl (PCB). C, hepatic microsomal proteins from a control, untreated rat; T, hepatic microsomal proteins from a rat treated with PCBs. From *A Textbook of Modern Toxicology*, 3rd ed., ed. E. Hodgson. New York: Wiley, 2004.

the formation of a colored insoluble product of an enzymatic reaction, the enzyme, such as alkaline phosphatase or horseradish peroxidase, being covalently linked to the secondary antibody.

Immunoaffinity purification involves the use of antibodies, bound to an insoluble matrix, for chromatography. The advantage of this method is that it is highly specific, often permitting purification in a single step. *Immunoprecipitation* is a variant of immunoaffinity purification and is a means to remove a protein from a complex mixture in a highly specific manner.

Western blotting is a widely used technique in which antibodies are used to detect proteins following electrophoresis, generally sodium dodecyl sulfate (SDS) poly-acrylamide gel electrophoresis that permits the separation of proteins on the basis of their molecular weights (Figure 2.6). Western blotting can be used to determine the presence and relative amount of a particular protein in a biological sample as well as its molecular weight.

Radioimmunoassay (RIA) is a very sensitive method used to measure minute quantities of an antigen. Since this method is most often used to measure drugs, toxicants, and other xenobiotics, the antigen used to produce the antibody is the small molecule (hapten) linked covalently to a protein. Among the techniques used in the actual measurement is the antigen capture method, in which the competition between radiolabeled antigen and the unlabeled antigen in the sample is the most common.

Depending upon the design of the method, *enzyme-linked immunosorbent assay* (ELISA) can be used to measure either antigens or antibodies in mixtures by using enzymatic-mediated detection of the corresponding immobilized immune complex. Even though this method has proven most useful for the rapid estimation of antibodies or antigens in complex biological mixtures, it has also been used for the quantitation of small molecules in a manner analogous to RIAs.

Inhibitory antibodies are frequently used in studies of xenobiotic metabolism, usually to estimate the contribution of particular enzymes in multienzyme mixtures. An important example is the use of antibodies to estimate the contribution of individual cytochrome P450 (CYP) isoforms to the overall metabolism of a xenobiotic in microsomal preparations.

2.5 PROTEOMICS

The proteome is defined as the protein complement present in the biological unit (e.g., cell, organ, organism) and represents that portion of the genome currently being expressed. Proteomics is represented by broad, inclusive techniques to separate, identify, and study the structure of the proteins of the proteome. Separation is usually by two-dimensional polyacrylamide gel electrophoresis and identification by a number of variants of mass spectrometry. Details are available in Merrick (2008).

2.6 METABOLOMICS

Genomics has the goal of determining, through analysis of mRNA, which genes are being expressed. Proteomics (Deighton, 2008) has the goal of determining whether expression of mRNA results in protein synthesis, while metabolomics has the goal of determining whether the expressed proteins are metabolically active. Metabolomics is, therefore, the identification and quantification of all of the metabolites in a biological system at some point in time. It is important to remember that the metabolites in question are the products of the normal endogenous metabolism of the cell, organ, or organism and not the metabolic products of toxicants or other xenobiotics although in the latter case, the techniques of metabolomics can be invaluable.

Given the large number, chemical diversity, and concentration range of the entire metabolome, of necessity, a number of techniques are needed to obtain the complete picture needed. Initially, an unbiased extraction technique must be selected or developed. Since no single extraction technique is likely to extract all metabolites, several techniques are usually employed. Metabolite identification depends on two sensitive techniques, mass spectrometry and nuclear magnetic resonance spectroscopy.

2.7 BIOINFORMATICS

In the narrow and original meaning, bioinformatics was the application of information technology to molecular biology. While this is still the most important aspect of bioinformatics, it is increasingly applied to other fields of biology, including molecular and other aspects of toxicology. It is characterized by computationally intensive methodology and includes the design of large databases and the development of techniques for their manipulation, including data mining.

2.8 SUMMARY AND CONCLUSIONS

Rapid and dramatic progress in development of new techniques based on molecular biology and analytical chemistry are stimulating new approaches and progress in many fields of toxicology. This progress is reflected not only in advances in the understanding of the fundamental mechanisms of toxicity but also in new approaches to toxicity testing. Both of these aspects have important implications for human health and for human health risk assessment. Further rapid progress is expected in the immediate future.

BIBLIOGRAPHY AND SUGGESTED READING

Deighton, N. Metabolomics. In *Molecular and Biochemical Toxicology*, 4th ed., eds. R. C. Smart and E. Hodgson. Hoboken, NJ: Wiley, 2008.

Hodgson, E., ed. *A Textbook of Modern Toxicology*, 3rd edition. Hoboken, NJ: Wiley, 2004.

LeBlanc, G. A. Immunochemical techniques in toxicology. In *Molecular and Biochemical Toxicology*, 4th ed., eds. R. C. Smart and E. Hodgson. Hoboken, NJ: Wiley, 2008.

Merrick, B. A. Proteomics. In *Molecular and Biochemical Toxicology*, 4th ed., eds. R. C. Smart and E. Hodgson. Hoboken, NJ: Wiley, 2008.

Meyer, S. A. Cellular techniques in toxicology. In *Molecular and Biochemical Toxicology*, 4th ed., eds. R. C. Smart and E. Hodgson. Hoboken, NJ: Wiley, 2008.

Oleksiac, M. F. Toxicogenomics. In *Molecular and Biochemical Toxicology*, 4th ed., eds. R. C. Smart and E. Hodgson. Hoboken, NJ: Wiley, 2008.

Smart, R. C. Overview of molecular techniques in toxicology: Genes/transgenes. In *Molecular and Biochemical Toxicology*, 4th ed., eds. R. C. Smart and E. Hodgson. Hoboken, NJ: Wiley, 2008.

Smart R. C. and E. Hodgson. eds. *Molecular and Biochemical Toxicology*. 4th ed. Hoboken, NJ: Wiley, 2008.

Stone, E. A and D. M. Nielsen. Bioinformatics. In *Molecular and Biochemical Toxicology*, 4th ed., eds. R. C. Smart and E. Hodgson. Hoboken, NJ: Wiley, 2008.

SAMPLE QUESTIONS

1. Briefly discuss the importance of cultured cell techniques in toxicology.

2. Define the following terms:
   a. Northern and Southern blot analysis;
   b. Polymerase chain reaction;
   c. Microarray.

3. Define the following terms:
   a. Western blot;
   b. Polyclonal and monoclonal antibodies;
   c. Immunoaffinity purification and immunoprecipitation.

CLASSES OF TOXICANTS

Exposure Classes, Toxicants in Air, Water, Soil, Domestic, and Occupational Settings

W. GREGORY COPE

3.1 AIR POLLUTANTS

3.1.1 History

Air pollution probably occurred as soon as humans started to use wood fires for heat and cooking. For centuries, fire was used in such a way that living areas were filled with smoke. After the invention of the chimney, combustion products and cooking odors were removed from living quarters and vented outside. Later, when soft coal was discovered and used for fuel, coal smoke became a problem in the cities. By the thirteenth century, records show that coal smoke had become a nuisance in London, and in 1273, Edward I made the first antipollution law, one that prohibited the burning of coal while Parliament was in session: "Be it known to all within the sound of my voice, whosoever shall be found guilty of burning coal shall suffer the loss of his head." Despite this and various other royal edicts, however, smoke pollution continued in London.

Increasing domestic and industrial combustion of coal caused air pollution to get steadily worse, particularly in large cities. During the twentieth century, the most significant change was the rapid increase in the number of automobiles, from almost none at the turn of the century to millions within only a few decades. During this time, few attempts were made to control air pollution in any of the industrialized countries until after World War II. Action was then prompted, in part, by two acute pollution episodes in which human deaths were caused directly by high levels of pollutants. One incident occurred in 1948 in Donora, a small steel mill town in western Pennsylvania. In late October, heavy smog settled in the area, and a weather inversion prevented the movement of pollutants out of the valley. Twenty-one deaths were attributed directly to the effects of the smog. The "Donora episode" helped focus attention on air pollution in the United States.

A Textbook of Modern Toxicology, Fourth Edition. Edited by Ernest Hodgson
Copyright © 2010 John Wiley & Sons, Inc.

In London in December 1952, the now infamous killer smog occurred. A dense fog at ground level coupled with smoke from coal fireplaces caused severe smog lasting more than a week. The smog was so heavy that daylight visibility was only a few meters, and bus conductors had to walk in front of the buses to guide the drivers through the streets. Two days after the smog began, the death rate began to climb, and between December 5 and December 9, there were an estimated 4000 deaths above the normal daily count. The chief causes of death were bronchitis, pneumonia, and associated respiratory complaints. This disaster resulted in the passage in Britain of the Clean Air Act in 1956.

In the United States, the smog problem began to occur in large cities across the country, becoming especially severe in Los Angeles. In 1955, federal air pollution legislation was enacted, providing federal support for air pollution research, training, and technical assistance. Responsibility for the administration of the federal program now lies with the U.S. Environmental Protection Agency (EPA). Technological interest since the mid-1950s has centered on automobile air pollution, pollution by oxides of sulfur and nitrogen, and the control of these emissions. Attention is also being directed toward the problems that are being caused by the greenhouse effect resulting from increased concentrations of carbon dioxide (CO_2) in the atmosphere, depletion of the stratospheric ozone layer, long-range transport of pollution, and acid deposition.

3.1.2 Types of Air Pollutants

What is clean air? Unpolluted air is a concept of what the air would be if humans and their works were not on earth, and if the air were not polluted by natural sources such as volcanoes and forest fires. The true composition of "unpolluted" air is unknown because humans have been polluting the air for thousands of years. In addition, there are many natural pollutants such as terpenes from plants, smoke from forest fires, fumes and smoke from volcanoes. Table 3.1 lists the components that, in the absence of such pollution, are thought to constitute clean air.

TABLE 3.1 Gaseous Components of Normal Dry Air

Compound	Percent by Volume	Concentration (ppm)
Nitrogen	78.09	780,900
Oxygen	20.94	209,400
Argon	0.93	9,300
Carbon dioxide	0.0325	325
Neon	0.0018	18
Helium	0.0005	5.2
Methane	0.0001	1.1
Krypton	0.0001	1.0
Nitrous oxide		0.5
Hydrogen		0.5
Xenon		0.008
Nitrogen dioxide		0.02
Ozone		0.01–0.04

Gaseous Pollutants These substances are gases at normal temperature and pressure as well as vapors evaporated from substances that are liquid or solid. Among pollutants of greatest concern are carbon monoxide (CO), hydrocarbons, hydrogen sulfide (H_2S), nitrogen oxides (N_xO_y), ozone (O_3), and other oxidants, sulfur oxides (S_xO_y), and CO_2. Pollutant concentrations are usually expressed as micrograms per cubic meter ($\mu g/m^3$) or for gaseous pollutants as parts per million (ppm) by volume in which 1 ppm = 1 part pollutant per million parts (10^6) of air.

Particulate Pollutants Fine solids or liquid droplets can be suspended in air. Some of the different types of particulates are defined as follows:

1. *Dust.* Relatively large particles about 100 μm in diameter that come directly from substances being used (e.g., coal dust, ash, sawdust, cement dust, grain dust).
2. *Fumes.* Suspended solids less than 1 μm in diameter usually released from metallurgical or chemical processes (e.g., zinc and lead oxides).
3. *Mist.* Liquid droplets suspended in air with a diameter less than 2.0 μm (e.g., sulfuric acid mist).
4. *Smoke.* Solid particles (0.05–1.0 μm) resulting from incomplete combustion of fossil fuels.
5. *Aerosol.* Liquid or solid particles (<1.0 μm) suspended in air or in another gas.

3.1.3 Sources of Air Pollutants

Natural Pollutants Many pollutants are formed and emitted through natural processes. An erupting volcano emits particulate matter as well as gases such as sulfur dioxide, hydrogen sulfide, and methane; such clouds may remain airborne for long periods of time. Forest and prairie fires produce large quantities of pollutants in the form of smoke, unburned hydrocarbons, CO, nitrogen oxides, and ash. Dust storms are a common source of particulate matter in many parts of the world, and oceans produce aerosols in the form of salt particles. Plants and trees are a major source of hydrocarbons on the planet, and the blue haze that is so familiar over forested mountain areas is mainly from atmospheric reactions with volatile organics produced by the trees. Plants also produce pollen and spores, which cause respiratory problems and allergic reactions.

Anthropogenic Pollutants These substances come primarily from three sources: (1) combustion sources that burn fossil fuel for heating and power, or exhaust emissions from transportation vehicles that use gasoline or diesel fuels; (2) industrial processes; and (3) mining and drilling.

The principal pollutants from combustion are fly ash, smoke, sulfur, and nitrogen oxides, as well as CO and CO_2. Combustion of coal and oil, both of which contain significant amounts of sulfur, yields large quantities of sulfur oxides. One effect of the production of sulfur oxides is the formation of acidic deposition, including acid rain. Nitrogen oxides are formed by thermal oxidation of atmospheric nitrogen at high temperatures; thus, almost any combustion process will produce nitrogen oxides. Carbon monoxide is a product of incomplete combustion; the more efficient the combustion, the higher is the ratio of CO_2 to CO.

Transportation sources, particularly automobiles, are a major source of air pollution and include smoke, lead particles from tetraethyl lead additives, CO, nitrogen oxides, hydrocarbons, and more recently, the platinum group metals, which are used in automobile catalytic converters. Since the mid-1960s, there has been significant progress in reducing exhaust emissions, particularly with the use of low-lead or no-lead gasoline as well as the use of oxygenated fuels—for example, fuels containing ethanol or methyl *t*-butyl ether (MTBE).

Industries may emit various pollutants relating to their manufacturing processes—acids (sulfuric, acetic, nitric, and phosphoric); solvents and resins; gases (chlorine and ammonia); and metals (cadmium, copper, lead, and zinc).

Indoor Pollutants   In general, the term indoor air pollution refers to home and nonfactory public buildings such as office buildings and hospitals, and results in a contamination often referred to as "sick building syndrome." The pollution can come from heating and cooking, pesticides, tobacco smoking, radon, gases, and most commonly, microbes such as bacteria and fungi (molds) that grow in the structure or the heating and cooling systems due to excessive moisture.

Although indoor air pollution has increased in developed nations because of tighter building construction and the use of building materials that may give off gaseous chemicals, indoor air pollution is a particular problem in developing countries. Wood, crop residues, animal dung, and other forms of biomass are used extensively for cooking and heating—often in poorly ventilated rooms. For women and children in particular, this leads to high exposures of air pollutants such as CO and polycyclic aromatic hydrocarbons.

3.1.4   Examples of Air Pollutants

Most of the information on the effects of air pollution on humans comes from acute pollution episodes such as the ones in Donora and London. Illnesses may result from chemical irritation of the respiratory tract, with certain sensitive subpopulations being more affected: (1) very young children, whose respiratory and circulatory systems are poorly developed; (2) the elderly, whose cardiorespiratory systems function poorly; and (3) people with cardiorespiratory diseases such as asthma, emphysema, and heart disease. Heavy smokers are also affected more adversely by air pollutants. In most cases, the health problems are attributed to the combined action of particulates and sulfur dioxides (SO_2); no one pollutant appears to be responsible. Table 3.2 summarizes some of the major air pollutants and their sources and effects.

Carbon Monoxide   Carbon monoxide combines readily with hemoglobin (Hb) to form carboxyhemoglobin (COHb), thus preventing the transfer of oxygen to tissues. The affinity of hemoglobin for CO is approximately 210 times its affinity for oxygen. A blood concentration of 5% COHb, equivalent to equilibration at approximately 45 ppm CO, is associated with cardiovascular effects. Concentrations of 100 ppm can cause headaches, dizziness, nausea, and breathing difficulties. An acute concentration of 1000 ppm is invariably fatal. Carbon monoxide levels during acute traffic

TABLE 3.2   Principal Air Pollutants, Sources, and Effects

Pollutant	Sources	Significance
Sulfur oxides, particulates	Coal and oil power plants Oil refineries, smelters Kerosene heaters	Main component of acid deposition Damage to vegetation, materials Irritating to lungs, chronic bronchitis
Nitrogen oxides	Automobile emissions Fossil fuel power plants	Pulmonary edema, impairs lung defenses Important component of photochemical smog and acid deposition
Carbon monoxide	Motor vehicle emissions Burning fossil fuels Incomplete combustion	Combines with hemoglobin to form carboxyhemoglobin, poisonous Asphyxia and death
Carbon dioxide	Product of complete combustion	May cause "greenhouse effect"
Ozone (O_3)	Automobile emissions Photochemical smog	Damage to vegetation Lung irritant
Hydrocarbons, C_xH_y	Smoke, gasoline fumes Cigarette smoke, industry Natural sources	Contributes to photochemical smog Polycyclic aromatic hydrocarbons; lung cancer
Radon	Natural sources	Lung cancer
Asbestos	Asbestos mines Building materials Insulation	Asbestosis Lung cancer, mesothelioma
Allergens	Pollen, house dust Animal dander	Asthma, rhinitis
Arsenic	Copper smelters	Lung cancer

congestion have been known to be as high as 400 ppm; in addition, people who smoke elevate their total body burden of CO as compared with nonsmokers. The effects of low concentrations of CO over a long period are not known, but it is possible that heart and respiratory disorders are exacerbated.

Sulfur Oxides   Sulfur dioxide is a common component of polluted air that results primarily from the industrial combustion of coal, with soft coal containing the highest levels of sulfur. The sulfur oxides tend to adhere to air particles and enter the inner respiratory tract, where they are not effectively removed. In the respiratory tract, SO_2 combines readily with water to form sulfurous acid, resulting in irritation of mucous membranes and bronchial constriction. This irritation in turn increases the sensitivity of the airway to other airborne toxicants.

Nitrogen Oxides   Nitrogen dioxide (NO_2), a gas found in photochemical smog, is also a pulmonary irritant and is known to lead to pulmonary edema and hemorrhage. The main issue of concern is its contribution to the formation of photochemical smog and ozone, although nitrogen oxides also contribute to acid deposition.

Ozone   A highly irritating and oxidizing gas is formed by photochemical action of ultraviolet (UV) light on nitrogen dioxide in smog. The resulting ozone can produce pulmonary congestion, edema, and hemorrhage.

$$NO_2 + UV\ light \rightarrow NO + O^{\bullet}$$

$$O^{\bullet} + O_2 \rightarrow O_3$$

At this point, it is worth distinguishing between "good" and "bad" ozone. *Tropospheric ozone* occurs from 0 to 10 mi above the earth's surface, and is harmful. *Stratospheric ozone*, located about 30 mi above the earth's surface, is responsible for filtering out incoming UV radiation and thus is beneficial. It is the decrease in the stratospheric ozone layer that has been of much concern recently. It is estimated that a 1% decrease in stratospheric ozone will increase the amount of UV radiation reaching the earth's surface by 2% and will cause a 10% increase in skin cancer. Major contributors to damage to stratospheric ozone are thought to be the chlorofluorocarbons (CFCs). Chlorine is removed from the CFC compounds in the upper atmosphere by reaction with UV light and is then able to destroy the stratospheric ozone through self-perpetuating free radical reactions.

$$Cl + O_3 \rightarrow ClO + O_2$$

$$ClO + O \rightarrow Cl + O_2$$

Before being inactivated by nitrogen dioxide or methane, each chlorine atom can destroy up to 10,000 molecules of ozone. The use of CFC compounds is now being phased out and banned by international agreements.

Hydrocarbons or Volatile Organic Compounds (VOCs)   These are derived primarily from two sources: approximately 50% are derived from trees as a result of the respiration process (biogenic); the other 45–50% comes from the combustion of fuel and from vapor from gasoline. Many gasoline pumps now have VOC recovery devices to reduce pollution.

Lead   One of the most familiar of the particulates in air pollutants is lead, with young children and fetuses being the most susceptible. Lead can impair renal function, interfere with the development of red blood cells, and impair the nervous system, leading to mental retardation and even blindness. The two most common routes of exposure to lead are inhalation and ingestion. It is estimated that approximately 20% of the total body burden of lead comes from inhalation.

Solid Particles   Dust and fibers from coal, clay, glass, asbestos, and minerals, can lead to scarring or fibrosis of the lung lining. Pneumoconiosis, a condition common among coal miners that breathe coal dust, silicosis caused by breathing silica-containing dusts, and asbestosis from asbestos fibers are all well-known industrial pollution diseases.

TABLE 3.3   Examples of Air Pollution Injury to Vegetation

Pollutant	Symptoms
Sulfur dioxide	Bleached spots, interveinal bleaching
Ozone	Flecking, stippling, bleached spotting
Peroxyacetylnitrate (PAN)	Glazing, silvering, or bronzing on lower leaf surfaces
Nitrogen dioxide	White or brown collapsed lesion near leaf margins
Hydrogen fluoride	Tip and margin burns, dwarfing

3.1.5 Environmental Effects

Vegetation   Pollutants may visibly injure vegetation by bleaching, other color changes, and necrosis, or by more subtle changes such as alterations in growth or reproduction. Table 3.3 lists some of the more common visual effects of air pollutants on vegetation. Air pollution can also result in measurable effects on forest ecosystems, such as reduction in forest growth, change in forest species, and increased susceptibility to forest pests. High-dose exposure to pollutants, which is associated with point source emissions such as smelters, frequently results in complete destruction of trees and shrubs in the surrounding area.

Domestic Animals   Although domestic animals can be affected directly by air pollutants, the main concern is chronic poisoning as a result of ingestion of forage that has been contaminated by airborne pollutants. Pollutants important in this connection are arsenic, lead, and molybdenum. Fluoride emissions from industries producing phosphate fertilizers and derivatives have damaged cattle throughout the world. The raw material, phosphate rock, can contain up to 4% fluoride, some of which is released into the air and water. Farm animals, particularly cattle, sheep, and swine, are susceptible to fluoride toxicity (fluorosis), which is characterized by mottled and soft teeth, and osterofluoritic bone lesions, which lead to lameness and, eventually, death.

Materials and Structures   Building materials have become soiled and blackened by smoke, and damage by chemical attack from acid gases in the air has led to the deterioration of many marble statues in western Europe. Metals are also affected by air pollution; for example, SO_2 causes many metals to corrode at a faster rate. Ozone is known to oxidize rubber products, and one of the effects of Los Angeles smog is cracking of rubber tires. Fabrics, leather, and paper are also affected by SO_2 and sulfuric acid, causing them to crack, become brittle, and tear more easily.

Atmospheric Effects   The presence of fine particles (0.1–1.0 mm in diameter) or NO_2 in the atmosphere can result in atmospheric haze or reduced visibility due to light scattering by the particles. The major effect of atmospheric haze has been degradation in visual air quality and is of particular concern in areas of scenic beauty, including most of the major national parks such as Great Smoky Mountain, Grand Canyon, Yosemite, and Zion Parks.

There is also concern over the increase in CO_2 in the atmosphere because CO_2 absorbs heat energy strongly and retards the cooling of the earth. This is often referred to as the greenhouse effect; theoretically, an increase in CO_2 levels would result in a global increase in air temperatures. In addition to CO_2, other gases contributing to the greenhouse effect include methane, CFCs, nitrous oxide, and ozone.

Acidic Deposition   Acidic deposition is the combined total of wet and dry deposition, with wet acidic deposition being commonly referred to as acid rain. Normal uncontaminated rain has a pH of about 5.6, but acid rain usually has a pH of less than 4.0. In the eastern United States, the acids in acid rain are approximately 65% sulfuric, 30% nitric, and 5% other; whereas in the western states, 80% of the acidity is due to nitric acid.

Many lakes in northeastern North America and Scandinavia have become so acidic that fish are no longer able to live in them. The low pH not only directly affects fish, but also contributes to the release of potentially toxic metals, such as aluminum, from the soil. The maximum effect occurs when there is little buffering of the acid by soils or rock components. Maximum fish kills occur in early spring due to the "acid shock" from the melting of winter snows. Much of the acidity in rain may be neutralized by dissolving minerals in the soil, such as aluminum, calcium, magnesium, sodium, and potassium, which are leached from the soil into surface waters. The ability of the soil to neutralize or buffer the acid rain is very dependent on the alkalinity of the soil. Much of the area in eastern Canada and the northeastern United States is covered by thin soils with low acid neutralizing capacity. In such areas, the lakes are more susceptible to the effects of acid deposition leading to a low pH and high levels of aluminum, a combination toxic to many species of fish.

A second area of concern is that of reduced tree growth in forests. The leaching of nutrients from the soil by acid deposition may cause a reduction in future growth rates or changes in the type of trees to those able to survive in the altered environment. In addition to the change in soil composition, there are the direct effects on the trees from sulfur and nitrogen oxides as well as ozone.

3.2   WATER AND SOIL POLLUTANTS

With three quarters of the earth's surface covered by water and much of the remainder covered by soil, it is not surprising that water and soil serve as the ultimate sinks for most anthropogenic chemicals. Until recently, the primary concern with water pollution was that of health effects due to pathogens and, in fact, this is still the case in most developing countries. In the United States and other developed countries, however, treatment methods have largely eliminated bacterial disease organisms from the water supply, and attention has been turned to chemical contaminants.

3.2.1   Sources of Water and Soil Pollutants

Surface water can be contaminated by *point* or *nonpoint* sources. An effluent pipe from an industrial plant or a sewage-treatment plant is an example of a point source; a field from which pesticides and fertilizers are carried by rainwater into a river is an example of a nonpoint source. Industrial wastes probably constitute the greatest single pollution problem in soil and water. These contaminants include organic wastes such as solvents, inorganic wastes, such as chromium, and many unknown chemicals. Contamination of soil and water results when by-product chemicals are not properly disposed of or conserved. In addition, industrial accidents may lead to severe local contamination. For a more in-depth discussion of sources and movements of water pollutants, see Chapter 26.

Domestic and municipal wastes, both from sewage and from disposal of chemicals, are another major source of chemical pollutants. At the turn of the twentieth century, municipal wastes received no treatment and were discharged directly into rivers or oceans. Even today, many older treatment plants do not provide sufficient treatment, especially plants in which both storm water and sewage are combined. In addition to organic matter, pesticides, fertilizers, detergents, and metals are significant pollutants discharged from urban areas.

Contamination of soil and water also results from the use of pesticides and fertilizers. Persistent pesticides applied directly to the soil have the potential to move from the soil into the water and thus enter the food chain from both soil and water. In a similar way, fertilizers leach out of the soil or runoff during rain events and flow into the natural water systems.

Pollution from petroleum compounds has been a major concern since the mid-1960s. In 1967, the first major accident involving an oil tanker occurred. The *Torrey Canyon* ran onto rocks in the English Channel, spilling oil that washed onto the shores of England and France. It is estimated that at least 10,000 serious oil spills occur in the United States each year. In addition, flushing of oil tankers plays a major role in marine pollution. Other sources, such as improper disposal of used oil by private car owners and small garages, also contribute to oil pollution.

3.2.2 Examples of Pollutants

Metals that are of environmental concern fall into three classes: (1) metals that are suspected carcinogens, (2) metals that move readily in soil, and (3) metals that move through the food chain.

Lead The heavy metals of greatest concern for health with regard to drinking water exposure are lead and arsenic. The sources of lead in drinking water that are most important are from lead pipes and lead solder. Also of concern is the seepage of lead from soil contaminated with the fallout from leaded gasoline and seepage of lead from hazardous waste sites. Lead poisoning has been common in children, particularly in older housing units and inner city dwellings, in which children may consume chips of lead contaminated paint. Lead and associated toxic effects are discussed more fully in Chapter 4.

Arsenic Drinking water is at risk for contamination by arsenic from the leaching of inorganic arsenic compounds formerly used in pesticide sprays, from the combustion of arsenic-containing fossil fuels, and from the leaching of mine tailings and smelter runoff. Chronic high-level exposures can cause abnormal skin pigmentation, hyperkeratosis, nasal congestion, and abdominal pain. At lower levels of chronic exposure, cancer is the major concern. Epidemologic studies have linked chronic arsenic exposure to various cancers, including skin, lungs, and lymph glands.

Cadmium   One of the most significant effects of metal pollution is that aquatic organisms can accumulate metals in their tissues, leading to increased concentrations in the food chain. Concern about long-term exposure to cadmium intensified after recognition of the disease Itai-Itai (painful-painful) in certain areas of Japan. The disease is a combination of severe kidney damage and painful bone and joint disease and occurs in areas where rice is contaminated with high levels of cadmium. This contamination resulted from irrigation of the soil with water containing cadmium released from industrial sources. Cadmium toxicity in Japan has also resulted from consumption of cadmium-contaminated fish taken from rivers near smelting plants.

Mercury   In Japan in the 1950s and 1960s, wastes from a chemical and plastics plant containing mercury were discharged into Minamata Bay. The mercury was converted to the readily absorbed methylmercury by bacteria in the aquatic sediments. Consumption of fish and shellfish by the local population resulted in numerous cases of mercury poisoning, or Minamata disease. By 1970, at least 107 deaths had been attributed to mercury poisoning, and 800 cases of Minamata disease were confirmed. Even though the mothers appeared healthy, many infants born to these mothers who had eaten contaminated fish developed cerebral palsy-like symptoms and mental deficiency.

Pesticides   are also a major source of concern as water and soil pollutants. Because of their stability and persistence, the most hazardous pesticides are the organochlorine compounds such as DDT (1,1,1-trichloro-2,2-di(4-chlorophenyl)ethane), aldrin, dieldrin, and chlordane. Persistent pesticides can accumulate in food chains; for example, shrimp and fish can concentrate some pesticides as much as 1000–10,000-fold. This bioaccumulation has been well documented with the pesticide DDT, which is now banned in many parts of the world. In contrast to the persistent insecticides, the organophosphorus (OP) pesticides, such as malathion, and the carbamates, such as carbaryl, are short-lived and generally persist for only a few weeks to a few months. Thus, these compounds do not usually present as serious a problem as the earlier insecticides. Herbicides, because of the large quantity used, are also of concern as potential toxic pollutants. Pesticides are discussed in more detail in Chapter 4.

Nitrates and phosphates   are two important nutrients that have been increasing markedly in natural waters since the mid-1960s. Sources of nitrate contamination include fertilizers, discharge from sewage treatment plants, and leachate from septic systems and manure. Nitrates from fertilizers leach readily from soils, and it has been estimated that up to 40% of applied nitrates enter water sources as runoff and leachate. Fertilizer phosphates, however, tend to be absorbed or bound to soil particles, so that only 20–25% of applied phosphates are leached into water. Phosphate detergents are another source of phosphate, one that has received much media attention in recent years.

The increase in these nutrients, particularly phosphates, is of environmental concern because excess nutrients can lead to "algal blooms" or eutrophication, as it is known, in lakes, ponds, estuaries, and very slow moving rivers. The algal bloom reduces light penetration and restricts atmospheric reoxygenation of the water. When the dense algal growth dies, the subsequent biodegradation results in anaerobic conditions and the death of many aquatic organisms. High phosphate concentrations and algal blooms are generally not a problem in moving streams, because such streams are continually flushed out and algae do not accumulate.

There are two potential adverse health effects from nitrates in drinking water: (1) nitrosamine formation and (2) methemoglobinemia. Ingested nitrates can be converted to nitrites by intestinal bacteria. After entering the circulatory system, nitrite ions combine with hemoglobin to form methemoglobin, thus decreasing the oxygen-carrying capacity of the blood and resulting in anemia or blue baby disease. It is particularly severe in young babies who consume water and milk formula prepared with nitrate-rich water. Older children and adults are able to detoxify the methemoglobin as a result of the enzyme methemoglobin reductase, which reverses the formation of methemoglobin. In infants, however, the enzyme is not fully functional. Certain nitrosamines are known carcinogens.

Oils and petroleum are ever-present pollutants in the modern environment, whether from the used oil of private motorists or spillage from oil tankers. At sea, oil slicks are responsible for the deaths of many birds. Very few birds that are badly contaminated recover, even after de-oiling and hand feeding. Oil is deposited on rocks and sand as well, thus preventing the beaches from being used for recreation until after costly clean up. Shore animals such as crabs, shrimp, mussels, and barnacles are also affected by the toxic hydrocarbons they ingest. The subtle and perhaps potentially more harmful long-term effects on aquatic life are not yet fully understood.

VOCs are common groundwater contaminants. They include halogenated solvents and petroleum products, collectively referred to as VOCs. Both groups of compounds are used in large quantities by a variety of industries, such as degreasing, dry cleaning, paint, and the military. Historically, petroleum products were stored in underground tanks that would erode, or were spilled onto soil surfaces. The EPA's National Priority List includes 11 VOCs: trichloroethylene; toluene; benzene; chloroform; tetrachloroethylene; 1,1,1-trichloroethane; ethylbenzene; *trans*-1,2-dichloroethane; xylene; dichloromethane; and vinyl chloride.

The physical and chemical properties of VOCs permit them to move rapidly into groundwater, and almost all of the previously mentioned chemicals have been detected in groundwater near contaminant sites. High levels of exposure can cause headache, impaired cognition, and kidney toxicities. At levels of exposure most frequently encountered, cancer and reproductive effects are of utmost concern, particularly childhood leukemia.

Low molecular weight chlorinated hydrocarbons are a by-product of the chlorination of municipal water. Chlorine reacts with organic substances commonly found in water to generate trihalomethanes (THMs), such as chloroform. The main organics that have been detected are chloroform, bromodichloromethane, dibromochloromethane, bromoform, carbon tetrachloride, and 1,2-dichloroethane. These compounds are associated with an increased risk of cancer. Studies in New Orleans in the mid-1970s showed that tap water in New Orleans contained more chlorinated hydrocarbons than did untreated Mississippi River water or well water. In addition, chlorinated hydrocarbons, including carbon tetrachloride, were detected in blood plasma from volunteers who drank treated tap water. Epidemiologic studies indicated that the cancer death rate was higher among white males who drank tap water than among those who drank well water.

Radioactive contamination as background radiation from natural sources, such as radon, occurs in some regions of the world, but there is particular concern over the contamination of surface water and groundwater by radioactive compounds generated by the production of nuclear weapons and by the processing of nuclear fuel. Many of these areas have remained unrecognized because of government secrecy.

Acids present in rain or drainage from mines are major pollutants in many fresh-water rivers and lakes. Because of their ability to lower the pH of the water to toxic levels and release toxic metals into solution, acids are considered particularly hazardous (see Chapter 4).

The number of organic compounds found as soil and water contaminants continue to grow each year. They include polychlorinated biphenyls (PCBs), phenols, cyanides, plasticizers, solvents, and numerous industrial chemicals. PCBs were historically used as coolants in electrical transformers and are also known by-products of the plastic, lubricant, rubber, and paper industries. They are stable, lipophilic, and break down only slowly in tissues. Because of these properties, they accumulate to high concentrations in fish and waterfowl; in 1969, PCBs were responsible for the death of thousands of birds in the Irish Sea.

Dioxins have contaminated large areas of water and soil, most notably with the extremely toxic TCDD (2,3,7,8-tetrachlorodibenzo-*p*-dioxin) through industrial accidents and through widespread use of the herbicide 2,4,5-T. Small amounts of TCDD were contained as contaminants in herbicide manufacturing. The U.S. Army used this herbicide, known as Agent Orange, extensively as a defoliant in Vietnam. TCDD is one of the most toxic synthetic substances known for laboratory animals: lethal dose 50 (LD_{50}) for male rats, 0.022 mg/kg; LD_{50} for female rats, 0.045 mg/kg; LD_{50} for female guinea pigs (the most sensitive species tested), 0.0006 mg/kg. In addition, it is fetotoxic to pregnant rats at a dose of only 1/400 of the LD_{50} and has been shown to cause birth defects at levels of 1–3 ng/kg. TCDD is a proven carcinogen in both mice and rats, with the liver being the primary target. Although TCDD does not appear to be particularly acutely toxic to humans, chronic low-level exposure is suspected of contributing to reproductive abnormalities and carcinogenicity.

3.3   OCCUPATIONAL TOXICANTS

Assessment of hazards in the workplace is a concern of occupational/industrial toxicology and has a history that dates back to ancient civilizations. The Greek historian Strabo, who lived in the first century AD, gave a graphic description of the arsenic mines in Pantus: "The air in mines is both deadly and hard to endure on account of the grievous odor of the ore, so that the workmen are doomed to a quick death." With the coming of the industrial revolution in the nineteenth century, industrial diseases increased, and new ones, such as chronic mercurialism caused by exposure to mercuric nitrate used in "felting" animal furs, were identified. Hatmakers, who were especially at risk, frequently developed characteristic tremors known as "hatters' shakes," and the expression "mad as a hatter" was coined. In recent years, concern has developed over the carcinogenic potential of many workplace chemicals.

3.3.1 Regulation of Exposure Levels

The goal of occupational toxicology is to ensure work practices that do not entail any unnecessary health risks. To do this, it is necessary to define suitable permissible levels of exposure to industrial chemicals, using the results of animal studies and epidemiological studies. These levels can be expressed by the following terms for allowable concentrations.

Threshold limit values (TLVs) refer to airborne concentrations of substances and represent conditions under which it is believed that nearly all workers may be repeatedly exposed day after day without adverse effect. Because of wide variation in individual susceptibility, a small percentage of workers may experience discomfort from some substances at or below the threshold limit; a smaller percentage may be affected more seriously by aggravation of a preexisting condition or by development of an occupational illness. Threshold limits are based on the best available information from industrial experience, from experimental human and animal studies and, when possible, from a combination of the three. The basis on which the values are established may differ from substance to substance; protection against impairment of health may be a guiding factor for some, whereas reasonable freedom from irritation, narcosis, nuisance, or other forms of stress may form the basis for others. Three categories of TLVs follow.

Threshold limit value–time-weighted average (TLV-TWA) is the TWA concentration for a normal 8-h workday or 40-h workweek to which nearly all workers may be repeatedly exposed, day after day, without adverse effect. TWAs allow certain permissible excursions above the limit provided they are compensated by equivalent excursions below the limit during the workday. In some instances, the average concentration is calculated for a workweek rather than for a workday.

Threshold limit value–short-term exposure limit (TLV-STEL) is the maximal concentration to which workers can be exposed for a period up to 15 min continuously without suffering from (1) irritation, (2) chronic or irreversible tissue change, or (3) narcosis of sufficient degree that would increase accident proneness, impair self-rescue, or materially work efficiency, provided that no more than four excursions per day are permitted, with at least 60 min between exposure periods, and provided that the daily TLV-TWA is not exceeded.

Threshold limit value-ceiling (TLV-C) is the concentration that should not be exceeded even instantaneously. For some substances—for instance, irritant gases—only one category, the TLV-C, may be relevant. For other substances, two or three categories may be relevant.

Biologic limit values (BLVs) represent limits of amounts of substances (or their affects) to which the worker may be exposed without hazard to health or well-being as determined by measuring the worker's tissues, fluids, or exhaled breath. The biologic measurements on which the BLVs are based can furnish two kinds of information useful in the control of worker exposure: (1) measure of the worker's overall exposure, and (2) measure of the worker's individual and characteristic response. Measurements of response furnish a superior estimate of the physiological status of the worker, and may consist of (1) changes in amount of some critical biochemical constituent, (2) changes in activity or a critical enzyme, and (3) changes in some physiological function. Measurement of exposure may be made by (1) determining in blood, urine, hair, nails, or body tissues and fluids the amount of substance to which the worker was exposed; (2) determining the amount of the metabolite(s) of the substance in tissues and fluids; and (3) determining the amount of the substance in the exhaled breath. The biologic limits may be used as an adjunct to the TLVs for air, or in place of them.

Immediately dangerous to life or health (IDLH) conditions pose a threat of severe exposure to contaminants, such as radioactive materials, that are likely to have adverse cumulative or delayed effects on health. Two factors are considered when establishing IDLH concentrations. The worker must be able to escape (1) without loss of life or without suffering permanent health damage within 30 min and (2) without severe eye or respiratory irritation or other reactions that could inhibit escape. If the concentration is above the IDLH, only highly reliable breathing apparatus is allowed.

3.3.2 Routes of Exposure

The principal routes of industrial exposure are dermal and inhalation. Occasionally, toxic agents may be ingested, if food or drinking water is contaminated. Exposure to the skin often leads to localized effects known as "occupation dermatosis" caused by either irritating chemicals or allergenic chemicals. Such effects include scaling, eczema, acne, pigmentation changes, ulcers, and neoplasia. Some chemicals may also pass through the skin; these include aromatic amines such as aniline, and solvents such as carbon tetrachloride and benzene.

Toxic or potentially toxic agents may be inhaled into the respiratory tract where they may cause localized effects such as irritation (e.g., ammonia, chlorine gas), inflammation, necrosis, and cancer. Chemicals may also be absorbed by the lungs into the circulatory system, thereby, leading to systemic toxicity (e.g., CO, lead).

3.3.3 Examples of Industrial Toxicants

Carcinogen exposure is largely due to lifestyle, such as cigarette smoking, but occupation is an important source of exposure to carcinogens. Table 3.4 lists some occupational chemical hazards and the cancers associated with them.

Cadmium is a cumulative toxicant with a biologic half-life of up to 30 years in humans. More than 70% of the cadmium in the blood is bound to red blood cells; accumulation occurs mainly in the kidney and the liver, where cadmium is bound to metallothionein. In humans, the critical target organ after long-term exposure to cadmium is the kidney, with the first detectable symptom of kidney toxicity being an increased excretion of specific proteins.

Chromium toxicity results from compounds of hexavalent chromium that can be readily absorbed by the lung and gastrointestinal (GI) tract and, to a lesser extent, by the skin. Occupational exposure to chromium (Cr^{6+}) causes dermatitis, ulcers on the hands and arms, perforation of the nasal septum (probably caused by chromic acid), inflammation of the larynx and liver, and bronchitis. Chromate is a carcinogen causing bronchogenic carcinoma; the risk to chromate plant workers for lung cancer is 20 times grater than that for the general population. Compounds of trivalent chromium are poorly absorbed. Chromium is not a cumulative chemical, and once absorbed, it is rapidly excreted into the urine.

TABLE 3.4 Some Occupational Hazards and Associated Cancers

Agent	Tumor Sites	Occupation
Asbestos	Lung, pleura, peritoneum	Miners, manufacturers, users
Arsenic	Skin, lung, liver	Miners and smelters, oil refinery, pesticide workers
Benzene	Hemopoietic tissue	Process workers, textile workers
Cadmium	Lung, kidney, prostate	Battery workers, smelters
Chloroethers	Lung	Chemical plant workers, process workers
Chromium	Lung, nasal cavity, sinuses	Process and production workers, pigment workers
Mustard gas	Bronchi, lung, larynx	Production workers
Naphthylamines	Bladder	Dyestuff makers and workers, chemical workers, printers
Nickel	Lung, nasal sinuses	Smelters and process workers
Polycyclic aromatic hydrocarbons	Respiratory system, bladder	Furnace, foundry, shale, and gas workers; chimney sweeps
Radon, radium, uranium	Skin, lung, bone tissue, bone marrow	Medical and industrial chemists, miners
UV radiation	Skin	Outdoor exposure
X-rays	Bone marrow, skin	Medical and industrial workers

Lead is a ubiquitous toxicant in the environment, and consequently, the normal body concentration of lead is dependent on environmental exposure conditions. Approximately 50% of lead deposited in the lung is absorbed, whereas usually less than 10% of ingested lead passes into the circulation. Lead is not a major occupational problem today, but environmental pollution is still widespread. Lead interferes in the biosynthesis of porphyrins and heme, and several screening tests for lead poisoning make use of this interaction by monitoring either inhibition of the enzyme δ-aminolevulinic acid dehydratase (ALAD) or appearance in the urine of aminolevulinic acid (ALA) and coproporphorin (UCP). The metabolism of inorganic lead is closely related to that of calcium, and excess lead can be deposited in the bone where it remains for years. Inorganic lead poisoning can produce fatigue, sleep disturbances, anemia, colic, and neuritis. Severe exposure, mainly of children who have ingested lead, may cause encephalopathy, mental retardation, and occasionally, impaired vision.

Organic lead has an affinity for brain tissue; mild poisoning may cause insomnia, restlessness, and GI symptoms, whereas severe poisoning results in delirium, hallucinations, convulsions, coma, and even death.

Mercury is widely used in scientific and electrical apparatus, with the largest industrial use of mercury being in the chlorine-alkali industry for electrolytic production of chlorine and sodium hydroxide. Worldwide, this industry has been a major source of mercury contaminations. Most mercury poisoning, however, has been due to methylmercury, particularly as a result of eating contaminated fish. Inorganic and organic mercury differ in their routes of entry and absorption. Inhalation is the principal route of uptake of metallic mercury in industry, with approximately 80% of the mercury inhaled as vapor being absorbed; metallic mercury is less readily absorbed by the GI route. The principal sites of deposition are the kidney and brain after exposure to inorganic mercury salts. Organic mercury compounds are readily absorbed by all routes. Industrial mercurialism produces features such as inflammation of the mouth, muscular tremors (hatters' shakes), psychic irritation, and a nephritic syndrome characterized by proteinuria. Overall, however, occupational mercurialism is not a significant problem today.

Benzene was used extensively in the rubber industry as a solvent for rubber latex in the latter half of the nineteenth century. The volatility of benzene, which made it so attractive to the industry, also caused high atmospheric levels of the solvent. Benzene-based rubber cements were used in the canning industry and in the shoe manufacturing industry. Although cases of benzene poisoning had been reported as early as 1897 and additional reports and warnings were issued in the 1920s, the excellent solvent properties of benzene resulted in its continued extensive use. In the 1930s, cases of benzene toxicity occurred in the printing industry in which benzene was used as an ink solvent. Today, benzene use exceeds 11 billion gallons per year.

Benzene affects the hematopoietic tissue in the bone marrow and also appears to be an immunosuppressant. There is a gradual decrease in white blood cells, red blood cells, and platelets, and any combination of these signs may be seen. Continued exposure to benzene results in severe bone marrow damage and aplastic anemia. Benzene exposure has also been associated with leukemia.

Asbestos and other fibers of naturally occurring silicates will separate into flexible fibers. Asbestos is the general name for this group of fibers. Chrysotile is the most important commercially and represents about 90% of the total used. The use of asbestos has been extensive, especially in roofing and insulation, asbestos cements, brake linings, electrical appliances, and coating materials. Asbestosis, a respiratory disease, is characterized by fibrosis, calcification, and lung cancer. In humans, not only is there a long latency period between exposure and development of tumors, but other factors also influence the development of lung cancer. Cigarette smoking, for example, enhances tumor formation. Recent studies have shown that stomach and bowel cancers occur in excess in workers (such as insulation workers) exposed to asbestos. Other fibers have been shown to cause a similar disease spectrum, for instance, zeolite fibers.

BIBLIOGRAPHY AND SUGGESTED READING

Air Pollutants

Costa, D. L. Air pollution. In *Casarett and Doull's Toxicology: The Basic Science of Poisons*, 6th ed., ed. C. D. Klaassen, pp. 979–1012. New York: McGraw-Hill, 2001.

Holgate, S. T., J. M. Samet, H. Koren, et al., eds. *Air Pollution and Health*. San Diego, CA: Academic Press, 1999.

Water and Soil Pollutants

Abel, P. D., ed. *Water Pollution Biology*. London: Taylor & Francis, 1996.

Hoffman, D. J., B. A. Rattner, G. A. Burton, eds. *Handbook of Ecotoxicology*, 2nd ed. Boca Raton, FL: Lewis, 2002.

Larson, S. J., P. D. Capel, and M. S. Majewski, eds. *Pesticides in Surface Waters*. Chelsea, MI: Ann Arbor Press, 1998.

Occupational Toxicants

Doull, J. Recommended limits for occupational exposure to chemicals. In *Casarett and Doull's Toxicology: The Basic Science of Poisons*, 6th ed., ed. C. D. Klaassen, pp. 1155–1176. New York: McGraw-Hill, 2001.

Thorne, P. S. Occupational toxicology. In *Casarett and Doull's Toxicology: The Basic Science of Poisons*, 6th ed., ed. C. D. Klaassen, pp. 1123–1140. New York: McGraw-Hill, 2001.

SAMPLE QUESTIONS

1. Of the current criteria, air pollutants that were discussed, name the one that is a colorless gas composed of three atoms of oxygen, and briefly describe how and why it is formed in the lower part of the atmosphere such that it results in a human health concern.

2. Describe the condition termed sick building syndrome and name the most common groups of organisms responsible for this problem.

3. Define and explain the distinction between point source pollution and nonpoint source pollution, and provide an example of each.

4. Describe the health condition (name of disease) and sensitive subpopulation of humans that occurs when excessive nitrate contaminates groundwater supplies.

Classes of Toxicants: Use Classes

W. GREGORY COPE and ERNEST HODGSON

4.1 INTRODUCTION

As indicated earlier, organisms are not exposed to one chemical at a time, but rather they are exposed to mixtures of chemicals, the composition of which changes over time. The information contained in this chapter is, because of its largely descriptive nature, closely similar to that in the corresponding chapter in the 3rd edition. It is nevertheless essential for understanding which toxicants are in use commercially, which have been in use so recently as to still be in the environment, and which are naturally occurring.

As discussed in Chapter 1, use classes include not only chemicals currently in use but also the toxicological aspects of the development of new chemicals for commercial use, chemicals produced as by-products of industrial processes, and chemicals resulting from the use and/or disposal of chemicals. Because any use class may include chemicals from several different chemical classes, this classification is not sufficient for mechanistic considerations. It is, however, essential for an understanding of the scope of toxicology and, in particular, is essential for many applied branches of toxicology such as exposure assessment, industrial hygiene, public health toxicology and regulatory toxicology. It also provides the information necessary for understanding why certain chemicals have greater priority for research, which have greater priority for the toxicity testing required for human and environmental risk analysis and which are likely to be components of the mixture of toxicants characteristic of particular exposure scenarios.

4.2 METALS

4.2.1 History

Although most metals occur in nature in rocks, ores, soil, water, and air, levels are usually low and widely dispersed. In terms of human exposure and toxicological

A Textbook of Modern Toxicology, *Fourth Edition*. Edited by Ernest Hodgson
Copyright © 2010 John Wiley & Sons, Inc.

significance, it is anthropogenic activities that are most important because they increase the levels of metals at the site of human activities.

Metals have been used throughout much of human history to make utensils, machinery, and so on, and mining and smelting supplied metals for these uses. These activities increased environmental levels of metals. More recently, metals have found a number of uses in industry, agriculture, and medicine. These activities have increased exposure not only to metal-related occupational workers but also to consumers of the various products.

Despite the wide range of metal toxicity and toxic properties, there are a number of toxicological features that are common to many metals. Some of the more important aspects are discussed briefly in the following sections. For a metal to exert its toxicity, it must cross the membrane and enter the cell. If the metal is in a lipid-soluble form such as methylmercury, it readily penetrates the membrane; when bound to proteins such as cadmium-metallothionein, the metal is taken into the cell by endocytosis; other metals (e.g., lead) may be absorbed by passive diffusion. The toxic effects of metals usually involve interaction between the free metal and the cellular target. These targets tend to be specific biochemical processes and/or cellular and subcellular membranes.

4.2.2 Common Toxic Mechanisms and Sites of Action

Enzyme Inhibition/Activation A major site of toxic action for metals is interaction with enzymes, resulting in either enzyme inhibition or activation. Two mechanisms are of particular importance: inhibition may occur as a result of interaction between the metal and sulfhydryl (SH) groups on the enzyme, or the metal may displace an essential metal cofactor of the enzyme. For example, lead may displace zinc in the zinc-dependent enzyme δ-aminolevulinic acid dehydratase (ALAD), thereby inhibiting the synthesis of heme, an important component of hemoglobin and heme-containing enzymes, such as the various cytochromes.

Subcellular Organelles Toxic metals may disrupt the structure and function of a number of organelles. For example, enzymes associated with the endoplasmic reticulum may be inhibited, metals may be accumulated in the lysosomes, respiratory enzymes in the mitochondria may be inhibited, and metal inclusion bodies may be formed in the nucleus.

Carcinogenicity A number of metals have been shown to be carcinogenic in humans or in animals. Arsenic, certain chromium compounds, and nickel are known human carcinogens; beryllium, cadmium, and cisplatin are probable human carcinogens. The carcinogenic action, in some cases, is thought to result from the interaction of the metallic ions with DNA (see Chapter 11 for a detailed discussion of carcinogenesis).

Kidney Because the kidney is the main excretory organ of the body, it is a common target organ for metal toxicity. Cadmium and mercury, in particular, are potent nephrotoxicants and are discussed more fully in the following sections on cadmium and mercury and in Part V on organ toxicity (Chapter 14).

Nervous System   The nervous system is also a common target of toxic metals, particularly organic metal compounds (see Chapter 15). For example, methylmercury, because it is lipid soluble, readily crosses the blood–brain barrier and enters the nervous system. By contrast, inorganic mercury compounds, which are more water soluble, are less likely to enter the nervous system and are primarily nephrotoxicants. Likewise, organic lead compounds are mainly neurotoxicants, whereas the first site of inorganic lead is enzyme inhibition (e.g., enzymes involved in heme synthesis).

Endocrine and Reproductive Effects   Because the male and female reproductive organs are under complex neuroendocrine and hormonal control, any toxicant that alters any of these processes can affect the reproductive system (see Chapter 16). In addition, metals can act directly on the sex organs. Cadmium is known to produce testicular injury after acute exposure, and lead accumulation in the testes is associated with testicular degeneration, inhibition of spermatogenesis, and Leydig cell atrophy.

Respiratory System   Occupational exposure to metals in the form of metal dust makes the respiratory system a likely target. Acute exposure may cause irritations and inflammation of the respiratory tract, whereas chronic exposure may result in fibrosis (aluminum) or carcinogenesis (arsenic, chromium, nickel). Respiratory toxicants are discussed more fully in Chapter 18.

Metal-Binding Proteins   The toxicity of many metals such as cadmium, lead, and mercury depends on their transport and intracellular bioavailabiltiy. This availability is regulated to a degree by high-affinity binding to certain cytosolic proteins. Such ligands usually possess numerous SH binding sites that can outcompete other intracellular proteins and thus mediate intracellular metal bioavailability and toxicity. These intracellular "sinks" are capable of partially sequestering toxic metals away from sensitive organelles or proteins until their binding capacity is exceeded by the dose of the metal. *Metallothionein* (MT) is a low molecular weight metal-binding protein (approximately 7000 Da) that is particularly important in regulating the intracellular bioavailability of cadmium, copper, mercury, silver, and zinc. For example, *in vivo* exposure to cadmium results in the transport of cadmium in the blood by various high molecular weight proteins and uptake by the liver, followed by hepatic induction of MT. Subsequently, cadmium can be found in the circulatory system bound to MT as the cadmium-metallothionein complex (CdMT).

4.2.3   Lead

Because of the long-term and widespread use of lead, it is one of the most ubiquitous of the toxic metals. Exposure may be through air, water, or food sources. In the United States, the major industrial uses, such as in fuel additives and lead pigments in paints, have been phased out, but other uses, such as in batteries, have not been reduced. Other sources of lead include lead from pipes and glazed ceramic food containers.

Inorganic lead may be absorbed through the gastrointestinal (GI) tract, the respiratory system, and the skin. Ingested inorganic lead is absorbed more efficiently from the GI tract of children than that of adults, readily crosses the placenta, and in children penetrates the blood–brain barrier. Initially, lead is distributed in the blood, liver, and kidney; after prolonged exposure, as much as 95% of the body burden of lead is found in bone tissue.

The main targets of lead toxicity are the hematopoietic system and the nervous system. Several of the enzymes involved in the synthesis of heme are sensitive to inhibition by lead, the two most susceptible enzymes being ALAD and heme synthetase (HS). Although clinical anemia occurs only after moderate exposure to lead, biochemical effects can be observed at lower levels. For this reason, inhibition of ALAD or appearance in the urine of aminolevulinic acid (ALA) can be used as an indication of lead exposure.

The nervous system is another important target tissue for lead toxicity, especially in infants and young children in whom the nervous system is still developing (Chapter 15). Even at low levels of exposure, children may show hyperactivity, decreased attention span, mental deficiencies, and impaired vision. At higher levels, encephalopathy may occur in both children and adults. Lead damages the arterioles and capillaries, resulting in cerebral edema and neuronal degeneration. Clinically, this damage manifests itself as ataxia, stupor, coma, and convulsions.

Another system affected by lead is the reproductive system (Chapter 16). Lead exposure can cause male and female reproductive toxicity, miscarriages, and degenerate offspring.

4.2.4 Mercury

Mercury exists in the environment in three main chemical forms: elemental (Hg^0), inorganic mercurous (Hg^+) and mercuric ($Hg2^+$) salts, and organic methylmercury (CH_3Hg) and dimethylmercury (CH_3HgCH_3) compounds. Elemental mercury, in the form of mercury vapor, is almost completely absorbed by the respiratory system, whereas ingested elemental mercury is not readily absorbed and is relatively harmless. Once absorbed, elemental mercury can cross the blood–brain barrier into the nervous system. Most exposure to elemental mercury tends to be from occupational sources.

Of more concern from environmental contamination is exposure to organic mercury compounds. Inorganic mercury may be converted to organic mercury through the action of sulfate-reducing bacteria, to produce methylmercury, a highly toxic form readily absorbed across membranes. Several large episodes of mercury poisoning have resulted from consuming seed grain treated with mercury fungicides or from eating fish contaminated with methylmercury. In Japan in the 1950s and 1960s, wastes from a chemical and plastics plant containing mercury were drained into Minamata Bay. The mercury was converted to the readily absorbed methylmercury by bacteria in the aquatic sediments. Consumption of fish and shellfish by the local population resulted in numerous cases of mercury poisoning or Minamata disease. By 1970, at least 107 deaths had been attributed to mercury poisoning, and 800 cases of Minamata disease were confirmed. Even though the mothers appeared healthy, many infants born to mothers who had eaten contaminated fish developed cerebral palsy-like symptoms and mental deficiency. Organic mercury primarily affects the nervous system, with the fetal brain being more sensitive to the toxic effects of mercury than that of adults.

Inorganic mercury salts, however, are primarily nephrotoxicants, with the site of action being the proximal tubular cells. Mercury binds to SH groups of membrane proteins, affecting the integrity of the membrane and resulting in aliguria, anuria, and uremia.

4.2.5 Cadmium

Cadmium occurs in nature primarily in association with lead and zinc ores and is released near mines and smelters processing these ores. Industrially, cadmium is used as a pigment in paints and plastics, in electroplating, and in making alloys and alkali storage batteries (e.g., nickel-cadmium batteries). Environmental exposure to cadmium is mainly from contamination of groundwater from smelting and industrial uses as well as the use of sewage sludge as a food crop fertilizer. Grains, cereal products, and leafy vegetables usually constitute the main source of cadmium in food. Reference has already been made to the disease Itai-Itai resulting from consumption of cadmium-contaminated rice in Japan (see Chapter 3, Section 3.2.2).

Acute effects of exposure to cadmium result primarily from local irritation. After ingestion, the main effects are nausea, vomiting, and abdominal pain. Inhalation exposure may result in pulmonary edema and chemical pneumonitis.

Chronic effects are of particular concern because cadmium is very slowly excreted from the body, with a half-life of about 30 years. Thus, low levels of exposure can result in considerable accumulation of cadmium. The main organ damaged following long-term exposure is the kidney, with the proximal tubules being the primary site of action. Cadmium is present in the circulatory system bound primarily to the metal-binding protein, MT, produced in the liver. Following glomerular filtration in the kidney, CdMT is reabsorbed efficiently by the proximal tubule cells, where it accumulates within the lysosomes. Subsequent degradation of the CdMT complex releases Cd^{+2}, which inhibits lysosomal function, resulting in cell injury.

4.2.6 Chromium

Because chromium occurs in ores, environmental levels are increased by mining, smelting, and industrial uses. Chromium is used in making stainless steel, various alloys, and pigments. The levels of this metal are generally very low in air, water, and food, and the major source of human exposure is occupational. Chromium occurs in a number of oxidation states from Cr^{+2} to Cr^{+6}, but only the trivalent (Cr^{+3}) and hexavalent (Cr^{+6}) forms are of biologic significance. Although the trivalent compound is the most common form found in nature, the hexavlent form is of greater industrial importance. In addition, hexavalent chromium, which is not water soluble, is more readily absorbed across cell membranes than is trivalent chromium. *In vivo*, the hexavalent form is reduced to the trivalent form, which can complex with intracellular macromolecules, resulting in toxicity. Chromium is a known human carcinogen and induces lung cancers among exposed workers. The mechanism of chromium (Cr^{+6}) carcinogenicity in the lung is believed to be its reduction to Cr^{+3} and generation of reactive intermediates, leading to bronchogenic carcinoma.

4.2.7 Arsenic

In general, the levels of arsenic in air and water are low, and the major source of human exposure is food. In certain parts of Taiwan and South America, however, the water contains high levels of this metalloid, and the inhabitants often suffer from dermal hyperkeratosis and hyperpigmentation. Higher levels of exposure result in a more serious condition: gangrene of the lower extremities or "blackfoot disease." Cancer of the skin also occurs in these areas.

Approximately 80% of arsenic compounds are used in pesticides. Other uses include glassware, paints, and pigments. Arsine gas is used in the semiconductor industry. Arsenic compounds occur in three forms: (1) pentavalent, As^{+5}, organic or arsenate compounds (e.g., alkyl arsenates); (2) trivalent, As^{+3}, inorganic or arsenate compounds (e.g., sodium arsenate, arsenic trioxide); and (3) arsine gas, AsH_3, a colorless gas formed by the action of acids on arsenic. The most toxic form is arsine gas with the threshold limit value–time-weighted average (TLV-TWA) of 0.05 ppm. Microorganisms in the environment convert arsenic to dimethylarsenate, which can accumulate in fish and shellfish, providing a source for human exposure. Arsenic compounds can also be present as contaminants in well water. Arsenite (As^{+3}) compounds are lipid soluble and can be absorbed following ingestion, inhalation, or skin contact. Within 24 h of absorption, arsenic distributes over the body, where it binds to SH groups of tissue proteins. Only a small amount crosses the blood–brain barrier. Arsenic may also replace phosphorus in bone tissue and be stored for years.

After acute poisoning, severe GI symptoms occur within 30 min to 2 h. These include vomiting, watery and bloody diarrhea, severe abdominal pain, and burning esophageal pain. Vasodilatation, myocardial depression, cerebral edema, and distal peripheral neuropathy may also follow. Later stages of poisoning include jaundice and renal failure. Death usually results from circulatory failure within 24 h to 4 days.

Chronic exposure results in nonspecific symptoms such as diarrhea, abdominal pain, hyperpigmentation, and hyperkeratosis. A symmetrical sensory neuropathy often follows. Late changes include gangrene of the extremities, anemia, and cancer of the skin, lung, and nasal tissue.

4.2.8 Treatment of Metal Poisoning

Treatment of metal exposure to prevent or reverse toxicity is done with chelating agents or antagonists. Chelation is the formation of a metal ion complex, in which the metal ion is associated with an electron donor ligand. Metals may react with O-, S-, and N-containing ligands (e.g., –OH, –COOH, –S–S–, and –NH₂). Chelating agents need to be able to reach sites of storage, form nontoxic complexes, not readily bind essential metals (e.g., calcium, zinc), and be easily excreted.

One of the first clinically useful chelating drugs was British anti-lewisite (BAL [2,3-dimercaptopropanol]), which was developed during World War II as an antagonist to arsenical war gases. BAL is a dithiol compound with two sulfur atoms on adjacent carbon atoms that compete with critical binding sites involved in arsenic toxicity. Although BAL will bind a number of toxic metals, it is also a potentially toxic drug with multiple side effects. In response to BAL's toxicity, several analogs have now been developed. Table 4.1 lists some of the more common chelating drugs in therapeutic use.

TABLE 4.1   Examples of Chelating Drugs Used to Treat Metal Toxicity

British anti-lewisite (BAL[2,3-dimercaptopropanol]), dimercaprol
DMPS (2,3-dimercapto-1-propanesulfonic acid)
DMSA (meso-2,3-dimercaptosuccinic acid)
EDTA (ethylenediaminetetraacetic acid, calcium salt)
DTPA (diethylenetriaminepentaacetic acid, calcium salt)
DTC (dithiocarbamate)
Penicillamine (β-β-dimethylcysteine), hydrolytic product of penicillin

4.3   AGRICULTURAL CHEMICALS (PESTICIDES)

4.3.1   Introduction

Chemicals have been used to kill or control pests for centuries. The Chinese used arsenic to control insects, the early Romans used common salt to control weeds and sulfur to control insects. In the 1800s, pyrethrin (i.e., compounds present in the flowers of the chrysanthemum, *Pyrethrum cinerariaefolium*) was found to have insecticidal properties. The roots of certain Derris species (*Derris species, D. elliptica* and *Lonchocarpus* spp.) were used by the Chinese and by South American natives as a fish poison. The active ingredient (AI), rotenone, was isolated in 1895 and was used for insect control. Another material developed for insect control in the 1800s was Paris Green, a mixture of copper and arsenic salts. Fungi were controlled with Bordeaux Mixture, a combination of lime and copper sulfate.

However, it was not until the 1900s that the compounds we identify today as having pesticidal properties came into being. Petroleum oils, distilled from crude mineral oils, were introduced in the 1920s to control scale insects and red spider mites. The 1940s saw the introduction of the chlorinated hydrocarbon insecticides such as DDT and the phenoxy acid herbicides such as 2,4-D). Natural compounds such as Red Squill, derived from the bulbs of red squill, *Urginea (Scilla) maritima* were effective in controlling rodents. Triazine herbicides, such as atrazine, introduced in the late 1950s, dominated the world herbicide market for years. Synthetic pyrethrins or pyrethroid insecticides (e.g., resmethrin) became and continue to be widely used insecticides due to their low toxicity, enhanced persistence compared to the pyrethrins, and low application rates. New families of fungicides, herbicides, and insecticides continue to be introduced into world markets as older compounds lose their popularity due to pest resistance or adverse health effects.

Pesticides are unusual among environmental pollutants in that they are used deliberately for the purpose of killing some form of life. Ideally, pesticides should be highly selective, destroying target organisms while leaving nontarget organisms unharmed. In reality, most pesticides are not so selective. In considering the use of pesticides, the benefits must be weighed against the risk to human health and environmental quality. Among the benefits of pesticides are control of vector-borne diseases, increased agricultural productivity, and control of urban pests. A major risk is environmental contamination, especially translocation within the environment where pesticides might enter both food chains and natural water systems. Factors to be considered in this regard are persistence in the environment and potential for bioaccumulation.

4.3.2 Definitions and Terms

The term, agricultural chemicals, has largely been replaced by the term pesticides or the two terms are used synonymously. It should be borne in mind, however, that all agricultural chemicals are not pesticides, for example, plant growth regulators and fertilizers. Pesticides are defined as economic poisons, regulated by federal and state laws that are used to control, kill, or repel pests. Depending upon what a compound is designed to do, pesticides have been subclassified into a number of categories (Table 4.2). The primary classes of pesticides in use today are fumigants,

TABLE 4.2 Classification of Pesticides, with Examples

Class	Principal Chemical Type	Example, Common Name
Algicide	Organotin	Brestar
Fungicide	Dicarboximide	Captan
	Chlorinated aromatic	Pentchlorophenol
	Dithiocarbamate	Maneb
	Mercurial	Phenylmercuric acetate
Herbicide	Amides, acetamides	Propanil
	Bipyridyl	Paraquat
	Carbamates, thiocarbamates	Barban
	Phenoxy	2,4-D
	Dinitrophenol	DNOC
	Dinitroaniline	Trifluralin
	Substitute urea	Monuron
	Triazine	Atrazine
Nematocide	Halogenated alkane	Ethylene dibromide (EDB)
Molluscicide	Chlorinated hydrocarbon	Bayluscide
Insecticide	Chlorinated hydrocarbons	
	DDT analogs	DDT
	Chlorinated alicyclic	BHC
	Cyclodiene	Aldrin
	Chlorinated terpenes	Toxaphene
	Organophosphorus	Chlorpyriphos
	Carbamate	Carbaryl
	Thiocyanate	Lethane
	Dinitrophenols	DNOC
	Fluoacetate	Nissol
	Botanicals	
	Nicotinoids	Nicotine
	Rotenoids	Rotenone
	Pyrethroids	Pyrethrin
	Synthetic pyrethroids	Fenvalerate
	Synthetic nicotinoids	Imidacloprid
	Fiproles	Fipronil
	Juvenile hormone analogs	Methroprene
	Growth regulators	Dimilin
	Inorganics	
	Arsenicals	Lead arsenate
	Fluorides	Sodium fluoride
	Microbials	Thuricide, avermectin

TABLE 4.2 *Continued*

Class	Principal Chemical Type	Example, Common Name
Insecticide synergists	Methylenedioxyphenyl	Piperonyl butoxide
	Dicarboximides	MGK-264
Acaricides	Organosulfur	Ovex
	Formamidine	Chlordimeform
	Dinitrophenols	Dinex
	DDT analogs	Chlorbenzilate
Rodenticides	Anticoagulants	Warfarin
	Botanicals	
	Alkaloids	Strychine sulfate
	Glycosides	Scillaren A and B
	Fluorides	Fluoroacetate
	Inorganics	Thallium sulfate
	Thioureas	ANTU

TABLE 4.3 Use Patterns of Pesticides Used in the United States[a]

Class	Percentage of Total Pesticide Use
Herbicides	47
Insecticides	19
Fungicides	13
Others[b]	21

[a] Most recent data: for 1997, published by U.S. EPA in 2001.
[b] Includes fumigants and wood preservatives.

fungicides, herbicides, and insecticides, with total U.S. production of 1.2 billion pounds (1997: United States Environmental Protection Agency's [U.S. EPA's] latest figures) and production of some 665 million pounds of wood preservatives. Table 4.3 describes the relative use of different classes of pesticides in the United States.

Generally, it takes some 5–7 years to bring a pesticide to market once its pesticidal properties have been verified. Many tests must be conducted to determine such things as the compound's chemical and physical properties and its efficacy. In addition, in order for registration for use by the U.S. EPA, numerous toxicity tests are undertaken, including those for acute toxicity, those for chronic effects such as reproductive anomalies, carcinogenesis, and neurological effects, and those for environmental effects; see Chapter 20).

The mandated pesticide label contains a number of specified items, including the concentration and/or percentage of both the AI and inert ingredients; proper mixing of the formulation with water to obtain the application rate of AI, what the AI will control, and how and when to apply. In addition, the label describes environmental hazards, proper storage of the material, reentry intervals (REIs) for application sites, and the personal protective equipment (PPE) that must be worn during application or harvesting.

Depending upon the toxicity, formulation concentration, and use patterns, pesticides can be classified as "general" or "restricted" use. A general use pesticide will cause no unreasonable, adverse effects when used according to the label and can be purchase and applied by anyone. A restricted use pesticide, defined as generally causing undesirable effects on the environment, applicator, or workers can only be purchased and applied by an individual who is licensed by the state.

The U.S. EPA has developed "category use" definitions based upon toxicity. Category I pesticides are highly hazardous, are classified as restricted use, and have an oral lethal dose 50 (LD_{50}) less than or equal to 1.0 mg/kg of body weight; Category II pesticides are moderately toxic and have an oral LD_{50} less than or equal to 500 mg/kg; Category III pesticides are generally nontoxic and have an oral LD_{50} less than or equal to 15,000 mg/kg. In addition, the U.S. EPA has developed a "Carcinogenicity Categorization" to classify pesticides for carcinogenicity.

Figure 4.1 Some examples of chemical structures of common pesticides.

4.3.3 Organochlorine Insecticides

The chlorinated hydrocarbon insecticides were introduced in the 1940s and 1950s and include familiar insecticides such as DDT (1,1,1-trichloro-2,2-di(4-chlorophenyl)ethane, methoxychlor, chlordane, heptachlor, aldrin, dieldrin, endrin, toxaphene, mirex, and lindane. The structures of two of the more familiar ones, DDT and dieldrin, are shown in Figure 4.1. The chlorinated hydrocarbons are neurotoxicants and cause acute effects by interfering with the transmission of nerve impulses. Although DDT was synthesized in 1874, its insecticidal properties were not noted until 1939, when Dr. Paul Mueller, a Swiss chemist, discovered its effectiveness as an insecticide and was awarded a Nobel Prize for his work. During World War II, the United States used large quantities of DDT to control vector-borne diseases, such as typhus and malaria, to which U.S. troops were exposed. After the war, DDT use became widespread in agriculture, public health, and households. Its persistence, initially considered a desirable attribute, later became the basis for public concern.

Permethrin   2,4-D

Atrazine   Paraquat

Maneb   Warfarin

Figure 4.1 *Continued*

The publication of Rachel Carson's book *Silent Spring* in 1962 stimulated this concern and eventually led to the ban of DDT and other chlorinated insecticides in the United States in 1972.

DDT, as well as other organochlorines, were used extensively from the 1940s through the 1960s in agriculture and mosquito control, particularly in the World Health Organization (WHO) malaria control programs. The cyclodiene insecticides, such as chlordane [4,7-methano1H-indene-1,2,3,4,5,6,7,7a,8,8-octachloro-2,3,3a,4,7,7a-hexahydro-(CAS9CI) were used extensively as termiticides into the 1980s but were removed from the market due to measurable residue levels penetrating into interiors and allegedly causing health problems. Residue levels of chlorinated insecticides continue to be found in the environment and, although the concentrations are now so low as to approach the limit of delectability, there continues to be concern.

4.3.4 Organophosphorus (OP) Insecticides

OPs are phosphoric acid esters or thiophosphoric acid esters (Figure 4.1) and are among the most widely used pesticides for insect control. During the 1930s and 1940s, Gerhard Schrader and coworkers began investigating OP compounds. They realized the insecticidal properties of these compounds and by the end of World War II had made many of the insecticidal OPs in use today, such as ethyl parathion [*O,O*-diethyl *O*-(4-nitrophenyl)phosphorothioate]. The first OP insecticide to find widespread use was tetraethylpyrophosphate (TEPP), approved in Germany in 1944 and marketed as a substitute for nicotine to control aphids. Because of its high mammalian toxicity and rapid hydrolysis in water, TEPP was replaced by other OP insecticides.

Chlorpyrifos [*O,O*-diethyl *O*-(3,5,6-trichloro-2-pyridinyl) phosphorothioate] became one of the largest selling insecticides in the world and had both agricultural and urban uses. The insecticide could be purchased for indoor use by homeowners, but health-related concerns caused U.S. EPA to cancel home indoor and lawn application uses in 2001. The only exception is its continued use as a termiticide.

Parathion was another widely used insecticide due to its stability in aqueous solutions and its broad range of insecticidal activity. However, its high mammalian toxicity through all routes of exposure led to the development of less hazardous compounds. Malathion [diethyl (dimethoxythiophosphorylthio) succinate], in particular, has low mammalian toxicity because mammals possess certain enzymes, the carboxylesterases, which readily hydrolyze the carboxyester link, detoxifying the compound. Insects, by contrast, do not readily hydrolyze this ester, and the result is its selective insecticidal action.

OPs are toxic because of their inhibition of the enzyme acetylcholinesterase. This enzyme inhibition results in the accumulation of acetylcholine in nerve tissue and effector organs, with the principal site of action being the peripheral nervous system (PNS) (see Chapter 15). In addition to acute effects, some OP compounds have been associated with delayed neurotoxicity, known as organophosphorus-induced delayed neuropathy (OPIDN). The characteristic clinical sign is bilateral paralysis of the distal muscles, predominantly of the lower extremities, occurring from 7 to 10 days following ingestion (see Chapter 15). Not all OP compounds cause delayed neuropathy. Among the pesticides associated with OPIDN are leptophos, mipafox, EPN, DEF (tribufos), and trichlorofon. Testing is now required for OP substances prior to their use as insecticides.

The OP and carbamate insecticides are relatively nonpersistent in the environment. They are applied to the crop or directly to the soil as systemic insecticides, and they generally persist from only a few hours to several months. Thus, these compounds, in contrast to the organochlorine insecticides, do not represent a serious problem as contaminants of soil and water and rarely enter the human food chain. Being esters, the compounds are susceptible to hydrolysis, and their breakdown products are generally nontoxic. Direct contamination of food by concentrated compounds has been the cause of poisoning episodes in several countries.

4.3.5 Carbamate Insecticides

The carbamate insecticides are esters of *N*-methyl (or occasionally *N,N*-dimethyl) carbamic acid (H_2NCOOH). The toxicity of the compound varies according to the phenol or alcohol group. One of the most widely used carbamate insecticides is carbaryl (1-napthyl methylcarbamate), a broad spectrum insecticide (Figure 4.1). It is used widely in agriculture, including home gardens where it generally is applied as a dust. Carbaryl is not considered to be a persistent compound because it is readily hydrolyzed. Based upon its formulation, it carries a toxicity classification of II or III with an oral LD_{50} of 250 mg/kg (rat) and a dermal lethal concentration 50 (LC_{50}) of >2,000 mg/kg.

An example of an extremely toxic carbamate is aldicarb [2-methyl-2-(methylthio) proprionaldehyde]. Both oral and dermal routes are the primary portals of entry, and it has an oral LD_{50} of 1.0 mg/kg (rat) and a dermal LD_{50} of 20 mg/kg (rabbit). For this reason, it is recommended for application to soils on crops such as cotton, citrus, and sweet potatoes. This compound moves readily through soil profiles and has contaminated groundwater supplies. Like the OP insecticides, the mode of action of the carbamates is acetylcholinesterase inhibition, with the important difference that the inhibition is more rapidly reversed than with OP compounds.

4.3.6 Botanical Insecticides

Extracts from plants have been used for centuries to control insects. Nicotine [(*S*)-3-(1-methyl-2-pyrrolidyl)pyridine] (Figure 4.1) is an alkaloid occurring in a number of plants and was first used as an insecticide in 1763. Nicotine is quite toxic orally as well as dermally. The acute oral LD_{50} of nicotine sulfate for rats is 83 mg/kg and the dermal LD_{50} is 285 mg/kg. Symptoms of acute nicotine poisoning occur rapidly, and death may occur with a few minutes. In serious poisoning cases, death results from respiratory failure due to paralysis of respiratory muscles. In therapy, attention is focused primarily on support of respiration.

Pyrethrin is an extract from several types of chrysanthemum, and is one of the oldest insecticides used by humans. There are six esters and acids associated with this botanical insecticide. Pyrethrin is applied at low doses and is considered to be nonpersistent.

Mammalian toxicity to pyrethrins is quite low, apparently due to its rapid breakdown by liver microsomal enzymes and esterases. The acute LD_{50} to rats is about 1500 mg/kg. The most frequent reaction to pyrethrins is contact dermatitis and allergic respiratory reactions, probably as a result of other constituents in the formulation. Synthetic mimics of pyrethrins, known as the pyrethroids, were developed to overcome the lack of persistence.

4.3.7 Pyrethroid Insecticides

As stated, pyrethrins are not persistent, which led pesticide chemists to develop compounds of similar structure having insecticidal activity but being more persistent. This class of insecticides, known as pyrethroids, have greater insecticidal activity and are more photostable than pyrethrins. There are two broad classes of pyrethroids depending upon whether the structure contains a cyclopropane ring [e.g., cypermethrin {(±)-α-cyano-3-phenoxybenzyl (±)-*cis,trans*-3-(2,2-dichlorovinyl_2,2-dimethyl cyclopropanecarboxylate)}] or whether this ring is absent in the molecule [e.g., fenvalerate{(RS)-α-cyano-3-phenoxybenzyl(RS)-2-(4-chlorophenyl)-3-methylbutyrate}]. They are generally applied at low doses (e.g., 30 g/Ha) and have low mammalian toxicities [e.g., cypermethrin, oral (aqueous suspension) LD_{50} of 4,123 mg/kg (rat) and dermal LD_{50} of >2,000 mg/kg (rabbit)]. Pyrethroids are used in both agricultural and urban settings (e.g., termiticide) (Figure 4.1).

Pyrethrins affect nerve membranes by modifying the sodium and potassium channels, resulting in depolarization of the membranes. Formulations of these insecticides frequently contain the insecticide synergist piperonyl butoxide [5-{2-(2-butoxyethoxy) ethoxymethyl}-6-propyl-1,3-benzodioxole], which acts to increase the efficacy of the insecticide by inhibiting the cytochrome P450 enzymes responsible for the breakdown of the insecticide.

4.3.8 New Insecticide Classes

There are new classes of insecticides that are applied at low dosages and are extremely effective but are relatively nontoxic to humans. One such class is the fiproles, and one of these receiving major attention is fipronil [(5-amino-1-(2,6-dichloro-4-(trifluoromethyl)phenyl)-4-((1,R,S)-(trifluoromethyl)su-1-H-pyrasole-3-carbonitrile)]. Although it is used on corn, it is becoming a popular termiticide because of its low application rate (ca 0.01%) and long-term effectiveness. Another class of insecticides, the chloronicotinoids, is represented by imidacloprid [1-(6-chloro-3-pyridin-3-ylmethyl)-N-nitroimidazolidin-2-ylidenamine] (Figure 4.1) that also is applied at low dose rates to soil and effectively controls a number of insect species, including termites.

4.3.9 Herbicides

Herbicides control weeds and are the most widely used class of pesticides. The latest U.S. EPA data show that some 578 million pounds of herbicides were used in the United States in 1997 and accounts for some 47% of pesticides used. This class of pesticide can be applied to crops using many strategies to eliminate or reduce weed populations. These include preplant incorporation, preemergent applications, and postemergent applications. New families of herbicides continue to be developed, and are applied at low doses, are relatively nonphytotoxic to beneficial plants, and are environmentally friendly. Some of the newer families such as the imidazolinones inhibit the action of acetohydroxyacid synthase that produces branched-chain amino acids in plants. Because this enzyme is produced only in plants, these herbicides have low acute toxicities to mammals, fish, insects, and birds.

The potential for environmental contamination continues to come from families of herbicides that have been used for years. The chlorophenoxy herbicides such as 2,4-D (2,4-dichlorophenoxy acetic acid) and 2,4,5-T (2,4,5-trichlorophenoxy-acetic acid) (Figure 4.1) are systemic acting compounds to control broadleaf plants and have been in use since the 1940s. The oral toxicities of these compounds are low.

A mixture of 2,4-D and 2,4,5-T, known as Agent Orange, was used by the U.S. military as a defoliant during the Vietnam conflict, and much controversy has arisen over claims by military personnel of long-term health effects. The chemical of major toxicological concern was identified as a contaminant, TCDD (2,3,7,8-tetrachlorodibenzo-p-dioxin), which was formed during the manufacturing process. TCDD is one of the most toxic synthetic substances known in laboratory animals. The LD_{50} for male rats is 0.022 mg/kg, and for female guinea pigs (the most sensitive species tested) the LD_{50} is 0.0006 mg/kg. In addition, it is toxic to developing embryos in pregnant rats at a dose of only 1/400 of the LD_{50}, and has been shown to cause birth defects at levels of 1–3 ng/kg of body weight. TCDD is a proven carcinogen in both mice and rats, with the liver being the primary target. This chemical has also been shown to alter the immune system and enhance susceptibility in exposed animals.

Another family of herbicides, the triazines, continues to cause concern to environmentalists and toxicologists because of the contamination of surface and groundwater supplies that become public drinking water. The herbicide, atrazine [6-chloro-N-ethyl-N-(1-methylethyl)-1,2,5-triazine-2,4-diamine (Figure 4.1) is used primarily on corn and has an MCL (maximum contaminant level) of 3.0 μg/L. This herbicide has been found in surface and groundwaters worldwide with widely varying concentrations (e.g., from 1 to >130 μg/L). Atrazine, along with two other triazines, cyanazine [2-{{4-chloro-6-(ethylamino)-1,3,5-triazin-2-yl}amino}-2-methylpropanenitrile] and simazine (6-chloro-N,N-diethyl-1,3,5-triazine-2,4-diamine) (MCL of 4.0 μg/L). The uses of cyanazine were canceled in 2001, and no further have been permitted after 2002. Although relatively nontoxic (e.g., atrazine, oral LD_{50} of 3,100 mg/kg [rat]), the major concern with these types of compounds is their carcinogenic effects, and U.S. EPA considers these three triazines as possible human carcinogens (Category C).

A member of the bipyridylium family of herbicides is the compound paraquat (1,1-dimethyl-4,4-bipyridinium ion as the chloride salt) (Figure 4.1), a very water-soluble contact herbicide that is active against a broad range of plants and is used as a defoliant on many crops. The compound binds tightly to soil particles following application and becomes inactivated. However, this compound is classified as a Class I toxicant with an oral LD_{50} of 150 mg/kg (rat). Most poisoning cases, which are often fatal, are due to accidental or deliberate ingestion of paraquat. Toxicity results from lung injury resulting from both the preferential uptake of paraquat by the lungs and the redox cycling mechanism.

4.3.10 Fungicides

Annually, fungi cause crop losses in the United States amounting to millions of dollars. In addition, recent studies have shown that toxins and other airborne organic compounds, released from fungi inhabiting the interior of dwellings, probably are responsible for a number of adverse health effects. Compounds produced to combat these losses and adverse health effects are called fungicides, and a number of these families have been around for years.

The fungicide, chlorothalonil (tetrachloroisophthalonitrile), is a broad-spectrum fungicide that is used widely in urban environments. It is relatively cheap and controls some 140 species of organisms. As a result of the popularity of this compound, it is found routinely in surface waters entering public drinking water supplies. In the formulation that can be purchased by the general public, it is relatively nontoxic.

One family of fungicides that is of concern are the dithiocarbamates, sulfur derivatives of dithiocarbamic acid and include the metallic dimethydithiocarbamates. The latter group includes mancozeb (a coordination product of zinc ion and manganese ethylene bisdithiocarbamate), maneb (manganese ethylenebisdithiocarbamate) (Figure 4.1), and zineb (zinc ethylenebisdithiocarbamate). All are effective fungicides and are used on a variety of crops including grapes, sugar beets, and ornamental plants. Although relatively nontoxic, they do hydrolyze producing known carcinogens such as ethylenethiourea (ETU).

4.3.11 Rodenticides

This class of compounds is used to control rodents that cause yearly losses of 20–30% in grain and other food storage facilities. These pests harbor diseases in the form of fleas that carry bacteria and other organisms. A number of rodenticides have been used for years and include warfarin [3-(α-acetonylbenzyl)-4-hydroxycoumarin] (Figure 4.1), an anticoagulant. This is a potent toxicant with an oral LD_{50} of 3.0 mg/kg (rat). As the rats navigate through narrow passages, they bruise themselves, developing small hemorrhages. Anticoagulants prevent the blood from clotting, and the animals bleed to death in about a week. Humans who are exposed to this class of compounds are given vitamin K and, if the poisoning is severe, blood transfusions as a treatment. Other rodenticides poison the animal and many times are applied along with an attractant such as peanut butter to overcome bait shyness. Fluoroacetamide is a fast acting poison with an oral LD_{50} (rat) of 15 mg/kg. This material is supplied as bait pellets or grains. ANTU (α-naphthylthiourea), strychnine, and thallium salts are other fast acting poisons, and have been on the market for many years. Most of the rodenticides are classified as restricted use and are applied only by licensed pest control operators. Human poisonings associated with rodenticides usually result from accidental or suicidal ingestion of the compounds.

4.3.12 Fumigants

Fumigants are extremely toxic gases used to protect stored products, especially grains, and to kill soil nematodes. These materials are applied to storage warehouses, to freight cars, and to houses infested with insects such as powder post beetles. They present a special hazard due to inhalation exposure and rapid diffusion into pulmonary blood; thus, extreme care must be taken when handling and applying this class of pesticides. All fumigants are classified as restricted use compounds and require licensed applicators to handle them.

One of the most effective fumigants is methyl bromide. It essentially sterilizes soil when applied under a ground covering, because it not only kills insects, nematodes, and weed seed but is also used to fumigate warehouses. Overexposure to this compound causes respiratory distress, cardiac arrest, and central nervous effects. The inhalation LC_{50} is 0.06 mg/L (15 min) of air (rat) and 7900 ppm (1.5 h) (human). Methyl bromide has been classified as an ozone depleter under the Clean Air Act and is due to be phased out of use.

Chloropicrin (trichloronitromethane) is another soil/space fumigant that has been used for many years. It has an inhalation LC_{50} of 150 ppm (15 min). Thus, it is highly toxic by inhalation, can injure the heart, and cause severe eye damage.

4.3.13 Conclusions

This section has covered only a few of the pesticides available today in the United States and world markets. An understanding of the basic chemical processes affected by pesticides has led to the discovery and production of new families of chemicals. Today's modern pesticide is generally safe to use if the directions on the label are followed. Advances in instrumentation and an understanding of how adverse health effects are produced have resulted in the production of many environmentally friendly but effective pesticides.

4.4 FOOD ADDITIVES AND CONTAMINANTS

Chemicals are added to food for a number of reasons: as preservatives, either antibacterial, antifungal, or antioxidants; to change physical characteristics, particularly for processing; to change taste; to change color; and to change odor. In general, food additives have proved to be safe and without chronic toxicity. Many were introduced when toxicity testing was relatively unsophisticated, however, and some of these have been subsequently shown to be toxic. Table 4.4 gives examples of different types of organic food additives. Inorganics, the most important of which are nitrate and nitrite, are discussed later. Certainly, hundreds, and possibly thousands, of food additives are in use worldwide, many with inadequate testing. The question of synergistic interactions between these compounds has not been explored adequately. Not all toxicants in food are synthetic; many examples of naturally occurring toxicants in the human diet are known, including carcinogens and mutagens.

TABLE 4.4   Examples of Organic Chemicals Used as Food Additives

Function	Class	Example
Preservatives	Antioxidants	Butylatedhydroxyanisole
		Ascorbic acid
	Fungistatic agents	Methyl *p*-benzoic acid
		Propionates
	Bactericides	Sodium nitrite
Processing aids	Anticaking agents	Calcium silicate
		Sodium aluminosilicate
	Emulsifiers	Propylene glycol
		Monoglycerides
	Chelating agents	EDTA
		Sodium tartrate
	Stabilizers	Gum ghatti
		Sodium alginate
	Humectants	Propylene glycol
		Glycerol
Flavor and taste Modification	Synthetic sweeteners	Saccharin
		Mannitol
		Aspartame
	Synthetic flavors	Piperonal
		Vanillin
Color modification	Synthetic dyes	Tartrazine (FD&C yellow5)
		Sunset yellow
Nutritional supplements	Vitamins	Thiamin
		Vitamin D3
	Amino acids	Alanine
		Aspartic acid
	Inorganics	Manganese sulfate
		Zinc sulfate

4.5   TOXINS

4.5.1   History

A discussion of toxins first necessitates the understanding and distinction between the toxicological terms toxicant and toxin. A *toxicant* is any chemical, of natural or synthetic origin, capable of causing a deleterious effect on a living organism. A *toxin* is a toxicant that is produced by a living organism and is not used as a synonym for toxicant—all toxins are toxicants, but not all toxicants are toxins. Toxins, whether produced by animals, plants, insects, or microbes are generally metabolic products that have evolved as defense mechanisms for the purpose of repelling or killing predators or pathogens. The action of natural toxins has long been recognized and understood throughout human history. For example, ancient civilizations used natural toxins for both medicinal (therapeutic) and criminal purposes. Even today, we continue to discover and understand the toxicity of natural products, some for beneficial pharmaceutical or therapeutic purposes whose safety and efficacy are tested, and some for other less laudable purposes like biological or chemical warfare. Toxins may be classified in various ways depending on interest and need, such as by target organ toxicity or mode of action, but are commonly classified according to source.

4.5.2   Microbial Toxins

The term "microbial toxin" is usually reserved by microbiologists for toxic substances produced by microorganisms that are of high molecular weight and have antigenic properties; toxic compounds produced by bacteria that do not fit these criteria are referred to simply as poisons. Many of the former are proteins or mucoproteins and may have a variety of enzymatic properties. They include some of the most toxic substances known, such as tetanus toxin, botulinus toxin, and diphtheria toxin. Bacterial toxins may be extremely toxic to mammals and may affect a variety of organ systems, including the nervous system and the cardiovascular system. A detailed account of their chemical nature and mode of action is beyond the scope of this volume.

The range of poisonous chemicals produced by bacteria is also large. Again, such compounds may also be used for beneficial purposes; for example, the insecticidal properties of *Bacillus thuringiensis*, due to a toxin, have been utilized in agriculture for some time.

4.5.3   Mycotoxins

The range of chemical structures and biologic activity among the broad class of fungal metabolites is large and cannot be summarized briefly. Mycotoxins do not constitute a separate chemical category, and they lack common molecular features.

Mycotoxins of most interest are those found in human food or in the feed of domestic animals. They include the ergot alkaloids produced by *Claviceps* sp., aflatoxins, and related compounds produced by *Aspergillus* sp., and the tricothecenes produced by several genera of fungi imperfecti, primarily *Fusarium* sp.

The ergot alkaloids are known to affect the nervous system and to be vasoconstrictors. Historically, they have been implicated in epidemics of both gangrenous and convulsive ergotism (St. Anthony's fire), although such epidemics no longer occur in humans due to increased knowledge of the cause and to more varied modern diets. Outbreaks of ergotism in livestock do still occur frequently, however. These compounds have also been used as abortifacients. The ergot alkaloids are derivatives of ergotine, the most active being, more specifically, amides of lysergic acid.

Aflatoxins are products of species of the genus *Aspergillus*, particularly *A. flavus*, a common fungus found as a contaminant of grain, maize, peanuts, and so on. First implicated in poultry diseases such as Turkey-X disease, they were subsequently shown to cause cancer in experimental animals and, from epidemiological studies, in humans. Aflatoxin B1, the most toxic of the aflatoxins, must be activated enzymatically to exert its carcinogenic effect.

Tricothecenes are a large class of sesquiterpenoid fungal metabolites produced particularly by members of the genera *Fusarium* and *Tricoderma*. They are frequently acutely toxic, displaying bactericidal, fungicidal, and insecticidal activity, as well as causing various clinical symptoms in mammals, including diarrhea, anorexia, and ataxia. They have been implicated in natural intoxications in both humans and animals, such as Abakabi disease in Japan and Stachybotryotoxicosis in the former USSR, and are at the center of a continuing controversy concerning their possible use as chemical warfare agents.

Mycotoxins may also be used for beneficial purposes. The mycotoxin avermectin is currently generating considerable interest both as an insecticide and for the control of nematode parasites of domestic animals.

4.5.4 Algal Toxins

Algal toxins are broadly defined to represent the array chemicals derived from many species of cyanobacteria (blue-green bacteria), dinoflagellates, and diatoms. The toxins produced by these freshwater and marine organisms often accumulate in fish and shellfish inhabiting the surrounding waters, causing both human and animal poisonings, as well as overt fish kills. Unlike many of the microbial toxins, algal toxins are generally heat stable and, therefore, not altered by cooking methods, which increases the likelihood of human exposures and toxicity. Many of the more common algal toxins responsible for human poisonings worldwide are summarized herein.

Amnesic shellfish poisoning (ASP) was first identified in 1987 from Prince Edward Island, Canada after four people died from eating contaminated mussels. It is caused by domoic acid produced by several species of *Pseudonitzschia* diatoms. The main contamination problems include mussels, clams, and crabs of the Pacific Northwest of the United States and Canada.

Paralytic shellfish poisoning (PSP) was first determined to be a problem in 1942 after three people and many seabirds died from eating shellfish on the west coast of the United States, near the Columbia River. It is caused by the saxitoxin family (saxitoxin + 18 related compounds) produced by several species of *Alexandrium* dinoflagellates. The main contamination problems include mussels, clams, crabs, and fish of the Pacific Northwest and Northeast Atlantic.

Neurotoxic shellfish poisoning (NSP) is caused by a red tide producer that was first identified in 1880 from Florida, with earlier historical references. It causes sickness in humans lasting several days. NSP is not fatal to humans; however, it is known to kill fish, invertebrates, seabirds, and marine mammals (e.g., manatees). It is caused by the brevetoxin family (brevetoxin + 10 related compounds) produced by the dinoflagellate *Karenia brevis* a.k.a. *Gymnodinium breve*. The main contamination problems include oysters, clams, and other filter feeders of the Gulf of Mexico and southeast Atlantic, including North Carolina.

Diarrheic shellfish poisoning (DSP) was first identified from human poisonings in the 1960s. It causes sickness in humans lasting several days, but is not fatal. It is caused by chemicals of the okadaic acid family (okadaic acid + four related compounds) produced by several species of *Dinophysis* dinoflagellates. The main contamination problems include mussels, clams, and other bivalves of the cold and warm temperate areas of the Atlantic and Pacific Oceans, mainly in Japan and Europe. Only two cases of DSP have been documented in North America.

Ciguatera fish poisoning (CFP) was first identified in 1511. CFP is a tropical-subtropical seafood poisoning that affects up to 50,000 people each year and is the most often reported foodborne disease of a chemical origin in the United States. Caused by consumption of reef fishes (e.g., grouper, snapper), sickness in humans lasts several days to weeks, but the human fatality rate is low. It is caused by the ciguatoxin family (ciguatoxin + three or more related compounds) and is produced by several species of dinoflagellates including *Gambierdiscus*, *Prorocentrum*, and *Ostreopsis*. The main contamination problems include herbivorous tropical reef fish worldwide.

Cyanobacterial (blue-green bacteria) toxin poisonings were first recognized in the late 1800s. Human poisonings are rare; however, kills of livestock, other mammals, birds, fish, and aquatic invertebrates are common. It is caused by a variety of bio-toxins and cytotoxins, including anatoxin, microcystin, and nodularin produced by several species of cyanobacteria including, *Anabaena*, *Aphanizomenon*, *Nodularia*, *Oscillatoria*, and *Microcystis*. The main contamination problems include all eutrophic freshwater rivers, lakes, and streams.

*Ambush predator (*Pfiesteria piscicida *and toxic* Pfiesteria *complex) toxins* come from members belonging to this group of organisms that were first identified in 1991 from estuaries in North Carolina. They were believed to produce a toxin that has been implicated in several large fish kills and is suspect in causing adverse human health effects. However, the toxin or toxins are not yet identified and toxicity tests are not universally conclusive. These toxins are produced by several dinoflagellate species including *Pfiesteria piscicida, Pfiesteria shumwayae*, and perhaps several other unidentified, unnamed dinoflagellates belonging to the potentially toxic *Pfiesteria* complex. The main problems include major fish kills in North Carolina and Maryland and potential human health effects. The range may extend from the Gulf of Mexico to the Atlantic estuarine waters, including Florida, North Carolina, Maryland, and Delaware, and possibly outward to Europe.

4.5.5 Plant Toxins

The large array of toxic chemicals produced by plants (phytotoxins), usually referred to as secondary plant compounds, are often held to have evolved as defense mechanisms against herbivorous animals, particularly insects and mammals. These compounds may be repellent, but not particularly toxic, or they may be acutely toxic to a wide range of organisms. They include sulfur compounds, lipids, phenols, alkaloids, glycosides, and many other types of chemicals. Many of the common drugs of abuse such as cocaine, caffeine, nicotine, morphine, and the cannabinoids are plant toxins. Many chemicals that have been shown to be toxic are constituents of plants that form part of the human diet. For example, the carcinogen safrole and related compounds are found in black pepper. Solanine and chaconine, which are cholinesterase inhibitors and possible teratogens, are found in potatoes, and quinines and phenols are widespread in food. Livestock poisoning by plants is still an important veterinary problem in some areas.

4.5.6 Animal Toxins

Some species from practically all phyla of animals produce toxins for either offensive or defensive purposes. Some are passively venomous, often following inadvertent ingestion, whereas others are actively venomous, injecting poisons through specially adapted stings or mouthparts. It may be more appropriate to refer to the latter group only as venomous and to refer to the former simply as poisonous. The chemistry of animal toxins extends from enzymes and neurotoxic and cardiotoxic peptides and proteins to many small molecules such as biogenic amines, alkaloids, glycosides, terpenes, and others. In many cases, the venoms are complex mixtures that include both proteins and small molecules and depend on the interaction of the components for the full expression of their toxic effect. For example, bee venom contains a biogenic amine, histamine, three peptides, and two enzymes (Table 4.5). The venoms and defensive secretions of insects may also contain many relatively simple toxicants or irritants such as formic acid, benzoquinone, and other quinines, or terpenes such as citronellal. Bites and stings from the hymenoptera (ants, bees, wasps, and hornets) result in 5–60 fatal anaphylactic reactions each year in the United States. According to experts, about 0.3–3.0% of the U.S. population experiences anaphylactic reactions from insect stings and bites.

Snake venoms have been studied extensively; their effects are due, in general, to toxins that are peptides with 60–70 amino acids. These toxins are cardiotoxic or neurotoxic, and their effects are usually accentuated by the phospholipases, peptidases, proteases, and other enzymes present in venoms. These enzymes may affect the blood-clotting mechanisms and damage blood vessels. Snake bites are responsible for less than 10 deaths per year in the United States, but many thousands worldwide.

Many fish species, over 700 species worldwide, are either directly toxic, or upon ingestion, are poisonous to humans. A classic example is the toxin produced by the puffer fishes (*Sphaeroides* spp.) called tetrodotoxin (TTX). TTX is concentrated in the gonads, liver, intestine, and skin, and poisonings occur most frequently in Japan and other Asian countries where the flesh, considered a delicacy, is eaten as "fugu." Death occurs within 5–30 min and the fatality rate is about 60%. TTX is an inhibitor of the voltage-sensitive Na channel (like saxitoxin); it may also be found in some salamanders and may be bacterial in origin.

TABLE 4.5 Some Components of Bee Venom

Compound	Effect
Biogenic amine	
Histamine	Pain, vasodilation, increased capillary permeability
Peptides	
Apamine	CNS effects
Melittin	Hemolytic, serotonin release, cardiotoxic
Mast cell degranulating peptide	Histamine release from mast cells
Enzymes	
Phospholipase A	Increased spreading and penetration of tissues
Hyaluronidase	

Toxins and other natural products generally provide great benefit to society. For example, some of the most widely used drugs and therapeutics like streptomycin, the aminoglycoside antibiotic from soil bacteria, and acetylsalicylic acid (aspirin), the nonsteroidal anti-inflammatory from willow tree bark, are used by millions of people everyday to improve health and well-being. On the other hand, adverse encounters with natural toxins like fish and shellfish toxins, plant, and insect toxins do result in harm to humans.

4.6   SOLVENTS

Although solvents are more a feature of the workplace, they are also found in the home. In addition to cutaneous effects, such as defatting and local irritation, many have systemic toxic effects, including effects on the nervous system or, as with benzene, on the blood-forming elements. Commercial solvents are frequently complex mixtures and may include nitrogen- or sulfur-containing organics— gasoline and other oil-based products are examples of this. The common solvents fall into the following classes:

1. *Aliphatic hydrocarbons*, such as hexane. These may be straight or branched chain compounds and are often present in mixtures.
2. *Halogenated aliphatic hydrocarbons*. The best-known examples are methylene dichloride, chloroform, and carbon tetrachloride, although chlorinated ethylenes are also widely used.
3. *Aliphatic alcohols*, such as methanol and ethanol.
4. *Glycols and glycol ethers* such as ethylene and propylene glycols. Use in antifreeze gives rise to considerable exposure of the general public. The glycol ethers, such as methyl cellosolve, are also widely used.
5. *Aromatic hydrocarbons*. Benzene is probably the one of greatest concern, but others, such as toluene, are also used.

4.7   THERAPEUTIC DRUGS

Although the study of the therapeutic properties of chemicals falls within the province of pharmacology, essentially all therapeutic drugs can be toxic, producing deleterious effects at some dose. The danger to the individual depends on several factors, including the nature of the toxic response, the dose necessary to produce the toxic response, and the relationship between the therapeutic dose and the toxic dose. Drug toxicity is affected by all of the factors that affect the toxicity of other xenobiotics, including individual (genetic) variation, diet, age, and the presence of other exogenous chemicals.

Even when the risk of toxic side effects from a particular drug has been evaluated, it must be weighed against the expected benefits. The use of a very dangerous drug with only a narrow tolerance between the therapeutic and toxic doses may still be justified if it is the sole treatment for an otherwise fatal disease. However, a relatively safe drug may be inappropriate if safer compounds are available or if the condition being treated is trivial.

The three principal classes of cytotoxic agents used in the treatment of cancer all contain carcinogens, for example, melphalan, a nitrogen mustard, adriamycin, an antitumor antibiotic, and methotrexate, an antimetabolite. Diethylstilbestrol (DES), a drug formerly widely used, has been associated with cancer of the cervix and vagina in the offspring of treated women.

Other toxic effects of drugs can be associated with almost every organ system. The stiffness of the joints accompanied by damage to the optic nerve (subacute myelo-optic neuropathy [SMON]) that was common in Japan in the 1960s was apparently a toxic side effect of chloroquinol, an antidiarrhea drug. Teratogenosis can also be caused by drugs, with thalidomide being the most alarming example. Skin effects (dermatitis) are common side effects of drugs, an example being topically applied corticosteroids.

A number of toxic effects on the blood have been documented, including agranulocytosis caused by chlorpromazine, hemolytic anemia caused by methyldopa, and megaloblastic anemia caused by methotrexate. Toxic effects on the eye have been noted and range from retinotoxicity caused by thioridazine to glaucoma caused by systemic corticosteroids.

4.8 DRUGS OF ABUSE

All drugs are toxic at some dose. Drugs of abuse, however, either have no medicinal function or are taken at dose levels higher than would be required for therapy. Although some drugs of abuse may affect only higher nervous functions—mood, reaction time, and coordination—many produce physical dependence and have serious physical effects, with fatal overdoses being a frequent occurrence.

The drugs of abuse include central nervous system depressants such as ethanol, methaqualone and secobarbital; central nervous system stimulants, such as cocaine, methamphetamine (speed), caffeine and, nicotine; opioids such as heroin, and mependine (demerol); and hallucinogens such as lysergic acid diethylamide (LSD), phencyclidine (PCP), and tetrahydrocannabinol, the most active principal of marijuana. A further complication of toxicological significance is that many drugs of abuse are synthesized in illegal and poorly equipped laboratories with little or no quality control. The resultant products are therefore often contaminated with compounds of unknown, but conceivably dangerous, toxicity. The structures of some of these chemicals are shown in Figure 4.2.

4.9 COMBUSTION PRODUCTS

While many air pollutants (see Chapter 3) are the products of natural or anthropomorphic combustion, some of the most important from the point of view of human health are polycyclic aromatic hydrocarbons. Although also found in natural products such as coal and crude oil, they are generally associated with incomplete combustion of organic materials and are found in smoke from wood, coal, oil, tobacco, and so on, as well as in broiled foods. Because some of them are car-

1. CNS Depressants

C₂H₅OH
Ethanol

Methaqualone

Secobarbital

2. CNS Stimulants

Cocaine

Methamphetamine

Nicotine

Caffeine

3. Opioids

Meperidine

Morphine
(Heroine, OHs = OCOCH₃)

4. Hallucinogens

Phencyclidine

Lysergic Acid Diethylamide
(LSD)

Δ⁹-Tetrahydrocannabinol
(THC)

Figure 4.2 Some common drugs of abuse.

cinogens, they have been studied intensively from the point of view of metabolic activation, interactions with DNA, and other aspects of chemical carcinogenesis. Some are heterocyclic, containing nitrogen in at least one of the rings. Some representative structures of the most studied polycyclic aromatic hydrocarbons are shown in Figure 4.3.

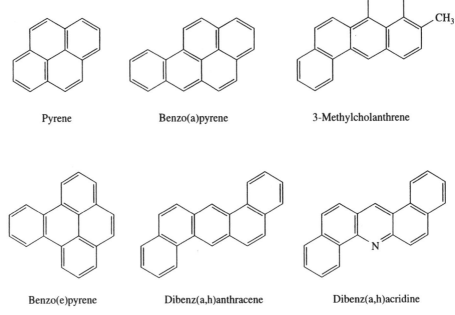

Pyrene   Benzo(a)pyrene   3-Methylcholanthrene

Benzo(e)pyrene   Dibenz(a,h)anthracene   Dibenz(a,h)acridine

Figure 4.3   Some common polycyclic aromatic hydrocarbons.

4.10 COSMETICS

The most common deleterious effects of modern cosmetics are occasional allergic reactions and contact dermatitis. The highly toxic and/or carcinogenic azo or aromatic amine dyes are no longer in use, nor are the organometallics, used in even earlier times. Bromates, used in some cold wave neutralizers, may be acutely toxic if ingested, as may the ethanol used as a solvent in hair dyes and perfumes. Thioglycolates and thioglycerol used in cold wave lotion and depilatories and sodium hydroxide used in hair straighteners are also toxic on ingestion. Used as directed, cosmetics appear to present little risk of systemic poisoning, due in part to the deletion of ingredients now known to be toxic and in part to the small quantities absorbed.

BIBLIOGRAPHY AND SUGGESTED READING

General

Hodgson, E., R. B. Mailman, and J. E. Chambers. *Dictionary of Toxicology*, 2nd ed. Norwalk, CT: Appleton & Lange, 1994.

Klaassen, C. D., ed. *Casarett and Doull's Toxicology: The Basic Science of Poisons*, 6th ed. New York: McGraw-Hill, 2001.

Metals

Goyer, R. A. and T. W. Clarkson. Toxic effects of metals. In *Casarett and Doull's Toxicology: The Basic Science of Poisons*, 6th ed., ed. C. D. Klaassen, pp. 811–867. New York: McGraw-Hill, 2001.

Pesticides

Ecobichon, D. J. Toxic effects of pesticides. In *Casarett and Doull's Toxicology: The Basic Science of Poisons*, 6th ed., ed. C. D. Klaassen, pp. 763–810. New York: McGraw-Hill, 2001.

Krieger, R. *Handbook of Pesticide Toxicology*. San Diego, CA: Academic Press, 2001.

Toxins

Burkholder, J. M. and H. B. Glasgow. *Pfiesteria piscicida* and other *Pfiesteria*-like dinoflagellates: Behavior, impacts, and environmental controls. *Limnol. Oceanogr.* **42**:1052–1075, 1997.

Falconer, I., ed. *Algal Toxins in Seafood and Drinking Water*. New York: Academic, 1993.

Kotsonis, F. N., G. A. Burdock, and W. G. Flamm. Food toxicology. In *Casarett and Doull's Toxicology: The Basic Science of Poisons*, 6th ed., ed. C. D. Klaassen, pp. 1049–1088. New York: McGraw-Hill, 2001.

Norton, S. Toxic effects of plants. In *Casarett and Doull's Toxicology: The Basic Science of Poisons*, 6th ed., ed. C. D. Klaassen, pp. 965–976. New York: McGraw-Hill, 2001.

Russell, F. E. Toxic effects of terrestrial animal venoms and poisons. In *Casarett and Doull's Toxicology: The Basic Science of Poisons*, 6th ed., ed. C. D. Klaassen, pp. 945–964. New York: McGraw-Hill, 2001.

Solvents

Bruckner, J. V. and A. Warren. Toxic effects of solvents and vapors. In *Casarett and Doull's Toxicology: The Basic Science of Poisons*, 6th ed., ed. C. D. Klaassen, pp. 945–964. New York: McGraw-Hill, 2001.

Therapeutic Drugs

Hardman, J. G., P. B. Limbird, R. W. Molinoff, eds. *Goodman and Gilman's The Pharmacological Basis of Therapeutics*, 9th ed. New York: McGraw-Hill, 1996.

SAMPLE QUESTIONS

1. Describe the process by which a toxic metal causes enzyme inhibition or enzyme activation.

2. Of the three forms or species of mercury, which one is the most toxicologically significant for the general population and state its primary route or source of exposure to humans.

3. Describe the primary site and mode of action for the organophosphorus insecticides in insect (and human) systems.

4. Name the most widely used class of pesticide that is designed to control weed species of plants.

5. Define and explain the distinction between the terms toxicant and toxin.

6. Would the puffer fish from which tetrodotoxin is isolated (consumed by humans as fugu) be considered as a poisonous animal or a venomous animal? Explain your choice.

TOXICANT PROCESSING *IN VIVO*

Absorption and Distribution of Toxicants

RONALD E. BAYNES and ERNEST HODGSON

5.1 INTRODUCTION

As illustrated in the previous chapter, the human body can be exposed to a variety of toxicants which may be present in various environmental media such as air, soil, water, or food. However, just simply being exposed to these hazardous chemicals does not necessarily translate into a toxicological response. The mammalian body has several inherent defense mechanisms and membrane barriers which tend to prevent the entry or absorption and distribution of these toxicants once an exposure event has occurred. However, if the toxicant is readily absorbed into the body, there are still other anatomical and physiological barriers which may prevent distribution to the target tissue to elicit a toxic response. As the toxicological response is often related to the exposed dose, interactions between the toxicant and the body's barriers and defense mechanisms will have an effect on toxicant movement in the body, and ultimately modulate the rate and extent of toxicant absorption and distribution to the target tissue.

The skin represents the largest organ in the human body, and one of its primary functions can be seen as a physical barrier to absorption of toxicants. The other major routes of toxicant entry into the body are through the respiratory and gastrointestinal tract (GIT) which can be seen to offer less resistance to toxicant absorption than the skin. In general, the respiratory tract offers the most rapid route of entry and the dermal the least rapid. One reason for this major difference is primarily because membrane thickness, which is really the physical distance between the external environment (skin surface, air in the lung, or lumen of the gut) and the blood capillaries, varies across these portals of entry. The overall entry depends on both the amount present and the saturability of the transport processes involved.

Liver metabolism will have the most significant effect on toxicant bioavailability following gastrointestinal absorption, but microbial activity and various enzymes in the GIT and the skin can play a significant role in oral and dermal absorption, respectively. Physicochemical characteristics of the toxicant such as the chemical form can be a useful indicator of whether the toxicant will be absorbed and distributed in the body. In this regard, toxicant molecular weight, ionization (pKa), and octanol/water partition coefficient (logP) are useful indices of predicting chemical transport from an environmental media across biological membranes to the bloodstream. The reader should also be aware that for those toxicants that are readily ionized, the pH gradient across membranes can determine the extent of toxicant transport and accumulation in tissues.

Once the toxicant has been absorbed, the toxicant molecules can move around the body in two ways: (1) by bulk flow transfer (i.e., in the bloodstream) and (2) by diffusional transfer (i.e., molecule by molecule over short distances). Disposition is the term often used to describe the simultaneous effects of distribution and elimination processes subsequent to absorption. The cardiovascular system provides distribution of all toxicant irrespective of their chemical nature to various organs and tissues with various levels of affinities for toxicants. It should be remembered that organ mass and blood perfusion can vary, which can account for differential distribution of toxicants. Toxicant disposition can also be influenced by plasma protein binding in the bloodstream. The nature of this toxicant-protein interaction is dependent on the chemical nature of the toxicant, the presence of other toxicants or drugs in the bloodstream, as well as plasma protein levels. However, what distinguishes one toxicant pharmacokinetically from another is its diffusional characteristics. That is, its ability to cross nonaqueous diffusional barriers (e.g., cell membranes) from an aqueous compartment. This usually involves movement across several compartments separated by lipid membranes. It is therefore important to understand the *mechanisms by which drugs cross membranes* and the *physiochemical properties* of molecules and membranes that influence the movement of drugs from the environment to the body via either oral, inhalation, or dermal routes. These factors also influence movement from one compartment to another within the body during distribution as well as metabolism and excretion.

We can quantitate this movement or transport from one compartment to another using mathematical models to describe transport rates. This in fact is what we do in pharmacokinetic analysis and modeling. *Pharmaco- or toxicokinetics* is therefore the quantitation of the time course of toxicants in the body during the various processes of absorption, distribution, and elimination or clearance (metabolism and/or excretion) of the toxicant. Stated differently, this is a study of how the body "handles" the toxicant as it is reflected in the plasma concentration at various time points. The two most important pharmacokinetic parameters that describe the disposition of a chemical are volume of distribution (V_d) and systemic (body) clearance. *Pharmaco- and toxicodynamics* is the study of the biochemical and physiological effects of drugs and toxicants and determines their mechanism of action. Physiologically based pharmaco- or toxicokinetic models are used to integrate this information and to predict disposition of toxicants for a given exposure scenario. These concepts will be introduced at the end of this chapter.

5.2 CELL MEMBRANES

During absorption, distribution, and elimination processes, the toxicant will encounter various cell membranes before interacting with the target tissue. Each step of these processes involves translocation of the chemical across various membrane barriers, from the skin or mucosa through the capillary membranes, and through

Figure 5.1 Schematic showing membranes that a chemical may need to cross during passage from the environment to the site of action. Redrawn from Hodgson and Levi, eds. *Introduction to Biochemical Toxicology*, 2nd ed. Norwalk, CT: Appleton & Lange, 1994, p. 12.

Figure 5.2 Schematic diagram of biological membrane. Head groups of lipids represented by spheres, tail ends by zigzag lines. Black, white, or stippled spheres indicate different kinds of lipids and illustrate asymmetry in certain cases. Large bodies are membrane-associated proteins. Modified from Singer and Nicolson. *Science* **175**:720, 1972.

the cellular and organelle membranes (Figure 5.1). These membrane barriers vary from the relatively thick areas of the skin to the relatively thin lung membranes. In all cases, however, the membranes of tissue, cell, and cell organelle are relatively similar.

The cell membranes are predominantly a lipid matrix or can be considered a lipid barrier with an average width of a membrane being approximately 75 Å. The membrane is described as the fluid mosaic model (Figure 5.2) which consists of: (1) a bilayer of phospholipids with hydrocarbons oriented inward (hydrophobic phase); (2) hydrophilic heads oriented outward (hydrophilic phase); and (3) associated intra- and extracellular proteins that transverse the membrane. The ratio of lipid to protein varies from 5:1 for the myelin membrane to 1:5 for the inner structure of the mitochondria. However, 100% of the myelin membrane surface is lipid bilayer, whereas the inner membrane of the mitochondria may have only 40% lipid bilayer surface. In this example, the proportion of membrane surface that is lipid will clearly influence distribution of toxicants of varying lipophilicity.

The lipid constituents in the membrane permit considerable movement of macro-molecules, and membrane constituents may move appreciably within membranes. Membrane fluidity, a function of lipid composition, can be altered by temperature and chemicals (e.g., anesthetics). Several types of lipids are found in membranes, with phospholipids and cholesterol predominating. Sphingolipids comprise the primary minor component. Phosphatidylcholine, phosphatidylserine, and phospha-tidylethanolamine are the primary phosphatides, and their two fatty acid hydro-carbon chains (typically from 16 to 18, but varying from 12 to 22) comprise the nonpolar region. Some of the fatty acids are unsaturated and contribute appreciably to the fluidity of the membrane.

Proteins, which have many physiological roles in normal cell function, are inti-mately associated with lipids and may be located throughout lipid bilayers. These proteins may be located on either the surface or traverse the entire structure. Hydrophobic forces are responsible for maintaining the structural integrity of pro-teins and lipids within membranes, but movement within the membranes may occur. External and internal membrane proteins can function as receptors. Many proteins that traverse the membrane are transport proteins and are involved in translocation of ligands; that is, they are involved in active and facilitated transport.

Complexes of intrinsic membrane proteins and lipids can form hydrophilic or hydrophobic channels that allow transport of molecules with different physico-chemical characteristics. The amphipathic nature of the membrane creates a barrier for ionized, highly polar drugs, although it does not completely exclude them. The presence of pores of approximately 4 Å are believed to allow for ready move-ment of small molecules such as water. Thus, certain molecules that ordinarily would be excluded can rapidly traverse the highly lipid membrane barrier.

It is worth noting that differences between membranes, such as the presence of different lipids, the amount of surface lipid, differences in size and shape of proteins, or physical features of bonding may cause differences in permeability between membranes. These biochemical and biophysical differences are thought to be responsible for permeability differences in skin from different anatomical regions of the body.

5.3 MECHANISMS OF TRANSPORT

In general, there are four main ways by which small molecules cross biological lipid membranes:

1. *Passive diffusion*, that is, by diffusing through the lipid membrane
2. *Filtration*, that is, by diffusing through aqueous pores
3. *Special transport*, that is, by combination with a carrier molecule, which acts as a "ferryboat"
4. *Endocytosis*, that is, by pinocytosis for liquids and phagocytosis for solids

The first and third routes are important in relation to pharmacokinetic mechanisms. The aqueous pores are too small in diameter for diffusion of most drugs and toxicant, although important for movement of water and small polar molecules (e.g., urea). Pinocytosis is important for some macromolecules (e.g., insulin crossing the blood–brain barrier [BBB]).

5.3.1 Passive Diffusion

Most drugs and toxicant pass through membranes by *simple diffusion* down a concentration gradient. The driving force being the concentration gradient across the membrane.

This diffusion process can continue until equilibrium, although in reality there is always movement, but the net flux is zero. Eventually, the concentration of unionized or unbound (free) toxicant is the same on either side of the membrane. Please note that there is no competition of molecules, and there is generally a lack of saturation. Solubility in the lipid bilayer is important, and the greater the partition coefficient, the higher the concentration in the membrane, and the greater the rate of diffusion across the membrane. For ionized toxicants, the steady state concentration is dependent on the differences in pH across the membrane. Most membranes are relatively permeable to water either by diffusion or by flow that results from hydrostatic or osmotic differences across the membrane, and bulk flow of water can also carry with it small and water-soluble molecules by this mechanism. These substances generally have a molecular weight of less than 200. Although inorganic ions are small and will readily diffuse across some membranes, their hydrated ionic radius is relatively large. In such cases, active transport is required (see below). Specific ion fluxes are also controlled by specific channels that are important in nerves, muscles, and signal transduction.

We can now quantitate the *rate* at which a toxicant can be transported by passive diffusion, and this can be described by *Fick's Law of Diffusion* in the following equation:

$$\text{Rate of Diffusion} = \frac{D \times \text{Sa} \times \text{Pc}}{d}(C_H - C_L)$$

D is the diffusion coefficient; Sa is the surface area of the membrane, Pc is the partition coefficient, d is the membrane thickness, and C_H and C_L are the concentrations at both sides of the membrane (high and low, respectively). The first part of this equation ($D\text{Pc}/d$) represents the permeability coefficient of the drug. The permeability expresses the ease of penetration of a chemical and has units of velocity, distance/time (cm/hr).

The diffusion coefficient or diffusivity of the toxicant, D, is primarily dependent on solubility of the toxicant in the membrane and its molecular weight and molecular conformation. Depending on the membrane, there is a functional molecular size and/or weight cutoff that prevents very large molecules from being passively absorbed across any membrane. One would expect small molecular weight molecules to diffuse more rapidly than larger molecular weight toxicants. Therefore, the magnitude of a toxicant's diffusion coefficient really reflects the ease with which it is able to diffuse through the membrane. The reader should also be aware that as a toxicant crosses from the donor or aqueous medium and through the membrane medium, there are really two diffusion environments and thus two diffusion coefficients to consider. Another important factor that can influence the diffusion coefficient is membrane viscosity. This physicochemical characteristic should remain constant in biological systems, but can be modified in skin membranes exposed to various pharmaceutical or pesticide formulations. Formulation additives or excipients may enter the membrane barrier and reversibly or irreversibly change viscosity and thus diffusion coefficient of the drug or pesticide in the barrier membranes of the skin.

The partition coefficient that will be described in more detail later in this chapter is the relative solubility of the compound in lipid and water, and really reflects the ability of the toxicant to move from a relatively aqueous environment across a lipid membrane. It is this factor that is often manipulated in pesticide and drug formulations to create a vehicle. Membrane permeability is therefore strongly correlated to the lipid solubility of the toxicant in the membrane as well as the aqueous environment surrounding the membrane. Please be aware that there are instances where partition coefficient or lipid solubility of the toxicant may be very large, and there may be a tendency for the drug to sequester in the membrane. Membrane surface area and membrane thickness can also vary across different organs in the body, but one does not expect these two factors in Fick's equation to vary considerably. The final component of Fick's equation is the concentration gradient $(C_H - C_L)$ across the membrane, which is the *driving force for diffusion*, and as will be demonstrated below in our discussion on first-order kinetics, is the most important factor dictating the rate of transport across most biological membranes.

First-Order Kinetics: When the rate of a process is dependent upon a *rate constant* and a concentration gradient, a linear or first-order kinetic process will be operative. The reader should be aware that there are numerous deviations from the first-order process when chemical transport *in vivo* is analyzed, and this can be deemed an *approximation* since in many barriers, penetration is slow, and a long period of time is required to achieve steady state.

The rate of movement of a toxicant across a membrane may be expressed as the *change in amount of toxicant, A, (dA) or toxicant concentration, C, (dC) per unit of time (dt) which equals dA/dt.* Calculus can be used to express instantaneous rates over very small time intervals *(dt)*. Thus, rate processes may then be generally expressed as:

$$dA/dt = K\ A^n$$

where dA/dt is the rate of chemical (X) movement (e.g., absorption, distribution, elimination); k is the rate constant of the process; and n is the kinetic order of the transport process (e.g., absorption). The "n" either equals 1 (first order) or 0 (zero order). Thus, a "first-order" rate equation is:

$$dA/dt = k\ A^1 = KA$$

and a "zero-order" rate equation is:

$$dA/dt = K\ A^0 = K.$$

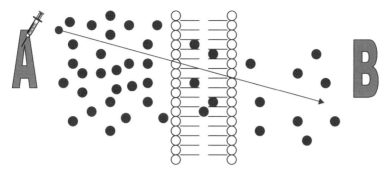

Figure 5.3   Illustration of concentration gradient generated by administration of a drug that can travel down this gradient from area A and across a biological membrane to area B.

We know from Fick's Law that the rate of diffusion (now expressed as dA/dt) is:

$$dA/dt = D \cdot \text{Sa} \cdot \text{Pc}(A_1 - A_2)/d.$$

Once a toxicant crosses a membrane, it is rapidly removed from the "receiving side" (compartment B in Figure 5.3) either by uptake into the bloodstream or elimination from the organism. Thus, it is A_1 that is the primary driving force, and if we replace this with "A" in all equations, then:

$$dA/dt = (D \cdot \text{Sa} \cdot \text{Pc}/d) A$$

If we let $\mathbf{K} = (D \cdot \text{Sa} \cdot \text{Pc}/d)$ and realize that since A is present in this equation, n must equal 1, we have a first-order rate process. Fick's Law of Diffusion, which is so important for quantitating rates of absorption, distribution, and elimination, is thus the basis for using first-order kinetics in most pharmacokinetic models.

Therefore, in a first-order process, the rate of drug movement is directly proportional to the amount of drug (A) in the body, which is usually a function of the dose. K is the *first-order fractional rate constant with units of l/time (time⁻¹)* and *represents the fraction of drug that is transported per unit of time.* Thus, in a *first-order* process, *the rate of drug movement is proportional to dose, but the fraction moved per unit of time is constant and independent of dose.*

When first-order kinetics hold, a simple relationship exists between the penetration rate constant, K, and $t_{0.5}$ (time necessary for one-half of the applied dose to penetrate):

$$K = \frac{0.693}{t_{0.5}} \quad \text{and the units of } K \text{ are percentage of change/time unit.}$$

We can also derive the concentration of the toxicant if we know the volume or V_d of the toxicant compartment as:

$$V_d \,(\text{volume}) = A\,(\text{mass})/C\,(\text{mass/volume})$$

V_d is discussed in more detail later in this chapter.

5.3.2 Carrier-Mediated Membrane Transport

This mechanism is important for compounds that lack sufficient lipid solubility to move rapidly across the membrane by simple diffusion. A membrane-associated protein is usually involved; specificity, competitive inhibition, and the saturation phenomenon and their kinetics are best described by *Michaelis–Menten enzyme kinetic models*. Membrane penetration by this mechanism is more rapid than simple diffusion and, in the case of active transport, may proceed beyond the point at which concentrations are equal on both sides of the membrane. Generally, there are two types of specialized carrier-mediated transport processes:

Passive facilitated diffusion involves movement down a concentration gradient without an input of energy. However, this mechanism, which may be highly selective for specific conformational structures, is necessary for transport of endogenous compounds whose rate of transport by simple diffusion would otherwise be too slow. The classical example of the latter is transport of glucose into red blood cells.

Active transport requires energy and transport is against a concentration. Maintenance against this gradient often requires energy. It is often coupled to energy-producing enzymes (e.g., ATPase) or to the transport of other molecules (e.g., Na^+, Cl, H^+) that generate energy as they cross the membranes. Carrier-mediated drug transport can occur in only a few sites in the body, and the main sites are:

1. BBB, neuronal membranes, choroid plexus;
2. renal tubular cells; and
3. hepatocytes, biliary tract.

There are instances in which toxicants have chemical or structural similarities to endogenous chemicals that rely on these special transport mechanisms for normal physiological uptake and can thus utilize the same system for membrane transport. Useful examples of drugs known to be transported by this mechanism include levodopa, which is used in treating Parkinson's disease, and fluorouracil, a cytotoxic drug. Levodopa is taken up by the carrier that normally transports phenylalanine, and fluorouracil is transported by the system that carries the natural pyrimidines, thymine, and uracil. Iron is absorbed by a specific carrier in the mucosal cells of the jejunum, and calcium by a vitamin D-dependent carrier system. Lead may be more quickly moved by a transport system that is normally involved in the uptake of calcium.

For carrier-mediated transport, the rate of movement across a membrane will now be *constant*, since flux is dependent upon the capacity of the membrane carriers and not the mass of the chemical to be transported. These processes are described by *zero-order* kinetic rate equations of the form:

$$dX/dt = K\ X^0 = K_0$$

K_0 is now the *zero-order* rate constant and is expressed in terms of *mass/time*. In an active carrier-mediated transport process following zero-order kinetics, the rate of drug transport is always equal to K once the system is fully loaded or saturated. At subsaturation levels, the rate is *initially first order* as the carriers become loaded

TABLE 5.1 An Example of Differences between First-Order and Zero-Order Processes although the System Is Exposed to the Same Initial Mass of Toxicant of 1000, 100, or 10 mg

Initial Toxicant Mass (mg)	Amount Transported/min (First Order)	Amount Transported/min (Zero Order)
1000	100	10
100	10	10
10	1	10

with the toxicant, but at concentrations normally encountered in pharmacokinetics, the rate becomes constant. Thus, as dose increases, the rate of transport does *not* increase in proportion to dose as it does with the fractional rate constant seen in first-order process. This is illustrated in Table 5.1 below, where it is assumed that the *first-order* rate constant is 0.1 (10% per min) and the *zero-order* rate is 10 mg/min.

In the case of first order, these amounts will subsequently diminish (10% of 900 is 90, and so on). In the case of zero order, the amount transported does not vary with time (constant rate of transport).

The plot below illustrates the differences in passive (linear) versus carrier-mediated (nonlinear) transport. At relatively low concentrations of drug, carrier-mediated processes may appear to be first order since the protein carriers are not saturated. However, at higher concentrations, zero-order behavior becomes evident. It is in plots such as this that the terms *linear* (first order) and *nonlinear* (zero order) come into existence.

5.4 PHYSICOCHEMICAL PROPERTIES RELEVANT TO DIFFUSION

The following physicochemical properties are important for chemical diffusion, and we have discussed several of these in previous sections of this chapter as it relates to passive diffusion mechanism and how it impacts rate of toxicant transport across membranes.

1. Molecular size and shape
2. Solubility at site of absorption
3. Degree of ionization
4. Relative lipid solubility of ionized and unionized forms

Although molecular weight is important, it is less important than the drug's *lipid solubility* when it comes to assessing the rate of passive diffusion across membranes. The permeability, P ($P = Pc \times D$), of a nonpolar substance through a cell membrane is dependent on two physicochemical factors: (1) *solubility in the membrane* (Pc), which can be expressed as a partition coefficient of the drug between the aqueous phase and membrane phase, and (2) *diffusivity or diffusion coefficient* (D), which is a measure of mobility of the drug molecules within the lipid. The latter may vary only slightly between toxicants, but the former is more important. Lipid solubility is therefore one of the most important determinants of the pharmacokinetic characteristics of a chemical, and it is important to determine whether a toxicant is readily ionized or not influenced by pH of the environment. If the toxicant is readily ionized, then one needs to understand its chemical behavior in various environmental matrices in order to adequately assess its transport mechanism across membranes.

5.4.1 Ionization

For the purposes of this discussion on membrane transport, chemicals can be broadly categorized into those that are ionized and those that are not ionized. Many drugs (e.g., antibiotics) and several toxicants (e.g., strychnine) are either weak acids or weak bases and can exist in solution as a mixture of nonionized and ionized forms. Generally, these drugs and toxicants must be in the uncharged or nonionized form to be transported by passive diffusion across biological membranes. This is because biological membranes are of a lipid nature and are less permeable to the ionized form of the chemical. The pH of the environment (e.g., lumen of the GIT and renal tubules) can influence transfer of toxicant that is ionizable by increasing or decreasing the amount of nonionized form of the toxicant. Aminoglycosides (e.g., gentamicin) are the exception to this general rule in that the uncharged species is insufficiently lipid soluble to cross the membrane appreciably. This is due to a preponderance of hydrogen-bonding groups in the sugar moiety that render the uncharged molecule hydrophilic. Note that some amphoteric drugs (e.g., tetracyclines) may be absorbed from both acidic and alkaline environments. In essence, the amount of drug or toxicant in ionized or nonionized form depends on the pKa (pH at which 50% of the drug is ionized) of the drug and the pH of the solution in which the drug is dissolved. *The pKa, which is the negative logarithm of the dissociation constant of a weak acid or weak base, is a physicochemical characteristic of the drug or toxicant.* When the pH of the solution is equal to the pKa, then 50% of the toxicant is in the ionized form and 50% is in the nonionized form. The ionized and nonionized fractions can be calculated according to the *Henderson–Hasselbach* equations listed below:

For weak acids $\quad$ pKa $-$ pH $=$ log[nonionized form/ionized form]

For weak bases $\quad$ pKa $-$ pH $=$ log[ionized form/nonionized form]

For an organic acid ($RCOOH \Leftrightarrow RCOO^- + H^+$), acidic conditions (pH less than the pKa of the compound) will favor the formation of the nonionized RCOOH, whereas alkaline conditions (pH greater than pKa) will shift the equilibrium to the right. For an organic base ($RNH_2 + H^+ \Leftrightarrow RNH_3^+$), the reverse is true, and decreasing the pH (increasing the concentration of H^+) will favor formation of the ionized form, whereas increasing the pH (decreasing the concentration of H^+) will favor formation of the nonionized form.

Memory aid. In general, weak organic acids readily diffuse across a biological membrane in an acidic environment, and organic bases can similarly diffuse in a basic environment. This is illustrated quite well in Table 5.2 below for chemical in the rat intestine. There are the usual exceptions to the generalizations concerning

TABLE 5.2   Effect of pH on Absorption of Weak Organic Acids and Bases from Rat Intestine

Compound	Percent Absorbed at Various pH Values				
	pKa	3.6–4.3	4.7–5.0	7.0–7.2	7.8–8.0
Acids					
Nitrosalicyclic	2.3	40	27	<02	<02
Salicyclic	3.0	64	35	30	10
Benzoic	4.2	62	36	35	05
Bases					
Aniline	4.6	40	48	58	61
Aminopyrene	5.0	21	35	48	52
Quinine	8.4	09	11	41	54

ionization and membrane transport, and some compounds, such as pralidoxime (2-PAM), paraquat, and diquat, are absorbed to an appreciable extent even in the ionized forms. The mechanisms allowing these exceptions are not well understood.

Ion trapping can occur when at equilibrium the total (ionized + nonionized) concentration of the drug will be different in each compartment, with an acidic drug or toxicant being concentrated in the compartment with the relatively high pH, and vice versa. The pH partition mechanism explains some of the qualitative effects of pH changes in different body compartment on the pharmacokinetics of weakly basic or acidic drugs or toxicant as it relates to renal excretion and penetration of the BBB. Alkalization of urine in the lumen of renal tubules can enhance elimination of weak acids. However, this phenomenon is not the main determinant of absorption of drugs or toxicants from the GIT. In the GIT, the enormous absorptive surface area of the villi and microvilli in the ileum compared to the smaller absorptive area of the stomach is of overriding importance.

5.4.2   Partition Coefficients

A second physicochemical parameter influencing chemical penetration through membranes is the relative lipid solubility of the potential toxicant which can be ascertained from its known partition coefficient. The partition coefficient is a measure of the ability of a chemical to separate between two immiscible phases. The phases consist of an organic phase (e.g., octanol or heptane) and an aqueous phase (e.g., water). The lipid solvent used for measurement is usually octanol because it best mimics the carbon chain of phospholipids, but many other systems have been reported (chloroform/water, ether/water, olive oil/water). The lipid solubility and the water solubility characteristics of the chemical will allow it to proportionately partition between the organic and water phase. The partition coefficients can be calculated using the following equation:

$$P = V_w/V_o \left[C_{wo} - C_w/C_w \right]$$

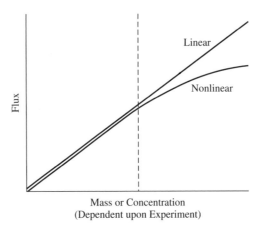

Figure 5.4   Plot depicting a linear relationship (first order) and nonlinear relationship (zero order) between chemical flux across a membrane and the initial mass or concentration of the chemical.

where P is the partition coefficient and is usually expressed in terms of its logarithmic value (log P); V_w and V_o are the volumes of aqueous and oil or organic phase, respectively; and C_{wo} and C_w are drug or toxic concentrations in the aqueous phase before and after shaking, respectively.

The lower the partition coefficient, the more water soluble, and the least permeable the toxicant is across a membrane. Regarding dermal absorption, partition coefficients can be predictive of absorption. However, toxicants with extremely high partition coefficients tend to remain in the membrane or skin. This explains why a strong correlation between permeability and partition coefficient can exist for a hypothetical series of analogous chemicals for a specific range of partition coefficients, but the correlation does not exist for log P greater values greater than 6 in many instances. A log P of around 1 is often taken as desirable for skin penetration. The reader should also recall that this parameter is operative as the chemical diffuses across membranes (Figure 5.1) of varying lipid content during absorption, distribution, and elimination processes.

5.5   ROUTES OF ABSORPTION

Primary routes of entry of toxicants to the human body are dermal, gastrointstinal, and respiratory. Methods for studying these different routes are numerous, but they are perhaps best developed for the study of dermal absorption because this route is subject to more direct methodology, whereas methods for studying respiratory or gastrointestinal absorption require more highly specialized instrumentation. Additional routes encountered in experimental studies include intraperitoneal, intramuscular, and subcutaneous routes. When direct entry into the circulatory system is desired, intravenous (IV) or intra-arterial injections can be used to bypass the absorption phase. Information from this more direct route of entry (e.g., IV) should, however, be used in addition to data from the extravascular route of interest to adequately assess the true extent of absorption of a toxicant.

5.5.1   Extent of Absorption

It is often useful to determine *how much of the drug* actually penetrates the membrane barrier (e.g., skin or GIT) and gets into the bloodstream. This is usually determined experimentally for oral and dermal routes of administration. The *area under the curve (AUC)* of the concentration-time profiles for oral or dermal routes is compared with the AUC for IV routes of administration. The AUC is determined by breaking the curve up into a series of trapezoids and summing all of the areas with the aid of an appropriate computer program (Figure 5.5).

The intravenous correction is *very* important if absolute bioavailability is desired. The ratio of these AUC values is absolute bioavailability, *F*

$$F = \frac{(\text{AUC})_{\text{route}}}{(\text{AUC})_{\text{IV}}}$$

The above relationship holds if the same doses were used with both routes; however, the bioavailability should be corrected if different doses were used.

$$F = \frac{\text{AUC}_{\text{route}} \times \text{Dose}_{\text{IV}}}{\text{AUC}_{\text{IV}} \times \text{Dose}_{\text{route}}}$$

Another technique is to monitor drug or toxicant excretion rather than blood concentrations, especially when blood or plasma concentrations are very low. Using the same equations, the AUC is now replaced by chemical concentrations in urine, feces, and expired air. Some chemicals are primarily excreted by the kidney, and

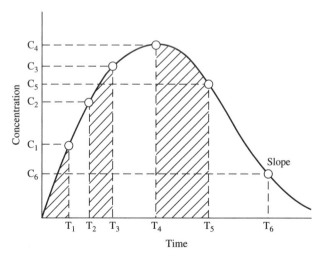

Figure 5.5 Plasma concentration time profile for oral exposure to a toxicant and depiction of AUCs determined by summation of trapezoids at several time periods.

urine data alone may be necessary. The rate and extent of absorption are clearly important for therapeutic and toxicological considerations. For example, different formulations of the same pesticide can change the absorption rate in skin or GIT, and not bioavailability, but can result in blood concentrations near the toxic dose. Also, different formulations can result in similar absorption rates, but different bioavailability.

5.5.2 Gastrointestinal Absorption

The GIT is a hollow tube (Figure 5.6a) lined by a layer of columnar cells, and is usually protected by mucous that offers minimal resistance to toxicant penetration. The distance from the outer membrane to the vasculature is about 40 μm, from which point further transport can easily occur. However, we should recognize that the cornified epithelium of the esophagus prevents absorption from this region of the GIT. Most of the absorption will therefore occur in the intestine (pH = 6) and, to some extent, in the stomach (pH = 1–3). Buccal and rectal absorption can occur in special circumstances. Note that secretions from the lachrymal duct, salivary gland, and nasal passages can enter the GIT via the buccal cavity. Therefore, following IV administration, a toxicant can enter the GIT if the drug is in these secretions.

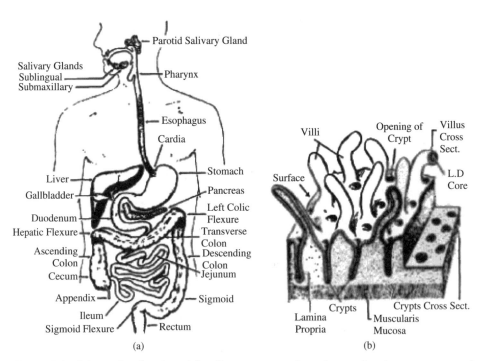

Figure 5.6 Schematic showing (a) alimentary canal and associated structures and (b) lining of the small intestine. From (a) Scholtelius and Scholtelius in *Textbook of Physiology*, St. Louis: Mosby, 1973, and (b) Ham and Cormack in *Histology*, 8th ed, Philadelphia: Lippincott, 1979.

The intestine can compensate the 2.5 log units difference between it and the stomach by the increased surface area in the small intestines. The presence of microvilli (Figure 5.6b) in the intestine is an increase of 600-fold in surface area compared to a hollow tube of comparable length. Note that there is no absorption, except for water, in the large intestine.

Most of the absorption in the GIT is by passive diffusion, except for nutrients; glucose, amino acids, and drugs that look like these substances are taken up by active transport. For toxicants with structural similarities to compounds normally taken up by these active transport mechanisms, entry is enhanced. For example, cobalt is absorbed by the same active transport mechanism that normally transports iron, and 5-bromouracil is absorbed by the pyrimidine transport system.

Very lipid-soluble toxicants and drugs, which are not miscible in the aqueous intestinal fluid, are presented as emulsions, and brought into solution through the action of detergent-like bile acids. The product of this mixing is large surface area micelles (hydrophobic interior) that deliver the lipids to the brush border of the intestine for diffusion across the membrane. As stated previously, the rate of passive transfer will be dependent on ionization and lipid solubility. Very strong bases (e.g., tubocurarine, succinylcholine) and strong acids are not readily absorbed in the GIT. These muscle relaxants therefore are given intravenously. The smaller the particle size of the toxicant, the greater the absorption, and a chemical must be in aqueous solution for it to be absorbed in the GIT. A feature of the GIT that seems to contradict basic assumptions of absorption is the penetration of certain very large molecules. Compounds such as bacterial endotoxins, large particles of azo dyes, and carcinogens are apparently absorbed by endocytotic mechanisms.

GIT motility has a significant effect on GIT absorption of a toxicant. For example, excessively rapid movement of gut contents can reduce absorption by reducing residence time in the GIT, while the presence of food in the stomach can delay the progress of drugs from the stomach to the small intestine where most of the absorption will occur. Increased splanchnic blood flow after a meal can result in absorption of several drugs (e.g., propranolol), but in hypovolemic states, absorption can be reduced.

Biotransformation in the GIT prior to absorption can have a significant impact on bioavailability of a toxicant. The resident bacterial population can metabolize drugs in the GIT. Because of microbial fermentation in the rumen of ruminants and large intestine and cecum of horses and rabbits, it is often difficult to compare drug absorption profiles with carnivores (e.g., dogs) and omnivores (e.g., humans, pigs). Acid hydrolysis of some compounds can also occur, and enzymes in the intestinal mucosa can also have an effect on oral bioavailability. If the toxicant survive these microbial and chemical reactions in the stomach and small intestine, it is absorbed in the GIT and carried by the hepatic portal vein to the liver, which is the major site of metabolism. Chapters 7, 8, and 9 will discuss liver metabolism of toxicants in more detail. In brief, this activity in the liver can result in detoxification and/or bioactivation. Some drugs and toxicant that are conjugated (e.g., glucuronidation) in the liver are excreted via the biliary system back into the GIT. Once secreted in bile by active transport and excreted from the bile duct into the small intestine, this conjugated toxicant can be subjected to microbial beta-glucuronidase activity which can result in regeneration of the parent toxicant that is more lipophilic than the conjugate. The toxicant can now be reabsorbed by the GIT, prolonging the presence of the drug or toxicant in the systemic circulation. This is called *enterohepatic circulation*, which will be covered in greater detail in subsequent chapters.

5.5.3 Dermal Absorption

The skin is a complex multilayered tissue with a large surface area exposed to the environment. Skin anatomy, physiology, and biochemistry vary between species, within species, and even between anatomic sites within an individual animal or human. Logically, these biological factors alone can influence dermal absorption. However, what is consistent is that the outer layer, the *stratum corneum (SC)*, can provide as much as 80% of the resistance to absorption to most ions as well as aqueous solutions. However, the skin is permeable to many toxicants, and dermal exposure to agricultural pesticides and industrial solvent can result in severe systemic toxicity.

The anatomy of the skin is depicted in the schematic diagram below (Figure 5.7a). In mammalian skin there are really three distinct layers which are the epidermis, dermis, and hypodermis or subcutaneous fat layer. Human skin is 3 mm thick, but it is the epidermis which is only 0.1–0.8 mm that provides the greatest resistance to toxicant penetration. The five layers of the epidermis starting from the outside are the SC, stratum lucidum, stratum granulosum, stratum spinosum, and stratum basale. The basal cells of the epidermis proliferate and differentiate as they migrate outward toward the surface of the skin. It requires about 2~28 days for cells to migrate from the basal layer to the SC, where they are eventually sloughed off. These dead, keratinized cells are, however, very water absorbant (hydrophilic), a property that keeps the skin soft and supple. Sebum, a natural oil covering the skin, functions in maintaining the water-holding ability of the epidermis. The SC is the primary barrier to penetration, and it consists primarily of these dead keratin-filled keratinocytes embedded in an extracellular lipid matrix. The lipids are primarily sterols, other

(a)   (b)

Figure 5.7 (a) Schematic diagram of the microstructure of mammalian skin and potential pathways for absorption by (A) intercellular, (B) transcellular, (C) transfollicular, or (D) sweat pore routes. (b) "Brick and Mortar" model of the stratum corneum depicting intercellular pathway (i.e., route A) between keratinocytes through the lipid domain of the stratum corneum.

neutral lipids, and ceramides. This association between lipids and dead keratinized cells is often referred to as the "Brick and Mortar" model as depicted in Figure 5.7b and is often used to simplify the composition of the SC that are integral to chemical transport through skin.

A number of appendages are associated with the skin, including hair follicles, sebaceous glands, eccrine and apocrine sweat glands, and nails. Recently, it was found that removal of the SC does not allow complete absorption; thus, it is apparent that some role, although of lesser importance, is played by other parts of the skin. The dermis and subcutaneous areas of the skin are less important in influencing penetration, and once a toxicant has penetrated the epidermis, the other layers are traversed rather easily. The dermis is highly vascular, a characteristic that provides maximal opportunity for further transport once molecules have gained entry through the epidermis or through skin appendages. Most of the systemic absorption occurs at the capillary loops located at the epidermis–dermis junction. The blood supply of the dermis is under neural and humoral influences whose temperature-regulating functions could thus affect penetration and distribution of toxicants. Vasoactive drugs or environmental temperature can also influence absorption by altering blood flow to these capillaries. The subcutaneous layer of the skin is highly lipid in nature and serves as a shock absorber, an insulator, and a reserve depot of energy. The pH of the skin varies between 4 and 7 and is markedly affected by hydration.

Cutaneous biotransformation is mostly associated with the stratum basale layer where there can be Phase I and Phase II metabolism. However, the skin is not very efficient, compared to the liver. The epidermal layer accounts for the major portion of biochemical transformations in the skin, although the total skin activity is low (2–6% that of the liver). If activity is based on epidermis alone, however, that layer is as active as the liver or, in the case of certain toxicants, several times more active. For some chemicals, metabolism can influence absorption, and transdermal delivery systems of drugs utilize this activity. For example, prodrug such as lipid esters are applied topically, and cutaneous esterases liberate the free drug. These basal cells and extracellular esterases have been shown to be involved in detoxification of several pesticides and bioactivation of carcinogens such as benzo(a)pyrene. For rapidly penetrating substances, metabolism by the skin is not presently considered to be of major significance; however, skin may have an important first-pass metabolic function, especially for compounds that are absorbed slowly.

The *intercellular pathway* is now accepted as the major pathway for absorption. Recall that the rate of penetration is often correlated with the partition coefficient. In fact, this is a very tortuous pathway, and the "h" (skin thickness) in Fick's First Law of Diffusion is really 10× the measured distance. By placing a solvent (e.g., ether, acetone) on the surface or a tape stripping the surface, the SC is removed, and absorption can be significantly increased by removing this outer barrier. This may not be the case for very lipophilic chemical. This is because the viable epidermis and dermis are regarded as aqueous layers compared to the SC. Note that the more lipophilic the drug, the more likely it will form depot in the SC, and slowly absorbed over time, leading to a prolonged half-life.

The *transcellular pathway* has been discredited as a major pathway, although some polar substances can penetrate the outer surface of the protein filaments of hydrated SC. The *transfollicular pathway* is really an invagination of the epidermis into the dermis, and the chemical still has to penetrate the epidermis to be absorbed into the bloodstream. This is also regarded as a *minor* route. *Sweat pores* are not lined with the SC layer, but the holes are small, and this route is still considered a minor route for chemical absorption. In general, the epidermal surface is 100–1000 times the surface area of skin appendages, and it is likely that only very small and/or polar molecules penetrate the skin via these appendages.

Variations in areas of the body cause appreciable differences in penetration of toxicants. The rate of penetration is in the following order:

Scrotal > Forehead > Axilla > = Scalp > Back = Abdomen > Palm and Plantar

The palmar and plantar regions are highly cornified and are 100–400 times thicker than other regions of the body. Note that there are differences in blood flow and, to a lesser extent, hair density, which may influence absorption of more polar toxicants.

Formulation additives used in topical drug or pesticide formulations can alter the SC barrier. Surfactants are least likely to be absorbed, but they can alter the lipid pathway by fluidization and delipidization of lipids, and proteins within the keratinocytes can become denatured. This is mostly likely associated with formulations containing anionic surfactants than nonionic surfactants. Similar effects can be observed with solvents. Solvents can partition into the intercelluar lipids, thereby changing membrane lipophilicity and barrier properties in the following order: ether/acetone > DMSO (dimethyl sulfoxide) > ethanol > water. Higher alcohols and oils do not damage the skin, but they can act as a depot for lipophilic drugs on the skin surface. The presence of water in several of these formulations can hydrate the skin. Skin occlusion with fabric or transdermal patches, creams, and ointments can increase epidermal hydration which can increase permeability.

The reader should be aware of the animal model being used to estimate dermal absorption of toxicants in humans. For many toxicants, direct extrapolation from a rodent species to human is not feasible. This is because of differences in skin thickness, hair density, lipid composition, and blood flow. Human skin is the least permeable compared to skin from rats, mice, and rabbits. Pig skin is, however, more analogous to human skin anatomically and physiologically, and pig skin is usually predictive of dermal absorption of most drugs and pesticides in human skin. Human skin is the best model, followed by skin from pigs, primates, and hairless guinea pigs, and rats, mice, and rabbits. In preliminary testing of a transdermal drug, if the drug does not cross rabbit or mice skin, it is very unlikely that it will cross human skin. There are several *in vitro* experimental techniques such as static diffusion (Franz) cells or flow-through diffusion (Bronough) cells. There are several *ex vivo* methods, including the isolated perfused porcine skin flap (IPPSF), which with its intact microvasculature makes this model unique. *In vivo* methods are the golden standard, but they are very expensive, and there are human ethical and animal rights issues to be considered.

There are other factors that can influence dermal absorption, and these can include environmental factors such as air flow, temperature, and humidity. Preexisting skin disease and inflammation should also be considered. The topical dose this is usually expressed in per unit surface area can vary, and relative absorption usually decreases with increase in dose.

5.5.4 Respiratory Penetration

As observed with the GIT and skin, the respiratory tract can be regarded as an external surface. However, the lungs, where gas/vapor absorption occurs, are preceded by protective structures (e.g., nose, mouth, pharynx, trachea, and bronchus) which can reduce the toxicity of airborne substances, especially particles. There is little or no absorption in these structures, and residual volume can occur in these sites. However, cells lining the respiratory tract may absorb agents which can cause a toxicological response. The absorption site which is the alveoli–capillary membrane is very thin (0.4–1.5 µm). The membranes to cross from the alveolar air space to the blood will include *Type I cells to basement membrane to capillary endothelial cells* (Figure 5.8). This short distance allows for rapid exchange of gases/vapors. The analogous absorption distance in skin is 100–200 µm, and in GIT it is about 30 µm. There is also a large surface area (50 times the area of skin) available for absorption as well as significant blood flow which makes it possible to achieve rapid adjustments in plasma concentration. Gases/vapors must get into solution in the thin fluid film in the alveoli for systemic absorption to occur. Please note that doses are often a measurement of partial pressures, which is important for gases/vapors.

The process of respiration involves the movement and exchange of air through several interrelated passages, including the nose, mouth, pharynx, trachea, bronchi, and successively smaller airways terminating in the alveoli, where gaseous exchange occurs. These alveoli consist mainly of type I pneumocytes that represent 40% of all cells but cover >90% of surface area, and type II pneumocytes that represent 60% of all cells but cover 5% of surface area. Macrophages make up 90% of cells in alveolar space. The amount of air retained in the lung despite maximum expiratory effort is known as the residual volume. Thus, toxicants in the respiratory air may not be cleared immediately because of slow release from the residual volume.

Figure 5.8 Schematic representation of the respiratory unit of the lung. From Bloom and Fawcett, *A Textbook of Histology*, Philadelphia: Saunders, 1975.

The rate of entry of vapor-phase toxicants is controlled by the alveolar ventilation rate, with the toxicant being presented to the alveoli in an interrupted fashion approximately 20 times/min.

Airborne toxicants can be simplified to two general types of compounds, namely, gases and aerosols. Compounds such as gases, solvents, and vapors are subject to gas laws and are carried easily to alveolar air. Much of our understanding of xenobiotic behavior is with anesthetics. Compounds such as aerosols, particulates, and fumes are not subject to gas laws because they are in particulate form.

The transfer of gas from alveoli to blood is the actual absorption process. Among the most important factor that will determine rate and extent of absorption of a gas in lungs is the solubility of that gas. Therefore, it is not membrane partition coefficient that necessarily affects absorption as has been described for skin and GIT membranes, but rather the blood:gas partition coefficient or blood/gas solubility of the gas. A high blood:gas partition coefficient indicates that the blood can hold a large amount of gas. Keeping in mind that it is the *partial pressure* at equilibrium that is important, and the more soluble the gas is in blood, the greater the amount of gas that is needed to dissolve in the blood to raise the partial pressure or tension in blood. For example, anesthetics such as diethyl ether and methoxyflurane which are soluble (Table 5.3), require a longer period for this partial pressure to be realized. Again, the aim is to generate the same tension in blood as in inspired air. Because these gases are very soluble, detoxification is a prolonged process. In practice, anesthetic induction is slower, and so is recovery from anesthesia. For less soluble gases (e.g., NO, isoflurane, halothane), the partial pressure or tension in blood can be raised a lot easier to that of inspired gases, and detoxification takes less time than those gases that are more soluble.

There are several other important factors that can determine whether the gas will be absorbed in blood and then transported from the blood to the perfused tissue. The concentration of the gas in inspired air influences gas tension, and partial pressure can be increased by overventilation. In gas anesthesiology, we know that the effects of respiratory rate on speed of induction are transient for gases that have low solubility in blood and tissues, but there is a significant effect for agents that are more soluble and take a longer time for gas tensions to equilibrate. In determining how much of the gas is absorbed, it is important to consider what fraction of the lung is ventillated and what fraction is perfused. However, one should be aware that due to diseased lungs, there can be differences between these fractions. For example, decreased perfusion will decrease absorption although there is agent in the alveoli and vice versa. The rate at which a gas passes into tissues is also

TABLE 5.3   Blood:Gas Partition Coefficient of Various Anesthetic Gases

Agent	Blood:Gas Partition Coefficient in Humans
Methoxyflorane	13.0–15.0
Halothane	2.3–2.5
Isoflurane	1.4
NO	0.5

dependent on gas solubility in the tissues, rate of delivery of the gas to tissues, and partial pressures of gas in arterial blood and tissues. After uptake of the gas, the blood takes the gas to other tissues. The mixed venous blood returned to the lungs progressively begins to have more of the gas, and differences between arterial (or alveolar) and mixed venous gas tensions decrease continuously.

While gases are more likely to travel freely through the entire respiratory tract to the alveoli, passage of aerosols and particles will be affected by the upper respiratory tract which can act as an effective filter to prevent particulate matter from reaching the alveoli. Mucous traps particles to prevent entry to alveoli, and the mucociliary apparatus in the trachea traps and pushes particles up the trachea to the esophagus where they are swallowed and possibly absorbed in the GIT.

In addition to upper pathway clearance, lung phagocytosis is very active in both upper and lower pathways of the respiratory tract and may be coupled to the mucus cilia. Phagocytes may also direct engulfed toxicants into the lymph, where the toxicants may be stored for long periods. If not phagocytized, particles $\leq 1\,\mu m$ may penetrate to the alveolar portion of the lung. Some particles do not desquamate but instead form a dust node in association with a developing network of reticular fibers. Overall, removal of alveolar particles is markedly slower than that achieved by the directed upper pulmonary mechanisms. This defense mechanism is not important for vapors/gases. The efficiency of the system is illustrated by the fact than on average, only 100 g of coal dust is found postmortem in the lungs of coal miners although they inhale approximately 6000 g during their lifetime.

The deposition site of particles in the respiratory tract is primarily dependent on the *aerodynamic behavior* of the particles. The particle size, density, shape, hygroscopicity, breathing pattern, and lung airway structure are also important factors influencing the deposition site and efficiency. The *aerodynamic-equivalent diameter* (for particle >0.5 μm) and *diffusion-equivalent diameter* (<0.5 μm) are *defined* as the diameter of a *unit density sphere* having the same *settling velocity* (aerodynamic-equivalent) or the same *diffusion rate* (diffusion-equivalent) as the *irregularly shaped particle of interest*. Deposition occurs by five possible mechanisms: electrostatic precipitation, interception, impaction, sedimentation, impaction, and diffusion. The latter three are most important. Only particle sizes less than 10–20 μm that get past the nasopharyngeal regions and reach the alveoli are of medical concern. As particle size decreases below 0.5 μm, the aerosol begins to behave like a gas (Figure 5.9). For these particles, diffusion becomes the primary mechanism of deposition in the respiratory tract before it finally reaches the alveoli.

5.6 TOXICANT DISTRIBUTION

5.6.1 Physicochemical Properties and Protein Binding

Absorption of toxicants into the blood needs to be high enough so that it will have a significant effect at the site of action in other areas of the body. The distribution process that takes the absorbed drug to other tissues is dependent on various physiological and physicochemical properties of the drug. This process is therefore a reversible movement of the toxicant between blood and tissues or between extracellular and intracellular compartments. There are, however, several complicating

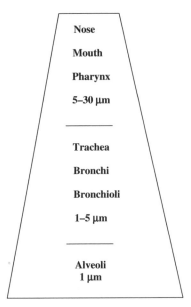

Figure 5.9   Schematic illustration of the regions where absorption may occur in the respiratory tract.

factors that can influence the distribution of a toxicant. For example, *perfusion* of tissues is an important physiological process, as some organs are better perfused (e.g., heart, brain) than others (e.g., fat). There can also be significant *protein binding* that affects delivery of drug to tissues. To further complicate the issue, elimination processes such as excretion and biotransformation (discussed at a later time) are occurring simultaneously to remove the toxicant from the blood as well as the target site.

There are several physiochemical properties of the toxicant that can influence its distribution. These include lipid solubility, pKa, and molecular weight that have been described earlier in this chapter (Section 5.4) and will not be described here. For many toxicants, distribution from the blood to tissues is by simple diffusion down a concentration gradient, and the absorption principles described earlier also apply here. The concentration gradient will be influenced by the partition coefficient or rather the ratio of toxicant concentrations in blood and tissue. Tissue mass and blood flow will also have a significant effect on distribution. For example, a large muscle mass can result in increased distribution to muscle, while limited blood flow to fat or bone tissue can limit distribution. The ratio of blood flow to tissue mass is also a useful indicator of how well the tissue is perfused. The well-perfused tissues include liver, kidney, and brain, and the low-perfused tissues include fat and bone where there is slow elimination from these tissues. Initial distribution to well-perfused tissues (e.g., heart, brain) occurs within the first few minutes, while delivery of drug to other tissues (e.g., fat, skin) is slower.

If the affinity for the target tissue is high, then the chemical will accumulate or form a depot. The advantage here is that if this were a drug, there is no need to load up the central compartment to get to the active site. However, if the reservoir for the drug has a large capacity and fills rapidly, it so alters the distribution of the drug that larger quantities of the drug are required initially to provide a therapeutic effective concentration at the target organ. If this were a toxicant, this may be an advantageous feature as toxicant levels at the target site will be reduced. In general, lipid-insoluble toxicants stay mainly in the plasma and interstitial fluids, while lipid-soluble toxicants reach all compartments and may accumulate in fat. There are numerous examples of cellular reservoirs for toxicants and drugs to distribute. Tetracycline antibiotics have a high affinity for calcium-rich tissues in the body. The bone can become a reservoir for the slow release of chemicals such as lead, and effects may be chronic or there may be acute toxicity if the toxicant is suddenly released or mobilized from these depots. The antimalaria drug quinacrine accumulates due to reversible intracellular binding, and the concentration in the liver can be several thousand times that of plasma. Another antimalaria drug, chloroquine, has a high affinity for melanin, and this drug can be taken up by tissues such as the retina that are rich in melanin granules and can cause retinitis with a drug overdose. Lipophilic pesticides and toxicants (e.g., polychlorinated biphenyls [PCBs]) and lipid-soluble gases can be expected to accumulate in high concentration in fat tissue.

There are unique anatomical barriers that can limit distribution of toxicants. A classical example of such a unique barrier is the BBB which can limit the distribution of toxicants into the central nervous system (CNS) and cerebrospinal fluid (CSF). There are three main processes or structures that keep drug or toxicant concentrations low in this region: (1) the BBB, which consists of capillary endothelial tight junctions and glial cells, surround the precapillaries, reduce filtration, and requires that the toxicant cross several membranes in order to get to the CSF. Please note that endothelial cells in other organs can have intercellular pores and pinocytotic vesicles; (2) active transport systems in the choroid plexus allow for transport of organic acids and bases from the CSF into blood; and (3) the continuous process of *CSF* production in the ventricles and venous drainage continuously dilutes toxicant or drug concentrations. Disease processes such as meningitis can disrupt this barrier and can allow for penetration of antibiotics (e.g., aminoglycosides) that would not otherwise readily cross this barrier in a healthy individual. Other tissue/blood barriers include prostate/blood, testicles/blood, and globe of eye/blood; inflammation or infection can also increase permeability of these barriers. Toxicants can cross the placenta primarily by simple diffusion, and this is most easily accomplished if the toxicants are lipid soluble (i.e., nonionized weak acids or bases). The view that the placenta is a barrier to drugs and toxicants is inaccurate. The fetus is, at least to some extent, exposed to essentially all drugs even if those with low lipid solubility are taken by the mother.

As indicated earlier, the circulatory system and components in the bloodstream are primarily responsible for transport of toxicants to target tissues or reservoirs. Erythrocytes and lymph can play important roles in transport of toxicants, but compared to plasma proteins, their role in toxicant distribution is relatively minor for most toxicants. Plasma protein binding can affect distribution because only the unbound toxicant is free or available to diffuse across the cell membranes. The toxicant–protein binding reaction is reversible and obeys the laws of mass action:

$$\textbf{Toxicant} + \textbf{Protein} \underset{k_2}{\overset{k_1}{\rightleftharpoons}} \textbf{Toxicant} - \textbf{Protein}$$
$$\small\text{(free)} \qquad\qquad\qquad\qquad \text{(bound)}$$

Usually, the ratio of unbound plasma concentration (Cu) of the toxicant to total toxicant concentration in plasma (C) is the fraction of drug unbound, fu, that is,

$$fu = Cu/C$$

The constants k_1 and k_2 are the specific rate constants for association and dissociation, respectively. The association constant K_a will be the ratio k_1/k_2, and conversely, the dissociation constant, K_d will be k_2/k_1. The constants and parameters are often used to describe and more importantly compare the relative affinity of xenobiotics for plasma proteins.

There are many circulating proteins, but those involved in binding xenobiotics include albumin, α_1-acid glycoprotein, lipoproteins, and globulins. Because many toxicants are lipophilic, they are more likely to bind to plasma α- and β-lipoproteins. There are mainly three classes of lipoproteins, namely, high-density lipoprotein (HDL), low-density lipoprotein (LDL), and very low-density lipoprotein (VLDL). Iron and copper are known to interact strongly with the metal-binding globulins transferin and ceruloplasmin, respectively. Acidic drugs bind primarily to albumin, and basic drugs are bound primarily to α_1-acid glycoprotein and β-globulin. Albumin makes up 50% of total plasma proteins, and it reacts with a wide variety of drugs and toxicants. The α_1-acid glycoprotein does not have as many binding sites as albumin, but has one high-affinity binding site. The amount of toxicant drug that is bound depends on free drug concentration, and its affinity for the binding sites, and protein concentration. Plasma protein binding is nonselective, and therefore, toxicants and drugs with similar physicochemical characteristics can compete with each other and endogenous substances for binding sites. Binding to these proteins does not necessarily prevent the toxicant from reaching the site of action, but slows the rate at which the toxicant reaches a concentration sufficient to produce a toxicological effect. Again, this is related to what fraction of the toxicant is free or unbound (fu).

Toxicants complex with proteins by various mechanisms. Covalent binding may have a pronounced effect on an organism due to the modification of an essential molecule, but such binding is usually a very minor portion of the total dose. Because covalently bound molecules dissociate very slowly, if at all, they are not considered further in this discussion. However, we should recognize that these interactions are often associated with carcinogenic metabolites. Noncovalent binding is of primary importance to distribution because the toxicant or ligand can dissociate more readily than it can in covalent binding. In rare cases, the noncovalent bond may be so stable that the toxicant remains bound for weeks or months, and for all practical purposes, the bond is equivalent to a covalent one. Types of interactions that lead to noncovalent binding under the proper physiological conditions include ionic binding, hydrogen bonding, van der Waals forces, and hydrophobic interactions. There are, however, some transition metals that have high association constants and dissociation is slow.

We know more about ligand–protein interactions today because of the numerous protein binding studies performed with drugs. The major difference between drugs and most toxicants is the frequent ionizability and high water solubility of drugs as compared with the nonionizability and high lipid solubility of many toxicants. Thus, experience with drugs forms an important background, but one that may not always be relevant to other potentially toxic compounds.

TABLE 5.4 Distribution of Insecticides into Albumin and Lipoproteins

Insecticide	Percent Distribution of Bound Insecticide			
	Percent Bound	Albumin	LOL	HDL
DDT	99.9	35	35	30
Deildrin	99.9	12	50	38
Lindane	98.0	37	38	25
Parathion	98.7	67	21	12
Diazinon	96.6	55	31	14
Carbaryl	97.4	99	<1	<1
Carbofuran	73.6	97	1	2
Aldicarb	30.0	94	2	4
Nicotine	25.0	94	2	4

Source: Adapted from Maliwal, B. P. and F. E. Guthrie. *Chem. Biol. Interact.* **35**:177–188, 1981.
LOL, low-density lipoprotein; HDL, high-density lipoprotein.

Variation in chemical and physical features can affect binding to plasma constituents. Table 5.4 shows the results of binding studies with a group of insecticides with greatly differing water and lipid solubilities. The affinity for albumin and lipoproteins is inversely related to water solubility, although the relation may be imperfect. Chlorinated hydrocarbons bind strongly to albumin but even more strongly to lipoproteins. Strongly lipophilic organophosphates bind to both protein groups, whereas more water-soluble compounds bind primarily to albumin. The most water-soluble compounds appear to be transported primarily in the aqueous phase. Chlordecone has partitioning characteristics that cause it to bind in the liver whereas dichlorodiphenyldichloroethylene (DDE), the metabolite of dichlorodiphenyltrichloroethane (DDT), partitions into fatty depots. Thus, the toxicological implications for these two compounds may be quite different.

Although highly specific (high-affinity, low-capacity) binding is more common with drugs, examples of specific binding for toxicants seem less common. It seems probable that low-affinity, high-capacity binding describes most cases of toxicant binding. The number of binding sites can only be estimated, often with considerable error, because of the nonspecific nature of the interaction. The number of ligand or toxicant molecules bound per protein molecule, and the maximum number of binding sites, n, define the definitive capacity of the protein. Another consideration is the binding affinity $K_{binding}$ (or $1/K_{diss}$). If the protein has only one binding site for the toxicant, a single value, $K_{binding}$, describes the strength of the interaction. Usually more than one binding site is present, each site having its intrinsic binding constant, $k_1, k_2, \ldots k_n$. Rarely does one find a case where $k_1 = k_2 = \ldots = k_n$, where a single value would describe the affinity constant at all sites. This is especially true when hydrophobic binding and van der Waals forces contribute to nonspecific, low-affinity binding. Obviously, the chemical nature of the binding site is of critical importance in determining binding. The three-dimensional molecular structure of the binding site, the environment of the protein, the general location in the overall protein molecule, and allosteric effects are all factors that influence binding. Studies with toxicants, and even more extensive studies with drugs, provided an adequate elucidation of these factors. Binding appears to be too complex a phenomenon to be accurately described by any one set of equations.

There are many methods for analyzing binding, but equilibrium dialysis is the most extensively used. Again, the focus of these studies is to determine the percentage of toxicant bound, the number of binding sites (n), and the affinity constant (Ka). The examples presented here are greatly simplified to avoid the undue confusion engendered by a very complex subject.

Toxicant–protein complexes that utilize relatively weak bonds (energies of the order of hydrogen bonds or less) readily associate and dissociate at physiological temperatures, and the law of mass action applies to the thermodynamic equilibrium.

$$K_{\text{binding}} = \frac{[TP]}{[T][P]} = \frac{1}{K_{\text{diss}}}$$

where K_{binding} is the equilibrium constant for association, $[TP]$ is the molar concentration of toxicant–protein complex, $[T]$ is the molar concentration of free toxicant, and $[P]$ is the molar concentration of free protein. This equation does not describe the binding site(s) or the binding affinity. To incorporate these parameters and estimate the extent of binding, double-reciprocal plots $1/[TP]$ *versus* $1/[T]$ may be used to test the specificity of binding. The $1/[TP]$ term can also be interpreted as moles of albumin per moles of toxicant. The slope of the straight line equals $1/n$Ka, and the intercept of this line with the x-axis equals $-$Ka. Regression lines passing through the origin imply infinite binding, and the validity of calculating an affinity constant under these circumstances is questionable. Figure 5.10a illustrates one such case with four pesticides, and the insert illustrates the low-affinity, "unsaturable" nature of binding in this example.

The two classes of toxicant–protein interactions encountered may be defined as (1) specific, high affinity, low capacity; and (2) nonspecific, low affinity, high capacity. The term high affinity implies an affinity constant (K_{binding}) of the order of 10^8/M, whereas low affinity implies concentrations of 10^4/M. Nonspecific, low-affinity binding is probably most characteristic of nonpolar compounds, although most cases are not as extreme as that shown in Figure 5.10.

An alternative and well-accepted treatment for binding studies is the Scatchard equation especially in situations of high-affinity binding:

$$v = \frac{nk[T]}{1+k[T]}$$

which is simplified for graphic estimates to

$$\frac{v}{[T]} = k(n-v)$$

where v is the moles of ligand (toxicant) bound per mole of protein, $[T]$ is the concentration of free toxicant, k is the intrinsic affinity constant, and n is the number of sites exhibiting such affinity. When v $[T]$ is plotted against v, a straight line is obtained if only one class of binding sites is evident. The slope is $-k$ and the intercept on the v axis becomes n (number of binding sites). If more than one class of sites

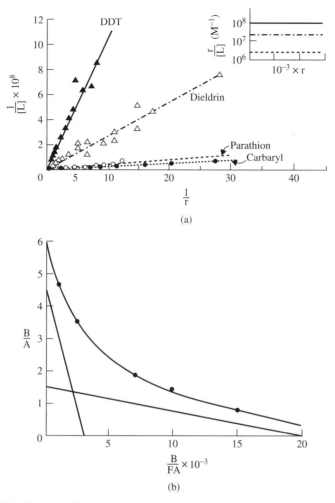

Figure 5.10 Binding of toxicants to blood proteins: (a) Double-reciprocal plot of binding of rat serum lipoprotein fraction with four insecticides. Insert illustrates magnitude of differences in slope with Scatchard plot. (b) Scatchard plot of binding of salicylate to human serum proteins. From (a) Skalsky and Guthrie, *Pest. Biochem. Physiol.* **7**:289, 1977 and (b) Moran and Walker, *Biochem. Pharmacol.* **17**:153, 1968.

occurs (probably the most common situation for toxicants), a curve is obtained from which the constants may be obtained. This is illustrated in Figure 5.10b, for which the data show not one but two species of binding sites: one with low capacity but high affinity and another with about three times the capacity but less affinity. Commonly used computer programs usually solve such data by determining one line for the specific binding and one line for nonspecific binding, the latter being an average of many possible solutions.

When hydrophobic binding of lipid toxicants occurs, as is the case for many environmental contaminants, binding is probably not limited to a single type of plasma protein. For example, the binding of the chlorinated hydrocarbon DDT is strongest for lipoproteins and albumin, but other proteins account for a significant part of overall transport. Similar results have been observed for several compounds with a range of physiochemical properties.

The presence of another toxicant and/or drug that can bind at the same site can also increase the amount of free or unbound drug. This is an example of drug inter-action, which can have serious toxicological or pharmacological consequences. *In general*, when bound concentrations are less than 90% of the total plasma con-centrations, plasma protein binding has little clinical importance. Plasma protein binding *becomes important when it is >90%*. For example, if a toxicant is 99% bound to plasma proteins, then 1% is free, but if there is toxicant interaction (e.g., com-petitive binding) that results in 94% bound and 6% is now free. Note that because of this interaction, the amount of available toxicant to cause a toxicological response *has increased sixfold*. Such a scenario may result in severe acute toxicity. Extensive plasma protein binding can influence renal clearance if glomerular filtration is the major elimination process in the kidney, but not if it is by active secretion in the kidney. Binding can also affect drug clearance if the extraction ratio (ER) in the liver is low, but not if the ER is high for that toxicant. Plasma protein binding can vary between and within chemical classes, and it is also species specific. For example, humans tend to bind acidic drugs more extensively than do other species.

There are several other variables that can alter plasma protein concentrations. These include malnutrition, pregnancy, cancer, liver abscess, renal disease, and age that can reduce serum albumin. Furthermore, α_1-glycoprotein concentrations can increase with age, inflammation, infections, obesity, renal failure, and stress. Small changes in body temperature or changes in acid–base balance may alter chemical protein binding characteristics. Although termination of drug or toxicant effect is usually by biotransformation and excretion, it may also be associated with redistri-bution from its site of action into other tissues. The classical example of this is when highly lipid-soluble drugs or toxicants that act on the brain or cardiovascular system are administered by IV or by inhalation.

5.6.2 V_d

Usually, after a toxicant or drug is absorbed, it can be distributed into various physi-ologic fluid compartments. The total body water represents 57% of total body mass (0.57 L/kg). The plasma, interstitial fluid, extracellular fluid, and intracellular fluid represent about 5, 17, 22, and 35% body weight, respectively (Table 5.5). The extra-cellular fluid comprises the blood plasma, interstitial fluid, and lymph. Intracellular

TABLE 5.5 Examples of Several Compartment Fluid Volumes

Compartment	Volume of Distribution Liters/Kilogram Body Weight (Liters/70 kg Body Weight)
Plasma	0.05 (3.5 L)
Interstitial fluid	0.18 (12.6 L)
Extracellular fluid	0.23 (16.1 L)
Intracellular fluid	0.35 (24.5 L)
Total body water	0.55 (39 L)

fluid includes the sum of fluid contents of all cells in the body. There is also transcellular fluid which represents 2% body weight, and this includes cerebrospinal, intraocular, peritoneal, pleural, and synovial fluids, and digestive secretions. Fat is about 20% body weight, while the GIT contents in monogastrics make up 1% body weight, and in ruminants it can constitute 15% body weight.

It is sometimes useful to quantitate how well a drug or toxicant is distributed into these various fluid compartments, and in this context, the apparent V_d can be a useful parameter. The apparent V_d is defined as the volume of fluid required to contain the total amount, (A), of drug in the body at the same concentration as that present in plasma, C_p,

$$V_d = A/C_p$$

In general, the V_d for a drug is, to some extent, descriptive of its distribution pattern in the body. For example, drugs or toxicants with relatively small V_d values may be confined to the plasma as diffusion across the capillary wall is limited. There are other toxicants that have a slightly larger V_d (e.g., 0.23 L/kg), and these toxicants may be distributed in the extracellular compartment. This includes many polar compounds (e.g., tubocurarine, gentamicin; V_d = 0.2–0.4 L/kg). These toxicants cannot readily enter cells because of their low lipid solubility. If the V_d for some of these toxicants is in excess of the theoretical value, this may be due to limited degree of penetration into cells or from the extravascular compartment. Finally, there are many toxicants that are distributed throughout the body water ($V_d \geq 0.55$ L/kg), and may have V_d values much greater than that for total body water. This distribution is achieved by relatively lipid-soluble toxicants and drugs that readily cross cell membranes (e.g., ethanol, diazepam; V_d = 1–2 L/kg). Binding of the toxicant anywhere outside of the plasma compartment, as well as partitioning into body fat can increase V_d beyond the absolute value for total body water. In general, toxicants with a large V_d can even reach the brain, fetus, and other transcellular compartments. In general, toxicants with large V_d are a consequence of extensive tissue binding. The reader should be aware that we are talking about tissue binding, and not plasma protein binding where distribution is limited to plasma for obvious reasons.

The fraction of toxicant located in plasma is dependent on whether a toxicant binds to both plasma and tissue components. Plasma binding can be measured directly, but not tissue binding. It can, however, be inferred from the following relationship:

Amount in Body = Amount in Plasma + Amount Outside Plasma

$$V_d \times C = V_p \times C + V_{TW} \times C_{TW}$$

Where V_d = apparent volume of distribution; V_p = volume of plasma; V_{TW} = apparent volume of tissue; and C_{TW} = tissue concentration. If the above equation is divided by C, it now becomes:

$$V_d = V_p + V_{TW} \times \frac{C_{TW}}{C}$$

Recall, $fu = Cu/C$ occurs with plasma. Then also fraction unbound in tissues, $fu_T = Cu_T/C_{TW}$.

Assuming at equilibrium that unbound concentration in tissue and plasma are equal, then the ratio of fu/fu_T replaces C_{TW}/C, and the V_d can be determined as follows:

$$V_d = V_p + V_{TW} \times (fu/fu_T)$$

One can now predict what happens to V_d when fu or fu_T changes as a result of physiological or disease processes in the body that can change plasma and/or tissue protein concentrations. For example, V_d can increase with increased unbound toxicant in plasma or with a decrease in unbound toxicant tissue concentrations. The above equation explains why, because of both plasma and tissue binding, some V_d values rarely correspond to a real volume such as plasma volume, extracellular space, or total body water. Finally, interspecies differences in V_d values can be due to differences in body composition of body fat and protein, organ size, and blood flow as alluded to earlier in this section of this chapter. The reader should also be aware that in addition to V_d, there are other volumes of distribution that can be obtained from pharmacokinetic analysis of a given data set. These include the volume of distribution at steady state ($V_{d_{ss}}$), volume of the central compartment (V_c), and the V_d that is operative over the elimination phase ($V_{d_{area}}$). The reader is advised to consult other relevant text for a more detailed description of these parameters and when it is appropriate to use these parameters.

5.7 TOXICOKINETICS

The explanation of the pharmacokinetics or toxicokinetics involved in absorption, distribution, and elimination processes is a highly specialized branch of toxicology, and is beyond the scope of this chapter. However, our focus here is to introduce a few basic concepts that are related to the several transport rate processes that have been described earlier in this chapter. Toxicokinetics is an extension of pharmacokinetics in that these studies are conducted at higher doses than pharmacokinetic studies, and the principles of pharmacokinetics are applied to xenobiotics. In addition, these studies are essential to provide information on the fate of the xenobiotic following exposure by a define route. This information is essential if one is to adequately interpret the dose–response relationship in the risk assessment process. In recent years, these toxicokinetic data from laboratory animals are utilized in physiologically based pharmacokinetic (PBPK) models to help extrapolations to low-dose exposures in humans. The ultimate aim in all of these analyses is to provide an estimate of tissue concentrations at the target site associated with the toxicity.

Immediately on entering the body, a chemical begins changing location, concentration, or chemical identity. It may be transported independently by several components of the circulatory system, absorbed by various tissues, or stored; the chemical may effect an action, be detoxified, or be activated; the parent compound or its metabolite(s) may react with body constituents, be stored, or be eliminated— to name some of the more important actions. Each of these processes may be

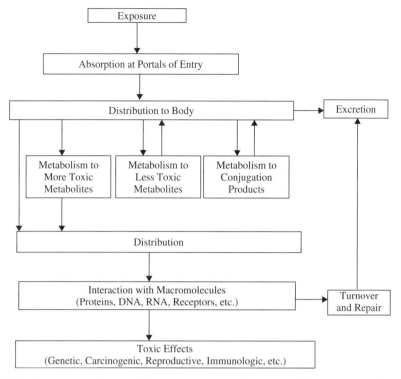

Figure 5.11 Sequence of events following exposure of an animal to exogenous chemicals.

described by rate constants similar to those described earlier in our discussion about first-order rate processes which are associated with toxicant absorption, distribution, and elimination that are occurring simultaneously. Thus, at no time is the situation stable but is constantly changing as indicated in Figure 5.11.

It should be noted however, that as the toxicant is being absorbed and distributed throughout the body, it is being simultaneously eliminated by various metabolism and/or excretion mechanisms which will be discussed in more detail in future chapters. However, one should mention here that an important pharmacokinetic parameter known as *clearance* (Cl) can be used to quantitatively assess elimination of a toxicant. Clearance is defined as the rate of toxicant excreted relative to its plasma concentration, C_p:

$$Cl = \text{Rate of Toxicant Excretion}/C_p$$

The rate of excretion is really the administered dose times the fractional elimination rate constant K_{el} described earlier. Therefore, we can express the above equation in terms of K_{el} and administered dose as well as V_d:

$$Cl = K_{el} \cdot \text{Dose}/C_p = K_{el} \cdot (V_d \cdot C_p)/C_p = K_{el} \cdot V_d$$

In physiological terms, we can also define clearance as the volume of blood cleared of the toxicant by an organ or body per unit time. Therefore, as the equations above indicate, the body clearance of a toxicant is expressed in units of volume per unit time (e.g., liter/hour), and can be derived if we know the V_d of the toxicant and fractional rate constant. In many instances, this can only be derived by appropriate pharmacokinetic analysis of a given data set following blood or urine sample collection and appropriate chemical analyses to determine toxicant concentrations in either of these biological matrices.

Each of the processes discussed thus far, absorption, distribution, and elimination, can be described as a rate process. In general, these are assumed to be first-order processes in which the rate of transfer at any time is proportional to the amount of drug in the body at that time. Recall that the rate of transport (dC/dt) is proportional to toxicant concentration (C) or stated mathematically:

$$dC/dt = KC$$

where K is the rate constant (fraction per unit time). Many pharmacokinetic analyses of a chemical are based primarily on toxicant concentrations in blood or urine samples. It is often assumed in these analyses that the rate of change of toxicant concentration in blood reflects quantitatively the change in toxicant concentration throughout the body (first-order principles). Because of the elimination/clearance process which is also assumed to be a first-order rate process, the above rate equation now needs a negative sign. This is really a decaying process that is observed as a decline of toxicant concentration in blood or urine after IV administration. The IV route is preferred in these initial analyses because there is no absorption phase, but only chemical depletion phase. However, one cannot measure infinitesimal change of C or time, t; therefore there needs to be integration after rearrangement of the above equation:

$$-dC/C = kdt \text{ becomes } \int -dC/C = k\int dt$$

or expressed as:

$$C = C^0 e^{-kt}$$

where e is the base of the natural logarithm, and we can remove e by taking the ln of both sides:

$$\ln C^t = \ln C^0 - kt.$$

Please note that K is the slope of the straight line for a semilog plot of toxicant concentration versus time (Figure 5.12), and in the above equation, it is the elimination rate constant that is related to half-life of the toxicant described earlier in this chapter. The derived C^0 can now be used to calculate the V_d of the toxicant as follows:

$$V_d = \text{Dose}/C^0.$$

However, toxicokinetic data for many toxicants do not always provide a straight line when plotted as described above, and more complicated equations with more than one exponential term with rate constants may be necessary to mathematically describe the concentration-time profile. These numerous rate constants are indicative of chemical transport between various compartments in the body and not

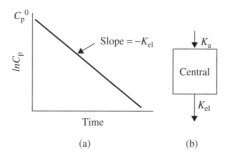

Figure 5.12 (a) Semilog plot of plasma concentration (C_p) versus time. C_p^o is the intercept on the y-axis, and K_{el} is the elimination rate constant. (b) Single compartment model with rate constants for absorption, K_a and for elimination, K_{el}.

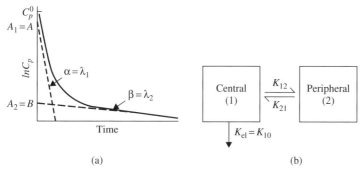

Figure 5.13 (a) Semilog plot of plasma concentration for (C_p) versus time representative of a two-compartment model. The curve can be broken down into an α or λ_1 distribution phase and β or λ_2 elimination phase. (b) Two-compartment model with transfer rate constants, K_{12} and K_{21}, and elimination rate constant, K_{el}.

a single central compartment as suggested in the simple equation and semilog plot described above in Figure 5.12. In some instances, the data may fit to a bi-exponential concentration-time profile Figure 5.13, and the equation to describe this model is:

$$C = Ae^{-\alpha t} + Be^{-\beta t}$$

In other instances, complex profiles may require a three- or multi-exponential concentration-time profile (Figure 5.14), and the equation to describe the former is:

$$C = Ae^{-\alpha t} + Be^{-\beta t} + Ce^{-\gamma t}$$

In the physiological sense, one can divide the body into "compartments" that may represent discrete parts of the whole—blood, liver, urine, and so on—or the mathematical model describing the process may be a composite representing the pooling of parts of tissues involved in distribution and bioactivation. Usually, pharmacokinetic compartments have no anatomical or physiological identity; they represent all locations within the body that have similar characteristics relative to the transport

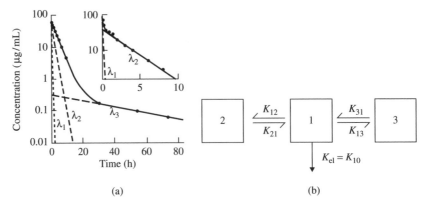

Figure 5.14   (a) Semilog plot of plasma concentration for (C_p) versus time representative of a three- or multicompartment model. The curve can be broken down into three phases, λ_1, λ_2, and λ_3. (b) Three-compartment model with transfer rate constants, K_{12}, K_{21}, K_{13}, K_{31}, and elimination rate constant, K_{el}. As these models can get more complicated; the α, β, and γ nomenclature may get replaced with λn as indicated in the profile.

rates of the particular toxicant. Simple first-order kinetics is usually accepted to describe individual rate processes for the toxicant after entry. The resolution of the model necessitates mathematical estimates (as a function of time) concerning the absorption, distribution, biotransformation, and excretion of the toxicant.

Drugs and toxicants with multi-exponential behavior depicted in Figure 5.14 require calculation of the various micro-constants. An alternative method involves using model-independent pharmacokinetic to arrive at relevant parameters. This would not be covered in any detail in this chapter, but very briefly it involves determination of AUC of the concentration-time profiles, and the emergence of micro-computers in recent years has greatly facilitated this approach.

In conclusion, pharmacokinetics is a study of the time course of absorption, distribution, and elimination of a chemical. We use pharmacokinetics as a tool to analyze plasma concentration time profiles after chemical exposure, and it is the derived rates and other parameters that reflect the underlying physiological processes that determine the fate of the chemical. There are numerous software packages available today to accomplish these analyses. The user should, however, be aware of the experimental conditions and time frame over which the data were collected and many of the assumptions embedded in the analyses. For example, many of the transport processes described in this chapter may not obey first-order kinetics and thus may be nonlinear especially at toxicological doses. The reader is advised to consult other text for more detailed descriptions of these nonlinear interactions and data analyses.

BIBLIOGRAPHY AND SUGGESTED READING

Bloom, D. and W. Fawcett. *Bloom and Fawcett: A Textbook of Histology*. Philadelphia: Saunders, 1975.

Bronaugh, R. and H. Maibach, eds. *Percutaneous Absorption*. New York: Marcel Dekker, 1989.

Gilman, A. G., T. W. Rall, A. S. Nies, eds. *Goodman and Gilman's The Pharmacological Basis of Therapeutics*, 8th ed. Elmsford, NY: Pergamon Press, 1990.

Grandjean, P., ed. *Skin Penetration: Hazardous Chemicals at Work*. Philadelphia: Taylor & Francis, 1990.

Ham, A.W. and D. H. Cormack. *Histology*, 8th ed. Philadelphia: Lippincott, 1979.

Hodgson, E. and P. E. Levi, eds. *Introduction to Biochemical Toxicology*, 2nd ed. Norwalk, CT: Appleton & Lange, 1994, p. 12.

Krieger, R., ed. *Handbook of Pesticide Toxicology*, 2nd ed. San Diego, CA: Academic Press, 2001.

Maliwal, B. P. and F. E. Guthrie. Interactions of insecticides with human liopproteins. *Chem. Biol. Interact.* **35**:177–188, 1981.

Moran, C. J. and W. H. Walker. The binding of salicylate to human serum. *Biochem. Pharmacol.* **17**:153–156, 1968.

Rowland, M. and T. N. Tozer, eds. *Clinical Pharmacokinetics: Concepts and Applications*, 3rd ed. Philadelphia: Lea and Febiger, 1995.

Schottelius, B. A. and D. D. Schottelius. *Textbook of Physiology*. St. Louis, MO: Mosby, 1973.

Shargel, L. and A. B. C. Yu, eds. *Applied Biopharmaceutics and Pharmacokinetics*, 4th ed. Norwalk, CT: Appleton & Lange, 1999.

Singer, S. J. and G. L. Nicolson. The fluid mosaic model of the structure of cell membranes. *Science* **175**:720–731, 1972.

Skalsky, H. L. and F. E Guthrie. Affinities of parathion, DDT, dieldrin, and carbaryl for macromolecules in the blood of the rat and American cockroach and the competitive interaction of steroids. *Pest. Biochem. Physiol.* **7**:289–296, 1977.

SAMPLE QUESTIONS

1. **(a)** In carrier-mediated transport, chemical transport can reach saturation.
   (b) In passive diffusion, energy is usually expended in chemical transport.
   (c) In general, a chemical is more readily absorbed across human skin than mouse skin.
   (d) The unionized form of a chemical is more readily absorbed than the ionized form.

2. **(a)** In zero-order transport, the amount moved per unit time is constant.
   (b) In first-order transport, the fraction moved is dependent on dose.
   (c) A chemical that binds extensively to plasma proteins will most likely have a high V_d.
   (d) A chemical with high lipid solubility most likely has a high V_d.
   (e) The half-life of any chemical can be increased by *increasing* the V_d and/or *decreasing* its clearance.

3. Explain how significant plasma protein binding can influence distribution of a toxicant in the body.

4. What are the differences between active and passive transport of toxicants?

5. Define the term partition coefficient. How is it related to Fick's law?

Metabolism of Toxicants

ERNEST HODGSON and RANDY L. ROSE*

6.1 INTRODUCTION

Since the publication of the 3rd edition of this textbook, there have been few, if any, additions to the roster of xenobiotic-metabolizing enzymes (XMEs) or to the reactions they catalyze. However, important advances have been made in the molecular biology of XMEs and to their action in humans, the latter aspect being of particular importance in human health risk analysis.

One of the most important determinants of xenobiotic persistence in the body and subsequent toxicity to the organism is the extent to which they can be metabolized and excreted. Several families of metabolic enzymes, often with broad substrate specificity, are involved in xenobiotic metabolism. Some of the more important families of enzymes involved in xenobiotic metabolism include the cytochrome P450 monooxygenases (CYPs), flavin-containing monooxygenases (FMOs), alcohol and aldehyde dehydrogenases, amine oxidases, cyclooxygenases, reductases, hydrolases, and a variety of conjugating enzymes such as glucuronidases, sulfotransferases, methyltransferases, glutathione transferases, and acetyl transferases.

The majority of xenobiotic metabolism occurs in the liver, an organ devoted to the synthesis of many important biologically functional proteins, which also has the capacity to mediate chemical transformations of xenobiotics. Most xenobiotics that enter the body are lipophilic, a property that enables them to bind to lipid membranes and to be transported by lipoproteins in the blood. After entrance into the liver, as well as in other organs, xenobiotics may undergo one or two phases of metabolism. In Phase I, a polar reactive group is introduced into the molecule, rendering it a suitable substrate for Phase II enzymes. Enzymes typically involved in Phase I metabolism include the CYPs, FMOs, and hydrolases as will be discussed later. In Phase II, following the introduction of a polar group, conjugating enzymes typically add endogenous substituents, such as sugars, sulfates, or amino acids which result in substantially increasing the water solubility of the xenobiotic, making it

*deceased

A Textbook of Modern Toxicology, *Fourth Edition*. Edited by Ernest Hodgson
Copyright © 2010 John Wiley & Sons, Inc.

easily excreted. Although this process is generally a detoxication sequence, reactive intermediates may be formed that are much more toxic than the parent compound. It is, however, usually a sequence that increases water solubility and hence decreases the biological half-life ($t_{0.5}$) of the xenobiotic *in vivo*. The role of the transport proteins, known collectively as *transporters* (see Chapter 9) is often referred to as Phase III.

Phase I monooxygenations are more likely to form reactive intermediates than Phase II metabolism because the products are usually potent electrophiles capable of reacting with nucleophilic substituents on macromolecules, unless detoxified by some subsequent reaction. In the following discussion, examples of both detoxication and intoxication reactions are given, although greater emphasis on activation products is provided in Chapter 7.

6.2   PHASE I REACTIONS

Phase I reactions include microsomal monooxygenations, cytosolic and mitochondrial oxidations, co-oxidations in the prostaglandin synthetase reaction, reductions, hydrolyses, and epoxide hydration. All of these reactions, with the exception of reductions, introduce polar groups to the molecule that, in most cases, can be conjugated during Phase II metabolism. The major Phase I reactions are summarized in Table 6.1.

6.2.1   The Endoplasmic Reticulum, Microsomes, and Monooxygenations

Monooxygenations of xenobiotics are catalyzed either by the CYP-dependent monooxygenase system or by the FMOs. Both are located in the endoplasmic reticulum of the cell and have been studied in many tissues and organisms. This is particularly true of CYPs, probably the most studied of all enzymes.

Microsomes are derived from the endoplasmic reticulum as a result of tissue homogenization and are isolated by centrifugation of the postmitochondrial supernatant fraction, described below. The endoplasmic reticulum is an anastomosing network of lipoprotein membranes extending from the plasma membrane to the nucleus and mitochrondria, whereas the microsomal fraction derived from it consists of membranous vesicles contaminated with free ribosomes, glycogen granules, and fragments of other subcellular structures such as mitochondria and Golgi apparatus. The endoplasmic reticulum, and consequently the microsomes derived from it, consists of two types, rough and smooth, the former having an outer membrane studded with ribosomes, which the latter characteristically lack. Although both rough and smooth microsomes have all the components of the CYP-dependent monooxygenase system, the specific activity of the smooth type is usually higher.

The preparation of the postmitochondrial fraction (S9) microsomes and cytosolic fractions from tissue homogenates involves the use of two or three centrifugation steps. Following tissue extraction, careful mincing, and rinses of tissue for blood removal, the tissues are typically homogenized in buffer and centrifuged at $10,000 \times g$ for 20 min. The resulting supernatant, the S9 fraction, can be used in studies where both microsomal and cytosolic enzymes are desired. More often, however, the S9

TABLE 6.1   Summary of Some Important Oxidative and Reductive Reactions of Xenobiotics

Enzymes and Reactions	Examples
Cytochrome P450	
Epoxidation/hydroxylation	Aldrin, benzo(a)pyrene, aflatoxin, bromobenzene
N-, O-, S-dealkylation	Ethylmorphine, atrazine, p-nitroanisole, methylmercaptan
N-, S-, P-oxidation	Thiobenzamide, chlorpromazine, 2-acetylaminofluorene
Desulfuration	Parathion, carbon disulfide
Dehalogenation	Carbon tetrachloride, chloroform
Nitro reduction	Nitrobenzene
Azo reduction	O-aminoazotoluene
Flavin-containing monooxygenase	
N-, S-, P-oxidation	Nicotine, imipramine, thiourea, methimazole
Desulfuration	Fonofos
Prostaglandin synthetase co-oxidation	
Dehydrogenation	Acetaminophen, benzidine, epinephrine
N-dealkylation	Benzphetamine, dimethylaniline
Epoxidation/hydroxylation	Benzo(a)pyrene, 2-aminofluorene, phenylbutazone
Oxidation	FANFT (N(4-(5-nitro-2-furyl)-2-thiazolyl formamide)), ANFT (2-amino-4-(5-nitro-2-furyl) thiazole), bilirubin
Molybdenum hydroxylases	
Oxidation	Purines, pteridine, methotrexate, 6-deoxycyclovir
Reductions	Aromatic nitrocompounds, azo dyes, nitrosoamines
Alcohol dehydrogenase	
Oxidation	Methanol, ethanol, glycols, glycol ethers
Reduction	Aldehydes and ketones
Aldehyde dehydrogenase	
Oxidation	Aldehydes resulting from alcohol and glycol oxidations
Esterases and amidases	
Hydrolysis	Parathion, paraoxon, dimethoate
Epoxide hydrolase	
Hydrolysis	Benzo(a)pyrene epoxide, styrene oxide

fraction is centrifuged at $100,000 \times g$ for 60 min to yield a microsomal pellet and a cytosolic supernatant. The pellet is typically resuspended in a volume of buffer, which will give 20–50 mg protein/mL and stored from −20 to −70°C. Often, the microsomal pellet is resuspended a second time and resedimented at $100,000 \times g$ for 60 min to further remove contaminating hemoglobin and other proteins. As described above, enzymes within the microsomal fraction (or microsomes) include CYPs, FMOs, cyclooxygenases, and other membrane-bound enzymes including necessary coenzymes such as reduced nicotine adenine dinucleotide phosphate (NADPH)-cytochrome P450 reductase for CYP. Enzymes found in the cytosolic fraction (derived from the supernatant of the first $1,000,000 \times g$ spin) include hydrolases and most of the conjugating enzymes such as glutathione transferases, glucuronidases, sulfotransferases, methyltransferases, and acetylases.

Monooxygenations, previously known as mixed-function oxidations, are those oxidations in which one atom of a molecule of oxygen is incorporated into the substrate while the other is reduced to water. Because the electrons involved in the reduction of CYPs or FMOs are derived from NADPH, the overall reaction can be written as follows (where RH is the substrate):

$$RH + O_2 + NADPH + H^+ \rightarrow NADP^+ + ROH + H_2O$$

6.2.2 The CYP-Dependent Monooxygenase System

The CYPs, the carbon monoxide-binding pigments of microsomes, are heme proteins of the b cytochrome type, containing protoporphyrin IX. Originally described as a single protein, over 7500 animal CYP isoforms in 781 gene families have been characterized across all taxa and genomic and protein sequences are known. As the list of CYPs is continually expanding, progress in this area can be readily accessed via the Internet at the website of the P450 Gene Superfamily Nomenclature Committee (http://drnelson.utmem.edu/nelsonhomepage.html) or at another excellent website (http://www.icgeb.trieste.it/p450).

A system of nomenclature based upon derived amino acid sequences was proposed in 1987 and entries are continuously updated (http://drnelson.utmem.edu/CytochromeP450.html). Degree of similarity in sequence classifies members to a CYP (cyp in the case of mice) numeric gene family, then letter subfamily such that individual isoforms have unique CYP number-letter-number annotations, for example, CYP1A1. Of the 110 animal CYP families, 18 are found in vertebrates.

In general, enzymes within a gene family share more than 40% amino acid sequence identity. Protein sequences within subfamilies have greater than 55% similarity in the case of mammalian genes, or 46% in the case of nonmammalian genes. So far, genes in the same subfamily have been found to lie on the same chromosome within the same gene cluster and are nonsegregating, suggesting a common origin through gene duplication events. Sequences showing less than 3% divergence are arbitrarily designated allelic variants unless other evidence exists to the contrary. Known sequences fit the classification scheme surprisingly well, with few exceptions found at the family, subfamily, or allelic variant levels, and in each case, additional information is available to justify the departure from the rules set out.

In some cases, a homologue of a particular CYP enzyme is found across species (e.g., CYP1A1). In other cases, the genes diverged subsequent to the divergence of the species and no exact analog is found in various species (e.g., the CYP2C subfamily). In this case, the genes are numbered in the order of discovery and the gene products from a particular subfamily may even have differing substrate specificity in different species (e.g., rodent vs. human).

The total number of functional CYP genes in any single mammalian species is thought to range from 60 to 200. Whereas some CYP isoforms are substrate specific, those involved in xenobiotic metabolism tend to be relatively nonspecific, although substrate preferences are usually evident. Although P450 is still appropriate as a prefix for the protein products, it is being rapidly replaced by the formal abbreviation, CYP. Unlike most cytochromes, the name is derived not from the absorption maximum of the reduced form in the visible region but from the unique wavelength of the absorption maximum of the carbon monoxide derivative of the reduced form, namely 450 nm.

The role of CYP as the terminal oxidase in monooxygenase reactions is supported by considerable evidence. The initial proof was derived from the demonstration of the concomitant light reversibility of the CO complex of CYP and the inhibition, by CO, of the C-21 hydroxylation of 17 α-hydroxy-progesterone by adrenal gland microsomes. This was followed by a number of indirect, but nevertheless convincing proofs involving the effects on both CYP and monooxygenase activity of CO, inducing agents, and spectra resulting from ligand binding and the loss of activity on degradation of CYP to cytochrome P420. Direct proof was subsequently provided by the demonstration that monooxygenase systems, reconstituted from apparently homogenous purified CYP, NADPH-CYP reductase, and phosphatidylcholine, can catalyze many monooxygenase reactions.

CYPs, like other hemoproteins, have characteristic absorptions in the visible region. The addition of many organic, and some inorganic, ligands results in perturbations of these spectra. Although the detection and measurement of these spectra requires a high-resolution spectrophotometer, these perturbations, measured as optical difference spectra, have been of tremendous use in the characterization of CYPs, particularly in the decades preceding the molecular cloning and expression of specific CYP isoforms.

The most important difference spectra of oxidized CYP are type I, with an absorption maximum at 385–390 nm. Type I ligands are found in many different chemical classes and include drugs, environmental contaminants, pesticides, and so on. They appear to be generally unsuitable, on chemical grounds, as ligands for the heme iron and are believed to bind to a hydrophobic site in the protein that is close enough to the heme to allow both spectral perturbation and interaction with the activated oxygen. Although most type I ligands are substrates, it has not been possible to demonstrate a quantitative relationship between K_S (concentration required for half-maximal spectral development) and K_m (Michaelis constant). Type II ligands, however, interact directly with the heme iron of CYP, and are associated with organic compounds having nitrogen atoms with sp^2 or sp^3 nonbonded electrons that are sterically accessible. Such ligands are frequently inhibitors of CYP activity.

The two most important difference spectra of reduced CYP are the well-known CO spectrum, with its maximum at or about 450 nm, and the type III spectrum, with two pH-dependent peaks at approximately 430 nm and 455 nm. The CO spectrum forms the basis for the quantitative estimation of CYP. The best-known type III ligands for CYP are ethyl isocyanide and compounds such as the methylenedioxyphenyl synergists and SKF 525A, the last two forming stable type III complexes that appear to be related to the mechanism by which they inhibit monooxygenations.

In the catalytic cycle of CYP, reducing equivalents are transferred from NADPH to CYP by a flavoprotein enzyme known as NADPH-CYP reductase. The evidence that this enzyme is involved in CYP monooxygenations was originally derived from the observation that cytochrome c, which can function as an artificial electron acceptor for the enzyme, is an inhibitor of such monooxygenations. This reductase is an essential component in CYP-catalyzed enzyme systems reconstituted from purified components. Moreover, antibodies prepared from purified reductase are inhibitors

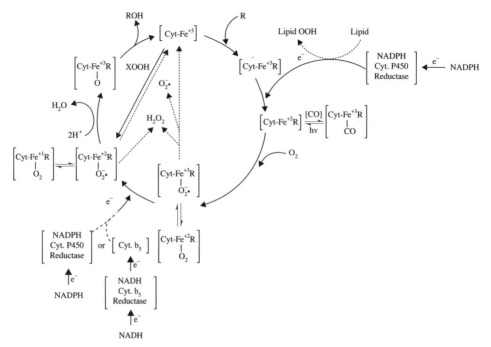

Figure 6.1 Generalized scheme showing the sequence of events for P450 monooxygenations.

of microsomal monooxygenase reactions. The reductase is a flavoprotein of approximately 80,000 Da that contain 2 mol each of flavin mononucleotide (FMN) and flavinadenine dinucleotied (FAD) per mole of enzyme. The only other component necessary for activity in the reconstituted system is a phospholipid, phosphatidylchloline. This is not involved directly in electron transfer but appears to be involved in the coupling of the reductase to the cytochrome and in the binding of the substrate to the cytochrome.

The generally recognized steps in the catalytic cycle are shown in Figure 6.1. The initial step consists of the binding of substrate to oxidized CYP followed by a one electron reduction catalyzed by NADPH-CYP reductase to form a reduced cytochrome–substrate complex. This complex can interact with CO to form the CO complex, which gives rise to the well-known difference spectrum with a peak at 450 nm and also inhibits monooxygenase activity. The next several steps are less well understood. They involve an initial interaction with molecular oxygen to form a ternary oxygenated complex. This ternary complex accepts a second electron, resulting in the further formation of one or more less understood complexes. One of these, however, is probably the equivalent of the peroxide anion derivative of the substrate-bound hemoprotein. Under some conditions, this complex may break down to yield hydrogen peroxide and the oxidized cytochrome–substrate complex. Normally, however, one atom of molecular oxygen is transferred to the substrate and the other is reduced to water, followed by dismutation reactions leading to the formation of the oxygenated product, water, and the oxidized cytochrome.

The possibility that the second electron is derived from NADH through cytochrome b_5 has been the subject of argument for some time and has yet to be completely resolved. Cytochrome b_5 is a widely distributed microsomal heme protein that is involved in metabolic reactions such as fatty acid desaturation that involve endogenous substrates. It is clear, however, that cytochrome b_5 is not essential for all CYP-dependent monooxygenations because many occur in systems reconstituted from NADPH, O_2, phosphatidylchloline, and highly purified CYP and NADPH-CYP reductase. Nevertheless, there is good evidence that many catalytic activities by isoforms including CYP3A4, CYP3A5, and CYP2E1 are stimulated by cytochrome b_5. In some cases, apocytochrome b_5 (devoid of heme) has also been found to be stimulatory, suggesting that an alternate role of cytochrome b_5 may be the result of conformational changes in the CYP/NADPH-CYP reductase systems. Thus, cytochrome b_5 may facilitate oxidative activity in the intact endoplasmic reticulum. The isolation of forms of CYP that bind avidly to cytochrome b_5 also tends to support this idea.

Distribution CYP In vertebrates, the liver is the richest source of CYP and is most active in the monooxygenation of xenobiotics. CYP and other components of the CYP-dependent monooxygenase system are also found in the skin, nasal mucosa, lung, and gastrointestinal tract, presumably reflecting the evolution of defense mechanisms at portals of entry. In addition to these organs, CYP has been demonstrated in the kidney, adrenal cortex and medulla, placenta, testes, ovaries, fetal and embryonic liver, corpus luteum, aorta, blood platelets, and the nervous system. In humans, CYP has been demonstrated in the fetal and adult liver, the placenta, kidney, testes, fetal and adult adrenal gland, skin, blood platelets, and lymphocytes.

Although CYPs are found in many tissues, the function of the particular subset of isoforms in a particular organ, tissue, or cell type does not appear to be the same in all cases. In the liver, CYPs oxidize a large number of xenobiotics as well as some endogenous steroids and bile pigments. The CYPs of the lung also appear to be concerned primarily with xenobiotic oxidations, although the range of substrates is more limited than that of the liver. The skin and small intestine also carry out xenobiotic oxidations, but their activities have been less well characterized. In normal pregnant females, the placental microsomes display little or no ability to oxidize foreign compounds, appearing to function as a steroid hormone metabolizing system. On induction of the CYP enzymes, such as occurs in pregnant women who smoke, CYP-catalyzed aryl hydrocarbon hydroxylase activity is readily apparent. The CYPs of the kidney are active in the ω-oxidation of fatty acids, such as lauric acid, but are relatively inactive in xenobiotic oxidation. Mitochondrial CYPs, such as those of the placenta and adrenal cortex, are active in the oxidation of steroid hormones rather than xenobiotics.

Distribution of CYPs within the cell has been studied primarily in the mammalian liver, where it is present in greatest quantity in the smooth endoplasmic reticulum and in smaller but appreciable amounts in the rough endoplasmic reticulum. The nuclear membrane has also been reported to contain CYP and to have detectable aryl hydrocarbon hydroxylase activity, a location that may be of importance in the metabolic activation of carcinogens.

Multiplicity of CYP, Purification, and Reconstitution of CYP Activity Even before appreciable purification of CYP had been accomplished, it was already apparent from indirect evidence that mammalian liver cells contained more than one CYP enzyme. Subsequent direct evidence on the multiplicity of CYPs included the separation and purification of CYP isozymes, distinguished from each other by chromatographic behavior, immunologic specificity, and/or substrate specificity after reconstitution and separation of distinct polypeptides by sodium dodecyl sulfate–polyacrylamide gel electrophoresis (SDS-PAGE), which could then be related to distinct CYPs present in the original microsomes.

Purification of CYP and its usual constituent isoforms was, for many years, an elusive goal; one, however, that was eventually resolved. However, the lengthy processes of CYP purification have now been largely replaced by the cloning and expression of transgenic isoforms.

Systems reconstituted from purified CYP, NADPH-CYP reductase, and phosphatidylchloline will, in the presence of NADPH and O_2, oxidize many xenobiotics, often at rates comparable to microsomes. Although systems reconstituted from this minimal number of components are enzymatically active, other microsomal components, such as cytochrome b_5, may facilitate activity either *in vivo* or *in vitro* or may even be essential for the oxidation of certain substrates.

One important finding from purification studies as well as cloning and expressing of individual isoforms is that the lack of substrate specificity of microsomes for monooxygenase activity is not an artifact caused by the presence of several specific cytochromes because it appears that many of the cytochromes isolated are still relatively nonspecific. The relative activity toward different substrates does, however, vary greatly from one CYP isoform to another even when both are relatively nonspecific. This lack of specificity is illustrated in Table 6.2, using human isoforms as examples.

Evolution of CYP Relationships between different CYP families and subfamilies are related to the rate and extent of CYP evolution.

Figure 6.2 demonstrates some of the evolutionary relationships between P450 genes between some of the earliest vertebrates and humans. This dendrogram compares P450 genes from the puffer fish (fugu) and eight other fish species with human CYPs (including three pseudogenes). The unweighted pair group method arithmetic averaging (UPGMA) phylogenetic tree demonstrates the presence of five CYP clans (clusters of CYPs that are consistently grouped together) and delineates the 18 known human CYPs. This data set demonstrates that the defining characteristics of vertebrate CYPs have not changed much in 420 million years. Of these 18 human CYPs, only 1 family was missing in fugu (CYP39), indicating that the mammalian diversity of CYPs likely predates the tetrapod-ray finned fish divergence. The fish genome also has new CYP1C, 3B, and 7C subfamilies that are not seen in mammals.

CYP Families with Xenobiotic Metabolizing Potential Although mammals are known to have 18 CYP families, only three families are primarily responsible for most xenobiotic metabolism. These families, families 1–3, are considered to be more recently derived from the "ancestral" CYP families. The remaining families are less promiscuous in their metabolizing abilities and are often responsible for specific metabolic steps. For example, members of the CYP4 family are responsible for the

TABLE 6.2   Some Important Human Cytochrome P450 Isozymes and Selected Substrates

P450	Drugs	Carcinogens/Toxicants/Endogenous Substrates	Diagnostic Substrates In Vivo [In Vitro]
1A1	Verlukast (very few drugs)	Benzo(a)pyrene, dimethylbenz(a) anthracene	[Ethoxyresorufin, benzo(a)pyrene]
1A2	Phenacetin, theophylline, acetaminophen, warfarin, caffeine, cimetidine	Aromatic amines, arylhydrocarbons, NNK[a], aflatoxin, estradiol	Caffeine, [acetanilide, methoxyresorufin, ethoxyresorufin]
2A6	Coumarin, nicotine	Aflatoxin, diethylnitrosamine, NNK[a]	Coumarin
2B6	Cyclophosphamide, ifosfamide, nicotine	6 Aminochrysene, aflatoxin, NNK[a]	[7-ethoxy-4-trifluoro-methyl coumarin]
2C8	Taxol, tolbutamide, carbamazepine	—	[Chloromethylfluorescein diethly ether]
2C9	Tienilic acid, tolbutamide, warfarin, phenytoin, tetrahydrocannabinol, hexobarbital, diclofenac	—	[Diclofenac (4'-OH)]
2C19	S-mephenytoin, diazepam, phenytoin, omeprazole, indomethacin, imipramine, propanolol, proguanil	—	[S-mephenytoin (4'-OH)]
2D6	Debrisoquine, sparteine, bufuralol, propanolol, thioridazine, quinidine, phenytoin, fluoxetine	NNK[a]	Dextromethorphan, [bufuralol (4'-OH)]
2E1	Chlorzoxazone, isoniazid, acetaminophen, halothane, enflurane, methoxyflurane	Dimethylnitrosamine, benzene, halogenated alkanes (e.g., CCl4), acylonitrile, alcohols, aniline, styrene, vinyl chloride	Chlorzoxazone (6-OH), [p-nitrophenol]
3A4	Nifedipine, ethylmorphine, warfarin, quinidine, taxol, ketoconazole, verapamil, erythromycin, diazepam	Aflatoxin, 1-nitropyrene, benzo(a) pyrene 7,8-diol, 6 aminochrysene, estradiol, progesterone, testosterone, other steroids, bile acids	Erythromycin, nifedipine [testosterone (6-β)]
4A9/11	(Very few drugs)	Fatty acids, prostaglandins, thromboxane, prostacyclin	[Lauric acid]

[a]NNK, 4(methylnitrosamino)-1-(3-pyridyl)-1-butanone, a tobacco-smoke specific nitrosamine.

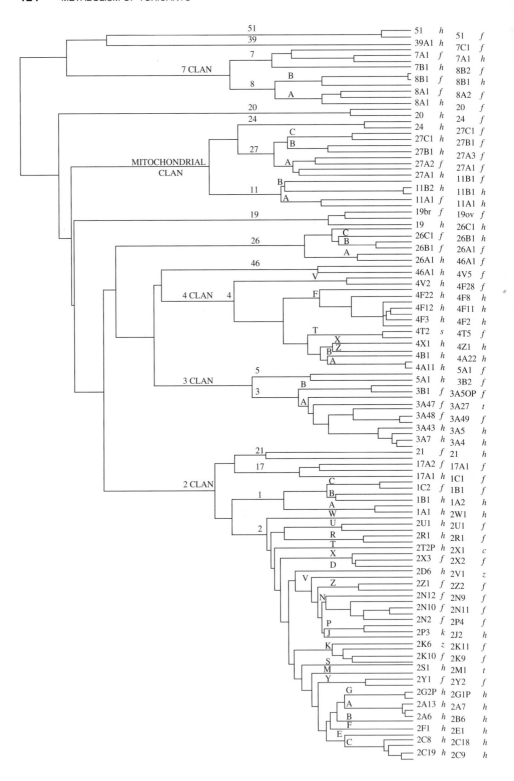

end-chain hydroxylation of long chain fatty acids. The remaining mammalian CYP families are involved in biosynthesis of steroid hormones. In fact, some of the nomenclature for some of these families is actually derived from the various positions in the steroid nucleus where the metabolism takes place. For example, CYP7 mediates hydroxylation of cholesterol at the 7 α-position, while CYP 17 and 21 catalyze the 17α and 21-hydroxylations of progesterone, respectively. CYP 19 is responsible for the aromatization of androgens to estrogen by the initial step of hydroxylation at the 19-position. Many of the CYPs responsible for steroidogenesis are found in the adrenal cortex, while those involved in xenobiotic metabolism are found predominantly in tissues that are more likely to be involved in exposure such as liver, kidneys, lungs, and olfactory tissues.

To simplify discussion of important CYP family members, the following discussion concentrates upon human CYP family members. However, since there is a great deal of homology among family members, many of the points of discussion are generally applicable to CYP families belonging to several species.

The CYP1 family contains three known human members, CYP1A1, CYP1A2, and CYP1B1. CYP1A1 and CYP1A2 are found in all classes of the animal kingdom. CYP2E1 is the only other CYP that retains the same gene designation in many different species.

CYP1A1 and CYP1A2 possess distinct but overlapping substrate specificities: CYP1A1 preferring neutral polycyclic aromatic hydrocarbons (PAHs), while the latter prefers polyaromatic and heterocyclic amines and amides. Because of the preference of this family for molecules with highly planar molecular structures, CYP1 family members are closely associated with metabolic activation of many procarcinogens and mutagens including benzo(a)pyrene, aflatoxin B1, dimethyl-benzanthracene, β-naphthylamine, 4-aminobiphenyl, 2-acetylaminofluorene, and benzidine. Figure 6.3 illustrates a typical reaction sequence that leads to the formation of epoxide and epoxide diols that are often implicated in the formation of carcinogenic metabolites formed by these enzymes.

Many of the planar PAH compounds induce their own metabolism by inducing transcription of the aryl hydrocarbon receptor (Ah receptor). Although expression of CYP1A1 and 1A2 is often coordinately induced, there are clear differences in regulation, not only with respect to substrate specificity but also in their biological expression. For example, CYP1A1 does not appear to be expressed in human liver unless induced, whereas CYP1A2 is endogenously expressed in the liver. CYP1A1, however, is present in many extrahepatic tissues including the lung, where there is a possible association between CYP-mediated activation of benzo(a) pyrene and other related chemicals present in cigarette smoke and lung cancer in man.

Figure 6.2   UPGMA tree of 54 puffer fish (fugu), 60 human, and 8 other fish P450s. Species are indicated by f, h, z, c, k, s, and t for fugu, human, zebrafish, catfish, killifish, sea bass, and trout, respectively. Reprinted from: Nelson, D. R. Comparison of P450s from human and fugu: 420 million years of vertebrate P450 evolution. *Arch. Biochem. Biophys.* **409**:18–24, 2003; with permission from Academic Press.

Figure 6.3   Examples of epoxidation reactions.

The CYP2 family consists of 10 subfamilies, five of which are present in mammalian liver. Some of the more important isoforms found in humans within this family are CYP2A6, CYP2B6, CYP2C8, CYP2C9, CYP2C19, CYP2D6, and CYP2E1. The enzyme CYP2A6 is expressed primarily in liver tissue, where it represents 1–10% of total CYP content. CYP2A6 is responsible for the 7-hydroxylation of the naturally occurring plant compound coumarin, and its activity is often phenotyped by monitoring this particular metabolic pathway. Other drugs metabolized by CYP2A6 include nicotine, 2-acetylaminofluorine, methoxyflurane, halothane, valproic acid, and disulfiram. Precarcinogens likely activated by CYP2A6 include aflatoxin B1, 1,3 butadiene, 2,6-dichlorobenzonitrile, and a number of nitrosamines. Because CYP2A6 is responsible for up to 80% of the human metabolism of nicotine, a number of studies have been conducted to determine whether individuals with CYP2A6 polymorphisms have reduced risk of lung cancers. Although theoretically, individuals lacking CYP2A6 would be expected to smoke less and be less likely to activate carcinogens found in tobacco smoke, studies have not conclusively demonstrated any clear associations between CYP2A6 polymorphisms and risk of lung cancer.

Like CYP2A6, the human isoform CYP2B6 has recently gained greater recognition for its role in metabolism of many clinical drugs. Some common pharmaceutical substrates for CYP2B6 include cyclophosphamide, nevirapine, *S*-mephobarbital, artemisinin, bupropion, propofol, ifosfamide, ketamine, selegiline, and methadone. CYP2B6 has also been demonstrated to have a role in the activation of the organophosphorus insecticide, chlorpyrifos, and in the degradation of the commonly used insecticide repellant, diethyl toluamide (DEET). Historically, it was thought that CYP2B6 is found in only a small proportion of livers (<25%), but more recent data using antibodies prepared from human proteins have demonstrated that most liver samples have detectable levels of CYP2B6, although greater than 20-fold differences in levels of protein have been observed. Recently, it has become apparent that CYP2B6 is important in the monooxidation of occupational chemicals, particularly agrochemicals. Since it is highly inducible, individual variation in concentration may be important in risk analysis and xenobiotic interactions.

In contrast with CYP2A6 and CYP2B6, members of the CYP2C family constitute a fairly large percentage of CYP in human liver (ca. 20%) and are responsible for the metabolism of several drugs. All four members of the subfamily in humans exhibit genetic polymorphisms, many of which have important clinical consequences in affected individuals. Genetic polymorphisms in CYP2C19 were shown to be responsible for one of the earliest described polymorphic effects, that involving mephenytoin metabolism. This particular polymorphism significantly reduces the metabolism of mephenytoin resulting in the classification of those individuals possessing this trait as poor metabolizers (PM). Among Caucasians, PMs represent only 3–5% of the populations, while in Asian and Polynesian populations 12–23% and 38–79% of the populations are represented, respectively. At least seven different mutations in this allele have been described, some of which negatively affect catalytic activity while others prevent expression of the protein. Other important drugs affected by these CYP2C19 polymorphisms include the antiulcer drug omeprazole, other important proton pump inhibitors, barbiturates, certain tricyclic antidepressants such as imipramine, and the antimalarial drug proguanil. Other important members of the CYP2C family in humans include CYP2C8, CYP2C9, and CYP2C18. Substrates metabolized exclusively by CYP2C8 include retinol, retinoic acid, taxol, and arachidonic acid. CYP2C9, the principal CYP2C in human liver, metabolizes several important drugs including the diabetic agent tolbutamide, the anticonvulsant phenytoin, the anticoagulant warfarin, and a number of anti-inflammatory drugs including ibuprofen, diclofenac, and others. Both CYP2C9 and CYP2C8, which are responsible for metabolism of the anticancer drug paclitaxel, have been demonstrated to be polymorphic.

CYP2E1 is the only member of the CYP2E family in most mammals with the exception of rabbits. Substrates for this family tend to be of small molecular weight and include ethanol, carbon tetrachloride, benzene, and acetaminophen. In contrast with many other inducible CYP families, CYP2E1 is regulated by a combination of increased transcription levels and increased message and protein stabilization.

Undoubtedly, the largest amount of CYP in human liver is that of the CYP3 family. CYP3A4 is the most abundant CYP in the human liver, on average accounting for approximately 30% of the total amount, and is known to metabolize many important drugs, including cyclosporin A, nifedipine, rapamycin, ethinyl \estradiol, quinidine, digitonin, lidocaine, erythromycin, midazolam, triazolam, lovastatin, and tamoxifen. Other important oxidations ascribed to the CYP3 family include many steroid hormones, macrolide antibiotics, alkaloids, benzodiazepines, dihydropyridines, organophosphorus insecticides and other insecticides, warfarin, polycyclic hydrocarbon-derived dihydrodiols, and aflatoxin B_1. Many chemicals are also capable of inducing CYPs of this family including phenobarbital, rifampicin, and dexamethasone. Because of potential difficulties arising from CYP induction, drugs metabolized by this family must be closely examined for the possibility of harmful drug–drug interactions.

CYP Reactions Although microsomal monooxygenase reactions are basically similar in the role played by molecular oxygen and in the supply of electrons, the many CYP isoforms can attack a large variety of xenobiotic substrates, with both substrates and products falling into many different chemical classes. In the following sections, therefore, enzyme activities are classified on the basis of the overall chemical reaction catalyzed; one should bear in mind, however, that not only do these classes often overlap, but often a substrate may also undergo more than one reaction. See Table 6.1 for a listing of important oxidation and reduction reactions of CYPs.

Epoxidation and Aromatic Hydroxylation Epoxidation is an extremely important microsomal reaction because not only can stable and environmentally persistent epoxides be formed (see aliphatic epoxidations below), but highly reactive intermediates of aromatic hydroxylations, such as arene oxides, can also be produced. These highly reactive intermediates are known to be involved in chemical carcinogesis as well as chemically induced cellular and tissue necrosis.

The oxidation of naphthalene was one of the earliest examples of an epoxide as an intermediate in aromatic hydroxylation. As shown in Figure 6.3, the epoxide can rearrange nonenzymatically to yield predominantly 1-naphthol, or interact with the enzyme epoxide hydrolase to yield the dihydrodiol, or interact with glutathione S-transferase (GST) to yield the glutathione conjugate, which is ultimately metabolized to a mercapturic acid.

More recent studies of the *in vitro* human metabolism of naphthalene have shown that the primary naphthalene metabolites from pooled human liver microsomes are *trans*-1,2-dihydro-1,2-naphthalenediol (dihydrodiol), 1-naphthol, and 2-naphthol. CYP1A2 was identified as the most efficient isoform for producing dihydrodiol and 1-naphthol while CYP3A4 was the most effective for 2-naphthol production. Further metabolism of the primary metabolites of naphthalene was also studied to identify secondary metabolites, and reactive metabolites such as 1,4-naphthoquinone were observed, and CYP1A2 and 2D6*1 were identified as the most active isoforms for the production of 1,4-naphthoquinone.

These aromatic epoxidation reactions are also of importance in the metabolism of other xenobiotics that contain an aromatic nucleus, such as the insecticide carbaryl and the carcinogen benzo(a)pyrene.

The ultimate carcinogens arising from the metabolic activation of benzo(a) pyrene are stereoisomers of benzo(a)pyrene 7,8-diol-9,10-epoxide (Figure 6.3). These metabolites arise by prior formation of the 7,8 epoxide, which gives rise to the 7,8-dihydrodiol through the action of epoxide hydrolase. This is further metabolized by the CYP to the 7,8-diol-9,10-epoxides, which are both potent mutagens and unsuitable substrates for the further action of epoxide hydrolase. Stereochemistry is important in the final product. Of the four possible isomers of the diol epoxide, the (+)-benzo(a)pyrene diol epoxide-2 is the most active carcinogen.

Aliphatic Hydroxylation Simple aliphatic molecules such as *n*-butane, *n*-pentane, *n*-hexane, and so on, as well as alicyclic compounds such as cyclohexane, are known to be oxidized to alcohols. However, alkyl side chains of aromatic compounds are more readily oxidized, often at more than one position, and provide good examples of this type of oxidation. For example, the *n*-propyl side chain of *n*-propyl benzene can be oxidized at any one of three carbons to yield 3-phenylpropan-1-ol ($C_6H_5CH_2CH_2CH_2OH$) by ω-oxidation, benzylmethylcarbinol ($C_6H_5CH_2CHOHCH_3$) by ω-1 oxidation, and ethyl-phenylcarbinol ($C_6H_5CHOHCH_2CH_3$) by α-oxidation. Further oxidation of these alcohols is also possible.

Aliphatic Epoxidation Many aliphatic and alicyclic compounds containing unsaturated carbon atoms are thought to be metabolized to epoxide intermediates (Figure 6.4). In the case of aldrin, the product, dieldrin, is an extremely stable epoxide and represents the principle residue found in animals exposed to aldrin. Epoxide formation in the case of aflatoxin is believed to be the final step in formation of the ultimate carcinogenic species and is, therefore, an activation reaction.

Dealkylation: O-, N-, and S-Dealkylation Probably the best-known example of O-dealkylation is the demethylation of *p*-nitroanisole. Due to the ease with which the product, *p*-nitrophenol, can be measured, it is a frequently used substrate for the demonstration of CYP activity. The reaction likely proceeds through formation of an unstable methylol intermediate (Figure 6.5).

The O-dealkylation of organophosphorus triesters differs from that of *p*-nitroanisole in that it involves the dealkylation of an ester rather than an ether. The reaction was first described for the insecticide chlorfenvinphos and is known to occur with a wide variety of vinyl, phenyl, phenylvinyl, and naphthyl phosphate and thionophosphate triesters (Figure 6.5).

Figure 6.4 Examples of aliphatic epoxidation.

Figure 6.5   Examples of dealkylation.

N-dealkylation is a common reaction in the metabolism of drugs, insecticides, and other xenobiotics. The drug ethylmorphine is a useful model compound for this reaction. In this case, the methyl group is oxidized to formaldehyde, which can be readily detected by the Nash reaction.

S-dealkylation is believed to occur with a number of thioethers, including methylmercaptan and 6-methylthiopurine, although with newer knowledge of the specificity of the FMO, it is possible that the initial attack is through sulfoxidation mediated by FMO rather than CYP.

N-Oxidation   N-oxidation can occur in a number of ways, including hydroxylamine formation, oxime formation, and N-oxide formation, although the latter is primarily dependent on the FMO enzyme. Hydroxylamine formation occurs with a number of amines such as aniline and many of its substituted derivatives. In the case of 2-acetylaminofluorene, the product is a potent carcinogen, and thus the reaction is an activation reaction (Figure 6.6).

Oximes can be formed by the N-hydroxylation of imines and primary amines. Imines have been suggested as intermediates in the formation of oximes from primary amines (Figure 6.6).

Oxidative Deamination   Oxidative deamination of amphetamine occurs in the rabbit liver but not to any extent in the liver of either the dog or the rat, which tend

(a) Hydroxylamine formation

2-Acetylaminofluorene

N-Hydroxy-
2-acetylaminofluorene

(b) Oxime formation

Trimethylacetophenone
imine

Trimethylacetophenone
oxime

Figure 6.6   Examples of N-oxidation.

to hydroxylate the aromatic ring. A close examination of the reaction indicates that it is probably not an attack on the nitrogen but rather on the adjacent carbon atom, giving rise to a carbinol amine, which eliminates ammonia, producing a ketone.

$$R_2CHNH_2 \xrightarrow{+O} R_2C(OH)NH_2 \xrightarrow{-NH_3} R_2C{=}O$$

The carbinol, by another reaction sequence, can also give rise to an oxime, which can be hydrolyzed to yield the ketone, which is thus formed by two different routes.

$$R_2C(OH)NH_2 \xrightarrow{-H_2O} R_2C{=}NH \xrightarrow{+O} R_2CNOH \xrightarrow{+H_2O} R_2C{=}O$$

S-Oxidation   Thioethers in general are oxidized by microsomal monooxygenases to sulfoxides, some of which are further oxidized to sulfones. This reaction is very common among insecticides of several different chemical classes, including carbamates, organophosphorus compounds, and chlorinated hydrocarbons. Recent work suggests that members of the CYP2C family are involved in sulfoxidation of several organophosphorus compounds including phorate, coumaphos, and demeton. The carbamate methiocarb is oxidized to a series of sulfoxides and sulfones, and among the chlorinated hydrocarbons, endosulfan is oxidized to endosulfan sulfate and methiochlor to a series of sulfoxides and sulfones, eventually yielding the bis-sulfone. Drugs, including chlorpromazine, and solvents, such as dimethyl sulfoxide, are also subject to S-oxidation. The fact that FMOs are versatile sulfur oxidation enzymes capable of carrying out many of the previously mentioned reactions raises important questions as to the relative role of this enzyme versus that of CYP.

P-Oxidation   P-oxidation, a little known reaction, involves the conversion of trisubstituted phosphines to phosphine oxides, for example, diphenylmethylphosphine to diphenylmethylphosphine oxide. Although this reaction is described as a typical CYP-dependent monooxygenation, it also is now known to be catalyzed by the FMO.

Figure 6.7 Desulfuration and oxidative dearylation.

Desulfuration and Ester Cleavage The phosphorothionates $[(R^1O)_2P(S)OR^2]$ and phosphorodithioates $[(R^1O)_2P(S)SR^2]$ owe their insecticidal activity and their mammalian toxicity to an oxidative reaction in which the P=S group is converted to P=O, thereby converting the compounds from chemicals relatively inactive toward cholinesterase into potent inhibitors (see Chapter 15 for a discussion of the mechanism of cholinesterase inhibition). This reaction has been described for many organophosphorus compounds. Much of the splitting of the phosphorus ester bonds in organophosphorus insecticides, formerly believed to be due to hydrolysis, is now known to be due to oxidative dearylation. This is a typical CYP-dependent monooxygenation, requiring NADPH and O_2 and being inhibited by CO. Current evidence supports the hypothesis that this reaction and oxidative desulfuration involve a common intermediate of the "phosphooxithirane" type (Figure 6.7). Some organophosphorus insecticides, all phosphonates such as fonofos, are activated by the FMO as well as the CYP.

Methylenedioxy (Benzodioxole) Ring Cleavage Methylenedioxyphenyl compounds, such as safrole or the insecticide synergist, piperonyl butoxide, many of which are effective inhibitors of CYP monooxygenations, are themselves metabolized to catechols. The most probable mechanism appears to be an attack on the methylene carbon, followed by elimination of water to yield a carbene. The highly reactive carbene either reacts with the heme iron to form a CYP-inhibitory complex or breaks down to yield the catechol (Figure 6.8).

6.2.3 The FMO

Tertiary amines such as trimethylamine and dimethylamine had long been known to be metabolized to N-oxides by a microsomal amine oxidase that was not dependent on CYP. This enzyme, now known as the microsomal FMO, is also dependent

Figure 6.8 Monooxygenation of methylenedioxyphenyl compounds.

on NADPH and O_2, and has been purified to homogeneity from a number of species. Isolation and characterization of the enzyme from liver and lung samples provided evidence of clearly distinct physicochemical properties and substrate specificities suggesting the presence of at least two different isoforms. Subsequent studies have verified the presence of multiple forms of the enzyme.

At least six different isoforms have been described by amino acid or cDNA sequencing, and are classified as FMO1 to FMO6. These isoforms share approximately 50–60% amino acid identity across species lines. The identity of orthologs is greater than 82%. Although each isoform has been characterized in humans, several are essentially nonfunctional in adults. For example, FMO1, expressed in the embryo, disappears relatively quickly after birth. FMO2 in most Caucasians and Asians contains a premature stop codon, preventing the expression of functional protein. Functional FMO2 is found in 26% of the African-American population and is, perhaps, also found in the Hispanic population. FMO3, the predominant human FMO, is poorly expressed in neonatal humans but is expressed in most individuals by 1 year of age. Gender-independent expression of FMO3 (contrasting with what is observed in other mammals) continues to increase through childhood, reaching maximal levels of expression at adulthood. Several polymorphic forms of FMO3 are responsible for the disease, trimethylaminuria, also known as "fish odor syndrome," characterized by the inability of some individuals to convert the malodorous trimethylamine, either from the diet or from metabolism, to its odorless N-oxide. Although the FMO4 transcript is found in several species, the protein has yet to be successfully expressed in any species. Although FMO5 is expressed in humans at low levels, the poor catalytic activity of FMO5 for most classical FMO substrates suggests that it has minimal participation in xenobiotic oxidation. No data is yet available on the role and abundance of the most recently discovered FMO, FMO6.

Substrates containing soft nucleopohiles (e.g., nitrogen, sulfur, phosphorus, and selenium) are good candidates for FMO oxidation (Figure 6.9). A short list of

Figure 6.9 Examples of oxidations catalyzed by the flavin-containing monooxygenase (FMO).

known substrates include drugs such as dimethylaniline, imipramine, thiobenzamide, chlorpromazine, promethazine, cimetidine, and tamoxifen; pesticides such as phorate, fonofos, and methiocarb; environmental agents including the carcinogen 2-aminofluorine, and the neurotoxicants nicotine and 1-methyl-4-phenyl-1,2,3,6-tetrahydropyridine (MPTP). Although there is no known physiologically relevant substrate for FMO, a few dietary and/or endogenous substrates have been identified, including trimethylamine, cysteamine, methionine, and several cysteine-S-conjugates. In most cases, metabolism by FMO results in detoxication products, although there are several examples of substrates that are bioactivated by FMO oxidation; particularly in the case of substrates involving sulfur oxidation.

Most FMO substrates are also substrates for CYP. Since both enzymes are microsomal and require NADPH and oxygen, it is difficult to distinguish which enzyme is responsible for oxidation without the use of techniques involving specific inactivation or inhibition or one or the other of these enzymes while simultaneously examining the metabolic contribution of the other. Since FMOs are generally heat labile, heating the microsomal preparation to 50°C for 1 min inactivates the FMOs while having minimal effects of CYPs. Alternatively, the contribution of FMO can be assessed by use of a general CYP inhibitor such as *N*-benzylimidazole or by an inhibitory antibody to NADPH-CYP reductase, a necessary CYP coenzyme. Typically, results of these tests are sought in combination so that the best estimates of CYP and FMO contribution can be obtained.

Toxicologically, it is of interest that the FMO enzyme is responsible for the oxidation of nicotine to nicotine-1'-N-oxide, whereas the oxidation of nicotine to cotinine is catalyzed by two enzymes acting in sequence: CYP followed by a soluble aldehyde dehydrogenase. Thus, nicotine is metabolized by two different routes, the relative contributions of which may vary with both the extrinsic and intrinsic factors outlined in Chapter 8.

6.2.4 Nonmicrosomal Oxidations

In addition to the microsomal monooxygenases, other enzymes are involved in the oxidation of xenobiotics. These enzymes are located in the mitochondria or in the soluble cytoplasm of the cell.

Alcohol Dehydrogenase Alcohol dehydrogenases catalyze the conversion of alcohols to aldehydes or ketones:

$$RCH_2OH + NAD^+ \rightarrow RCHO + NADH + H^+$$

This reaction should not be confused with the monooxygenation of ethanol by CYP that occurs in the microsomes. The alcohol dehydrogenase reaction is reversible, with the carbonyl compounds being reduced to alcohols.

Alcohol dehydrogenase is found in the soluble fraction of the liver, kidney, and lung and is probably the most important enzyme involved in the metabolism of foreign alcohols. Alcohol dehydrogenase is a dimer, the subunits of which can occur in several forms that are under genetic control, thus giving rise to a large number of variants of the enzyme. In mammals, six classes of enzymes have been described. Alcohol dehydrogenase can use either NAD or NADP as a coenzyme, but the reaction proceeds at a much slower rate with NADP. In the intact organism, the reaction proceeds in the direction of alcohol consumption, because aldehydes are further oxidized to acids. Because aldehydes are toxic and are not readily excreted because of their lipophilicity, alcohol oxidation may be considered an activation reaction, the further oxidation to an acid being a detoxication step.

Although short chain aliphatic alcohols, particularly ethanol, are the most studied substrates for alcohol dehydrogenase, larger molecules, such as phenoxybenzyl alcohol, the hydrolysis product of permethrin, have Km values up to two orders of magnitude lower than that for ethanol.

Primary alcohols are oxidized to aldehydes, *n*-butanol being the substrate oxidized at the highest rate. Although secondary alcohols are oxidized to ketones, the rate is less than for primary alcohols, and tertiary alcohols are not readily oxidized. Alcohol dehydrogenase is inhibited by a number of heterocyclic compounds such as pyrazole, imidazole, and their derivatives.

Aldehyde Dehydrogense Aldehydes are generated from a variety of endogenous and exogenous substrates. Endogenous aldehydes may be formed during metabolism of amino acids, carbohydrates, lipids, biogenic amines, vitamins, and steroids. Metabolism of many drugs and environmental agents produce aldehydes. Aldehydes are highly reactive electrophilic compounds and may react with thiol and amino groups to produce a variety of effects. Some aldehydes produce therapeutic effects, but more often effects are cytotoxic, genotoxic, mutagenic, or carcinogenic. Aldehyde dehydrogenases are important in helping to alleviate some of the toxic effects of aldehyde generation. This enzyme catalyzes the formation of acids from aliphatic and aromatic aldehydes; the acids are then available as substrates for conjugating enzymes:

$$RCHO + NAD^+ \rightarrow RCOOH + NADH + H^+$$

The aldehyde dehydrogenase gene superfamily is large, with more than 330 aldehyde dehydrogenase genes in prokaryote and eukaryotic species. The eukaryotic aldehyde dehydrogenase gene superfamily consists of 20 gene families, 9 of which contain 16 human genes and 3 pseudogenes. The importance of some of these genes in detoxication pathways is underscored by the fact that identified polymorphisms are associated with several metabolic diseases.

One especially interesting polymorphism is that which occurs at the aldehyde dehydrogenase 2 locus. When inherited as the homozygous trait, this aldehyde dehydrogenase polymorphism results in a 20-fold greater generation of acetaldehyde from ethanol, resulting in the flushing syndrome characteristic of many Asian individuals after ethanol consumption. Alcoholics are not likely to be found among individuals expressing this particular polymorphism.

Other enzymes in the soluble fraction of liver that oxidize aldehydes are aldehyde oxidase and xanthine oxidase, both flavoproteins that contain molybdenum; however, their primary role seems to be the oxidation of endogenous aldehydes formed as a result of deamination reactions.

Amine Oxidases The most important function of amine oxidases appears to be the oxidation of amines formed during normal processes. Two types of amine oxidases are concerned with oxidative deamination of both endogenous and exogenous amines. Typical substrates are shown in Figure 6.10.

Monoamine Oxidases The monomine oxidases are a family of flavoproteins found in the mitochondria of a wide variety of tissues: liver, kidney, brain, intestine, and blood platelets. They are a group of similar enzymes with overlapping specificities and inhibition. Although the enzyme in the central nervous system is concerned primarily with neurotransmitter turnover, that in the liver will deaminate primary, secondary, and tertiary aliphatic amines, reaction rates with the primary amines being faster. Electron-withdrawing substitutions on an aromatic ring increase the reaction rate, whereas compounds with a methyl group on the α-carbon such as amphetamine and ephedrine are not metabolized.

Diamine Oxidases Diamine oxidases are enzymes that also oxidize amines to aldehydes. The preferred substates are aliphatic diamines in which the chain length *p*-Chlorobenzylamine   *p*-Chlorobenzaldehyde

(a) Monoamine oxidase

$$H_2N(CH_2)_5NH_2 + O_2 + H_2O \longrightarrow H_2N(CH_2)_4CHO + NH_3 + H_2O_2$$
Cadaverine

(b) Diamine oxidase

Figure 6.10 Examples of oxidations catalyzed by amine oxidases.

Arachidonic acid

1. Cyclooxygenase
2. Peroxidase

PGG_2 OOH

Xenobiotic
co-oxidation

PGH_2 OH

Figure 6.11 Co-oxidation during prostaglandin biosynthesis.

is four (putrescine) or five (cadaverine) carbon atoms. Diamines with carbon chains longer than nine will not serve as substrates but can be oxidized by monoamine oxidases. Secondary and tertiary amines are not metabolized. Diamine oxidases are typically soluble pyridoxal phosphate-containing proteins that also contain copper. They have been found in a number of tissues, including liver, intestine, kidney, and placenta.

6.2.5 Co-oxidation by Cyclooxygenase (COX)

During the biosynthesis of prostaglandins, a polyunsaturated fatty acid, such as arachidonic acid, is first oxygenated to yield a hydroperoxy endoperoxide, prostaglandin G2. This is then further metabolized to prostaglandin H2, both reactions being catalyzed by the same enzyme, COX, also known as prostaglandin synthase (Figure 6.11). This enzyme is located in the microsomal membrane and is found in greatest levels in respiratory tissues such as the lung. It is also common in the kidney and seminal vesicle. It is a glycoprotein with a subunit molecular mass of about 70,000 Da, containing one heme per subunit. During the second step of the previous sequence (peroxidase), many xenobiotics can be co-oxidized, and investigations of the mechanism have shown that the reactions are hydroperoxide-dependent reactions catalyzed by a peroxidase that uses prostaglandin G as a substrate. In at least some of these cases, the identity of this peroxidase has been established as a prostaglandin synthase. Many of the reactions are similar or identical to those catalyzed by other peroxidases and also by microsomal monooxygenases; they include both detoxication and activation reactions. This mechanism is important in xenobiotic metabolism, particularly in tissues that are low in CYP and/or the FMO but high in prostaglandin synthase.

The COX enzyme is known to exist as two distinct isoforms. COX-1 is a constitutively expressed housekeeping enzyme found in nearly all tissues and mediates physiological responses. COX-2 is an inducible form expressed primarily by cells involved in the inflammatory response. Several tissues low in CYP expression are rich in COX, which is believed to have significance in the carcinogenic effects of aromatic amines in these organs.

During co-oxidation, some substrates are activated to become more toxic than they were originally. In some cases, substrate oxidation results in the production of free radicals, which may initiate lipid peroxidation or bind to cellular proteins or DNA. Another activation pathway involves the formation of a peroxyl radical from subsequent metabolism of prostaglandin G2. This reactive molecule can epoxidize many substates including polycyclic aromatic hydrocarbons, generally resulting in increasing toxicity of the respective substrates.

To differentiate between xenobiotic oxidations by COX and CYP, *in vitro* microsomal incubations of the xenobiotic may be performed in the presence of either arachidonic acid (COX catalyzed), or in the presence of NADPH (CYP catalyzed). In the presence of arachidonic acid while in the absence of NADPH, substrates co-oxidized by COX will be formed while those requiring CYP will not. Specific inhibitors of PG synthase (indomethacin) and CYP (metyrapone or SKF 525A) have also been used.

6.2.6 Reduction Reactions

A number of functional groups, such as nitro, diazo, carbonyl, disulfide sulfoxide, alkene, pentavalent arsenic, and so on, are susceptible to reduction, although in many cases it is difficult to tell whether the reaction proceeds enzymatically or nonenzymatically by the action of such biologic reducing agents as reduced flavins or reduced pyridine nucleotides. In some cases, such as the reduction of the double bond in cinnamic acid ($C_6H_5CH=CHCOOH$), the reaction has been attributed to the intestinal microflora. Examples of reduction reactions are shown in Figure 6.12.

Nitro Reduction   Aromatic amines are susceptible to reduction by both bacterial and mammalian nitroreductase systems. Convincing evidence has been presented that this reaction sequence is catalyzed by CYP. It is inhibited by oxygen, although NADPH is still consumed. Earlier workers had suggested a flavoprotein reductase was involved, and it is not clear if this is incorrect or if both mechanisms occur. It

(a) Nitro reduction

(b) Azo reduction

(c) Disulfide reduction

(d) Aldehyde reduction

(e) Sulfoxide reduction

Figure 6.12   Examples of metabolic reduction reactions.

is true, however, that high concentration of FAD or FMN will catalyze the nonenzymatic reduction of nitro groups.

Azo Reduction   Requirements for azo reduction are similar to those for nitroreduction, namely, anaerobic conditions and NADPH. They are also inhibited by CO, and presumably they involve CYP. The ability of mammalian cells to reduce azo bonds is rather poor, and intestinal microflora may play a role.

Disulfide Reduction   Some disulfides, such as the drug disulfiram (Antabuse), are reduced to their sulfhydryl constituents. Many of these reactions are three-step sequences, the last reaction of which is catalyzed by glutathione reductase, using glutathione (GSH) as a cofactor.

$$RSSR + GSH \rightarrow RSSG + RSH$$

$$RSSG + GSH \rightarrow GSSG + RSH$$

$$GSSG + NADPH + H^+ \rightarrow 2GSH + NADP^+$$

Ketone and Aldehyde Reduction In addition to the reduction of aldehyde and ketones through the reverse reaction of alcohol dehydrogenase, a family of aldehyde reductases also reduces these compounds. These reductases are NADPH-dependent, cytoplasmic enzymes of low molecular weight, and have been found in liver, brain, kidney, and other tissues.

Sulfoxide Reduction The reduction of sulfoxides has been reported to occur in mammalian tissues. Soluble thioredoxin-dependent enzymes in the liver are responsible in some cases. It has been suggested that oxidation in the endoplasmic reticulum followed by reduction in the cytoplasm may be a form of recycling that could extend the *in vivo* half-life of certain toxicants.

6.2.7 Hydrolysis

Enzymes with carboxylesterase and amidases activity are widely distributed in the body, occurring in many tissues and in both microsomal and soluble fractions. They catalyze the following general reactions:

$RC(O)OR' + H_2O \rightarrow RCOOH + HOR'$ Carboxylester hydrolysis

$RC(O)NR'R'' + H_2O \rightarrow RCOOH + HNR'R''$ Carboxyamide hydrolysis

$RC(O)SR' + H_2O \rightarrow RCOOH + HSR'$ Carboxythioester hydrolysis

Although carboxylesterases and amidases were thought to be different, no purified carboxylesterase has been found that does not have amidase activity toward the corresponding amide. Similarly, enzymes purified on the basis of their amidase activity have been found to have esterase activity. Thus, these two activities are now regarded as different manifestations of the same activity, specificity depending on the nature of R, R', and R" groups and, to a lesser extent, on the atom (O, S, or N) adjacent to the carboxyl group.

In view of the large number of esterases in many tissues and subcellular fractons, as well as the large number of substrates hydrolyzed by them, it is difficult to derive a meaningful classification scheme. The division into A-, B-, and C-esterases on the basis of their behavior toward such phosphate triesters as paraoxon, first devised by Aldridge, is still of some value, although not entirely satisfactory.

A-esterases, also referred to as arylesterases, are distinguished by their ability to hydrolyze esters derived from aromatic compounds. Organophosphates, such as the paraoxon, the active metabolite of the insecticide, parathion, are often used to characterize this group. B-esterases, the largest and most important group, are inhibited by organophosphates. All the B-esterases have a serine residue in their

Figure 6.13 Examples of esterase/amidase reactions involving xenobiotics.

active site that is phosphorylated by these inhibitors. This group includes a number of different enzymes and their isozymes, many of which have quite different substrate specificities. For example, the group contains carboxylesterase, amidases, cholinesterases, monoacylglycerol lipases, and arylamidases. Many of these enzymes hydrolyze physiological (endogenous) substrates as well as xenobiotics. Several examples of their activity toward xenobiotic substrates are shown in Figure 6.13. C-esterases, or acetylesterases, are defined as those esterases that prefer acetyl esters as substrates, and for which paraoxon serves as neither substrate nor inhibitor.

Two esterases that have received considerable attention in recent years are the carboxylesterase that hydrolyzes pyrethroid insecticides and PON1 (paraoxonase 1). In the first case, the esterase that hydrolyzes pyrethroids into acid and alcohol moieties has been of particular interest, particularly the human forms, hCE1 and hCE2. While the levels of each are only slightly variable in human liver microsomes, the level of hCE1 is some 46 times higher than of hCE2. While pyrethroids such as bioresmethrin are substrates for hCE1 but not hCE2, procaine is a substrate for hCE2 but not hCE1. It has also been shown that these carboxylases have endogenous substrates such as cholesterol esters and triglycerides and are important in cardiovascular function and health. Cholesteryl ester hydrolase and CE1 appear to be the same enzyme.

Paraxonase, originally named for its ability to hydrolyze paraoxon, has been shown to have broad substrate specificity as well as a role in the risk of vascular disease in humans. The form known as PON1 has been characterized with regard to its ability to hydrolyze paraoxon, diazoxon, chlorpyrifos oxon, and the chemical warfare agents, sarin and soman, all substrates. PON1 is a plasma enzyme associated with high-density lipoproteins that is also found in the liver. PON1 is polymorphic, and the GLn/Arg substitution at position 192 affects the catalytic activity. However, in addition to the variable catalytic activity determined by this polymorphism, the levels of PON1 vary as much as 15-fold between individuals. As a result, PON1 status cannot be assessed by genotyping alone; the determination of the rates of hydrolysis of different substrates (e.g., diazoxon vs. paraxon) relative to each other is also essential.

6.2.8 Epoxide Hydration

Epoxide rings of alkene and arene compounds are hydrated by enzymes known as epoxide hydrolases, the animal enzyme forming the corresponding *trans*-diols, although bacterial hydrolases are known that form *cis*-diols. Although in general, the hydration of the oxirane ring results in detoxication of the very reactive epoxide, in some cases, such as benzo(a)pyrene, the hydration of an epoxide is the first step in an activation sequence that ultimately yields highly toxic *trans*-dihydrodiol intermediates. In others, reactive epoxides are detoxified by both glutathione transferase and epoxide hydrolase. The reaction probably involves a nucleophilic attack by -OH on the oxirane carbon. The most studied epoxide hydrolase is microsomal, and the enzyme has been purified from hepatic microsomes of several species. Although less well-known, soluble epoxide hydrolases with different substrate specificities have also been described. Examples of epoxide hydrolase reactions are shown in Figure 6.14.

Styrene 7,8-oxide   + H₂O ⟶   Styrene 7,8-glycol

Naphthalene 1,2-oxide   + H₂O ⟶   Naphthalene dihydrodiol

Figure 6.14 Examples of epoxide hydrolase reactions.

Figure 6.15   DDT-dehydrochlorinase.

6.2.9 DDT Dehydrochlorinase

DDT   (1,1,1-trichloro-2,2-bis(4-chlorophenylethane))-dehydrochlorinase   is   an enzyme that occurs in both mammals and insects and has been studied most intensively in DDT-resistant houseflies. It catalyzes the dehydrochlorination of DDT to DDE (1,1-dichloro-bis(4-chlorophenyl) ethane) and occurs in the soluble fraction of tissue homogenates. Although the reaction requires GSH, it apparently serves in a catalytic role because it does not appear to be consumed during the reaction. The Km for DDT is 5×10^{-7} mol/L with optimum activity at pH 7.4. The monomeric form of the enzyme has a molecular mass of about 36,000 Da, but the enzyme normally exists as a tetramer. In addition to catalyzing the dehydrochlorination of DDT to DDE and DDD (2,2-bis(p-chlorophenyl)-1,1 -dichloroethane) to TDE (2,2-bis(p-chlorophenyl)-1-chlorothylen), DDT dehydrochlorinase also catalyzes the dehydrohalogenation of a number of other DDT analogs. In all cases, the p,p configuration is required, o,p and other analogs are not utilized as substrates. The reaction is illustrated in Figure 6.15.

6.3 PHASE II REACTIONS

Products of Phase I metabolism and other xenobiotics containing functional groups such as hydroxyl, amino, carboxyl, epoxide, or halogen can undergo conjugation reactions with endogenous metabolites, these conjugations being collectively termed Phase II reactions. The endogenous metabolites in question include sugars, amino acids, GSH, sulfate, and so on. Conjugation products, with rare exceptions, are more polar, less toxic, and more readily excreted than are their parent compounds.

Conjugation reactions usually involve activation by some high-energy intermediate and have been classified into two general types: type I, in which an activated conjugating agent combines with the substrate to yield the conjugated product, and type II, in which the substrate is activated and then combines with an amino acid to yield a conjugated product. The formation of sulfates and glycosides are examples of type I, whereas type II consists primarily of amino acid conjugation.

6.3.1 Glucuronide Conjugation

The glucuronidation reaction is one of the major pathways for elimination of many lipophilic xenobiotics and endobiotics from the body. The mechanism for this conjugation involves the reaction of one of many possible functional groups (R-OH, Ar-OH, R-NH2, AR-NH2, R-COOH, Ar-COOH) with the sugar derivative, uridine

5′-diphosphoglucuronic acid (UDPGA). Homogeneous glucuronosyl transferase has been isolated as a single polypeptide chain of about 59,000 Da, apparently containing carbohydrate, the activity of which appears to be dependent on reconstitution with microsomal lipid. There appears to be an absolute requirement for UDPGA; related UDP-sugars will not suffice. This enzyme, as it exists in the microsomal membrane, does not exhibit its maximal capacity for conjugation; activation by some means (e.g., detergents) is required. The reaction involves a nucleophilic displacement (SN^2 reaction) of the functional group of the substrate with Walden inversion. UDPGS is in the α-configuration whereas, due to the inversion, the glucuronide formed is in the β-configuration. The enzyme involved, the UDP glucuronosyl transferase (UGT), is found in the microsomal fraction of liver, kidney, intestine, and other tissues. Examples of various types of glucuronides are shown in Figure 6.16.

Glucuronide conjugation generally results in the formation of products that are less biologically and chemically reactive. This, combined with their greater polarity and greater susceptibility to excretion, contributes greatly to the detoxication of most xenobiotics. However, there are examples of glucuronide conjugation resulting in greater toxicity. Perhaps the best-known example involves the bioactivation

1-Naphthol   UDPGA   Naphthol glucuronide

(a) Reaction sequence

Coumarin   2-Naphthylamine   Thiophenol

Propanolol   Oxazapam   Imipramine

(b) Substrate examples

Figure 6.16 Reaction sequences of uridine diphospho glucuronosyl transferase and chemical structures of compounds that form glurcuronides. Arrows indicate the position on each molecule where glucuronidation occurs.

of *N*-hydroxy-2-acetylaminofluorine. This substrate, unlike 2-acetylaminofluorine, is unable to bind to DNA without metabolic activation. However, following glucuronide conjugation by linkage of the oxygen through the *N*-hydroxy group, this substrate becomes equipotent as a hepatocarcinogen with 2-acetylaminofluorine based on its ability to bind to DNA. Another relatively large class of xenobiotics that are often activated by glucuronide conjugation are the acyl glucuronides of carboxylic acids. Useful therapeutic drugs within this class include nonsteroidal anti-inflammatory drugs (NSAIDS), hypolipidemic drugs (clofibrate), and anticonvulsants (valproic acid). The various syndromes associated with the clinical use of some of these drugs (including cytotoxic, carcinogenic, and various immunologic effects) are thought to be the result of the ability of the glucuronide conjugates to react with nucleophilic macromolecules (protein and DNA).

A wide variety of reactions are mediated by glucuronosyltransferases. *O*-glucuronides, *N*-glucuronides, and *S*-glucuronides have all been identified. At this time, over 35 different UGT gene products have been described from several different species. These are responsible for the biotransformation of greater than 350 different substrates. Evidence from molecular cloning suggests that the UGTs belong to one of two large superfamilies, sharing less than 50% amino acid identity. Nomenclature of these genes is similar to that of the CYP superfamily. The UGT1 gene family consists of a number of UGTs that arise from alternate splicing of multiple first exons and share common exons 2–5. Members of the UGT2 family catalyze the glucuronidation of a wide variety of substrates including steroids, bile acids, and opioids.

There are nine known human isozymes within the UGT1 family and six within the UGT2 family. Polymorphic forms of some of these enzymes are associated with diseases and significant adverse effects to some drugs.

Jaundice, a condition resulting from the failure of either transport or conjugation of bilirubin, becomes clinically evident when serum bilirubin levels exceed $35\,\mu M/L$. Although the human UGT1A locus encompasses nine functional transferase genes, only one isoform, UGT1A1, is involved in inherited diseases of bilirubin metabolism. All three inheritable hyperbilirubinemias are the result of either mutant UGT1A1 alleles or UGT1A1 promoter polymorphisms. To date, 33 mutant UGT1A1 alleles have been identified. For the disease to be clinically manifest, one must either be homozygous for the mutant allele or have multiple heterozygous mutant alleles.

6.3.2 Glucoside Conjugation

Although rare in vertebrates, glucosides formed from xenobiotics are common in insects and plants. Formed from UDP-glucose, they appear to fall into the same classes as the glucuronides.

6.3.3 Sulfate Conjugation

Sulfation and sulfate conjugate hydrolysis, catalyzed by various members of the sulfotransferases (SULT) and sulfatase enzyme superfamilies, play important roles in the metabolism and disposition of many xenobiotics and endogenous substrates. Reactions of the sulfotransferase enzyme with various xenobiotics including

(a) Reaction sequence

(b) Substrate examples

Figure 6.17 Reaction sequence of sulfotransferases and chemical structures of compounds that form sulfates. Arrows indicate positions on each molecule where sulfotransferases may attack.

alcohols, arylamines, and phenols result in the production of water-soluble sulfate esters, which often are readily eliminated from the organism. Although generally these reactions are important in detoxication, they have also been shown to be involved in carcinogen activation, prodrug processing, cellular signaling pathways, and the regulation of several potent endogenous chemicals including thyroid hormones, steroids, and catechols. The overall sulfation pathway shown in Figure 6.17 consists of two enzyme systems: the SULTs, which catalyze the sulfation reaction, and the sulfatases, which catalyze the hydrolysis of sulfate esters formed by the action of the SULTs.

Sulfation is expensive in energy terms for the cell, since two molecules of ATP are necessary for the synthesis of one molecule of 3'-phosphoadenosine 5'-phosphosulfate (PAPS). Both enzymes involved in the synthesis of PAPS, ATP sulfurylase and adenosine phosphosulfate (APS) kinase, reside within a single bifunctional cytosolic protein of approximately 56 kDa, where substrate channeling of APS from ATP sulfurylase to APS kinase occurs. Several group VI anions other than sulfate can also serve as substrates, although the resultant anhydrides are unstable. Because this instability would lead to the overall consumption of ATP, these other anions can exert a toxic effect by depleting the cell of ATP.

In humans, there are five well-characterized SULT genes, each possessing widely different amino acid sequences and with widely different substrate specificities. Based on amino acid sequence identity as well as substrate preference, these can be separated into two families, phenol SULTs (P-PST, SULT1A2, M-PST, and EST) and hydroxysteroid sulfotransferase (HST). Phenol SULTs from rat liver have been separated into four distinct forms, each of which catalyzes the sulfation of various phenols and catecholamines. They differ, however, in pH optimum, relative substrate specificity, and immunologic properties. The molecules of all of them are in the range of 61,000–64,000 Da.

HST also appears to exist in several forms. This reaction is now known to be important, not only as a detoxication mechanism, but also in the synthesis and possibly the transport of steroids. HST will react with hydroxysterols and primary and secondary alcohols but not with hydroxyl groups in the aromatic rings of steroids.

6.3.4 Methyltransferases

A large number of both endogenous and exogenous compounds can be methylated by several *N*-, *O*-, and *S*-methyltransferases. The most common methyl donor is *S*-adenosylmethionine (SAM), formed from methionine and ATP. Even though these reactions may involve a decrease in water solubility, they are generally detoxication reactions. Examples of biologic methylation reactions are seen in Figure 6.18.

Figure 6.18 Examples of methyltransferase reactions.

N-*Methylation* Several enzymes are known that catalyze N-methylation reactions. They include histamine N-methyltransferase, a highly specific enzyme that occurs in the soluble fraction of the cell, phenylethanolamine N-methyltransferase, which catalyzes the methylation of noradrenaline to adrenaline as well as the methylation of other phenylethanolamine derivatives. A third N-methyltransferase is the indoethylamine N-methyltansferase, or nonspecific N-methyltransferase. This enzyme occurs in various tissues. It methylates endogenous compounds such as serotonin and tyrptamine and exogenous compounds such as nornicotine and norcodeine. The relationship between this enzyme and phenylethanolamine N-methyltransferase is not yet clear.

O-*Methylation* Catechol O-methyltransferase occurs in the soluble fraction of several tissues and has been purified from rat liver. The purified form has a molecular weight 23,000 Da, requires S-adenosylmethionine and Mg^+, and catalyzes the methylation of epinephrine, norepinephrine, and other catechol derivatives. There is evidence that this enzyme exists in multiple forms.

A microsomal O-methyltransferase that methylates a number of alkyl-, methoxy-, and halophenols has been described from rabbit liver and lungs. These methylations are inhibited by SKF-525, N-ethylmaleimide and p-chloromercuribenzoate. A hydroxyindole O-methyltransferase, which methylates N-acetylserotonin to melatonin and, to a lesser extent, other 5-hydroxyindoles and 5,6-dihydroxyindoles, has been described from the pineal gland of mammals, birds, reptiles, amphibians, and fish.

S-*Methylation* Thiol groups of some foreign compounds are also methylated, the reaction being catalyzed by the enzyme, thiol S-methyltransferase. This enzyme is microsomal and, as with most methyltransferases, utilizes S-adenosylmethionine. It has been purified from rat liver and is a monomer of about 28,000 Da. A wide variety of substrates are methylated, including thioacetanilide, mercaptoethanol, and diphenyl sulfide. This enzyme may also be important in the detoxication of hydrogen sulfide, which is methylated in two steps, first to the highly toxic methanethiol and then to dimethyl sulfide.

Methylthiolation, or the transfer of a methylthio (CH_3S-) group to a foreign compound may occur through the action of another recently discovered enzyme, cysteine conjugate β-lyase. This enzyme acts on cysteine conjugates of foreign compounds as follows:

$$RSCH_2CH(NH_2)COOH \rightarrow RSH + NH_3 + CH_3C(O)COOH$$

The thiol group can then be methylated to yield the methylthio derivative of the original xenobiotic.

Biomethylation of Elements The biomethylation of elements is carried out principally by microorganisms and is important in environmental toxicology, particularly in the case of heavy metals, because the methylated compounds are absorbed through the membranes of the gut, the blood–brain barrier, and the placenta more readily than are the inorganic forms. For example, inorganic mercury can be methylated first to monomethylmercury and subsequently, to dimethylmercury.

$$Hg^{2+} \rightarrow CH_3HG^+ \rightarrow (CH_3)_2 Hg$$

The enzymes involved are reported to use either *S*-adenosylmethionine or vitamin B_{12} derivatives as methyl donors and, in addition to mercury, the metals, lead, tin, and thallium as well as the metalloids, arsenic, selenium, tellurium, and sulfur, are methylated. Even the unreactive metals, gold and platinum, are reported as substrates for these reactions.

6.3.5 GSTs and Mercapturic Acid Formation

Although mercapturic acids, the *N*-acetylcysteine conjugates of xenobiotics, have been known since the early part of the twentieth century, only since the early 1960s has the source of the cysteine moiety (GSH) and the enzymes required for the formation of these acids been identified and characterized. The overall pathway is shown in Figure 6.19.

The initial reaction is the conjugation of xenobiotics having electrophilic substituents with GSH, a reaction catalyzed by one of the various forms of GST. This is followed by transfer of the glutamate by γ-glutamyltanspeptidase, by loss of glycine through cysteinyl glycinase, and finally by acetylation of the cysteine amino group. The overall sequence, particularly the initial reaction is extremely important in toxicology because, by removing reactive electrophiles, vital nucleophilic groups in macromolecules such as proteins and nucleic acids are protected. The mercapturic acids formed can be excreted either in the bile or in the urine.

The GSTs, the family of enzymes that catalyzes the initial step, are widely distributed, being found in essentially all groups of living organisms. Although the best-known examples have been described from the soluble fraction of mammalian liver, these enzymes have also been described in microsomes. All forms appear to be highly specific with respect to GSH but nonspecific with respect to xenobiotic

Figure 6.19   Glutathione transferase reaction and formation of mercapturic acids.

Figure 6.20   Examples of glutathione transferases reactions.

substrates, although the relative rates for different substrates can vary widely from one form to another. The types of reactions catalyzed include the following: alkyl-transferase, aryltransferase, aralkyltransferase, alkenetransferase, and epoxide-transferase. Examples are shown in Figure 6.20.

Multiple forms of GST have been demonstrated in the liver of many mammalian species; multiple forms also occur in insects. Most GSTs are soluble dimeric proteins with molecular weights ranging between 45,000–50,000 Da. All forms appear to be nonspecific with respect to the reaction types described, although the kinetic constants for particular substrates vary from one form to another. They are usually named from their chromatographic behavior. At least two are membrane bound glutathione transferases, one of which is involved in metabolism of xenobiotics and is designated the microsomal GST. The cytosolic GSTs are divided into six families (historically called classes): the α (alpha), κ (kappa), μ (mu), π (pi), σ (sigma), and θ (theta) families. A new system of nomenclature proposes the term GST for the enzyme, preceded by the use of a small roman letter for the species (m for mouse, h for humans, etc.) followed by a capital roman letter for the family (A for α, K for κ, etc.). Subunits are to be designated by Arabic numbers, with the two subunits represented with a hyphen between them. For example, hGSTM1-2 designates a heterodimer of the human family mu which possesses subunits one and two.

Glutathione conjugation dramatically increases the water solubility of the metabolites compared to the parent compounds. The metabolites are released from the cell by an active transport system belonging to the multidrug resistance protein (mdr). Prior to excretion, the metabolites are usually processed by multiple enzymes to release the substrate conjugated to a mercapturic acid (Figure 6.19). The enzymes involved in this process are γ-glutamyltranspeptidase, cysteinyl glycinase, and N-acetyltransferase.

γ-Glutamyltranspeptidase is a membrane-bound glycoprotein that has been purified from both the liver and kidney of several species. Molecular weights for the kidney enzyme are in the range of 68,000–90,000 Da, and the enzyme appears to consist of two unequal subunits; the different forms appear to differ in the degree of sialylation. This enzyme, which exhibits wide specificity toward γ-glutamyl peptides and has a number of acceptor amino acids, catalyzes two types of reactions:

Hydrolysis	γ-Glu-R + H₂O	Glu + HR
Transpeptidation	γ-Glu-R + Acceptor	γ-Glu-Acceptor + HR
	γ-Glu-r + γ-Glu-R	γ-Glu γ-Glu-R + HR

Aminopeptidases that catalyze the hydrolysis of cysteinyl peptides are known. The membrane-bound aminopeptidases are glycoproteins, usually with molecular weights of about 100,000 Da. They appear to be metalloproteins, one of the better known being a zinc-containing enzyme. Other enzymes, such as the leucine aminopeptidase, are cytosolic, but, at least in this case, are also zinc containing. The substrate specificity of these enzymes varies, but most are relatively nonspecific.

Little is known of the N-acetyltransferase(s) responsible for the acetylation of the S-substituted cysteine. It is found in the microsomes of the kidney and the liver, however, and is specific for acetyl CoA as the actyl donor. It is distinguished from other N-acetyltransferases by its substrate specificity and subcellular location.

6.3.6 Cysteine Conjugate β-Lyase

This enzyme uses cysteine conjugates as substrates, releasing the thiol of the xenobiotic, pyruvic acid, and ammonia; with subsequent methylation giving rise to the methylthio derivative, the enzyme from the cytosolic fraction of rat liver is pyridoxal phosphate requiring protein of about 175,000 Da. Cysteine conjugates of aromatic compounds are the best substrates, and it is necessary for the cysteine amino and carboxyl groups to be unsubstituted for enzyme activity.

6.3.7   Acylation

Acylation reactions are of two general types, the first involving an activated conjugation agent, coenzyme A (CoA), and the second involving activation of the foreign compounds and subsequent acylation of an amino acid. This type of conjugation is commonly undergone by exogenous carboxylic acids and amides and, although the products are often less water soluble than the parent compound, they are usually less toxic. Examples of acylation reactions are shown in Figure 6.21.

Acetylation   Acetylated derivatives of foreign exogenous amines are acetylated by *N*-acetyltransferase, the acetyl donor being CoA. This enzyme is cytosolic, has been purified from rat liver, and is known to occur in several other organs. Evidence exists for the existence of multiple forms of this enzyme. Although endogenous amino, hydroxy, and thiol compounds are acetylated *in vivo*, the acetylation of exogenous hydroxy and thiol groups is presently unknown.

Acetylation of foreign compounds is influenced by both development and genetics. Newborn mammals generally have a low level of the transferase whereas, due to the different genes involved, fast and slow acetylators have been identified in both rabbit and human populations. Slow acetylators are more susceptible to the effects of compounds detoxified by acetylation.

N, O-*acyltransferase***   The *N*-acyltransferase enzyme is believed to be involved in the carcinogenicity of arylamines. These compounds are first N-oxidized, and then, in species capable of their N-acetylation, acetylated to arylhydroxamic acids. The effect of *N*-, *O*-transacetylation is shown in Figure 6.22. The *N*-acyl group of

(a) Acetylation

(b) Amino acid conjugation

Figure 6.21   Examples of acylation reactions.

Figure 6.22   *N-, O*-acyltransferase reactions of arylhydroxamic acid. Ar, aryl group.

the hydroxamic acid is first removed and is then transferred, either to an amine to yield a stable amide, or to the oxygen of the hydroxylamine to yield a reactive *N*-acyloxyarylamine. These compounds are highly reactive in the formation of adducts with both proteins and nucleic acids, and *N, O*-acyltransferase, added to the medium in the Ames test, increases the mutagenicity of compounds such as *N*-hydroxy-2-acetylaminofluorene. In spite of its great instability, this enzyme has been purified from the cytosolic fraction of the rat liver.

Amino Acid Conjugation   In the second type of acylation reaction, exogenous carboxylic acids are activated to form S-CoA derivatives in a reaction involving ATP and CoA. These CoA derivatives then acylate the amino group of a variety of amino acids. Glycine and glutamate appear to be the most common acceptor of amino acids in mammals. In other organisms, other amino acids are involved. These include ornithine in reptiles and birds and taurine in fish.

The activating enzyme occurs in the mitochondria and belongs to a class of enzymes known as the ATP-dependent acid: CoA ligases but has also been known as acyl CoA synthetase and acid-activating enzyme. It appears to be identical to the intermediate chain length fatty acyl–CoA-synthetase.

Two acyl-CoA:amino acid *N*-acyltransferases have been purified from liver mitochondria of cattle, Rhesus monkeys, and humans. One is a benzoyltransferase CoA that utilizes benzyl-CoA, isovaleryl-CoA, and tiglyl-CoA, but not phenylacetyl CoA, malonyl-CoA, or indolacetyl-CoA. The other is a phenylacetyl transferase that utilizes phenylacetyl-CoA and indolacetyl-CoA but is inactive toward benzoyl-CoA. Neither is specific for glycine, as had been supposed from studies using less defined systems; both also utilize asparagine and glutamine, although at lesser rates than glycine.

Bile acids are also conjugated by a similar sequence of reactions involving a microsomal bile acid: CoA ligase and a soluble bile acid *N*-acyltransferase. The latter has been extensively purified, and differences in acceptor amino acids, of which taurine is the most common, have been related to the evolutionary history of the species.

Deacetylation   Deacetylation occurs in a number of species, but there is a large difference between species, strains, and individuals in the extent to which the reaction occurs. Because acetylation and deacetylation are catalyzed by different enzymes, the levels of which vary independently in different species, the importance of deacetylation as a xenobiotic-metabolizing mechanism also varies between species. This can be seen in a comparison of the rabbit and the dog. The rabbit, which has high acetyltransferase activity and low deacetylase, excretes significant amounts of acetylated amines. The dog, in which the opposite situation obtains, does not.

A typical substrate for the aromatic deacetylases of the liver and kidney is acet-anilide, which is deacylated to yield aniline.

6.3.8 Phosphate Conjugation

Phosphorylation of xenobiotics is not a widely distributed conjugation reaction, insects being the only major group of animals in which it is found. The enzyme from the gut of cockroaches utilizes ATP, requires MG^+, and is active in the phosphoryla-tion of 1-naphthol and p-nitrophenol.

BIBLIOGRAPHY AND SUGGESTED READING

Benedetti, M. S. Biotransformation of xenobiotics by amine oxidases. *Fundam. Clin. Pharmacol.* **15**:75–84, 2001.

Cashman, J. R. and J. Zhang. Human flavin-containing monooxygenases. *Annu. Rev. Pharmacol. Toxicol.* **46**:65–100, 2006.

Coughtrie, M. W. H., S. Sharp, K. Maxwell, et al. Biology and function of the reversible sul-fation pathway catalyzed by human sulfotransferases and sulfatases. *Chem. Biol. Interact.* **109**:3–27, 1998.

Duffel, M. W., A. D. Marshall, P. McPhie, et al. Enzymatic aspects of the phenol (aryl) sulfo-transferases. *Drug Metab. Rev.* **33**:369–395, 2001.

Guengerich, F. P. Cytochrome P450 and chemical toxicology. *Chem. Res. Toxicol.* **21**:70–83, 2008.

Hayes, J. D. and D. J. Pulford. The glutathione S-transferase supergene family: Regulation of GST and the contribution of the isoenzymes to cancer chemoprotection and drug resistance. *Crit. Rev. Biochem. Mol. Biol.* **30**:445–600, 1995.

Hodgson, E., P. K. Das, T. M. Cho, et al. Phase 1—Metabolism of toxicants and metabolic interactions. In *Molecular and Biochemical Toxicology*, 4th ed., ed. R. C. Smart and E. Hodgson. Hoboken, NJ: John Wiley and Sons, 2008.

Koukouritaki, S. B., P. Simpson, C. K. Yeung, et al. Human hepatic flavin-containing mono-oxygenases 1 (FMO1) and 3 (FMO3) developmental expression. *Pediatric. Res.* **51**:236–243, 2002.

Lawton, M. P., J. R. Cashman, T. Cresteil, et al. A nomenclature for the mammalian flavin-containing monooxygenase gene family based on amino acid sequence identities. *Arch. Biochem. Biophys.* **308**:254–257, 1994.

LeBlanc, G. A. Phase II—Conjugation of toxicants. In *Molecular and Biochemical Toxicology*, 4th ed., ed. R. C. Smart and E. Hodgson. Hoboken, NJ: John Wiley and Sons, 2008.

Nebert, D. W., D. R. Nelson, M. J. Coon, et al. Superfamily—update on new sequences, gene-mapping, and recommended nomenclature. *DNA Cell Biol.* **10**:1–4, 1991.

Nelson, D. R. Comparison of P450s from human and fugu: 420 million years of vertebrate P450 evolution. *Arch. Biochem. Biophys.* **409**:18–24, 2003.

Ortiz de Montellano, P. R., ed. *Cytochrome P450: Structure, Mechanism and Biochemistry.* New York: Springer, 2004.

Richter, R., G. P. Jarvik, and C. E. Furlong. Paraoxonase 1 (PON1) status and substrate hydrolysis. *Toxicol. Appl. Pharmacol.* **235**:1–9, 2009.

Ritter, J. K. Roles of glucuronidation and UDP-glucuronosyltransferases in xenobiotic bio-activation reactions. *Chem. Biol. Interact.* **129**:171–193, 2001.

Ross, M. K. and J. A. Crow. Human carboxylases and their role in xenobiotic and endobiotic metabolism. *J. Biochem. Mol. Toxicol.* **21**:187–196, 2007.

Smart, R. C. and E. Hodgson, eds. *Molecular and Biochemical Toxicology*, 4th ed. Hoboken, NJ: John Wiley and Sons, 2008.

Tukey, R. H. and C. P. Strassburg. Human UDP-glucuronosyltransferases: Metabolism, expression, and disease. *Annu. Rev. Pharmacol. Toxicol.* **40**:581–616, 2000.

Vasiliou, V., A. Pappa, and D. R. Petersen. Role of aldehyde dehydrogenases in endogenous and xenobiotic metabolism. *Chem. Biol. Interact.* **129**:1–19, 2000.

Zeldin, D. C. and J. M. Seubert. Structure, mechanism and regulation of cytochromes P450. In *Molecular and Biochemical Toxicology*, 4th ed., ed. R. C. Smart and E. Hodgson. Hoboken, NJ: John Wiley and Sons, 2008.

SAMPLE QUESTIONS

1. Compare and contrast the following terms:
   a. Cytochrome P450 (CYP) and flavin-containing monooxygenase (FMO);
   b. Induction and activation of xenobiotic metabolizing enzymes (XMEs);
   c. Phase I and Phase II xenobiotic metabolism.

2. Give examples of the following:
   a. Epoxidation;
   b. O-Dealkylation;
   c. Desulfuration and ester cleavage.

3. Name the cofactors involved in the following xenobiotic conjugations:
   a. Mercapturic acid formation;
   b. Glucuronic acid formation;
   c. Sulfation;
   d. Methylation.

4. Outline the combined action of alcohol dehydrogenase and aldehyde dehydrogenase on ethanol.

5. Describe how the action of cysteine conjugate β-lyase can lead to methylthiolation of a xenobiotic.

Reactive Metabolites

ERNEST HODGSON and RANDY L. ROSE*

7.1 INTRODUCTION

Many xenobiotics that are relatively inert undergo metabolism to highly reactive intermediates. These metabolites may interact with cellular constituents in numerous ways, such as binding covalently to macromolecules and/or stimulating lipid peroxidation to produce toxic effects, or they may be detoxified and the products excreted. This biotransformation of relatively inert chemicals to highly reactive intermediary metabolites is commonly referred to as metabolic activation or bioactivation and is often the initial event in chemically induced toxicities. Some toxicants are direct acting and require no activation, whereas other chemicals may be activated nonenzymatically. The focus of this chapter, however, relates to toxicants requiring metabolic activation and to the processes involved in activation.

In the 1940s and 1950s, the pioneering studies of James and Elizabeth Miller provided early evidence for *in vivo* conversion of chemical carcinogens to reactive metabolites. They found that reactive metabolites of the aminoazo dye *N,N*-dimethyl-4-aminoazobenzene, a hepatocarcinogen in rats, would bind covalently to proteins and nucleic acids. The term, metabolic activation, was coined by the Millers to describe this process. Moreover, they demonstrated that covalent binding of these chemicals was an essential part of the carcinogenic process.

The overall scheme of metabolism for potentially toxic xenobiotics is outlined in Figure 7.1. As illustrated by this diagram, xenobiotic metabolism can produce not only nontoxic metabolites, which are more polar and more readily excreted (detoxication), but also highly reactive metabolites, which can interact with vital intracellular macromolecules, resulting in toxicity. In addition, reactive metabolites can be detoxified—for example, by interaction with glutathione or by epoxide hydration. In general, reactive metabolites are electrophiles (molecules containing positive centers). Those electrophiles not detoxified can in turn can react with cellular nucleophiles (molecules containing negative centers) such as proteins and nucleic acids. Other reactive metabolites may be free radicals or act as radical generators

*deceased

A Textbook of Modern Toxicology, Fourth Edition. Edited by Ernest Hodgson
Copyright © 2010 John Wiley & Sons, Inc.

Figure 7.1   The relationship between metabolism, activation, detoxication, and toxicity of a chemical.

that interact with oxygen to produce reactive oxygen species that are capable of causing damage to membranes, DNA, and other macromolecules.

Although a chemical can be metabolized by several routes, the activation pathway is often a minor route with the remainder of the pathways resulting in detoxication. Activation, however, may become a more dominant pathway in certain situations, thus leading to toxicity. Several examples illustrating these situations are discussed later in this chapter. Some important terms that are often used when discussing activation include parent compound, sometimes referred to as procarcinogen in the case of a carcinogen or prodrug for pharmaceutical compounds; proximate toxic metabolite or proximate carcinogen for one or more of the intermediates; and ultimate toxic metabolite or ultimate carcinogen for the reactive species that binds to macromolecules such as DNA and protein.

7.2 ACTIVATION ENZYMES

Whereas most, if not all, of the enzymes involved in xenobiotic metabolism can form reactive metabolites (examples are summarized in Table 7.1 and are described in more detail in Sections 7.7.1–7.7.10), the enzyme systems most frequently involved in the activation of xenobiotics are those that catalyze oxidation reactions. The cytochrome P450 (CYPs) monooxygenases are by far the most important enzymes involved in the oxidation of xenobiotics. This is because of the abundance of CYPs (especially in the liver), the numerous CYP isoforms, and the ability of CYPs to be induced by xenobiotics.

Although CYPs are most abundant in the liver, they are also present in other tissues including the skin, kidney, intestine, lung, placenta, and nasal mucosa. Because CYP exists as multiple isoforms with different substrate specificities, the presence or absence of a particular CYP isozyme may contribute to tissue-specific toxicities. Many drugs and other xenobiotics are known to induce one or more of the CYP isoforms, resulting in an increase, decrease, or an alteration in the metabolic pathway of chemicals metabolized by the CYP isoforms involved. Specific examples of these types of interactions are given later in this section.

TABLE 7.1   Some Examples of Activation Enzymes, Substrates, and Reactive Intermediates

Enzyme	Reaction	Substrate	Reactive Intermediate
Alcohol dehydrogenase	Oxidation	Methanol	Formaldehyde
Aldehyde dehydrogenase	Oxidation	Formaldehyde	Formic acid
CYP	Oxidation	Acetaminophen	N-acetylbenzoquinoneimine
CYP	Epoxidation	Aflatoxinb1	2,3-epoxide
CYP	N-hydroxylation	Acetylaminofluorene	N-hydroxyacetylaminofluorene
Sulfotransferase	Sulfate conjugation	N-hydroxyacetylaminofluorene	Sulfate conjugate
CYP	Epoxidation	Benzo(a)pyrene	Benzo(a)pyrene 7,8-epoxide
Epoxide hydrolase	Hydrolysis	Benzo(a)pyrene , 7,8-epoxide	Benzo(a)pyrene 7,8-dihydrodiol
CYP	Epoxidation	Benzo(a)pyrene 7,8-dihydrodiol	Benzo(a)pyrene 7,8-diol-9,10-epoxide
CYP	Oxidative dechlorination	Carbon tetrachloride	Trichloromethyl radical
CYP	Oxidative desulfuration	Chlorpyrifos	Chlorpyrifos oxon
			Reactive sulfur
CYP	Methylenedioxy ring cleavage	Piperonyl butoxide	Carbene derivative on methylene carbon
CYP	Epoxidation	Vinyl chloride	Epoxide
B-glucosidase (gut flora)	Hydrolysis	Cycasin	Methylazoxymethanol

159

In addition to activations catalyzed by CYPs and flavin-containing monooxygenases (FMOs), Phase II conjugations, co-oxidation by cyclooxygenase (COX) during prostaglandin biosynthesis, and metabolism by intestinal microflora may also lead to the formation of reactive toxic products. With some chemicals, only one enzymatic reaction is involved, whereas with other compounds, several reactions, often involving multiple pathways, are necessary for the production of the ultimate reactive metabolite.

7.3   NATURE AND STABILITY OF REACTIVE METABOLITES

Reactive metabolites include such diverse groups as epoxides, quinones, free radicals, reactive oxygen species, and unstable conjugates. Figure 7.2 gives some examples of activation reactions, the reactive metabolites formed, and the enzymes catalyzing their bioactivation.

As a result of their high reactivity, reactive metabolites are often considered to be short-lived. This is not always true, however, because reactive intermediates can be transported from one tissue to another, where they may exert their deleterious effects. Thus, reactive intermediates can be seen as occurring in several stability categories, or better as a continuum consisting of a continuous range of stabilities

Figure 7.2   Examples of some activation reactions.

that reflect their half-life under physiological conditions and how far they may be transported from the site of activation.

Some of these metabolites are ultrashort-lived and bind primarily to the enzyme involved in their formation. This category includes intermediates that form enzyme-bound complexes with the active site of the enzyme, hence the parent substrate is often referred to as a mechanism-based inhibitor or a "suicide substrate." A number of compounds are known to react in this manner with CYP, and such compounds are often used experimentally as CYP inhibitors (see discussion of piperonyl butoxide, Section 7.7.1 and of chlorpyrifos, Section 7.7.2). Other compounds, although not true suicide substrates, produce reactive metabolites that bind primarily to the activating enzyme or adjacent proteins altering the function of the protein.

Short-lived metabolites (as opposed to ultrashort lived) remain in the cell or travel only to nearby cells. In this case, covalent binding is restricted to the cell of origin and to adjacent cells. Many xenobiotics fall into this group and give rise to localized tissue damage occurring close to the sites of activation. For example, in the lung, the Clara cells contain high concentrations of CYP and several lung toxicants that require activation often result in damage primarily to Clara cells.

Longer-lived metabolites may be transported to other cells and tissues so that although the site of activation may be the liver, the target site may be in a distant organ. Reactive intermediates may also be transported to other tissues, not in their original form but as conjugates, which then release the reactive intermediate under the specific conditions of the target tissue. For example, carcinogenic aromatic amines are metabolized in the liver to the N-hydroxylated derivatives that, following glucuronide conjugation, are transported to the bladder, where the N-hydroxy derivative is released under the acidic conditions of urine.

7.4   FATE OF REACTIVE METABOLITES

While the production of reactive metabolites is the initial process in the sequence of events leading to toxicity, this sequence is not inevitable since the reactive metabolites are also subject to detoxication. Thus, a variety of reactions may occur depending on the nature of the reactive species and the physiology of the organism.

7.4.1   Binding to Cellular Macromolecules

As mentioned previously, most reactive metabolites are electrophiles that can bind covalently to nucleophilic sites on cellular macromolecules such as proteins, polypeptides, RNA, and DNA. This covalent binding is considered to be the initiating event for many toxic processes such as mutagenesis, carcinogenesis, and cellular necrosis, and is discussed in greater detail in chapters dealing with modes of toxic action.

7.4.2   Lipid Peroxidation

Radicals such as $CCl_3 \cdot$, produced during the oxidation of carbon tetrachloride, may induce lipid peroxidation and subsequent destruction of lipid membranes (Figure 7.3). Because of the critical nature of various cellular membranes (nuclear, mitochondrial, lysosomal, etc.), lipid peroxidation can be a pivotal event in cellular necrosis.

Figure 7.3 Metabolism of tetrachloromethane. Upon metabolic activation a CCl_3 radical is formed. This radical extracts protons from unsaturated fatty acids to form a free fatty acid radical. This leads to diene conjugates. At the same time, O_2 forms a hydroperoxide with the C radical. Upon its decomposition, malondialdehyde and other disintegration products are formed. In contrast, the CCl_3 radical is converted to chloroform, which undergoes further oxidative metabolism. Reprinted from: Bolt, H. M. and J. T. Borlak. Halogenated hydrocarbons. *Toxicol.* 645–657, 1999; with permission from Elsevier.

7.4.3 Trapping and Removal: Role of Glutathione

Once reactive metabolites are formed, mechanisms within the cell may bring about their rapid removal or inactivation. Toxicity then depends primarily on the balance between the rate of metabolite formation and the rate of removal. With some compounds, reduced glutathione plays an important protective role by trapping electrophilic metabolites and preventing their binding to hepatic proteins and enzymes. Although conjugation reactions occasionally result in bioactivation of a compound, the acetyl-, glutathione-, glucuronyl-, or sulfo-transferases usually result in the formation of a nontoxic, water-soluble metabolite that is easily excreted. Thus, availability of the conjugating chemical is an important factor in determining the fate of the reactive intermediates.

7.4.4 Trapping and Removal: Role of Epoxide Hydration

Reactive epoxides, in addition to being conjugated by glutathione S-transferase, may also be hydrolyzed by epoxide hydrolases to yield less toxic diols. Examples include styrene 7,8-oxide and naphthalene 1,2-oxide (see Chapter 6, Figure 6.14).

7.5 FACTORS AFFECTING TOXICITY OF REACTIVE METABOLITES

A number of factors can influence the balance between the rate of formation of reactive metabolites and the rate of removal, thereby affecting toxicity. The major factors discussed in this chapter are summarized in the following subsections.

A more in-depth discussion of other factors affecting metabolism and toxicity are presented in Chapter 8.

7.5.1 Levels of Activating Enzymes

Specific isozymes of CYPs are often important in determining metabolic activation of a foreign compound. As mentioned previously, many xenobiotics induce specific CYP isoforms. Frequently, the CYP isoforms induced are those involved in the metabolism of the inducing agent. Thus, a carcinogen or other toxicant has the potential for inducing its own activation. In addition, there are species and gender differences in enzyme levels as well as specific differences in the expression of particular isozymes.

7.5.2 Levels of Conjugating Enzymes

Levels of conjugating enzymes, such as glutathione transferases, are also known to be influenced by gender and species differences as well as by drugs and other environmental factors. All of these factors will, in turn, affect the detoxication process.

7.5.3 Levels of Cofactors or Conjugating Chemicals

Treatment of animals with *N*-acetylcysteine, a precursor of glutathione, protects animals against acetaminophen-induced hepatic necrosis, possibly by reducing covalent binding to tissue macromolecules. However, depletion of glutathione potentiates covalent binding and hepatotoxicity.

7.6 REACTIVE OXYGEN SPECIES

Although molecular oxygen normally exists in a relatively unreactive triplet state ($3O_2$), reactive oxygen species are formed *in vivo*, either during, or as a consequence of, aerobic metabolism. These reactive species include superoxide anion, hydrogen peroxide, singlet oxygen, nitric oxide and the highly reactive hydroxyl radical. There is a great deal of evidence that these reactive oxygen species are linked to a number of toxic end points, and this phemomenon is known as oxidative stress.

Oxygen is first converted to the oxidizing agent superoxide anion ($O_2^{\cdot-}$) by cellular nicotine adenine dinucleophosphate (NADPH) oxidase systems or by xanthine oxidase and, subsequently, hydrogen peroxide (H_2O_2) is formed by the further oxidation of $O_2^{\cdot-}$ by the enzyme, superoxide dismutase:

$$2O_2^{\cdot-} + 2H^+ \rightarrow H_2O_2 + O_2$$

H_2O_2 can then be converted to the highly toxic hydroxyl radical ($\cdot OH$) via the iron-catalyzed Fenton reaction:

$$Fe^{2+} + H_2O_2 \rightarrow Fe^{3+} + \cdot OH + OH^-$$

In addition, in the presence of $O_2^{\cdot-}$ and a divalent metal, H_2O_2 can produce $\cdot OH$ via the iron-catalyzed Haber–Weiss reaction:

$$Fe^{2+} + H_2O_2 \rightarrow Fe^{3+} \cdot OH + OH^-$$

$$Fe^{3+} + O_2^{\cdot-} \rightarrow Fe^{2+} + O_2$$

$$O_2^{\cdot-} + H_2O_2 \rightarrow \cdot OH + OH^-$$

Further details of the deleterious effects of these reactive oxygen species can be found in the various chapters dealing with modes of toxic action.

7.7   EXAMPLES OF ACTIVATING REACTIONS

The following examples have been selected to illustrate the various concepts of activation and detoxication discussed in the previous sections. They are also summarized in Table 7.1.

7.7.1   Piperonyl Butoxide

Methylenedioxyphenyl compounds, such as the insecticide synergist, piperonyl butoxide, are effective inhibitors of CYP monooxygenations, and are themselves metabolized to catechols. The most probable mechanism for inhibition and metabolism to a catechol appears to be oxidation at the methylene carbon, followed by elimination of water to yield a carbene. The highly reactive carbene either reacts with the heme iron to form a CYP-inhibitory complex or breaks down to yield the catechol (Chapter 6, Figure 6.8).

7.7.2   Chlorpyrifos

Chlorpyrifos is one of several organophosphorus insecticides of economic importance. Like all of the organophosphorus cholinesterase inhibitors containing the P=S moiety, chlorpyrifos must be metabolized to the reactive oxon (the P=O derivative). Oxon toxicity is the result of excessive stimulation of cholinergic nerves, which is dependent upon their ability to inhibit acetylcholinesterases (Figure 7.2). The activation reaction, which is CYP-catalyzed, is known as oxidative desulfuration. However, oxons are not the only activated products of oxidative desulfuration, as *in vitro* studies of rat and human liver have demonstrated that CYP isoforms are inactivated by the electrophilic sulfur atom released during oxidation of chlorpyrifos to chlorpyrifos oxon as well as other organophosphorus insecticides such the oxidation of parathion to paraoxon. The specific isoforms responsible for the metabolic activation are the ones destroyed in the process. For example, preincubations of NADPH-supplemented human liver microsomes with either chlorpyrifos or parathion resulted in the inhibition of some isoform-specific reactions including testosterone and estradiol oxidation (CYP3A4). These losses of metabolic activity are also associated with the loss of CYP content as measured by the CO-difference spectra. Thus, chlorpyrifos acts as a suicide substrate, in that its metabolism results in the destruction of the particular isoforms involved in its metabolism. This becomes

particularly important because the principal CYP involved in chlorpyrifos metabolism is CYP3A4, which is the dominant CYP in human liver; accounting for
between 30% and 50% of the total liver CYP. Because of this enzyme's importance
in drug and steroid hormone metabolism, the strong potential for inhibition by
organophosphorus compounds may have serious consequences for individuals
undergoing drug therapy.

7.7.3 Vinyl Chloride

Another example of a suicide inhibitor is vinyl chloride. The first step in the biotransformation of vinyl chloride involves the CYP-mediated oxidation of the double
bond leading to the formation of an epoxide, or oxirane, which is highly reactive
and can easily bind to proteins and nucleic acids. Following activation by CYP,
reactive metabolites such as those formed by vinyl chloride bind covalently to the
pyrrole nitrogens present in the heme moiety, resulting in destruction of the heme
and loss of CYP activity. The interaction of the oxirane structure with nucleic acids
results in mutations and cancer. The first indications that vinyl chloride was a human
carcinogen involved individuals who cleaned reactor vessels in polymerization
plants who were exposed to high concentrations of vinyl chloride and developed
angiosarcomas of the liver as a result of their exposure (Figure 7.2).

7.7.4 Methanol

Ingestion of methanol, particularly during the prohibition era, resulted in significant
illness and mortality. Where epidemics of methanol poisoning have been reported,
one-third of the exposed population recovered with no ill effects, one-third had
severe visual loss or blindness, and one-third had died. Methanol itself is not responsible for the toxic effects but is rapidly metabolized in humans by alcohol dehydrogenase to formaldehyde that is subsequently metabolized by aldehyde dehydrogenase
to form the highly toxic formic acid (Figure 7.2). The aldehyde dehydrogenase is
so efficient in its metabolism of formaldehyde that it is actually difficult to detect
formaldehyde in postmortem tissues. Accumulation of formic acid in the tissues
results first in blindness through edema of the retina, and eventually to death as a
result of acidosis. Successful treatment of acidosis by treatment with base was often
still unsuccessful in preventing mortality due to subsequent effects on the central
nervous system. Treatment generally consists of hemodialysis to remove the methanol, but where this option is not available, administration of ethanol effectively
competes with methanol for the alcohol dehydrogenase pathway, a competition that
is therapeutic inasmuch as acetaldehyde is much less toxic than formaldehyde.

7.7.5 Aflatoxin B$_1$

Aflatoxin B$_1$ (AFB1) is one of the mycotoxins produced by *Aspergillus flavus* and
A. parasiticus and is a well-known hepatotoxicant and hepatocarcinogen. It is generally accepted that the activated form of AFB1 that binds covalently to DNA is the
2,3-epoxide (Figure 7.2). AFB1-induced hepatotoxicity and carcinogenicity is
known to vary among species of livestock and laboratory animals. The selective
toxicity of AFB1 appears to be dependent on quantitative differences in formation

of the 2,3-epoxide, which is related to the particular enzyme complement of the organism. Because the epoxides of foreign compounds are frequently further metabolized by epoxide hydrolases or are nonenzymatically converted to the corresponding dihydrodiols, existence of the dihydrodiol is considered as evidence for prior formation of the epoxide. Because epoxide formation is catalyzed by CYP enzymes, the amount of AFB1-dihydrodiol produced by microsomes is reflective of the CYP isozyme complement involved in AFB1 metabolism. It has been shown that in rat microsomes in which specific CYP isozymes have been induced by phenobarbital (PB), dihydrodiol formation is considerably higher than that in control microsomes.

7.7.6 Carbon Tetrachloride (Tetrachloromethane)

Carbon tetrachloride has long been known to cause fatty acid accumulation and hepatic necrosis. Extraction of a chlorine atom from carbon tetrachloride by CYP results in the formation of a trichloromethyl radical which extracts protons from esterified desaturated fatty acids resulting in the production of chloroform (Figure 7.3). Chloroform also undergoes subsequent metabolism by CYP leading to the production of phosgene which covalently binds to sulfhydryl-containing enzymes and proteins, leading to toxicity. Differences between hepatic and renal effects of carbon tetrachloride and chloroform toxicity suggest that each tissue produces its own toxic metabolites from these chemicals.

In the case of hepatic toxicity due to carbon tetrachloride, the extraction of protons from fatty acids by the trichloromethyl radical results in the formation of highly unstable lipid radicals that undergo a series of transformations, including rearrangement of double bonds to produce conjugated dienes (Figure 7.3). Lipid radicals also readily react with oxygen, with the subsequent process, termed lipid peroxidation, producing damage to the membranes and enzymes. The resulting lipid peroxyl radicals decompose to aldehydes, the most abundant being malondialdehyde and 4-hydroxy-2,3-nonenal (Figure 7.3).

Since desaturated fatty acids are highly susceptible to free radical attack, neighboring fatty acids are readily affected and the initial metabolic transformation results in a cascade of detrimental effects on the tissue. The initial production of the trichloromethyl radical from carbon tetrachloride also results in irreversible covalent binding to CYP, resulting in its inactivation. In cases of carbon tetrachloride poisoning, preliminary sublethal doses actually become protective to an organism in the event of further poisoning since the metabolic activating enzymes are effectively inhibited by the first dose.

7.7.7 Acetylaminofluorene

In the case of the hepatocarcinogen, 2-acetylaminofluorene (2-AAF), two activation steps are necessary to form the reactive metabolites (Figure 7.4). The initial reaction, N-hydroxylation, is a CYP-dependent Phase I reaction, whereas the second reaction, resulting in the formation of the unstable sulfate ester, is a Phase II conjugation reaction that results in the formation of the reactive intermediate. Another Phase II reaction, glucuronide conjugation, is a detoxication step, resulting in a readily excreted conjugation product.

Figure 7.4 Bioactivation of 2-acetylaminofluorene.

In some animal species, 2-AAF is known to be carcinogenic, whereas in other species, it is noncarcinogenic. The species- and sex-specific carcinogenic potential of 2-AAF is correlated with the ability of the organism to sequentially produce the N-hydroxylated metabolite followed by the sulfate ester. Therefore, in an animal such as the guinea pig, which does not produce the N-hydroxylated metabolite, 2-AAF is not carcinogenic. In contrast, both male and female rats produce the N-hydroxylated metabolite, but only male rats have high rates of tumor formation. This is because male rats have up to 10-fold greater expression of sulfotransferase 1C1 than female rats, which has been implicated in the sulfate conjugation of 2-AAF resulting in higher production of the carcinogenic metabolite.

7.7.8 Benzo(a)pyrene

The polycyclic aromatic hydrocarbons are a group of chemicals consisting of two or more condensed aromatic rings that are formed primarily from incomplete combustion of organic materials including wood, coal, mineral oil, motor vehicle exhaust, and cigarette smoke. Early studies of cancer in the 1920s involving the fractionation of coal tar identified the carcinogenic potency of pure polycyclic aromatic hydrocarbons, including dibenz(a,h)anthracene and benzo(a)pyrene. Although several hundred different polycyclic aromatic hydrocarbons are known, environmental monitoring usually only detects a few compounds, one of the most important of which is benzo(a)pyrene. Benzo(a)pyrene is also one of the most prevalent polycyclic aromatic hydrocarbons found in cigarette smoke.

Extensive studies of metabolism of benzo(a)pyrene have identified at least 15 Phase I metabolites. The majority of these are the result of CYP1A1 and epoxide hydrolase reactions. Many of these metabolites are further metabolized by Phase II enzymes to produce numerous different metabolites. Studies examining the carcinogenicity of this compound have identified the 7,8-oxide and 7,8-dihydrodiol as proximate carcinogens and the 7,8-diol-9,10 epoxide as a strong mutagen and ultimate carcinogen. Because of the stereoselective metabolizing abilities of CYP isoforms, the reactive 7,8-diol-9,10-epoxide can appear as four different isomers (Figure 7.5). Interestingly, only one of these isomers (+)-benzo(a)pyrene 7,8-diol-9,10 epoxide-2 has significant carcinogenic potential. Comparative studies with

Benzo(a)pyrene

7,8 epoxide of benzo(a)pyrene

HO

OH

7,8 dihydrodiol of benzo(a)pyrene

HO

OH

7,8-diol-9,10-epoxides of benozo(a)pyrene

Figure 7.5 Selected stages of biotransformation of benzo(a)pyrene. The diol epoxide can exist in four diastereoisomeric forms of which the key carcinogenic metabolite is (+)-benzo(a) pyrene 7,8-diol-9,10-epoxide.

several other polycyclic aromatic hydrocarbons have demonstrated that only those substances that are epoxidized in the bay region of the ring system possess carcinogenic properties.

7.7.9 Acetaminophen

A good example of the importance of tissue availability of the conjugating chemical is found with acetaminophen. At normal therapeutic doses, acetaminophen is safe, but can be hepatotoxic at high doses. The major portion of acetaminophen is conjugated with either sulfate or glucuronic acid to form water-soluble, readily excreted metabolites, and only small amounts of the reactive intermediate, believed to be a quinoneimine (*N*-acetylbenzoquinoneimine), are formed by the CYP isoforms, principally by CYP2E1 (Figure 7.6).

When therapeutic doses of acetaminophen are ingested, the small amount of reactive intermediate formed is efficiently deactivated by conjugation with glutathione. When large doses are ingested, however, the sulfate and glucuronide cofactors (phosphoadenosine phosphosulfate and uridinediphosphoglucuronic acid) become depleted, resulting in more of the acetaminophen being metabolized to the reactive intermediate.

As long as glutathione (GSH) is available, most of the reactive intermediate can be detoxified. When the concentration of GSH in the liver also becomes depleted,

Figure 7.6 Metabolism of acetaminophen and formation of reactive metabolites.

however, covalent binding to sulfhydryl (-SH) groups of various cellular proteins increases, resulting in hepatic necrosis. If sufficiently large amounts of acetaminophen are ingested, as in drug overdoses and suicide attempts, extensive liver damage and death may result.

A further complication may result from the fact that ethanol is an inducer of CYP2E1. Thus, prior consumption of ethanol increases the isoform responsible for the generation of the reactive metabolite, *N*-acetylbenzoquinoneimine.

7.7.10 Cycasin

When flour from the cycad nut, which is used extensively among residents of South Pacific Islands, is fed to rats, it leads to cancers of the liver, kidney, and digestive tract. The active compound in cycasin, is the β-glucoside of methylazoxymethanol (Figure 7.7). If this compound is injected intraperitoneally rather than given orally, or if the compound is fed to germ-free rats, no tumors occur. Intestinal microflora possess the necessary enzyme, β-glucosidase, to form the active compound methylazoxymethanol, which is then absorbed into the body. The parent compound, cycasin, is carcinogenic only if administered orally because β-glucosidases are not present in mammalian tissues but are present in the gut. However, it can be

Figure 7.7 Bioactivation of cycasin by intestinal microflora to the carcinogen methylazoxymethanol.

demonstrated that the metabolite, methylazoxymethanol, will lead to tumors in both normal and germ-free animals regardless of the route of administration.

7.8 SUMMARY AND CONCLUSIONS

The current procedures for assessing safety and carcinogenic potential of chemicals using whole animal studies are expensive as well as becoming less socially acceptable. Moreover, the scientific validity of such tests for human risk assessment is also being questioned. Currently, a battery of short-term mutagenicity tests are used extensively as early predictors of mutagenicity and possible carcinogenicity.

Most of these systems use test organisms—for example, bacteria, that lack suitable enzyme systems to bioactivate chemicals, and therefore, an exogenous activating system is used. Usually, the post-mitochondrial fraction from rat liver, containing both Phase I and Phase II enzymes, is used as the activating system. The critical question is to what extent does this rat system represent the true *in vivo* situation, especially in humans? If not this system, then what is the better alternative? As some of the examples in this chapter illustrate, a chemical that is toxic or carcinogenic in one species or gender may be inactive in another, and this phenomenon is often related to the complement of enzymes, either activation or detoxication, expressed in the exposed organism.

Another factor to consider is the ability of many foreign compounds to selectively induce the CYP enzymes involved in their metabolism, especially if this induction results in the activation of the compound. With molecular techniques now available, considerable progress is being made in defining the enzyme and isozyme complements of human and laboratory species and understanding their mechanisms of control. Another area of active research is the use of *in vitro* expression systems to study the oxidation of foreign chemicals (e.g., bacteria containing genes for specific human CYP isozymes).

In summary, in studies of chemical toxicity, pathways and rates of metabolism as well as effects resulting from toxicokinetic factors and receptor affinities are critical in the choice of the animal species and experimental design. Therefore, it is important that the animal species chosen as a model for humans in safety evaluations metabolize the test chemical by the same routes as humans and, furthermore, that quantitative differences are considered in the interpretation of animal toxicity data. Risk assessment methods involving the extrapolation of toxic or carcinogenic potential of a chemical from one species to another must consider the metabolic and toxicokinetic characteristics of both species.

BIBLIOGRAPHY AND SUGGESTED READING

Anders, M. W., W. Dekant, and S. Vamvakas. Glutathione-dependent toxicity. *Xenobiotica* **22**:1135–1145, 1992.

Bolt, H. M. and J. T. Borlak. Halogenated hydrocarbons. In *Toxicology*, ed. H. Marquardt. San Diego, CA: Academic, 1999.

Bonner, J. C. Respiratory toxicology. In *Molecular and Biochemical Toxicology*, 4th ed., eds. R. C. Smart and E. Hodgson. Hoboken, NJ: Wiley, 2008.

Coughtrie, M. W. H., S. Sharp, K. Maxwell, et al. Biology and function of the reversible sulfation pathway catalysed by human sulfotransferases and sulfatases. *Chem. Biol. Interact.* **109**:3–27, 1998.

Gonzalez, F. J. Role of cytochromes P450 in chemical toxicity and oxidative stress: Studies with CYP2E1. *Mutat. Res.* **569**:101–110, 2005.

Gonzalez, F. J. and H. V. Gelboin. Role of human cytochromes P450 in the metabolic activation of chemical carcinogens and toxins. *Drug Metab. Rev.* **26**:165–183, 1994.

Guengerich, F. P. Bioactivation and detoxication of toxic and carcinogenic chemicals. *Drug Metab. Dispos.* **21**:1–6, 1993.

Guengerich, F. P. Metabolic activation of carcinogens. *Pharmacol. Ther.* **54**:17–61, 1992.

MacKenzie, E. L. Reactive oxygen species/reactive metabolites and toxicity. In *Molecular and Biochemical Toxicology*, 4th ed., eds. R. C. Smart and E. Hodgson. Hoboken, NJ: Wiley, 2008.

Omiecinski, C. J., R. P. Remmel, and V. P. Hosagrahara, Concise review of the cytochrome P450s and their roles in toxicology. *Toxicol. Sci.* **48**:151–156, 1999.

Park, B., N. Kitteringham, J. Maggs, et al. The role of metabolic activation in drug-induced hepatoxicity. *Annu. Rev. Pharmacol. Toxicol.* **45**:177–202, 2005.

Rinaldi, R., E. Eliasson, S. Swedmark, et al. Reactive intermediates and the dynamics of glutathione transferases. *Drug Metab. Dispos.* **30**:1053–1058, 2002.

Ritter, J. K. Roles of glucuronidation and UDP-glucuronosyltransferases in xenobiotic bioactivation reactions. *Chem. Biol. Interact.* **129**:171–193, 2000.

Vasiliou, V., A. Pappa, and D. R. Petersen. Role of aldehyde dehydrogenases in endogenous and xenobiotic metabolism. *Chem. Biol. Interact.* **129**:1–19, 2000.

SAMPLE QUESTIONS

1. Which xenobiotic-metabolizing enzymes are involved in the production of reactive metabolites? Which are most important?

2. Name the reactive metabolite(s) formed from piperonyl butoxide; chlorpyrifos; methanol; carbon tetrachloride; cycasin.

3. What are the principal ways in which reactive metabolites exert their toxic effects?

4. What are the principal ways that reactive metabolites may be detoxified before exerting their toxic effects?

5. Outline the process, starting with molecular oxygen, by which reactive oxygen species can be formed *in vivo*.

Chemical and Physiological Effects on Xenobiotic Metabolism

ANDREW D. WALLACE and ERNEST HODGSON

8.1 INTRODUCTION

Xenobiotic metabolism can be modified by many factors both extrinsic and intrinsic to the normal functioning of the organism, and it is probable that many changes in toxicity are due to changes in metabolism of the toxicant, because most sequences of events that lead to overt toxicity involve either activation or detoxication of the toxicant in question. However, due to the difficulty of relating single events, measured *in vitro*, to the complex and interrelated effects that occur *in vivo*, the chain of cause and effect is not always entirely clear. Nevertheless, the relationship between *in vitro* and *in vivo* studies is important and is one of the themes of this chapter. It is important to note that the chemical, nutritional, physiological, and other effects noted herein have been described primarily from experiments carried out on experimental animals. While these studies indicate that similar effects may occur in humans or other animals, they do not indicate that they must occur, or that they occur at the same magnitude in all species, if they occur at all.

8.2 NUTRITIONAL EFFECTS

Many nutritional effects on xenobiotic metabolism have been noted, but the information is scattered and often appears contradictory. This section (8.2) is concerned only with the effects of nutritional constituents of the diet; the effects of other xenobiotics in the diet are discussed under chemical effects (Section 8.5).

8.2.1 Protein

Low-protein diets generally decrease monooxygenase activity in rat liver microsomes, and gender and substrate differences may be seen in the effect. For example, aminopyrine N-demethylation, hexobarbital hydroxylation, and aniline

A Textbook of Modern Toxicology, Fourth Edition. Edited by Ernest Hodgson

hydroxylation are all decreased, but the effect on the first two is greater in males than in females. In the third case, aniline hydroxylation, the reduction in males is equal to that in females. Tissue differences may also be seen. These changes are presumably related to the reductions in the levels of cytochrome P450 (CYP) and Nicotinamide adenine dinucleotide phosphate (NADPH)-CYP reductase that are also noted. One might speculate that the gender and other variations are due to differential effects on specific CYP isoforms. Even though enzyme levels are reduced by low-protein diets, they can still be induced to some extent by compounds such as phenobarbital. Such changes may also be reflected in changes in toxicity. Changes in the level of azoreductase activity in rat liver brought about by a low-protein diet are reflected in an increased severity in the carcinogenic effect of dimethylamino-azobenzene. The liver carcinogen dimethylnitrosamine, which must be activated metabolically, is almost without effect in protein-deficient rats.

Strychnine, which is detoxified by microsomal monooxygenase action, is more toxic to animals on low-protein diets, whereas octamethylpyrophosphoramide, carbon tetrachloride, and heptachlor, which are activated by CYP monooxygenases, are less toxic. Phase II reactions may also be affected by dietary protein levels. Chloramphenicol glucuronidation is reduced in protein-deficient guinea pigs, although no effect is seen on sulfotransferase activity in protein-deficient rats.

8.2.2 Carbohydrates

High dietary carbohydrate levels in the rat tend to have much the same effect as low dietary protein, decreasing such activities as aminopyrine N-demethylase, pentobarbital hydroxylation, and *p*-nitrobenzoic acid reduction along with a concomitant decrease in the enzymes of the CYP-dependent monooxygenase system. Because rats tend to regulate total caloric intake, this may actually reflect low-protein intake.

In humans, it has been demonstrated that increasing the ratio of protein to carbohydrate in the diet stimulates oxidation of antipyrine and theophylline, while changing the ratio of fat to carbohydrate had no effect. In related studies, humans fed charcoal-broiled beef (food high in polycyclic hydrocarbon content) for several days had significantly enhanced activities of CYP1A1 and CYP1A2, resulting in enhanced metabolism of phenacetin, theophylline, and antipyrine. Studies of this nature indicate that there is significant interindividual variability in these observed responses.

8.2.3 Lipids

Dietary deficiencies in linoleic or in other unsaturated fats generally bring about a reduction in CYP and related monooxygenase activities in the rat. The increase in effectiveness of breast and colon carcinogens brought about in animals on high fat diets, however, appears to be related to events during the promotion phase rather than the activation of the causative chemical. Increases in dietary lipids have been shown to elevate CYP2E1 and CYP4A levels in humans. Lipids also appear to be necessary for the effect of inducers, such as phenobarbital, to be fully expressed.

8.2.4 Micronutrients

Vitamin deficiencies in general bring about a reduction in monooxygenase activity, although exceptions can be noted. Riboflavin deficiency causes an increase in CYP and aniline hydroxylation, although at the same time it causes a decrease in CYP reductase and benzo(a)pyrene hydroxylation. Ascorbic acid deficiency in the guinea pig not only causes a decrease in CYP and monooxygenase activity but also causes a reduction in microsomal hydrolysis of procaine. Deficiencies in vitamins A and E cause a decrease in monooxygenase activity, whereas thiamine deficiency causes an increase. The effect of these vitamins on different CYP isoforms has not been investigated. Changes in mineral nutrition have also been observed to affect monooxygenase activity. In the immature rat, calcium or magnesium deficiency causes a decrease, whereas, quite unexpectedly, iron deficiency causes an increase. This increase is not accompanied by a concomitant increase in CYP, however. An excess of dietary cobalt, cadmium, manganese, and lead all cause an increase in hepatic glutathione levels and a decrease in CYP content.

8.2.5 Starvation and Dehydration

Although in some animals starvation appears to have effects similar to those of protein deficiency, this is not necessarily the case. For example, in the mouse, monooxygenation is decreased, but reduction of p-nitrobenzoic acid is unaffected. In male rats, hexobarbital and pentobarbital hydroxylation as well as aminopyrine N-demethylation are decreased, but aniline hydroxylation is increased. All of these activities are stimulated in the female. Water deprivation in gerbils causes an increase in CYP and a concomitant increase in hexobarbital metabolism, which is reflected in a shorter sleeping time.

8.2.6 Nutritional Requirements in Xenobiotic Metabolism

Because xenobiotic metabolism involves many enzymes with different cofactor requirements, prosthetic groups, or endogenous co-substrates, it is apparent that many different nutrients are involved in their function and maintenance. Determination of the effects of deficiencies, however, is more complex because reductions in activity of any particular enzyme will be effective only if it affects a change in a rate-limiting step in a process. In the case of multiple deficiencies, the nature of the rate-limiting step may change with time.

Phase I Reactions   Nutrients involved in the maintenance of the CYP-dependent monooxygenase system are shown in Figure 8.1. The B complex vitamins niacin and riboflavin are both involved, the former in the formation of NADPH, the latter in the formation of flavin adenine dinucleotied (FAD) and flavin mononucleotide (FMN). Essential amino acids are, of course, required for the synthesis of all of the proteins involved. The heme of the cytochrome requires iron, an essential inorganic nutrient. Other nutrients required in heme synthesis include pantothenic acid, needed for the synthesis of the coenzyme A (CoA) used in the formation of acetyl CoA, pyridoxine, a cofactor in heme synthesis and copper, required in the

Figure 8.1 Nutritional requirements with potential effects on the cytochrome P450 monooxygenase system From: Donaldson, W. E. Nutritional factors. In *Introduction to Biochemical Toxicology,* 3rd ed., eds. E. Hodgson and R. C. Smart. New York: Wiley, 2001.

ferroxidase system that converts ferrous to ferric iron prior to its incorporation into heme. Although it is clear that dietary deficiencies could reduce the ability of the CYP system to metabolize xenobiotics, it is not clear how this effect would be manifested *in vivo* unless there is an understanding of the rate-limiting factors involved a considerable task in such a complex of interrelated reactions, and there is little reliable information on the effect of deficiencies in specific micronutrients.

Phase II As with Phase I reactions, Phase II reactions usually depend on several enzymes with different cofactors and different prosthetic groups and, frequently, different endogenous co-substrates. All of these many components can depend on nutritional requirements, including vitamins, minerals, amino acids, and others. Mercapturic acid formation can be cited to illustrate the principles involved. The formation of mercapturic acids starts with the formation of glutathione conjugates, reactions catalyzed by the glutathione S-transferases.

This is followed by removal of the glutamic acid and the glycine residues followed by acetylation of the remaining cysteine. Essential amino acids are required for the synthesis of the proteins involved, pantothenic acid for coenzyme A synthesis, and phosphorus for synthesis of the ATP needed for glutathione synthesis. Similar scenarios can be developed for glucuronide and sulfate formation, acetylation, and other Phase II reaction systems. As with Phase I enzymes, there is little reliable information on the effect of deficiencies in specific micronutrients.

8.3 PHYSIOLOGICAL EFFECTS

8.3.1 Development

Birth in mammals initiates an increase in the activity of many hepatic enzymes, including those involved in xenobiotic metabolism. The ability of the liver to carry out monooxygenation reactions appears to be very low during gestation and to increase after birth, with no obvious differences being seen between immature males and females. This general trend has been observed in many species, although the developmental pattern may vary according to gender and genetic strain. The component enzymes of the CYP-dependent monooxygenase system both follow the same general trend although there may be differences in the rate of increase. In the rabbit, the postnatal increase in CYP and its reductase is parallel; in the rat, the increase in the reductase is slower than that of the cytochrome.

Phase II reactions may also be age dependent. Glucuronidation of many substrates is low or undetectable in fetal tissues but increases with age. The inability of newborn mammals of many species to form glucuronides is associated with deficiencies in both glucuronosyltransferase and its cofactor, uridine diphosphate glucuronic acid (UDPGA). A combination of this deficiency, as well as slow excretion of the bilirubin conjugate formed, and the presence in the blood of pregnanediol, an inhibitor of glucuronidation, may lead to neonatal jaundice. Glycine conjugations are also low in the newborn, resulting from a lack of available glycine, an amino acid that reaches normal levels at about 30 days of age in the rat and 8 weeks in the human. Glutathione conjugation may also be impaired, as in fetal and neonatal guinea pigs, because of a deficiency of available glutathione. In the serum and liver of perinatal rats, glutathione transferase is barely detectable, increasing rapidly until adult levels are reached at about 140 days (Figure 8.2). This pattern is not followed in all cases, because sulfate conjugation and acetylation appear to be fully functional and at adult levels in the guinea pig fetus. Thus, some compounds that are glucuronidated in the adult can be acetylated or conjugated as sulfates in the young.

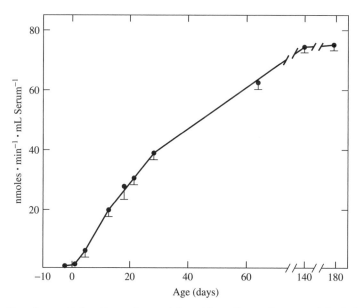

Figure 8.2 Developmental pattern of serum glutathione S-transferase activity in female rats. Adapted from Mukhtar and Bend. *Life Sci.* **21**:1277, 1977.

An understanding of how these effects may be related to the expression of individual isoforms is now beginning to emerge. It is known that in immature rats of either gender, CYP2A1, CYP2D6, and CYP3A2 predominate, whereas in mature rats, the males show a predominance of CYP2C11, CYP2C6, and CYP3A2, and the females CYP2A1, CYP2C6, and CYP2C12.

The effect of senescence on the metabolism of xenobiotics has yielded variable results. In rats, monooxygenase activity, which reaches a maximum at about 30 days of age, begins to decline some 250 days later, a decrease that may be associated with reduced levels of sex hormones. Glucuronidation also decreases in old animals, whereas monoamine oxidase activity increases. These changes in the monooxygenase activities are often reflected by changes in drug efficacy or overall toxicity.

In pediatric human populations, often there is a delayed increase in CYP enzyme expression and adult levels are not reached until 10 years of age or older. In elderly humans, age-related impairment of enzyme activity is highly controversial. Conflicting findings have found either age-related declines in or no changes with respect to the activity of CYP2C and CYP3A isoforms. Studies involving an erythromycin breath test in humans also suggested that there were no age-related declines associated with CYP3A4 activity. However, a study of CYP content and antipyrine clearance in liver biopsies obtained from 226 closely matched subjects indicated that subjects older than 70 had significantly less activity and clearance than younger subjects. Likewise, in older subjects, clearance of the drug omeprazole, a CYP2C19 substrate, was nearly half the rates observed in younger subjects. The effects of age on xenobiotic-metabolizing enzymes (XMEs) are complicated by other factors including liver mass, blood flow changes, and interindividual variability.

8.3.2 Gender Differences

Metabolism of xenobiotics may vary with the gender of the organism. Gender differences become apparent at puberty and are usually maintained throughout adult life. Adult male rats metabolize many compounds at rates higher than females, for example, hexobarbital hydroxylation, aminopyrine N-demethylation, glucuronidation of o-aminophenol, and glutathione conjugation of aryl substrates; however, with other substrates, such as aniline and zoxazolamine, no gender differences are seen. In other species, including humans, the gender difference in xenobiotic metabolism is less pronounced. The differences in microsomal monooxygenase activity between males and females have been shown to be under the control of sex hormones, at least in some species. Some enzyme activities are decreased by castration in the male, and administration of androgens to castrated males increases the activity of these sex-dependent enzyme activities without affecting the independent ones. Procaine hydrolysis is faster in male than female rats, and this compound is less toxic to the male. Gender differences in enzyme activity may also vary from tissue to tissue. Hepatic microsomes from adult male guinea pigs are less active in the conjugation of p-nitrophenol than are those from females, but no such gender difference is seen in the microsomes from lung, kidney, and small intestines.

Many differences in overall toxicity between males and females of various species are known (Table 8.1). Although it is not always known whether metabolism is the only or even the most important factor, such differences may be related to gender-related differences in metabolism. Hexobarbital is metabolized faster by male rats;

TABLE 8.1 Gender-Related Differences in Toxicity

Species	Toxicant	Susceptibility
Rat	EPN, warfarin, strychnine, hexobarbital, parathion	F > M
	Aldrin, lead, epinephrine, ergot alkaloids	M > F
Cat	Dinitrophenol	F > M
Rabbit	Benzene	F > M
Mouse	Folic acid	F > M
	Nicotine	M > F
Dog	Digitoxin	M > F

thus, female rats have longer sleeping times. Parathion is activated to the cholinesterase inhibitor paraoxon more rapidly in female than in male rats, and thus is more toxic to females. Presumably, many of the gender-related differences, as with the developmental differences, are related to quantitative or qualitative differences in the isoforms of the XMEs that exist in multiple forms, but this aspect has not been investigated extensively.

In the rat, sexually dimorphic CYPs appear to arise by programming, or imprinting, that occurs in neonatal development. This imprinting is brought about by a surge of testosterone that occurs in the male, but not the female, neonate and appears to imprint the developing hypothalamus so that in later development, growth hormone is secreted in a gender-specific manner. Growth hormone production is pulsatile in adult males with peaks of production at approximately 3-h intervals and more continuous in females, with smaller peaks. This pattern of growth hormone production and the higher level of circulating testosterone in the male maintain the expression of male-specific isoforms such as CYP2C11. The more continuous pattern of growth hormone secretion and the lack of circulating testosterone appears to be responsible for the expression of female-specific isoforms such as CYP2C12. The high level of sulfotransferases in the female appears to be under similar control, raising the possibility that this is a general mechanism for the expression of gender-specific XMEs or their isoforms. A schematic version of this proposed mechanism is seen in Figure 8.3.

Gender-specific expression is also seen in the flavin-containing monooxygenases. In mouse liver, FMO1 is higher in the female than in the male, and FMO3, present at high levels in female liver, is not expressed in male liver (Figure 8.4). No gender-specific differences are observed for FMO5. The important role of testosterone in the regulation of FMO1 and FMO3 was demonstrated in gonadectomized animals with and without testosterone implants. In males, castration increased FMO1 and FMO3 expression to levels similar to those observed in females, and testosterone replacement to castrated males resulted in ablation of FMO3 expression. Similarly, administration of testosterone to females caused ablation of FMO3 expression. Although these results clearly indicate a role for testosterone in the regulation of these isoforms, the physiological reasons for their gender-dependent expression remain unknown.

In humans, gender differences in the metabolism of some commonly used drugs have been observed, but overall, not much is known. These differences are thought to be due to expression levels of CYPs, such as CYP3A4, which leads to women generally being more susceptible to drug–drug interactions.

Figure 8.3 Hypothetical scheme for neonatal imprinting of the hypothalamus-pituitary-liver axis resulting in sexually dimorphic expression of hepatic enzymes in the adult rat. Neonatal surges of testosterone appear to play a role in imprinting. From Ronis, M. J. J. and H. C. Cunny. Physiological (endogenous) factors affecting the metabolism of xenobiotics. In *Introduction to Biochemical Toxicology*, 2nd ed., eds. E. Hodgson and P. E. Levi, p. 136. Norwalk, CT: Appleton & Lange, 1994.

8.3.3 Hormones

Hormones other than sex hormones are also known to affect the levels of XMEs, but these effects are much less studied or understood.

Thyroid Hormone Treatment of rats with thyroxin increases hepatic microsomal NADPH oxidation in both male and female rats, with the increase being greater in females. CYP content decreases in the male but not in the female.

Hyperthyroidism causes a decrease in gender-dependent monooxygenase reactions and appears to interfere with the ability of androgens to increase the activity of the enzymes responsible. Gender differences are not seen in the response of mice and rabbits to thyroxin. In mice, aminopyrine N-demethylase, aniline hydroxylase, and hexobarbital hydroxylase are decreased, whereas *p*-nitrobenzoic acid reduction is unchanged. In rabbits, hexobarbital hydroxylation is unchanged, whereas aniline hydroxylation and *p*-nitrobenzoic acid reduction increase. Thyroid hormone can also affect enzymes other than microsomal monooxygenases. For example, liver monoamine oxidase activity is decreased whereas the activity of the same enzymes in the kidney is increased.

Figure 8.4 Immunoreactivity of liver microsomes from sexually intact control, sham control, gonadectomized mice, or mice undergoing gonadectomy and/or receiving testosterone implants (5 mg). From Falls, J. G., D.-Y. Ryu, Y. Cao, et al. *Arch. Biochem. Biophys.* **342**:212–223, 1997.

Adrenal Hormones Removal of adrenal glands from male rats results in a decrease in the activity of hepatic microsomal enzymes, impairing the metabolism of aminopyrine and hexobarbital, but the same operation in females has no effect on their metabolism of xenobiotics. Cortisone or prednisolone restores activity to normal levels.

Insulin The effect of diabetes on xenobiotic metabolism is quite varied and, in this regard, alloxan-induced diabetes may not be a good model for the natural disease. The *in vitro* metabolism of hexobarbital and aminopyrine is decreased in alloxan-diabetic male rats, but is increased in similarly treated females. Aniline hydroxylase is increased in both males and females with alloxan diabetes. The induction of CYP2D1 in diabetes (and in fasting) is believed to be due to the high circulating levels of endogenously generated ketones. Studies of activity of the enzymes mentioned show no gender differences in the mouse; both sexes show an increase. Some Phase II reactions, such as glucuronidation, are decreased in diabetic animals. This appears to be due to a lack of UDPGA caused by a decrease in UDPG dehydrogenase, rather than a decrease in transferase activity, and the effect can be reversed by insulin.

Other Hormones Pituitary hormones regulate the function of many other endocrine glands, and hypophysectomy in male rats results in a decrease in the activity of XMEs. Administration of adrenocorticotropic hormone (ACTH) also results in a decrease of those oxidative enzyme activities that are gender dependent. In contrast, ACTH treatment of female rats causes an increase in aminopyrine N-demethylase but no change in other activities.

8.3.4 Pregnancy

Many XME activities decrease during pregnancy. Catechol O-methyltransferase and monoamine oxidase decrease, as does glucuronide conjugation. The latter may be related to the increasing levels of progesterone and pregnanediol, both known to be inhibitors of glucuronosyltransferase *in vitro*. A similar effect on sulfate conjugation has been seen in pregnant rats and guinea pigs. In some species, liver microsomal monooxygenase activity may also decrease during pregnancy, this decrease being accompanied by a concomitant decrease in CYP levels. An increased level of FMO2 is seen in the lung of pregnant rabbits.

8.3.5 Disease

Quantitatively, the most important site for xenobiotic metabolism is the liver; thus, effects on the liver are likely to have a pronounced effect on the organism's overall capacity in this regard. At the same time, effects on other organs can have consequences no less serious for the organism. Patients with acute hepatitis frequently have an impaired ability to oxidize drugs, with a concomitant increase in plasma half-life. Impaired oxidative metabolism has also been shown in patients with chronic hepatitis or cirrhosis. The decrease in drug metabolism that occurs in obstructive jaundice may be a consequence of the accumulation of bile salts, which are known inhibitors of some of the enzymes involved. Phase II reactions may also be affected, decreases in acetylation, glucuronidation, and a variety of esterase activities having been seen in various liver diseases. Hepatic tumors, in general, have a lower ability to metabolize foreign compounds than does normal liver tissue, although in some cases, the overall activity of tumor bearing livers may be no lower than that of controls. Kidney diseases may also affect the overall ability to handle xenobiotics because this organ is one of the main routes for elimination of xenobiotics and their metabolites. The half-lives of tolbutamide, thiopental, hexobarbital, and chloramphenicol are all prolonged in patients with renal impairment.

8.3.6 Diurnal Rhythms

Diurnal rhythms, both in CYP levels and in the susceptibility to toxicants, have been described, especially in rodents. Although such changes appear to be related to the light cycle, they may, in fact, be activity dependent because feeding and other activities in rodents are themselves markedly diurnal.

8.4 COMPARATIVE AND GENETIC EFFECTS

Comparative toxicology is the study of the variation in toxicity of exogenous chemicals toward different organisms, either of different genetic strains or of different taxonomic groups. Thus, the comparative approach can be used in the study of any aspect of toxicology, such as absorption, metabolism, mode of action, and acute or chronic effects. Most comparative data for toxic compounds exist in two areas— acute toxicity and metabolism. Until recently, comparative toxicology was largely a descriptive listing of differences between species and strains. Given the recent advances in molecular biology, comparative toxicology is entering a new phase in which the reasons for this variation will be elucidated. However, the value of the comparative approach to date can be summarized under three headings:

1. *Selective Toxicity.* Comparative toxicology is necessary for the development of selective biocides, toxic to the target organism but less toxic to other organisms, particularly humans.
2. *Experimental Models.* Comparative studies of toxic phenomena are necessary to select the most appropriate surrogate for extrapolation to humans for the testing and development of drugs and biocides and for human health risk analysis.
3. *Environmental Xenobiotic Cycles.* Different organisms in the complex ecological food webs metabolize compounds at different rates and to different products; the metabolic end products are released back to the environment, either to be further metabolized by other organisms or to exert toxic effects of their own. In addition to field studies, laboratory micro ecosystems have been developed, and chemicals and their metabolites can be followed through the plants and terrestrial and aquatic animals involved.

It might also be noted that many aspects of comparative toxicology are also comparative biochemistry in the broader sense. If the proper role of comparative biochemistry is to put evolution on a molecular basis, detoxication enzymes, like other enzymes, are suitable subjects for study. XMEs were probably essential in the early stages of animal evolution because secondary plant products, even those of low toxicity, are frequently lipophilic and, as a consequence would, in the absence of such enzymes, accumulate in lipid membranes and lipid depots. The evolution of CYP isoforms, with more than 7000 cDNA sequences known, is proving a useful tool for the study of biochemical evolution.

8.4.1 Variations among Taxonomic Groups

There are few differences in xenobiotic metabolism that are specific for large taxonomic groups. The formation of glucosides by insects and plants rather than the glucuronides of other animal groups is one of the most distinct. Although differences between species are common and are of toxicological significance, they are usually quantitative rather than qualitative in nature and tend to occur within as well as between taxonomic groups. Although the ultimate explanation of such differences must be at the level of gene expression, they are manifested at many levels, the most important of which may be summarized as follows.

In Vivo *Toxicity* An example may be seen in the toxicity of pesticides. Toxicity is a term used to describe the adverse effects of chemicals on living organisms. Available data on the toxicity of selected pesticides to rats suggest that herbicide use, in general, provides the greatest human safety factor by selectively killing plants. As the evolutionary position of the target species approaches that of humans, however, the human safety factor is narrowed considerably. Thus, as far as direct toxicity to humans and other mammals is concerned, pesticide toxicity seems to follow the progression: herbicides = fungicides < molluscicides < acaricides < nematocides < insecticides < rodenticides. However, this relationship is oversimplified because marked differences in lethality are observed when different members of a group of biocides are tested against laboratory test animals and target species.

Interspecific differences are also known for some naturally occurring poisons. Nicotine, for instance, is used as an insecticide and kills many insect pests at low doses, yet tobacco leaves constitute a normal diet for several species. As indicated earlier, most strains of rabbit eat Belladonna leaves without ill effects, whereas other mammals are easily poisoned. Natural tolerance to cyanide poisoning in millipedes and the high resistance to the powerful axonal blocking tetrodotoxin in puffer fish are examples of the tolerance of animals to the toxins they produce.

The specific organ toxicity of chemicals also exhibits wide species differences. Carbon tetrachloride, a highly potent hepatotoxicant, induces liver damage in many species, but chickens are almost unaffected by it. The eggshell thinning associated with dichlorodiphenyltrichloroethane (DDT) poisoning in birds is observed in falcons and mallard ducks, but not in gallinaceous species. Delayed neurotoxicity caused by organophosphates such as leptophos and tri-*o*-cresyl phosphate occurs in humans and chickens but not in most common laboratory mammals.

In Vivo *Metabolism* Many factors affect the rates of penetration, distribution, biotransformation, and excretion of chemicals, and thus govern their biological fate in the body. In general, the absorption of xenobiotics, their tissue distribution, and penetration across the blood–brain barrier and other barriers are dictated by their physicochemical nature and, therefore, tend to be similar in different animal species. The biological effect of a chemical depends on the concentration of its binding to tissue macromolecules. The biologic half-life is governed by the rates of metabolism and excretion and thus reflects the most important variable explaining interspecies differences in toxic response. Striking differences between species can be seen in the biologic half-lives of various drugs with humans, in general, metabolizing xenobiotics more slowly than various experimental animals. For example, phenylbutazone is metabolized slowly in humans, with a half-life averaging 3 days. In the monkey, rat, guinea pig, rabbit, dog, and horse, however, this drug is metabolized readily, with half-lives ranging between 3 and 6 h. The interdependence of metabolic rate, half-life, and pharmacological action is well illustrated in the case of hexobarbital. The duration of sleeping time is directly related to the biologic half-life and is inversely proportional to the *in vitro* degradation by liver enzymes from the respective species. Thus, mice inactivate hexobarbital readily, as reflected in a brief biologic half-life *in vivo* and short sleeping time, whereas the reverse is true in dogs.

Xenobiotics, once inside the body, undergo a series of biotransformations catalyzed by one or more XMEs acting either simultaneously or consecutively (see Chapter 6 for information on Phase I and Phase II XMEs). Because biotransformations are catalyzed by a large number of enzymes, it is to be expected that they will vary between species, with qualitative differences implying the involvement of different enzymes, quantitative differences implying differences in enzyme levels, in the extent of competing reactions, or in the efficiency of enzymes capable of reversing the reaction.

Even in the case of a xenobiotic undergoing oxidation primarily by a single reaction, there may be remarkable species differences in relative rates. Aromatic hydroxylation of aniline is an example. In this case, both *ortho* and *para* positions are susceptible to oxidative attack yielding the respective aminophenols. The biological fate of aniline has been studied in many species, and striking selectivity in hydroxylation position has been noted. These data show a trend, in that carnivores generally display a high aniline *ortho*-hydroxylase ability with a *para/ortho* ratio of ≤1, whereas rodents exhibit a striking preference for the *para* position, with a *para/ortho* ratio from 2.5 to 15. 2-Acetylaminofluorene may be metabolized in mammals by two alternative routes: N-hydroxylation, yielding the carcinogenic N-hydroxy derivative; and aromatic hydroxylation, yielding the noncarcinogenic 7-hydroxy metabolite. The former is the metabolic route in the rat, rabbit, hamster, dog, and in humans in which the parent compound is known to be carcinogenic. In contrast, the monkey carries out aromatic hydroxylation and the guinea pig appears to deacetylate the N-hydroxy derivative; thus, both escape the carcinogenic effects of this compound.

The hydrolysis of esters by esterases and of amides by amidases constitutes one of the most common enzymatic reactions of xenobiotics in humans and other animal species. Because both the number of enzymes involved in hydrolytic attack and the number of substrates for them is large, it is not surprising to observe interspecific differences in the disposition of xenobiotics due to variations in these enzymes. In mammals, the presence of a carboxylesterase that hydrolyzes malathion but is generally absent in insects explains the remarkable selectivity of this insecticide. As with esters, wide differences exist between species in the rates of hydrolysis of various amides *in vivo*. The insecticide dimethoate is susceptible to the attack of both esterases and amidases, yielding nontoxic products. In the rat and mouse, both reactions occur, whereas sheep liver contains only the amidases and that of guinea pig only the esterases. The relative rates of these degradative enzymes in insects are very low as compared with those of mammals, however, and this correlates well with the high selectivity of dimethoate.

Phase II reactions are concerned with the conjugation of primary metabolites of xenobiotics produced by Phase I reactions, and factors that alter or govern the rates of Phase II reactions may play a role in interspecific differences in xenobiotic metabolism.

Interspecific differences in the magnitude of biliary excretion of a xenobiotic excretion product largely depend on molecular weight, the presence of polar groups in the molecule, and the extent of conjugation. Conjugates with molecular weights of less than 300 are poorly excreted in bile and tend to be excreted with urine, whereas the reverse is true for those with molecular weights higher than 300. The critical molecular weight appears to vary between species, and marked species differences are noted for biliary excretion of chemicals with molecular weights of about 300. Thus, the biliary excretion of succinylsulfathiozole is 20- to 30-fold greater in the rat and the dog than in the rabbit and the guinea pig, and more than 100-fold greater than in the pig and the rhesus monkey. The cat and sheep are intermediate and excrete about 7% of the dose in the bile.

TABLE 8.2 Occurrence of Common and Unusual Conjugation Reactions

Conjugating Group	Common	Unusual
Carbohydrate	Glucuronic acid (animals)	N-acetylglucosamine (rabbits)
	Glucose (insects, plants)	Ribose (rats, mice)
Amino acids	Glycine	Glutamine (insects, humans)
	Glutathione	Ornithine (birds)
	Methionine	Arginine (ticks, spiders)
		Glycyltaurine (cats)
		Glycylglycine (cats)
		Serine (rabbits)
Acetyl	Acetyl group from acetyl-0CoA	
Formyl		Formylation (dogs, rats)
Sulfate	Sulfate group from PAPS	
Phosphate		Phosphate monoester formation (dogs, insects)

Source: Modified from Kulkarni, A. P. and E. Hodgson. Comparative toxicology. In *Introduction to Biochemical Toxicology*, eds. E. Hodgson and F. E. Guthrie, p. 115. New York: Elsevier, 1980.

Available evidence suggests some relationship between the evolutionary position of a species and its conjugation mechanisms. In humans and most mammals, the principal mechanisms involve conjugations with glucuronic acid, glycine, glutamine, and sulfate, mercapturic acid synthesis, acetylation, methylation, and thiocyanate synthesis. In some species of birds and reptiles, ornithine conjugation replaces glycine conjugation; in plants, bacteria, and insects, conjugation with glucose instead of glucuronic acid results in the formation of glucosides. In addition to these predominant reactions, certain other conjugative processes are found involving specific compounds in only a few species. These reactions include conjugation with phosphate, taurine, N-acetylglucosamine, ribose, glycyltaurine, serine, arginine, and formic acids (Table 8.2).

From the standpoint of evolution, similarity might be expected between humans and other primate species as opposed to the nonprimates. This phylogenic relationship is apparent from the relative importance of glycine and glutamine in the conjugation of arylacetic acids. The conjugating agent in humans is exclusively glutamine, and the same is essentially true with Old World monkeys while New World monkeys use both the glycine and glutamine pathways. Most nonprimates and lower primates carry out glycine conjugation selectively. A similar evolutionary trend is also observed in the N-glucuronidation of sulfadimethoxine and in the aromatization of quinic acid; both reactions occur extensively in humans, and their importance decreases with increasing evolutionary divergence from humans.

Defective operation of Phase II reactions usually causes a striking species difference in the disposition pattern of a xenobiotic. The origin of such species variations is usually either the absence or a low level of the enzyme(s) in question and/or its cofactors. Glucuronide synthesis is one of the most common detoxication mechanisms in most mammalian species. The cat and closely related species have a defective glucuronide-forming system, however. Although cats form little or no glucuronide from *o*-aminophenol, phenol, *p*-nitrophenol, 2-amino-4-nitrophenol, 1- or 2-naphthol, and morphine, they readily form glucuronides from phenolphthalein, bilirubin, thyroxine, and certain steroids. Recently, polymorphisms of UDP-glucuronosyltransferase have been demonstrated in rat and guinea pig liver preparations; thus, defective glucuronidation in the cat is probably related to the absence of the appropriate transferase rather than that of the active intermediate, UDPGA or UDP glucose dehydrogenase, which converts UDP glucose into UDPGA.

Studies on the metabolic fate of phenol in several species have indicated that four urinary products are excreted. Although extensive phenol metabolism takes place in most species, the relative proportions of each metabolite produced varies from species to species. In contrast to the cat, which selectively forms sulfate conjugates, the pig excretes phenol exclusively as the glucuronide. This defect in sulfate conjugation in the pig is restricted to only a few substrates, however, and may be due to the lack of a specific phenyl sulfotransferase because the formation of substantial amounts of the sulfate conjugate of 1-naphthol clearly indicates the occurrence of other forms of sulfotransferases.

In Vitro *Metabolism* Numerous variables simultaneously modulate the *in vivo* metabolism of xenobiotics; therefore, their relative importance cannot be studied easily. This problem is alleviated to some extent by *in vitro* studies of the underlying enzymatic mechanisms responsible for qualitative and quantitative species differences. Quantitative differences may be related directly to the absolute amount of active enzyme present and the affinity and specificity of the enzyme toward the substrate in question. Because many other factors alter enzymatic rates *in vitro*, caution must be exercised in interpreting data in terms of species variation. In particular, enzymes are often sensitive to the experimental conditions used in their preparation. Because this sensitivity varies from one enzyme to another, their relative effectiveness for a particular reaction can be sometimes miscalculated.

Species variation in the oxidation of xenobiotics, in general, is quantitative whereas qualitative differences, such as the apparent total lack of parathion oxidation by lobster hepatopancreas microsomes, are seldom observed. Although the amount of CYP or the activity of NADPH-CYP reductase seems to be related to the oxidation of certain substrates, this explanation is not always satisfactory because the absolute amount of CYP is not necessarily the rate-limiting characteristic. It is clear that there are multiple forms of CYP isoforms in each species, and that these forms differ from one species to another. Presumably, both quantitative and qualitative variations in xenobiotic metabolism depend in variations on the particular isoforms expressed and the extent of this expression.

Reductive reactions, like oxidation reactions, are carried out at different rates by enzyme preparations from different species. Microsomes from mammalian liver are 18 times or more higher in azoreductase activity and more than 20 times higher in nitroreductase activity than those from fish liver. Although relatively inactive in nitroreductase, fish can reduce the nitro group of parathion, suggesting multiple forms of reductase enzymes.

Hydration of epoxides catalyzed by epoxide hydrolase is involved in both detoxication and intoxication reactions. With high concentrations of styrene oxide as a substrate, the relative activity of hepatic microsomal epoxide hydrolase in several animal species is rhesus monkey > human = guinea pig > rabbit > rat > mouse. With some substrates, such as epoxidized lipids, the cytosolic hydrolase may be much more important than the microsomal enzyme.

The toxicity of the organophosphorus insecticide dimethoate depends on the rate at which it is hydrolyzed *in vivo*. This toxicant undergoes two main metabolic detoxication reactions, one catalyzed by an esterase and the other by an amidase. Although rat and mouse liver carry out both reactions, only the amidase occurs in sheep liver, and the esterase in guinea pig liver. The ability of liver preparations from different animal species to degrade dimethoate is as follows: rabbit > sheep > dog > rat > cattle > hen > guinea pig > mouse > pig, these rates being roughly inversely proportioned to the toxicity of dimethoate to the same species. Insects degrade this compound much more slowly than do mammals and hence are highly susceptible to dimethoate.

Glutathione S-transferase in liver cytosol from different animal species also shows a wide variation in activity. Activity is low in humans, whereas the mouse and guinea pig appear to be more efficient than other species. The ability of the guinea pig to form the initial glutathione conjugate contrasts with its inability to readily N-acetylate cysteine conjugates; consequently, mercapturic acid excretion is low in guinea pigs.

8.4.2 Selectivity

Selective toxic agents have been developed to protect crops, animals of economic importance, and humans from the vagaries of pests, parasites, and pathogens. Such selectivity is conferred primarily through distribution and comparative biochemistry.

Selectivity through differences in uptake permits the use of an agent toxic to both target and nontarget cells provided that lethal concentrations accumulate only in target cells, leaving nontarget cells unharmed. An example is the accumulation of tetracycline by bacteria, but not by mammalian cells, the result being drastic inhibition of protein synthesis in the bacteria, leading to death.

Certain schistosome worms are parasitic in humans and their selective destruction by antimony is accounted for by the differential sensitivity of phosphofructokinase in the two species, the enzyme from schistosomes being more susceptible to inhibition by antimony than is the mammalian enzyme.

Sometimes, both target and nontarget species metabolize a xenobiotic by the same pathways, but differences in rate determine selectivity. Malathion, a selective insecticide, is metabolically activated by CYPs to the cholinesterase inhibitor malaoxon. In addition to this activation reaction, several detoxication reactions also occur. Carboxylesterase hydrolyzes malathion to form the monoacid, phosphatases hydrolyze the P–O–C linkages to yield nontoxic products, and gluthathione S-alkyltransferase converts malathion to desmethylmalathion. Although all of these reactions occur in both insects and mammals, activation is rapid in both insects and mammals, whereas hydrolysis to the monoacid is rapid in mammals but slow in insects. As a result, malaoxon accumulates in insects but not in mammals, resulting in selective toxicity.

A few examples are also available, in which the lack of a specific enzyme in some cells in the human body has enabled the development of a therapeutic agent. For example, guanine deaminase is absent from the cells of certain cancers but is abundant in healthy tissue; as a result, 8-azaguanine can be used therapeutically.

Distinct differences in cells with regard to the presence or absence of target structures or metabolic processes also offer opportunities for selectivity. Herbicides such as phenylureas, simazine, and so on, block the Hill reaction in chloroplasts, thereby killing plants without harm to animals. This is not always the case, because paraquat, which blocks photosynthetic reactions in plants, is a pulmonary toxicant in mammals, due apparently to analogous free-radical reactions involving enzymes different from those involved in photosynthesis.

8.4.3 Genetic Differences

Just as the xenobiotic-metabolizing ability in different animal species seems to be related to evolutionary development and therefore to different genetic constitution, different strains within a species may differ from one another in their ability to metabolize xenobiotics. One reason for differences among strains is due to the fact that many genes are polymorphic or exist in multiple forms. A polymorphism is defined as an inherited monogenetic trait that exists in the population in at least two genotypes (two or more stable alleles) and is stably inherited. They arise as the result of a mutational event and generally result in an altered gene product. In humans, many polymorphisms are somewhat race specific, arising with greater frequency in one race than in another. Observed differences between strains of rats and mice as described below may also be the result of gene polymorphisms.

The effects of genetic polymorphisms can be observed at several different levels, as discussed below:

In Vivo *Toxicity*   The toxicity of organic compounds has been found to vary between different strains of laboratory animals. For example, mouse strain C3H is resistant to histamine, the LD_{50} being 1523 mg/kg in C3H/Jax mice as compared with 230 in Swiss/ICR mice; that is, the animals of the former strain are 6.6 times less susceptible to the effects of histamine. Striking differences in the toxicity of thiourea, a compound used in the treatment of hyperthyroidism, are seen in different strains of the Norway rat. Harvard rats were 11 times more resistant, and wild Norway rats were 335 times more resistant than were rats of the Hopkins strain.

The development of strains resistant to insecticides is an extremely widespread phenomenon that is known to have occurred in several hundred species of insects and mites, and resistance of up to several hundredfold has been noted. The different biochemical and genetic factors involved have been studied extensively and well characterized. Relatively few vertebrate species are known to have developed pesticide resistance and the level of resistance in vertebrates is low compared to that often found in insects. Susceptible and resistant strains of pine voles exhibit a 7.4-fold difference in endrin toxicity. Similarly, pine mice of a strain resistant to endrin were reported to be 12-fold more tolerant than a susceptible strain. Other examples include the occurrence of organochlorine insecticide-resistant and susceptible strains of mosquito fish, and resistance to Belladonna in certain rabbit strains.

Several genetic polymorphisms have been described and characterized with respect to CYP enzymes. The first and best-known example involves CYP2D6. In the course of a clinical trial for debrisoquine, a potential drug for use in lowering blood pressure, Dr. Robert Smith, one of the investigators who used himself as a volunteer, developed severe orthostatic hypotension with blood pressure dropping to 70/50. The effects of the drug persisted for 2 days, while in other volunteers no adverse effects were noted. Urine analysis demonstrated that in Dr. Smith, debrisoquine was excreted unchanged, while in the other volunteers the primary metabolite was 4-hydroxy debrisoquine. Subsequent studies demonstrated that CYP2D6 was responsible for the formation of 4-hydroxy debrisoquine and that the polymorphic form of CYP2D6 is prevalent in Caucasians and African Americans, in which approximately 7% are poor metabolizers. In Asian populations, the frequency of poor metabolizers is only 1%.

Another well-known genetic polymorphism has been described in the metabolism of drugs such as isoniazid. "Slow acetylators" are homozygous for a recessive gene; this is believed to lead to the lack of the hepatic enzyme acetyltransferase, which in normal homozygotes or heterozygotes (rapid acetylators) acetylates isoniazid as a step in the metabolism of this drug. This effect is seen also in humans, the gene for slow acetylation showing marked differences in distribution between different human populations. It is very low in Eskimos and Japanese, with 80–90% of these populations being rapid acetylators, whereas 40–60% of Blacks and some European populations are rapid acetylators. Rapid acetylators often develop symptoms of hepatotoxicity and polyneuritis at the dosage necessary to maintain therapeutic blood levels of isoniazid.

Many other significant polymorphisms in XMEs have been described including those for several CYP genes, alcohol and aldehyde dehydrogenases, epoxide hydrolase, and paraoxonase. One interesting polymorphism affecting metabolism of dietary trimethylamines involves FMO3. Individuals with FMO3 polymorphisms have a condition known as fish odor syndrome, or trimethylaminurea. Individuals with this syndrome exhibit an objectionable body odor resembling rotting fish due to their inability to N-oxidize trimethylamine which are found in many foods including meat, eggs, and soybeans. This syndrome often leads to social isolation, clinical depression, and even suicide. Other toxicological implications of this polymorphism are still not known.

Metabolite Production   Strain variations with respect to hexobarbital are often dependant on its degradation rate. For example, male mice of the AL/N strain are long sleepers, and this trait is correlated with slow inactivation of the drug. The reverse is true in CFW/N mice, which have short sleeping time due to rapid hexobarbital oxidation. This close relationship is further evidenced by the fact that the level of brain hexobarbital at awakening is essentially the same in all stains. Similar strain differences have been reported for zoxazolamine paralysis in mice.

Studies on the induction of aryl hydrocarbon hydroxylase by 3-methylcholanthrene have revealed several responsive and nonresponsive mouse strains, and it is now well established that the induction of this enzyme is controlled by a single gene. In the accepted nomenclature, aryl hydrocarbon (Ah)[b] represents the allele for responsiveness, whereas Ah[d] denotes the allele for nonresponsiveness.

In rats, both age and gender seem to influence strain variation in xenobiotic metabolism. Male rats exhibit about twofold variation between strains in hexobarbital metabolism, whereas female rats may display up to sixfold variation. In either gender, the extent of variations depends on age. The ability to metabolize hexobarbital is related to the metabolism of other substrates, and the interstrain differences are maintained.

A well-known interstrain difference in Phase II reactions is that of glucuronidation in Gunn rats. This is a mutant strain of Wistar rats that is characterized by a severe, genetically determined defect of bilirubin glucuronidation. Their ability to glucuronidate o-aminophenol, o-aminobenzoic acid, and a number of other substrates is also partially defective. This deficiency does not seem to be related to an inability to form UDPGA but rather to the lack of a specific UDP-glucuronosyltransferase. It has been demonstrated that Gunn rats can conjugate aniline by N-glucuronidation and can form the O-glucuronide of p-nitrophenol.

Rabbit strains may exhibit up to 20-fold variation, particularly in the case of hexobarbital, amphetamine, and aminopyrine metabolism. Relatively smaller differences between strains occur with chlorpromazine metabolism. Wild rabbits and California rabbits display the greatest differences from other rabbit strains in hepatic drug metabolism.

Enzyme Differences Variation in the nature and amount of constitutively expressed microsomal CYPs have not been studied extensively in different strains of the same vertebrate, although after induction, it has been shown that in addition to quantitative differences in the amount of CYP in different strains of mice, there may also be a qualitative difference in the CYP isoforms induced. (See Section 8.5.2, Induction.)

8.5 CHEMICAL EFFECTS

With regard to both logistics and scientific philosophy, the study of the metabolism and toxicity of xenobiotics must be initiated by considering single compounds. Unfortunately, humans and other living organisms are not exposed in this way; rather, they are exposed to many xenobiotics simultaneously, involving different portals of entry, modes of action, and metabolic pathways. An estimate of the number of chemicals in use in the United States is given in Table 8.3. Because it bears directly on the problem of toxicity-related interaction between different

TABLE 8.3   Estimates of the Number of Chemicals in Use in the United States

Number	Type	Source of Estimate[a]
1500	Active ingredients of pesticides	EPA
4000	Active ingredients of drugs	FDA
2000	Drug additives (preservatives, stabilizers, etc.)	FDA
2500	Food additives (nutritional value)	FDA
3000	Food additives (preservatives, stabilizers, etc.)	FDA
50,000	Additional chemicals in common use	EPA

[a] EPA, Environmental Protection Agency; FDA, Food and Drug Administration.

xenobiotics, the effect of chemicals on the metabolism of other exogenous compounds is one of the more important areas of biochemical toxicology.

Xenobiotics, in addition to serving as substrates for a number of enzymes, may also serve as inhibitors or inducers of these or other enzymes. Furthermore, many examples are known of compounds that first inhibit and subsequently induce enzymes such as the microsomal monooxygenases. The situation is even further complicated by the fact that although some substances have an inherent toxicity and are detoxified in the body, others without inherent toxicity can be metabolically activated to potent toxicants. The following examples are illustrative of the situations that might occur involving two compounds:

- Compound A, without inherent toxicity, is metabolized to a potent toxicant. In the presence of an inhibitor of its metabolism, there would be a reduction in toxic effect or, in the presence of an inducer, compound A would appear to be more toxic.
- Compound B, a toxicant, is metabolically detoxified. In the presence of an inhibitor of the detoxifying enzymes, there would be an increase in the toxic effect or, in the presence of an inducer, compound B would appear to be less toxic.

In addition, the toxicity of the inhibitor or inducer, as well as the time dependence of the effect, must also be considered because, as mentioned, many xenobiotics that are initially enzyme inhibitors ultimately become inducers. As the number of patients receiving multiple drugs has increased dramatically, there has been an increase in reports of drug–drug interactions resulting in adverse effects that has lead in some cases to the removal of drugs from the market. The U.S. Food and Drug Administration (FDA) has established guidelines for determining the ability of a drug to induce or inhibit xenobiotic enzymes and transporters. In the following sections, inhibition and induction will be discussed.

8.5.1 Inhibition

As previously indicated, inhibition of XMEs can cause either an increase or a decrease in toxicity. Several well-known inhibitors of such enzymes are shown in Figure 8.5 and are discussed in this section. Inhibitory effects can be demonstrated in a number of ways at different organizational levels.

Types of Inhibition: Experimental Demonstration

In Vivo *Symptoms* The measurement of the effect of an inhibitor on the duration of action of a drug *in vivo* was formerly a common method of demonstrating its action. These methods are open to criticism, however, because effects on duration of action can be mediated by systems other than those involved in the metabolism of the drug. Furthermore, they cannot be used for inhibitors that have pharmacological activity similar or opposite to the compound being used. At the present time, as a consequence of the availability of single expressed isoforms for direct studies of inhibitory mechanisms, these methods are used much less often although they are still valuable for human health risk analysis.

$(C_6H_5)_2CCO(CH_2)_2N(C_2H_5)_2$
with $\overset{O}{\underset{\|}{}}$ above the first C, and C_3H_7 below

SKF-525A [P450]
2-(Diethylamino)ethyl-
2,2-diphenylpentanoate

Piperonyl Butoxide [P450]
3,4-Methylenedioxy-6-propylbenzyl
N-butyl diethyleneglycol ether
(ring substituents C_3H_7 and $CH_2(OC_2H_4)_2OC_4H_9$)

$CH_2{=}CHCH_2$
$CHCNH_2$ with $\overset{O}{\underset{\|}{}}$
$(CH_3)_2CH$

Allylisopropylacetamide
[P450]

1-Aminobenzotriazole (1-ABT)
[P450]
(N, N, N triazole ring with NH_2)

C_6H_5
$PO{-}{\bigcirc}{-}NO_2$ with $\overset{O}{\underset{\|}{}}$
C_2H_5O

EPN [esterases]
O-Ethyl-O-p-nitrophenyl
phenylphosphonothioate

Metyrapone
[P450]
$CC(CH_3)_2$ with $\overset{O}{\underset{\|}{}}$, between two pyridine rings

$C_2H_5OCCH{=}CHCOC_2H_5$ with $\overset{O}{\underset{\|}{}}\ \overset{O}{\underset{\|}{}}$

Diethyl maleate
[glutathione S-transferase]

$(C_2H_5)_2NCSSCN(C_2H_5)_2$ with $\overset{S}{\underset{\|}{}}\ \overset{S}{\underset{\|}{}}$

Disulfiram (Antabuse)
[aldehyde dehydrogenase]

Figure 8.5 Some common inhibitors of xenobiotic-metabolizing enzymes.

Previously, the most used and most reliable of these tests involved the measurement of effects on the hexobarbital or pentobarbital sleeping time and the zoxazolamine paralysis time. Both of these drugs are fairly rapidly deactivated by the hepatic microsomal monooxygenase system; thus, inhibitors of this system prolong their action. For example, treatment of mice with chloramphenicol 0.5–1.0 h before pentobarbital treatment prolongs the duration of the pentobarbital sleeping time in a dose-related manner; it is effective at low doses (<5 mg/kg) and has a greater than 10-fold effect at high doses (100–200 mg/kg). The well-known inhibitor of drug metabolism, SKF-525A, causes an increase in both hexobarbital sleeping time and zoxazolamine paralysis time in rats and mice, as do the insecticide synergists piperonyl butoxide and tropital, the optimum pretreatment time being about 0.5 h before the narcotic is given. In the case of activation reactions, such as the activation of the insecticide azinphosmethyl to its potent anticholinesterase oxon derivative, a decrease in toxicity is apparent when rats are pretreated with the CYP inhibitor SkF-525A.

Cocarcinogenicity may also be an expression of inhibition of a detoxication reaction, as in the case of the cocarcinogenicity of piperonyl butoxide, a CYP inhibitor, and the carcinogens, freons 112 and 113.

Distribution and Blood Levels   Treatment of an animal with an inhibitor of xenobiotic metabolism may cause changes in the blood levels of an unmetabolized toxicant and/or its metabolites. This procedure may be used in the investigation of the inhibition of detoxication pathways; it has the advantage over *in vitro* methods of yielding results of direct physiological or toxicological interest because it is carried out in the intact animal. For example, if animals are first treated with either SKF-525A, glutethimide, or chlorcyclizine, followed in 1 h or less by pentobarbital, it can be shown that the serum level of pentobarbital is considerably higher in treated animals than in controls within 1 h of its injection. Moreover, the time sequence of the effects can be followed in individual animals, a factor of importance when inhibition is followed by induction—a not uncommon event.

Effects on Metabolism In Vivo   A further refinement of the previous technique is to determine the effect of an inhibitor on the overall metabolism of a xenobiotic *in vivo*, usually by following the appearance of metabolites in the urine and/or feces. In some cases, the appearance of metabolites in the blood or tissue may also be followed. Again, the use of the intact animal has practical advantages over *in vitro* methods, although little is revealed about the mechanisms involved.

Studies of antipyrine metabolism may be used to illustrate the effect of inhibition on metabolism *in vivo*; in addition, these studies have demonstrated variation between species in the inhibition of the metabolism of xenobiotics. In the rat, a dose of piperonyl butoxide of at least 100 mg/kg was necessary to inhibit antipyrine metabolism, whereas in the mouse, a single intraperitoneal (IP) or oral dose of 1 mg/kg produced a significant inhibition. In humans, an oral dose of 0.71 mg/kg had no discernible effect on the metabolism of antipyrine.

Disulfiram (Antabuse(Odyssey Pharmaceuticals, Florham Park, NJ, USA)) inhibits aldehyde dehydrogenase irreversibly, causing an increase in the level of acetaldehyde, formed from ethanol by the enzyme alcohol dehydrogenase. This results in nausea, vomiting, and other symptoms in the human—hence its use as a deterrent in alcoholism. Inhibition by disulfiram appears to be irreversible, the level returning to normal only as a result of protein synthesis.

Use of specific metabolic enzyme inhibitors may often provide valuable information with respect to the metabolism of a particular drug. For example, quinidine is a potent and selective inhibitor of CYP2D6. This drug has been used in clinical studies as a pharmacological tool to mimic the lack of CYP2D6 in humans. By demonstrating that quinidine substantially slows the metabolism of trimipramine (a tricyclic antidepressant), investigators have implicated CYP2D6 in its metabolism.

Effects on In Vitro *Metabolism Following* In Vivo *Treatment*   This method of demonstrating inhibition is of variable utility. The preparation of enzymes from animal tissues usually involves considerable dilution with the preparative medium during homogenization, centrifugation, and resuspension. As a result, inhibitors not tightly bound to the enzyme in question are lost, either in whole or in part, during the preparative processes. Therefore, negative results can have little utility because failure to inhibit and loss of the inhibitor give identical results. Positive results, however, not only indicate that the compound administered is an inhibitor but also provide a clear indication of excellent binding to the enzyme, most probably due to the formation of a covalent or slowly reversible inhibitory complex.

The inhibition of esterases following treatment of the animal with organophosphorus compounds, such as paraoxon, is a good example, because the phosphorylated enzyme is stable and is still inhibited after the preparative procedures. Inhibition by carbamates, however, is greatly reduced by the same procedures, because the carbamylated enzyme is unstable and, in addition, the residual carbamate is highly diluted.

Microsomal monooxygenase inhibitors that form stable inhibitory complexes with CYP, such as SKF-525A, piperonyl butoxide and other methylenedioxyphenyl compounds, and amphetamine and its derivatives, can be readily investigated in this way because the microsomes isolated from pretreated animals have a reduced capacity to oxidize many xenobiotics.

Another form of chemical interaction, resulting from inhibition *in vivo*, that can then be demonstrated *in vitro*, involves those xenobiotics that function by causing destruction of the enzyme in question, so-called suicide substrates. Exposure of rats to vinyl chloride results in a loss of CYP and a corresponding reduction in the capacity of microsomes subsequently isolated to metabolize foreign compounds. Allyl isopropylacetamide and other allyl compounds have long been known to have a similar effect.

In Vitro *Effects* *In vitro* measurement of the effect of one xenobiotic on the metabolism of another is by far the most common type of investigation of interactions involving inhibition. Although it is the most useful method for the study of inhibitory mechanisms, particularly when purified enzymes are used, it is of more limited utility in assessing the toxicological implications for the intact animal. The principal reason for this is that *in vitro* measurement does not assess the effects of factors that affect absorption, distribution, and prior metabolism, all of which occur before the inhibitory event under consideration. The use of human hepatocytes to study CYP inhibition can take into account some, but not all of these confounding factors, and has proven to provide data that is in good agreement with microsomal studies.

Although the kinetics of inhibition of XMEs can be investigated in the same ways as any other enzyme mechanism, a number of problems arise that may decrease the value of this type of investigation. They include the following:

- Inhibition of microsomal CYP-dependent oxidations and CYP isoforms have been investigated many times. However, as a result of using methods on particulate systems that were developed for single soluble enzymes, Lineweaver–Burk or other reciprocal plots are frequently curvilinear, and the same reaction may appear to have quite a different characteristic from laboratory to laboratory, species to species, and organ to organ.

- The nonspecific binding of substrate and/or inhibitor to membrane components is a further complicating factor affecting inhibition kinetics.

- Both substrates and inhibitors are frequently lipophilic, with low solubility in aqueous media.

- XMEs commonly exist in multiple forms (e.g., glutathione S-transferases and CYPs). These isoforms are all relatively nonspecific but differ from one another in the relative affinities of the different substrates.

The primary considerations in studies of inhibition mechanisms are reversibility and selectivity. The inhibition kinetics of reversible inhibition give considerable insight into the reaction mechanisms of enzymes and, for that reason, have been well studied. In general, reversible inhibition involves no covalent binding, occurs rapidly, and can be reversed by dialysis or, more rapidly, by dilution. Reversible inhibition is usually divided into competitive inhibition, uncompetitive inhibition, and noncompetitive inhibition. Because these types are not rigidly separated, many intermediate classes have been described.

Competitive inhibition is usually caused by two substrates competing for the same active site. Following classic enzyme kinetics, there should be a change in the apparent Michaelis constant (K_m), but not the maximum reaction rate (V_{max}). In microsomal monooxygenase reaction, type I ligands, which often appear to bind as substrates but do not bind to the heme iron, might be expected to be competitive inhibitors, and this frequently appears to be the case. Examples are the inhibition of the *O*-demethylation of *p*-nitroanisole by aminopyrine, aldrin epoxidation by dihydroaldrin, and *N*-demethylation of aminopyrene by nicotinamide. More recently, some of the polychlorinated biphenyls (PCBs), notably dichlorbiphenyl have been shown to have a high affinity as type I ligands for rabbit liver CYP and to be competitive inhibitors of the *O*-demethylation of *p*-nitroanisole.

Uncompetitive inhibition has seldom been reported in studies of xenobiotic metabolism. It occurs when an inhibitor interacts with an enzyme–substrate complex but cannot interact with free enzyme. Both K_m and V_{max} change by the same ratio, giving rise to a family of parallel lines in a Lineweaver–Burk plot.

Noncompetitive inhibitors can bind to both the enzyme and enzyme–substrate complex to form either an enzyme–inhibitor complex or an enzyme–inhibitor–substrate complex. The net result is a decrease in V_{max} but no change in K_m. Metyrapone (Figure 8.5), a well-known inhibitor of monooxygenase reactions, can also, under some circumstances, stimulate metabolism *in vitro*. In either case, the effect is noncompetitive, in that the K_m does not change, whereas V_{max} does, decreasing in the case of inhibition and increasing in the case of stimulation.

Irreversible inhibition, which is much more important toxicologically, can arise from various causes. In most cases, the formation of covalent or other stable bonds is involved or, more rarely, the disruption of the enzyme structure. In either of these cases, the effect cannot be readily reversed *in vitro* by either dialysis or dilution. The formation of stable inhibitory complexes may involve the prior metabolic formation of a reactive intermediate that then interacts with the enzyme, giving rise to the terms "mechanism-based inhibitor" or "suicide substrate." An excellent example of this type of inhibition is the effect of the insecticide synergist piperonyl butoxide (Figure 8.5) on hepatic microsomal monooxygenase activity. This methylenedioxyphenyl compound can form a stable inhibitory complex that blocks CO binding to CYP and also prevents substrate oxidation. This complex results from the formation of a reactive intermediate, which is shown by the fact that the type of inhibition changes from competitive to irreversible as metabolism, in the presence of NADPH and oxygen, proceeds. It appears probable that the metabolite in question is a carbene formed spontaneously by elimination of water following hydroxylation of the methylene carbon by the cytochrome (see Figure 6.8 for metabolism of methylenedioxyphenyl compounds). Piperonyl butoxide inhibits the *in vitro* metabolism of many substrates of the monooxygenase system, including aldrin, ethylmorphine, aniline, and aminopyrine, as well as carbaryl, biphenyl, hexobarbital, *p*-nitroanisole, and many others. Although most of the studies carried out on piperonyl butoxide have involved rat or mouse liver microsomes, they have also been carried out on pig, rabbit, and carp liver microsomes, and in various preparations from houseflies, cockroaches, and other insects. Certain classes of monooxygenase inhibitors, in addition to methylenedioxyphenyl compounds, are now known to form "metabolite inhibitory complexes," including amphetamine and its derivatives, and SKF-525A and its derivatives.

The inhibition of the carboxylesterase that hydrolyzes malathion by organophosphorus compounds, such as EPN, is a further example of xenobiotic interaction resulting from irreversible inhibition because in this case, the enzyme is phosphorylated by the inhibitor.

Another class of irreversible inhibitors of toxicological significance consists of those compounds that bring about the destruction of the XMEs, hence such inhibitors are also designated "suicide substrates." The drug allylisopropylacetamide (Figure 8.5), as well as other allyl compounds, has long been known to cause the breakdown of CYP and the resultant release of heme. More recently, the hepatocarcinogen vinyl chloride has also been shown to have a similar effect, probably also mediated through the generation of a highly reactive intermediate. Much information has accumulated since the mid 1970s on the mode of action of the hepatotoxicant carbon tetrachloride, which affects a number of irreversible changes in both liver protein structure, such as urea, detergents, strong acids, and so on, are probably of significance only in *in vitro* experiments.

The importance of irreversible inhibition as opposed to competitive inhibition is illustrated in Table 8.4, which summarizes the inhibition of human Phase I reactions of both xenobiotic and endogenous metabolites by pesticides. Apart from inhibition of estradiol metabolism by pyrethroids and carbamates, all other examples are the result of mechanism-based irreversible inhibition.

Synergism and Potentiation   The terms synergism and potentiation have been used and defined in various ways but, in any case, they involve a toxicity that is greater when two compounds are given simultaneously or are in close sequence than would be expected from a consideration of the toxicities of the compounds given alone. In medicine, synergism generally refers to the ability of two drugs to have a combined greater effect than the sum of the individual effects.

In an attempt to resolve the semantic difficulties and to make uniform the use of these terms, it is suggested that insofar as toxic effects are concerned, the terms be used as follows: *Both synergism and potentiation involve toxicity greater than would be expected from the toxicities of the compounds administered separately, but in the case of synergism, one compound has little or no intrinsic toxicity when administered alone, whereas in the case of potentiation, both compounds have appreciable toxicity when administered alone. It is further suggested that no special term is needed for simple additive toxicity of two or more compounds.*

An example of synergism has already been mentioned. Piperonyl butoxide, sesamex, and related compounds increase the toxicity of insecticides to insects by inhibiting insect CYP. Other insecticide synergists that interact with CYP include aryloxyalkylamines such as SKF-525A, Lilly 18947, and their derivatives; compounds containing acetylenic bonds such as aryl-2-propynyl phosphate esters

TABLE 8.4 Inhibition of Human Hepatic Phase I Metabolism by Pesticides

Substrate Reference	Enzyme	Inhibitor(s)
Xenobiotic substrates		
Carbaryl	Liver microsomes	Chlorpyrifos
Carbaryl	CYP2B6	Chlorpyrifos
Carbofuran	Liver microsomes	Chlorpyrifos
DEET	Liver microsomes	Chlorpyrifos
Fipronil	Liver microsomes	Chlorpyrifos
Fipronil	CYP3A4	Chlorpyrifos
Imipramine	Liver microsomes	Chlorpyrifos, azinphosphos methyl, parathion
Imipramine	CYPs 1A2, 3A4, 2C19	Chlorpyrifos, azinphosphos methyl, parathion
Nonane	Liver microsomes	Chlorpyrifos
Nonane	CYP2B6	Chlorpyrifos
Permethrin	Liver cytosol	Chlorpyrifos oxon, carbaryl
Endogenous substrates		
Estradiol	Liver microsomes	Chlorpyrifos, fonofos, carbaryl, naphthalene
Estradiol	CYP1A2	Chlorpyrifos, fonofos, carbaryl, naphthalene
Estradiol	CYP3A4	Chlorpyrifos, fonofos, deltamethrin, permethrin
Testosterone	Liver microsomes	Chlorpyrifos, phorate, fonofos
Testosterone	CYP3A4	Chlorpyrifos

containing propynyl functions; phosphorothionates; benzothiadiazoles; and some imidazole derivatives.

Antagonism In toxicology, antagonism may be defined as that situation in which the toxicity of two or more compounds administered together or sequentially is less than would be expected from a consideration of their toxicities when administered individually. Strictly speaking, this definition includes those cases in which the lowered toxicity results from induction of detoxifying enzymes (this situation is considered separately in Section 8.5.2). A part from the convenience of treating such antagonistic phenomena together with the other aspects of induction, they are frequently considered separately because of the significant time that must elapse between treatment with the inducer and subsequent treatment with the toxicant. The reduction of hexobarbital sleeping time and the reduction of zoxazolamine paralysis time by prior treatment with phenobarbital to induce drug-metabolizing enzymes are obvious examples of such induction effects at the acute level of drug action, whereas protection from the carcinogenic action of benzo(a)pyrene, aflatoxin B1, and diethylnitrosamine by phenobarbital treatment are examples of inductive effects at the level of chronic toxicity. In the latter case, the CYP isoforms induced by phenobarbital metabolize the chemical to less toxic metabolites.

Antagonism not involving induction is a phenomenon often seen at a marginal level of detection and is consequently both difficult to explain and is of marginal significance. In addition, several different types of antagonism of importance to toxicology that do not involve xenobiotic metabolism are known but are not appropriate for discussion in this chapter. They include competition for receptor sites, such as the competition between CO and O_2 in CO poisoning or situations in which one toxicant combines nonenzymatically with another to reduce its toxic effects, such as in the chelation of metal ions. Physiological antagonism, in which two agonists act on the same physiological system but produce opposite effects, is also of importance.

8.5.2 Induction

In the early 1960s, during investigations on the *N*-demethylation of aminoazo dyes, it was observed that pretreatment of mammals with the substrate or, more remarkably, with other xenobiotics, caused an increase in the ability of the animal to metabolize these dyes. It was subsequently shown that this effect was due to an increase in the microsomal enzymes involved. A symposium in 1965 and a landmark review by Conney in 1967 established the importance of induction in xenobiotic interactions. Since then, it has become clear that this phenomenon is widespread and nonspecific and involves enhanced gene transcription due to activation of receptors. Several hundred compounds of diverse chemical structure have been shown to induce monooxygenases and other enzymes. These compounds include drugs, pesticides, hydrocarbons, industrial chemicals, and many others; the only obvious common denominator is that they are organic and lipophilic. It has also become apparent that, even though all inducers do not have the same effects, the effects tend to be nonspecific to the extent that any single chemical can induce more than one xenobiotic-metabolizing gene. Other enzyme genes can also be induced, such as glutathione *S*-transferase, UDP-glucuronosyltransferases (UGTs), and epoxide hydrolases.

Specificity of Monooxygenase Induction   The majority of studies involving *CYP* induction have been conducted in mammals. Mammals have at least 17 distinct *CYP* families, coding for as many as 50–60 individual *CYP* genes in any given species. Many of these *CYP* families are fairly specific for endogenous metabolic pathways and are not typically involved in metabolism of foreign chemicals. As discussed in Chapter 6, *CYP* families 1–4 are the predominant families involved in xenobiotic metabolism. These CYP enzyme families are also known for their ability to respond to xenobiotic challenges by increasing their protein levels. Many of the genes within families 1–4 are transcriptionally activated through one of four receptor-dependent mechanisms. Others, such as *CYP2E1* are regulated at the level of mRNA stabilization and/or protein stabilization. These mechanisms of regulation are discussed in Section 8.5.2.

Inducers of CYP genes act by similar mechanisms and are exemplified by 2,3,7,8-tetrachlorodibenzo-*p*-dioxin (TCDD) (inducer of *CYP1A1*), phenobarbital (inducer of the *CYP2B* and *CYP3A* families), rifampicin (inducer of *CYP3A* and *2C* families), and ethanol (inducer of *CYP2E1*). Inducers of the phenobarbital type tend to share few structural features other than lipophilicity, while TCDD-like inducers are primarily polycyclic hydrocarbons. Other inducers, such as ethanol, dexamethasone, and clofibrate are more specific. Many inducers require either fairly high dose levels or repeated dosing to be effective, frequently >10 mg/kg and some as high as 100–200 mg/kg. Some insecticides, however, such as mirex, can induce at dose levels as low as 1 mg/kg, while the most potent inducer known, TCDD, is effective at 1 µg/kg in some species.

In the liver, phenobarbital-type inducers cause a marked proliferation of the smooth endoplasmic reticulum as well as an increase in the amount of CYP content. Often, these changes are sufficient to result in significant liver weight increases. Phenobarbital induction induces a wide range of oxidative activities including O-demethylation of *p*-nitroanisole, N-demethylation of benzphetamine, pentobarbital hydroxylation, and aldrin hydroxylation. *CYP* gene families that are primarily induced by phenobarbital and phenobarbital-like inducers include *CYP2B*, *CYP2C*, and *CYP3A* subfamilies.

In contrast with phenobarbital, induction by TCDD and polycyclic hydrocarbons does not cause proliferation of the endoplasmic reticulum although the CYP content is increased. CYP1A1 is the primary isoform induced, although other non-CYP proteins such as UGTs are also induced. Induction of CYP1A1 by polycyclic hydrocarbons results in the induction of a relatively narrow range of oxidative activities, consisting primarily of reactions involving aryl hydrocarbon hydroxylase, the best known reaction being the hydroxylation of benzo(a)pyrene.

Rifampicin and pregnenolone-16α-carbonitrile (PCN) induce members of the *CYP3A* family and represent a third type of inducer, in that the substrate specificity of the microsomes from treated animals differs from that of the microsomes from either phenobarbital-treated or TCDD-treated animals. Inducing substrates of this class include endogenous and synthetic glucocorticoids (e.g., dexamethasone (DEX)), pregnane compounds (e.g., PCN), and macrolide antibiotics (e.g., rifampicin).

Ethanol and a number of other chemicals, including acetone and certain imidazoles, induce CYP2E1. Piperonyl butoxide, isosafrole, and other methylenedioxyphenyl compounds are known to induce CYP1A2 by a non-aryl hydrocarbon receptor (AhR)-dependent mechanism. Peroxisome proliferators, including the drug clofibrate, and the herbicide synergist tridiphane induce a CYP4A isozyme that catalyzes the ω-oxidation of lauric acid.

All inducers do not fall readily into one or the other of these classes. Some oxidative processes can be induced by either type of inducer, such as the hydroxylation of aniline and the *N*-demethylation of chlorcyclizine. Some inducers, such as the mixture of PCBs designated Arochlor 1254, can induce a broad spectrum of CYP isoforms. Many variations also exist in the relative stimulation of different oxidative activities within the same class of inducer, particularly of the phenobarbital type.

It appears reasonable that because several types of CYP are associated with the endoplasmic reticulum, various inducers may induce one or more of them. Because each of these types has a relatively broad substrate specificity, differences may be caused by variations in the extent of induction of different CYP isoforms. Now that methods are available for gel electrophoresis of microsomes and identification of specific isoforms by immunoblotting and isoforms-specific antibodies, the complex array of inductive phenomena is being more logically explained in terms of specific isozymes.

Although the bulk of published investigations of the induction of monooxygenase enzymes have dealt with the mammalian liver, induction has been observed in other mammalian tissues and in nonmammalian species, both vertebrate and invertebrate. Many induced CYPs have now been cloned and/or purified from a variety of species. It is clear that many of these induced CYPs represent only a small percentage of the total CYP in the uninduced animal. For this reason, the "constitutive" isoforms, those already expressed in the uninduced animal, must be fully characterized because they represent the available xenobiotic-metabolizing capacity of the normal animal.

Mechanism and Genetics of Induction in Mammals   Many different mechanisms may be involved in CYP induction. These include increased transcription of DNA, increased mRNA translation to protein, mRNA stabilization, and protein stabilization. Induction can only occur in intact cells and cannot be achieved by the addition of inducers directly to cell fractions such as microsomes. It has been known for some time that in most, but not necessarily all, cases of increase in monooxygenase activity, there is a true induction involving synthesis of new mRNA, and not an increase in activity of an enzyme already synthesized, since induction is generally prevented by inhibitors of protein synthesis. For example, the protein synthesis inhibitors such as puromycin, ethionine, and cycloheximide inhibit aryl hydrocarbon hydroxylase activity. A simplified scheme for gene expression and protein synthesis is shown in Figure 8.6.

Perhaps the best understood example of induction involves ligand activation of the AhR by compounds such as TCDD and 3-methylcholanthrene. The use of suitable inhibitors of RNA polymerase activity has shown that an inhibitor such as actinomycin D block increases in aryl hydrocarbon hydroxylase activity. Thus, it appears that the increase in enzyme activity is due to the induction of *CYP* genes and requires new RNA synthesis.

These findings indicate these compounds act as ligands for receptors that mediate the induction of genes encoding XMEs in a manner analogous to steroid hormones—namely, combining with a cytosolic receptor followed by movement into the nucleus and then increased transcription of the appropriate gene. In the case of the AhR, TCDD or some other appropriate ligand enters the cell through the plasma membrane and binds to the cytosolic Ah receptor protein (Figure 8.7). After ligand binding, the receptor translocates to the nucleus where it forms a dimer with another protein known as aryl hydrocarbon receptor nuclear translocator (ARNT).

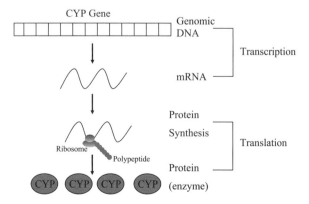

Figure 8.6   Simplified scheme for gene expression in animals of a *CYP* gene.

Figure 8.7   Proposed mechanism for TCDD-activated AhR translocation and DNA binding. Upon TCDD binding, activated AhR sheds chaperone proteins such as heat shock protein 90 (Hsp 90), translocates to the nucleus, and heterodimerizes with the ARNT protein. In the nucleus, the AhR–Arnt complex binds to dioxin response elements (DREs) such as those found in the *CYP1A1* gene promoter region. Activation of transcription of the *Cyp1a1* gene leads to formation of new RNA and an increase in *CYP1A1* protein levels.

Figure 8.8   The mouse *Cyp1a1* and *Cyp1a2* genes are located on chromosome 9 and contain 8 dioxin response elements (DREs) that are responsible for the AhR-mediated induction. A DRE cluster (DREC) is located 1.4 kb upstream of the *Cyp1a1* gene which also plays a role in the induction of *Cyp1a2*. (Source: Adapted from Nukaya, M., S. Moran, and C. A. Bradfield. The role of the dioxin-response element cluster between Cyp1a1 and Cyp1a2 loic in aryl hydrocarbon receptor biology. *Proc. Natl. Acad. Sci. USA* **106**:4923–4928, 2009.)

In the nucleus, the activated receptor acts as a transcription factor and interacts with specific sequences of DNA known as dioxin responsive elements (DREs) such as those found near the mouse *CYP1A1* gene (Figure 8.7). On mouse chromosome 9 multiple DREs, known as a DRE cluster (DREC), are located upstream from the transcriptional start site in the 5′ flanking region of the mouse *CYP1A1* gene, which also have been shown to influence CYP1A2 expression (Figure 8.8). The protein–DNA interaction that occurs at the DREs and along with interaction with coactivator proteins results in and increased transcription followed by increased protein synthesis. Activation of the AhR signaling pathway by TCDD-type inducers has been determined to induce a number of Phase I and Phase II genes.

Although phenobarbital induction has been studied for many years, the mechanism for induction has only recently been established.

The major advance in understanding phenobarbital induction came from a study using rat primary hepatocytes where phenobarbital responsiveness was demonstrated to be associated with a 163 bp DNA sequence at −2318 through −2155 bp of the *CYP2B2* gene. Subsequent studies using *in situ* transfection of *CYP2B2* promoter-luciferase constructs into rat livers confirmed this, as did similar studies involving the mouse *CYP2b10* gene. Additional deletion assays have narrowed phenobarbital responsiveness down to a minimum sequence of 51 bp from −2339 through −2289 of the *Cyp2b10* gene, now known as the phenobarbital-responsive enhancer module (PBREM). The PBREM sequence has also been found in rat *CYP2B1*, *CYP2B2*, and human *CYP2B6* genes. Multiple cis-acting elements within this fragment cooperate to bring about increased DNA transcription that include two receptor DR4 sites (NR1 and NR2).

The transcription factor that interacts with the PBREM is the nuclear receptor (NR) known as constitutive androstane receptor (CAR) or NR1I3 (NR subfamily 1, group I, member 3). CAR binds to each of the PBRE NR sites as a heterodimer with the retinoid X receptor (RXR), a common heterodimerization partner for many orphan NRs. Mouse CAR is normally found within the cytoplasm, but phenobarbital exposure results in a change in CAR's phosphorylation state, dissociation from chaperone proteins, and subsequently, translocation to the nucleus. Translocation to the nucleus leads to RXR association, DNA binding to the PBREM, and association coactivator proteins to form an activate transcriptional complex. Initial studies indicated that CAR was constitutively active, but this was most likely due to cell culture systems that did not contain adequate levels of co-chaperones to maintain CAR with the cytoplasm. Recent studies using CAR knockout mice indicate that many drug metabolizing genes are under CAR regulation, including *CYP2B*, *CYP3A*, NADPH-CYP reductase, and enzymes involved in sulfotransferase metabolism.

In the early 1980s, a distinct group of *CYPs* was described by several groups, which was characterized principally by its inducibility by glucocorticoids, by a mechanism not involving the glucocorticoid receptor, and by a wide variety of other chemicals with no structural similarities. Specifically, the *CYP3A* subfamily, which is well-known for the diversity of substrates which it is capable of metabolizing, was highly induced. In humans, the specific isoform, CYP3A4, is responsible not only for the metabolism of endogenous compounds such as testosterone, but also is credited for the metabolism of the largest number of currently used drugs. Many CYP3A substrates are also known for their ability to induce their own metabolism as well as the metabolism of other CYP3A substrates, resulting in patients with the potential for dangerous drug–drug type interactions. Regulation of the CYP3A family is primarily through enhanced transcription as demonstrated in studies using the RNA inhibitor actinomycin D and the translation inhibitor cycloheximide. Several studies in rats and human hepatocytes have identified several elements on the 5′ upstream promoter region as well as receptors involved in CYP3A regulation (Figure 8.9). Deletion studies involving transient transfections of various chimeric reporter gene constructs into primary cultures of rat hepatocytes demonstrated the presence of a dexamethasone/PCN response element within the first 164 bp at the start of transcription. Subsequent studies demonstrated that several CYP3A isoforms from different species contained NR binding sites that are activated by DEX/PCN but exhibit low activation by rifampicin. Further work identified an

−11.5 kB Far Enhancer   −7.0 kB Distal Enhancer   −250 Proximal Promoter   *CYP3A4* Gene

Figure 8.9 Illustration depicting the multiple pregnane X receptor (PXR) response elements found in the human *CYP3A4* gene. The activation of the human pregnane X receptor (PXR) by binding to the ligand rifampicin (R) leads to heterodimerization with the retinoid X receptor (RXR), subsequent binding to PXR response elements located in the proximal promoter, distal enhancer, and far enhancer. Transcriptional activation by PXR is aided by interaction with co-activators, such as the steroid receptor co-activator-1 (SRC-1), that help facilitate formation of the general transcriptional machinery (GTM), which includes RNA polymerase and associated proteins.

additional 230 bp distal element called the xenobiotic-responsive enhancer module (XREM) located at −7836 through −7607 of the CYP3A4 that conferred responsiveness to both rifampicin and dexamethasone when combined with the proximal promoter region. The XREM contains two PXR binding sites, one DR3 motif (dNR1) and one ER6 motif (dNR2), neither of which is solely responsible for the activity of XREM. An additional proximal promoter PXR binding site, known as dNR3, contains an ER6 motif and, most recently, a far distal enhancer was identified that contains aER6 site located 11.5 kb from the start of transcription, and both appear to have important roles in CYP3A4 induction.

The nuclear orphan receptor, pregnane X receptor (PXR), also known as NR1I2 (NR subfamily 1, group I, member 2), is the major determinant of *CYP3A* gene regulation by xenobiotics. Several lines of evidence support PXR involvement with *CYP3A* induction. First, both PXR and CYP3A isoforms are predominantly expressed in liver and intestine, with less expression found in lung and kidneys. Second, PXR binds to human and rat *CYP3A* promoter regions and can activate expression of *CYP3A4* promoter in transfection assays. Third, many of the same inducers of *CYP3A* isoforms also activate PXR. Fourth, interspecies differences in response to CYP3A inducers have been demonstrated to be due to the ability of these inducers to activate PXR in these species. Fifth, disruption of the mouse PXR gene eliminated induction of *CYP3A* by PCN, and transgenic mice "humanized" with the PXR gene were able to respond to rifampicin induction. PXR has a unique ligand binding domain (LBD) compared with other receptors, and X-ray crystallography of the LBD has shown that it is much larger, and thus can accommodate a wide variety of ligands. PXR has been shown to play a key role in the induction of a growing number of metabolizing genes including Phase I, Phase II, and xenobiotic transporters.

Peroxisome proliferators, including hypolipidemic drugs such as clofibrate, phthalate plasticizers, and herbicides bring about the induction of a CYP4A isoform that catalyzes the oxidation of many biologically important fatty acids, including arachidonic acid and other eicosanoids. *CYP4A* expression is part of a pleiotropic response in the rodent liver which includes increased liver weight, proliferation of peroxisomes, and the elevation of several peroxisomal enzymes such as catalase. Peroxisome proliferators are often epigenetic carcinogens in rodents but, since the effect is primarily seen in rodents, its significance for other species such as humans is unclear. The receptor protein, peroxisome proliferator-activated receptor-α (PPARα), was first cloned in 1990. PPARα knockout mice exposed to chemicals which normally induce CYP4A as well as peroxisome proliferation do not exhibit these characteristics, demonstrating the essential nature of PPARα for these responses. Like PXR, PPARα also binds to DNA as a heterodimer with RXR in response to peroxisome proliferating chemicals.

Other CYPs are induced by similar mechanisms involving other transcription factors including farnesoid X receptor (FXR), liver X receptor (LXR), hepatic nuclear factor (HNF) family members, GR, and CCAAT/enhancer-binding proteins (C/EBPs). Many of the previously mentioned inductions of *CYP1A1*, *CYP2B*, and *CYP3A* family members also involve these transcription factors and overlap exists in these regulatory pathways as genes may contain multiple transcription factor binding sites.

CYP2E1 catalyzes metabolism of several low molecular weight xenobiotics including drugs (e.g., acetaminophen), solvents (e.g., ethanol, carbon tetrachloride), and procarcinogens (e.g., *N*-nitrosodimethylamine). Induction of *CYP2E1* can occur as a result of exposure to several xenobiotics including ethanol, acetone, and imidazole, or alternatively, as a result of physiological conditions such as starvation and diabetes. Its induction by either fasting or diabetes is believed to be due to the high levels of ketones likely to be present in either of these conditions. It might also be noted that although *CYP2E1* is in the same family as *2B1* and *2B2*, it is not induced by phenobarbital-type inducers. In contrast to many other inducible *CYPs*, *CYP2E1* induction is not accompanied by high levels of *CYP2E1* mRNA, suggesting that regulation is by means of a posttranscriptional mechanism.

The regulation of *CYP2E1* gene expression involves several mechanisms that do not primarily include increased transcription. Recent studies demonstrated that rapid increases in CYP2E1 protein levels following birth are due to stabilization of preexisting proteins by ketone bodies released at birth. Rats treated with ethanol or acetone can have three- to sixfold increases in CYP2E1 protein in the absence of increased *CYP2E1* mRNA. Other studies have demonstrated that substrates including ethanol, imidazole, and acetone had little effects on *CYP2E1* transcript content and that these substrates tend to prevent protein degradation. Thus, increased protein expression levels in response to these substrates is due to enhanced translation and also protein stabilization a result of the inhibition of the proteasome ubiquitin degradation pathway of proteolysis. The ubiquitination process normally tags proteins with a chain of multiple ubiquitin moieties, thus targeting proteins for selective degradation by a cytosolic 26S protease, known as the proteasome. In recent studies, an antibody prepared against a putative ubiquitination-target site on the CYP2E1 protein quenched ubiquitination in a concentration-dependent manner. These results provide a plausible mechanistic explanation for the observation that substrate binding protects the CYP2E1 protein from ubiquitin-dependent proteolysis.

In other observations, diabetes is known to increase CYP2E1 expression at both the mRNA and protein levels in both chemically induced and spontaneous diabetic rats. Elevation of mRNA levels as a result of diabetes has been attributed to mRNA stabilization which can be reversed by daily insulin treatment. Recent research has shown that insulin destabilizes *CYP2E1* mRNA by binding to a 16 bp sequence within the 5′ coding sequence of *CYP2E1*. The mechanism for regulation by this means is still uncertain although other genes have also been reported with similar destabilizing sequences within their coding sequences.

Effects of Induction   The effects of inducers are usually the opposite of those of inhibitors; thus, their effects can be demonstrated by much the same methods, that is, by their effects on pharmacological or toxicological properties *in vivo* or by the effects on enzymes *in vitro* following prior treatment of the animal with the inducer. *In vivo* effects are frequently reported; the most common ones are the reduction of the hexobarbital sleeping time or zoxazolamine paralysis time. Also, *in vivo* effects of CYP3A inducers can be determined by pharmacokinetic studies by administering mice the CYP3A substrate midazolam (MDZ). After removing small amounts of blood over a time coure, the disappearance of midazolam, or the formation of MDZ metabolite can be measured by liquid chromatography mass spectrometry/mass spectrometry (LC-MS/MS) methods. These effects have been reported for numerous inducers and can be quite dramatic. For example, in the rat, the paralysis time resulting from a high dose of zoxazolamine can be reduced from 11 h to 17 min by treatment of the animal with benzo(a)pyrene 24 h before the administration of zoxazolamine.

The induction of CYP enzymes may be protective, as demonstrated by the creation of a DREC-deficient mouse model where loss of Cyp1a1 and Cyp1a2 induction lead to enhanced dioxin-induced hepatotoxicity. The induction of monooxygenase activity may also protect an animal from the effect of carcinogens by increasing the rate of detoxication. This has been demonstrated in the rat with a number of carcinogens including benzo(a)pyrene, *N*-2-fluorenylacetamide, and aflatoxin B_1. Effects on carcinogenesis may be expected to be complex because some carcinogens are both activated and detoxified by monooxygenase enzymes, while epoxide hydrolase, which can also be involved in both activation and detoxication, may also be induced. For example, the toxicity of the carcinogen 2-naphthylamine, the hepatotoxic alkaloid monocrotaline, and the cytotoxin cyclophosphamide are all increased by phenobarbital induction—an effect mediated by the increased population of reactive intermediates.

Organochlorine insecticides are also well-known inducers. Treatment of rats with either DDT or chlordane, for example, will decrease hexobarbital sleeping time and offer protection from the toxic effect of warfarin. Persons exposed to DDT and lindane metabolized antipyrine twice as fast as a group not exposed, whereas those exposed to DDT alone had a reduced half-life for phenylbutazone and an increased excretion of 6-hydroxycortisol.

Effects on xenobiotic metabolism *in vivo* are also widely known in both humans and animals. Cigarette smoke, as well as several of its constituent polycyclic hydrocarbons, is a potent inducer of aryl hydrocarbon hydroxylase in the placenta, liver, and other organs. The average content of CYP1A1 in liver biopsies from smokers was approximately fourfold higher than that from nonsmokers. Hepatic activity of CYP1A1 as measured by phenacetin *O*-demethylation, was also increased from 54 pmol/min/mg of protein in nonsmokers to 230 nmol/min/mg of protein in smokers. Examination of the term placentas of smoking human mothers revealed a marked stimulation of aryl hydrocarbon hydroxylase and related activities—remarkable in an organ that, in the uninduced state, is almost inactive toward foreign chemicals. These *in vitro* differences in metabolism are also observed *in vivo*, as smokers have been demonstrated to have increased clearance rates for several drugs metabolized principally by CYP1A1 including theophylline, caffeine, phenacetin, fluvoxamine, clozapine, and olanzapine.

Induction of XMEs Other than Monooxygenases Although less well studied, XMEs other than those of the CYP system have been shown to be induced in recent studies, frequently by the same inducers that induce the oxidases. These include glutathione S-transferases, epoxide hydrolase, UDP-glucuronosyltransferase, and sulfotransferases, and the mechanisms of induction involve receptors such as AhR, PXR, and CAR. The selective induction of one pathway over another can greatly affect the metabolism of a xenobiotic.

8.5.3 Biphasic Effects: Inhibition and Induction

Many inhibitors of mammalian monooxygenase activity can also act as inducers. Inhibition of microsomal monooxygenase activity is fairly rapid and involves a direct interaction with the cytochrome, whereas induction is a slower process. Therefore, following a single injection of a suitable compound, an initial decrease due to inhibition would be followed by an inductive phase. As the compound and its metabolites are eliminated, the levels would be expected to return to control values. Some of the best examples of such compounds are the methylenedioxyphenyl synergists, such as piperonyl butoxide. Because CYP combined with methylenedioxyphenyl compounds in an inhibitory complex cannot interact with CO, the CYP titer, as determined by the method of Omura and Sato (dependent upon CO-binding to reduced cytochrome), would appear to follow the same curve.

It is apparent from extensive reviews of the induction of monooxygenase activity by xenobiotics that many compounds other than methylenedioxyphenyl compounds have the same effect. It may be that any synergist that functions by inhibiting microsomal monooxygenase activity could also induce this activity on longer exposure, resulting in a biphasic curve as described previously for methylenedioxyphenyl compounds. This curve has been demonstrated for NIA 16824 (2-methylpropyl-2-propynyl phenylphosphonate) and WL 19255 (5,6-dichloro-1,2,3-benzothiadiazole), although the results were less marked with R05-8019 [2,(2,4,5-trichlorophenyl)-propynyl ether] and MGK 264 [*N*-(2-ethylhexyl)-5-norbornene-2,3-dicarboximide].

8.6 ENVIRONMENTAL EFFECTS

Because the *in vitro* effects of light, temperature, and so on, on XMEs are not different from their effects on other enzymes or enzyme systems, we are not concerned with them at present. This section deals with the effects of environmental factors on the intact animal as they relate to *in vivo* metabolism of foreign compounds.

8.6.1 Temperature

Although it might be expected that variations in ambient temperature would not affect the metabolism of xenobiotics in animals with homeothermic control, this is not the case. Temperature variations can be a form of stress and thereby produce changes mediated by hormonal interactions. Such effects of stress require an intact pituitary-adrenal axis and are eliminated by either hypophysectomy or adrenalectomy. There appears to be two basic types of temperature effect on toxicity; either with increase in toxicity at both high and low temperature, or an increase in toxicity with an increase in temperature. For example, both warming and cooling increases the toxicity of caffeine to mice, whereas the toxicity of *D*-amphetamine is lower at reduced temperatures and shows a regular increase with increases in temperature.

In many studies, it is unclear whether the effects of temperature are mediated through metabolism of the toxicant or via some other physiological mechanism. In other cases, however, temperature clearly affects metabolism. For example, in cold-stressed rats, there is an increase in the metabolism of 2-naphthylamine to 2-amino-1-naphthol.

8.6.2 Ionizing Radiation

In general, ionizing radiation reduces the rate of metabolism of xenobiotics both *in vivo* and in enzyme preparations subsequently isolated. This has occurred in hydroxylation of steroids, in the development of desulfuration activity toward azinphosmethyl in young rats, and in glucuronide formation in mice. Pseudocholinesterase activity is reduced by ionizing radiation in the ileum of both rats and mice.

8.6.3 Light

Because many enzymes, including some of those involved with xenobiotic metabolism, show a diurnal pattern that can be keyed to the light cycle, light cycles rather than light intensity would be expected to affect these enzymes. In the case of hydroxyindole-*O*-methyltransferase in the pineal gland, there is a diurnal rhythm with greatest activity at night; continuous darkness causes maintenance of the high level. CYP and the microsomal monooxygenase system show a diurnal rhythm in both the rat and the mouse, with greatest activity occurring at the beginning of the dark phase.

8.6.4 Moisture

No moisture effect has been shown in vertebrates, but in insects, it was noted that housefly larvae reared on diets containing 40% moisture had four times more activity for the epoxidation of heptachlor than did larvae reared in a similar medium saturated with water.

8.6.5 Altitude

Altitude can either increase or decrease toxicity. It has been suggested that these effects are related to the metabolism of toxicants rather than to the physiological mechanisms involving the receptor system, but in most examples, this has not been demonstrated clearly. Examples of altitude effects include the observations that at altitudes of ≥5000 ft, the lethality of digitalis or strychnine to mice is decreased, whereas that of *D*-amphetamine is increased.

8.6.6 Other Stress Factors

Noise has been shown to affect the rate of metabolism of 2-napthylamine, causing a slight increase in the rat. This increase is additive with that caused by cold stress.

8.7 SUMMARY AND CONCLUSIONS

It is apparent from the material presented in this chapter and the previous chapters related to metabolism that the metabolism of xenobiotics is complex, involving many enzymes; that it is susceptible to a large number of modifying factors, both physiological and exogenous; and that the toxicological implications of metabolism are important. In spite of the complexity, summary statements of considerable importance can be abstracted:

1. Phase I metabolism generally introduces a functional group into a xenobiotic, which enables conjugation to an endogenous metabolite to occur during Phase II metabolism.
2. The conjugates produced by Phase II metabolism are considerably more water soluble than either the parent compound or the Phase I metabolite(s) and hence are more easily excreted.
3. During the course of metabolism, and particularly during Phase I reactions, reactive intermediates that are much more toxic than the parent compound may be produced. Thus, xenobiotic metabolism may be either a detoxication or an activation process.
4. Because the number of enzymes involved in Phase I and Phase II reactions is large and many different sites on organic molecules are susceptible to metabolic attack, the number of potential metabolites and intermediates that can be derived from a single substrate is frequently very large.
5. Because both qualitative and quantitative differences exist between species, strains, individual organs, and cell types, a particular toxicant may have different effects in different circumstances.
6. Because exogenous chemicals can be inducers and/or inhibitors of the XMEs of which they are substrates; such chemicals may interact to bring about toxic sequelae different from those that might be expected from any of them administered alone.
7. Because endogenous factors also affect the enzymes of xenobiotic metabolism, the toxic sequelae to be expected from a particular toxicant will vary with developmental stage, nutritional statue, health or physiological status, stress, or environment.
8. It has become increasingly clear that most enzymes involved in xenobiotic metabolism occur as several forms, which coexist within the same individual and, frequently, within the same subcellular organelle. An understanding of the biochemistry and molecular genetics of these isoforms may lead to an understanding of the variation between species, individuals, organs, sexes, developmental stages, and so on.

BIBLIOGRAPHY AND SUGGESTED READING

Anderson, K. E. and A. Kappas. Dietary regulation of cytochrome P450. *Annu. Rev. Nutr.* **11**:141–167, 1991.

Benedetti, M. S., R. Whomsley, and M. Canning. Drug metabolism in the paediatric population and in the elderly. *Drug Discov. Today* **12**:599–610, 2007.

Brown, H. S., A. Chadwick, and J. B. Houston. Use of isolated hepatocyte preparations for cytochrome P450 inhibition studies: Comparison with microsomes for ki determination. *Drug Metab. Dispos.* **35**:2119–2126, 2007.

Conney, A. H. Induction of drug-metabolizing enzymes: A path to the discovery of multiple cytochromes P450. *Annu. Rev. Pharmacol. Toxicol.* **43**:1–30, 2003.

Donaldson, W. E. Nutritional factors. In *Introduction to Biochemical Toxicology*, 3rd ed., eds. E. Hodgson and R. C. Smart, pp. 255–276. Hoboken, NJ: Wiley, 2001.

Falls, J. G., D.-Y. Ryu, Y. Cao, et al. Regulation of mouse liver flavin-containing mono-oxygenases 1 and 3 by sex steroids. *Arch. Biochem. Biophys.* **342**:212–223, 1997.

Goodwin, B., M. R. Redinbo, and S. A. Kliewer. Regulation of CYP3A gene transcription by the pregnane X receptor. *Annu. Rev. Pharmacol. Toxicol.* **42**:1–23, 2002.

Guengerich, F. P. Cytochrome P-450 3A4: Regulation and role in drug metabolism. *Annu. Rev. Pharmacol. Toxicol.* **39**:1–17, 1999.

Hodgson, E. and E. Croom. Phase I—Toxicogenetics. In *Molecular and Biochemical Toxicology*, 4th ed., eds. E. Hodgson and R. C. Smart. Hoboken, NJ: Wiley, 2008.

Huang, S. M., J. M. Strong, L. Zhang, et al. New era in drug interaction evaluation: U.S. Food and Drug Administration update on CYP enzymes, transporters, and the guidance process. *J. Clin. Pharmacol.* **48**:662–670, 2008.

Kakizaki, S., Y. Yamamoto, A. Ueda, et al. Phenobarbital induction of drug/steroid-metabolizing enzymes and nuclear receptor CAR. *Biochim. Biophys. Acta.* **1619**:239–242, 2003.

Kocarek, T. A., R. C. Zanger, and R. F. Novak. Post-transcriptional regulation of rat CYP2E1 expression: Role of CYP2E1 mRNA untranslated regions in control of translational efficiency and message stability. *Arch. Biochem. Biophys.* **376**:180–190, 2000.

Kulkarni, A. P. and E. Hodgson. Comparative toxicology. In *Introduction to Biochemical Toxicology*, eds. E. Hodgson and F. E. Guthrie, p. 115. New York: Elsevier, 1980.

Lee, C. and S. L. Werlin. The induction of hepatic cytochrome P450 3A in rats: Effects of age. *Proc. Soc. Exp. Biol. Med.* **210**:134–139, 1995.

Lin, J. H. CYP induction-mediated drug interactions: In vitro assessment and clinical implications. *Pharm. Res.* **23**:1089–1116, 2006.7

Mukhtar, H. and J. R. Bend. Serum glutathione S-transferases: perinatal development, sex difference, and effect of carbon tetrachloride administration on enzyme activity in the rat. *Life Sci.* **21**:1277–1286, 1977.

Murray, M. Altered CYP expression and function in response to dietary factors: Potential roles in disease pathogenesis. *Curr. Drug Metab.* **7**:67–81, 2006.

Newton, D. J., R. W. Wang, and A. Y. H. Lu. Cytochrome P450 inhibitors: Evaluation of specificities in the in vitro metabolism of therapeutic agents by human liver microsomes. *Drug Metab. Dispos.* **23**:154–158, 1995.

Nukaya, M., S. Moran, and C. A. Bradfield. The role of the dioxin-responsive element cluster between the Cyp1a1 and Cyp1a2 loci in aryl hydrocarbon receptor biology. *Proc. Natl. Acad. Sci. USA* **106**:4923–4928, 2009.

Parkinson, A., D. R. Mudra, C. Johnson, et al. The effects of gender, age, ethnicity, and liver cirrhosis on cytochrome P450 enzyme activity in human liver microsomes and inducibility in cultured human hepatocytes. *Toxicol. Appl. Pharmacol.* **199**:193–209, 2004.

Ronis, M. J. J. and H. C. Cunny. Physiological (endogenous) factors affecting the metabolism of xenobiotics. In *Introduction to Biochemical Toxicology*, 2nd ed., eds. E. Hodgson and P. E. Levi, p. 136. Norwalk, CT: Appleton & Lange, 1994.

Ronis, M. J. J. and H. C. Cunny. Developmental effects on xenobiotic metabolism. In *Molecular and Biochemical Toxicology*, 4th ed., eds. E. Hodgson and R. C. Smart, Hoboken, NJ: Wiley, 2008.

Sueyoshi, T. and M. Negishi. Phenobarbital response elements of cytochrome P450 genes and nuclear receptors. *Annu. Rev. Pharmacol. Toxicol.* **41**:123–143, 2001.

Timsit, Y. E. and M. Negishi. CAR and PXR: The xenobiotic-sensing receptors. *Steroids* **72**:231–246, 2007.

Yao, M., M. Zhu, M. W. Sinz, et al. Development and full validation of six inhibition assays for five major cytochrome P450 enzymes in human liver microsomes using an automated 96-well microplate incubation format and LC-MS/MS analysis. *J. Pharm. Biomed. Anal.* **44**:211–223, 2007.

SAMPLE QUESTIONS

1. Chemical A has no inherent toxicity but is metabolized to a potent toxicant while chemical B, an inherently potent toxicant is metabolically detoxified. What would be the effect of treatment of the animal with either an inhibitor or an inducer of the enzymes involved in their metabolism?

2. What are the advantages and disadvantages of demonstrating inhibition of the metabolism of a toxicant by observation of (a) *in vivo* symptoms or (b) *in vitro* effects of the inhibitor on the xenobiotic-metabolizing enzyme involved.

3. Define "mechanism-based inhibitor" (also known as "suicide inhibitor") citing two examples.

4. Discuss the time course in the expected change in enzyme activity (i.e., the shape of the curve) following treatment of an animal with a synergist such as piperonyl butoxide.

5. Compound Z is being studied for its ability to alter the metabolism of the cytochrome P450 (CYP) substrate testosterone. You are presented with the *in vitro* metabolism data below. The rate of testosterone metabolism to the metabolite 6β-OH testosterone was studied using increasing concentrations of testosterone (0–8 μM) with a constant concentration of Compound Z (2 μM). What does the data tell you about how compound Z is altering testosterone metabolism? Briefly, describe how you would study the *in vivo* effects of Compound Z on liver metabolism.

6. After eating two charcoal-broiled hotdogs and one very well done hamburger at a picnic, you have significantly induced the levels of the metabolizing enzyme CYP1A1 in your liver. Name and briefly discuss the receptor pathway inducing CYP1A1 in the liver due to the polycyclic hydrocarbons (TCDD-like compounds) in the cooked meat ingested. Name two nutritional effects and two physiological effects that can alter the levels of xenobiotic metabolism.

Elimination of Toxicants

GERALD A. LEBLANC

9.1 INTRODUCTION

The ability to efficiently eliminate toxic materials is critical to the survival of a species. The complexity of toxicant elimination processes has increased commensurate with the increased complexity associated with animal form. For unicellular organisms, passive diffusion can suffice for the elimination of toxic metabolic wastes produced by the organism. Similarly, as exogenous toxic materials derived from the environment diffuse into a unicellular organism, they can also readily diffuse out of the organism. The large surface area to mass ratio of these organisms ensures that a toxic chemical within the cell is never significantly distanced from a surface membrane across which it can diffuse.

As organisms evolved in complexity, several consequences of increased complexity compromised the efficiency of the passive diffusion of toxic chemicals.

As organisms increased in complexity:

1. They increased in size
2. Their surface area to body mass decreased
3. Their bodies compartmentalized (i.e., cells, tissues, organs)
4. They generally increased in lipid content
5. They developed barriers to the external environment.

9.1.1 Size

With increased size of an organism, a toxic chemical has greater distance to traverse before reaching a membrane across which it can diffuse to the external environment. Thus, overall retention of the chemical will increase as will propensity for the chemical to elicit toxicity.

A Textbook of Modern Toxicology, Fourth Edition. Edited by Ernest Hodgson
Copyright © 2010 John Wiley & Sons, Inc.

9.1.2 Surface Area to Body Mass Ratio

Increased size of an organism is associated with a decrease in the surface area to body mass ratio. Accordingly, the availability of surface membranes across which a chemical can passively diffuse to the external environment decreases, and propensity for retention of the chemical increases.

9.1.3 Compartmentalization

With increased complexity comes increased compartmentalization. Cells associate to form tissues, and tissues associate to form organs. Compartmentalization increases the number of barriers across which chemicals must traverse before sites of elimination are reached. As different compartments often have different physicochemical characteristics (e.g., adipose tissue contains a significant amount of lipid; whereas blood contains a significant amount of water), chemicals are faced with the challenge to be mobile in these various environments.

9.1.4 Lipid Content

As a general though not universal rule, organisms have the ability to store energy as fat increases with increased size of the organism. Thus, large organisms tend to have significant lipid stores into which lipophilic chemicals can be stored for extended periods of time. These stored chemicals tend to be largely immobile and difficult to release from the adipose tissue.

9.1.5 Barriers to the Environment

Through evolution, increased complexity of organisms led to increased exploitation of various environments. In order to survive in these environments, organisms developed barriers such as skin and scales that protect the organisms from harsh conditions on the outside and minimize loss of vital constituents such as water on the inside. Likewise, these barriers impede the elimination of toxic constituents by the organisms, requiring the development of specialized membranes and organs through which toxic materials can be eliminated.

A consequence of this hindrance to elimination of toxic materials by complex organisms was the development of specialized routes of elimination. These routes generally evolved in concert (i.e., coevolved) with biotransformation processes that render chemicals amenable to these modes of elimination (see Chapter 6).

Three major routes of elimination culminate in the specialized organs of elimination: the liver, kidneys, and lungs. The liver serves as a major organ at which lipophilic materials are collected from the blood, biotransformed to generally less toxic and more polar derivatives, then eliminated into the bile or returned to the blood for renal elimination. The kidneys complement the liver in that these organs collect wastes and other chemicals in the blood through a filtration process and eliminate these wastes in the urine. The respiratory membranes of the lungs are ideal for the removal of volatile materials from the blood into expired air. In addition to these major routes of elimination, several quantitatively minor routes exist through which toxic materials can be eliminated from the body. These include the following.

1. *Skin:* Skin constitutes the largest organ in the human body, and it spans the interface between the body and the external environment. While the skin epidermis constitutes a relatively impervious membrane across which chemical elimination is difficult, the shear surface area involved requires consideration of this organ as a route of elimination. Volatile chemicals are particularly adept at traversing the skin and exiting the body through this route.

2. *Sweat:* Humans lose an average of 0.7 L of water per day due to sweating. This loss of fluid provides a route for the elimination of water-soluble chemicals.

3. *Milk:* Mother's milk is rich in lipids and lipoproteins. Milk thus serves as an ideal route for the elimination of both water-soluble and fat-soluble chemicals from the mother's body. For example, the dichlorodiphenyltrichloroethane (DDT) metabolite dichlorodiphenyldichloroethylene (DDE), the flame retardant mirex, and the polychlorinated biphenyls (PCBs) have been measured at significant levels in mother's milk from Arctic Inuit populations. While lactation may provide a benefit to the mother by the elimination of toxic chemicals, transfer of these toxicants to the suckling infant may pose health risk to the infant.

4. *Hair:* Growing hair can serve as a limited route through which chemicals can escape the body. Pollutants such as mercury and drugs such as cocaine have been measured in human hair, and hair analysis is often used as a marker of exposure to such materials.

9.2 TRANSPORT

For a chemical to be eliminated from the body at a site of elimination (e.g., kidney) that is distant from the site of storage (e.g., adipose tissue) or toxicity (e.g., brain), the chemical must be transported from the site of origin to the site of elimination. Chemicals are transported to the site of elimination largely via the circulatory system. Sufficiently, water-soluble chemicals can freely dissolve into the aqueous component of blood and be transported by both diffusion and blood circulation to sites of elimination. With decreasing water solubility and increasing lipid solubility, chemicals are less likely to freely diffuse into blood, and extraction of these chemicals from sites of toxicity or storage can be more challenging. These materials generally associate with transport proteins in the blood which either contain binding sites for chemical attachment or lipophilic cores (lipoproteins) into which lipophilic chemicals can diffuse. The blood contains various transport proteins that are typically suited for the transport of specific endogenous chemicals. These include albumin, sex steroid-binding globulin, and lipoproteins. Often, xenobiotics can utilize these proteins, particularly the nonspecific transporters, to facilitate mobilization and transport in the aqueous environment of the blood. At the site of elimination, xenobiotics may diffuse from the transport protein to the membranes of the excretory organ, or the transport protein may bind to surface receptors on the excretory organ, undergo endocytosis and intracellular processing, where the xenobiotic is released and undergoes processing leading to elimination.

9.3   RENAL ELIMINATION

The kidneys are the sites of elimination of water-soluble chemicals that are removed from the blood by the process of reverse filtration. Two characteristics are primarily responsible for determining whether a chemical will be eliminated by the kidneys: size and water solubility.

9.3.1   Size

The reverse filtration process requires that chemicals to be removed from the blood are able to pass through 70–100 A pores. As a general rule, chemicals having a molecular mass of less than ~65,000 are sufficiently small to be subject to reverse filtration.

9.3.2   Water Solubility

Non-water-soluble chemicals will be transported to the kidneys in association with transport proteins. Thus, in association with these proteins, the chemicals will not be able to pass through the pores during reverse filtration. Lipophilic chemicals are generally subject to renal elimination after they have undergone hydroxylation or conjugation reactions (Chapter 6) in the liver or elsewhere.

Blood is delivered to the human kidney by the renal artery. Blood flows to the kidneys of the adult human at a rate of roughly 1 L/min. The adult human kidney contains approximately 1 million functional units, called nephrons, to which the blood is delivered for removal of solutes. Collected materials are excreted from the body in the urine.

Blood entering the nephron passes through a network of specialized capillaries called the glomerulus (Figure 9.1). These capillaries contain the pores through which materials to be eliminated from the blood pass. Blood in the capillaries is maintained under high positive pressure from the heart coupled with the small diameter of the vessels. As a result, these sufficiently small solutes and water are forced through the pores of the glomerulus. This filtrate is collected in the glomerular (or Bowman's) capsule in which the glomerulus is located (Figure 9.1). Included in this filtrate are water, ions, small molecules such as glucose, amino acids, urate, and foreign chemicals. Large molecules such as proteins and cells are not filtered and are retained in the blood.

Following glomerular filtration, molecules important to the body are reabsorbed from the filtrate and returned to the blood. Much of this reabsorbtion occurs in the proximal tubules (Figure 9.1). Cells lining the proximal tubules contain fingerlike projections that extend into the lumen of the tubule. This provides an expanse of cell surface area across which water and ions can diffuse back into the cells and, ultimately, be returned to the blood. The proximal tubules also contain active transport proteins that recover small molecules such as glucose and amino acids from the filtrate. From the proximal tubules, the filtrate passes through the Loop of Henle. Significant water reabsorption occurs in the descending portion of the loop resulting in concentration of the filtrate. Water reabsorption does not occur in the ascending portion of the loop. Rather, the remaining, concentrated ions such as sodium, chloride, and potassium are reabsorbed. Those materials retained in the

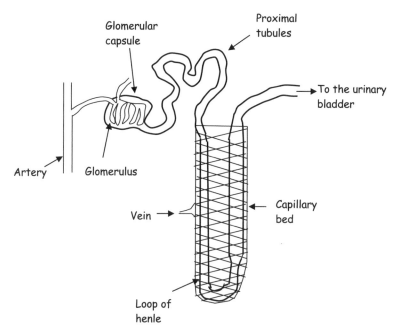

Glomerular capsule

Proximal tubules

To the urinary bladder

Artery   Glomerulus

Vein →   Capillary bed

Loop of henle

Figure 9.1 The nephron of the kidney. The nephron is the functional unit of the kidney that is responsible for the removal of water-soluble wastes and foreign compounds from the blood.

filtrate during passage through the nephron constitute the urine. The urine is transported through the ureters to the bladder and retained until excretion occurs.

The kidneys are a common site of chemical toxicity since the nephron functions to concentrate the toxicant and thus increase levels of exposure to the materials. This increased exposure can result from the concentration of the toxicant in the tubules. It also can occur by concentration within the cells of the nephrons when a chemical is capable of utilizing one of the active transport proteins and is shuttled from the lumen of the tubules into the renal cells.

9.4 HEPATIC ELIMINATION

The liver serves many vital functions to the body. It has a large capacity to hold blood and thus serves as a blood storage site. The liver synthesizes and secretes many substances that are necessary for normal bodily function. It cleanses the blood of various endogenous and foreign molecules. It biotransforms both endogenous and exogenous materials, typically reducing their bioreactivity and preparing them for elimination. It eliminates wastes and foreign chemicals through biliary excretion. Three of these functions occur coordinately in a manner that makes the liver a major organ of chemical elimination: chemical uptake from blood, chemical biotransformation, and biliary elimination of chemicals.

Blood is delivered to the liver from two sources. Oxygen-rich blood is delivered through the hepatic artery. In addition, blood is shunted from the capillaries that service the intestines and spleen to the liver by the hepatic portal vein. These two vessels converge, and the entire hepatic blood supply is passaged through sinusoids (Figure 9.2). Sinusoids are cavernous spaces among the hepatocytes which are the functional units of the liver. Hepatocytes are bathed in blood as the blood passes through the sinusoids as 70% of the hepatocyte surface membrane contacts the blood in the sinusoid. This provides for a tremendous surface area across which chemicals can diffuse to gain entry into the hepatocytes. Chemicals may passively diffuse across the sinusoidal membrane of the hepatocytes, they may be exchanged between blood transport proteins and the sinusoidal membranes, or their carrier proteins may bind to sinusoidal membrane receptors, then undergo endocytosis (Figure 9.3).

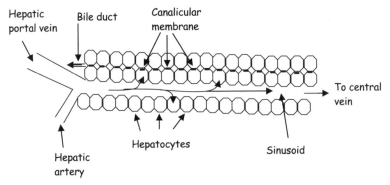

Figure 9.2   Diagrammatic representation of the basic architecture of the liver.

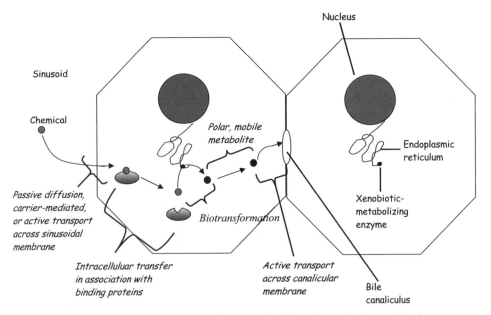

Figure 9.3   Vectorial transport of a chemical from the liver sinusoid, through the hepatocyte, to the cannalicular space.

Lipophilic materials require intracellular carrier proteins to be optimally mobilized, just as they required transport proteins in the blood (Figure 9.3). Several intracellular carrier proteins that mobilize specific endogenous chemical have been characterized. Many of these proteins have been shown to bind xenobiotics; however, less is known of their actual contribution to intracellular xenobiotic mobilization. The intracellular lipid binding proteins are a major family in intracellular carrier proteins. Within this family are the fatty acid binding proteins (FABP), the cellular retinoic acid binding proteins (CRABP), and the bile acid binding proteins (BABP). Some of the cytosolic glutathione S-transferase proteins also have been shown to non-catalytically bind xenobiotics and to be coordinately induced along with xenobiotic biotransformation enzymes and efflux transporters, suggesting that these proteins may function to mobilize xenobiotics.

Once mobilized in the hepatocyte, chemicals can contact and interact with biotransformation enzymes (Chapter 6). These enzymes generally increase the polarity of the chemical, thus reducing its ability to passively diffuse across the sinusoidal membrane back into the blood. Biotransformation reactions also typically render the xenobiotics susceptible to active transport across the canalicular membrane into the bile canaliculus and, ultimately, the bile duct (Figure 9.3). The bile duct delivers the chemicals, along with other constituents of bile, to the gall bladder that excretes the bile into the intestines for fecal elimination.

9.4.1 Entero-Hepatic Circulation

Upon biliary elimination into the gastrointestinal tract, chemicals that have undergone conjugation reactions in the liver may be subject to the action of hydrolytic enzymes that deconjugate the molecule. Deconjugation results in increased lipophilicity of the molecule and renders them once again subject to passive uptake. Reabsorbed chemicals reenter the circulation via the hepatic portal vein which shunts the chemical back to the liver, where the chemical can be reprocessed (i.e., biotransformed) and eliminated. This process is called entero-hepatic circulation (Figure 9.4). A chemical may undergo several cycles of entero-hepatic circulation

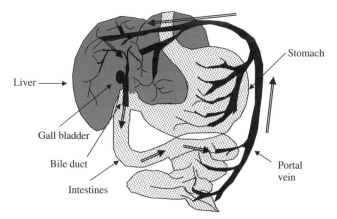

Figure 9.4 Enterohepatic circulation (as indicated by ➡). Polar xenobiotic conjugates are secreted into the intestine via the bile duct and gall bladder. Conjugates are hydrolyzed in the intestines, released xenobiotics are reabsorbed and transported back to the liver via the portal vein.

resulting in a significant increase in the retention time for the chemical in the body and increased toxicity.

The liver functions to collect chemicals and other wastes from the body. Accordingly, high levels of chemicals may be attained in the liver resulting in toxicity to this organ. Biotransformation of chemicals that occur in the liver sometimes results in the generation of reactive compounds that are more toxic than the parent compound resulting in damage to the liver (Chapter 7). Chemical toxicity to the liver is discussed elsewhere (Chapter 13).

9.4.2   Active Transporters of the Bile Canaliculus

The bile canaliculi are minute tubes that form among hepatocytes. These tubes are the product of aggregates of specialized membrane (canalicular membrane) on the basolateral surfaces the hepatocytes (Figure 9.3). Bile passes from the hepatocytes into these tubes which merge to form larger bile ductules and eventually the bile duct. The bile canaliculus constitutes only ~13% of the contiguous surface membrane of the hepatocyte yet must function in the efficient transfer of chemical from the hepatocyte to the bile duct. Active transport proteins located on the canalicular membrane are responsible for the efficient shuttling of chemicals across this membrane. These active transporters are members of a multigene superfamily of proteins know as the ATP-binding cassette transporters. Two subfamilies are currently recognized as having major roles in the hepatic elimination of xenobiotics, as well as endogenous materials. The P-glycoprotein (ABC B) subfamily is responsible for the elimination of a variety of structurally diverse compounds. P-glycoprotein substrates typically have one or more cyclic structures, a molecular weight of 400 or greater, moderate to low lipophilicity (log Kow < 2), and high hydrogen (donor)-bonding potential. Parent xenobiotics that meet these criteria and hydroxylated derivatives of more lipophilic compounds are typically transported by P-glycoproteins.

The multidrug-resistance associated protein (ABC C) subfamily of proteins largely recognizes anionic chemicals. ABC C substrates are commonly conjugates of xenobiotics (e.g., glutathione, glucuronic acid, sulfate conjugates). Thus, conjugation not only restricts passive diffusion of a lipophilic chemical but actually targets the xenobiotic for active transport across the canalicular membrane.

9.5   RESPIRATORY ELIMINATION

The lungs are highly specialized organs that function in the uptake and elimination of volatile materials (i.e., gasses). Accordingly, the lungs can serve as a primary site for the elimination of chemicals that have a high vapor pressure. The functional unit of the lung is the alveolus. These small, highly vascularized, membranous sacs serve to exchange oxygen from the air to the blood (uptake), and conversely, exchange carbon dioxide from the blood to the air (elimination). This exchange occurs through passive diffusion. Chemicals that are sufficiently volatile also may diffuse across the alveolar membrane, resulting in removal of the chemical from the blood and elimination into the air.

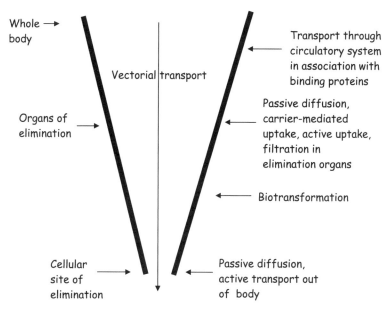

Figure 9.5   Processes involved in the vectorial transport of xenobiotics from the whole body point of origin to the specific site of elimination.

9.6   CONCLUSION

Many processes function coordinately to ensure that chemicals that are distributed throughout the body are efficiently eliminated at distinct and highly specialized locations. This directional transfer of chemicals from the site of origin (i.e., site of absorption, storage, and toxicity) to the site of elimination is a form of vectorial transport (Figure 9.5). The coordinate action of blood binding proteins, active transport proteins, blood filtration units, intracellular binding proteins, and bio-transformation enzymes ensures the unidirectional flow of chemicals ultimately resulting in their elimination. The evolution of this complex interplay of processes results in the efficient clearance of toxicants and has provided the way for the coevolution of complexity in form from unicellular to multiorgan organisms.

BIBLIOGRAPHY AND SUGGESTED READING

Kester, J. E. Liver. In *Encyclopedia of Toxicology*, Vol. 2, ed. P. Wexler, pp. 253–261. New York: Academic Press, 1998.

Miller, D. S. Cellular transport and elimination. In *Molecular and Biochemical Toxicology*, 4th ed., eds. R. C. Smart and E. Hodgson, pp. 273–286. Hoboken, NJ: Wiley, 2008.

Rankin, G. O. Kidney. In *Encyclopedia of Toxicology*, Vol. 2, ed. P. Wexler, pp. 198–225. New York: Academic Press, 1998.

Rozman, K. K. and C. D. Klassen. Absorption, distribution, and excretion of toxicants. In *Casarett & Doull's Toxicology*, 6th ed., ed. C. D. Klaassen, pp. 107–132. New York: McGraw-Hill, 2001.

SAMPLE QUESTIONS

1. What is meant by "vectorial transport" with respect to the elimination of toxicants from the body?

2. How do hepatic metabolic conjugation reactions contribute to the vectorial elimination of toxicants?

3. What are the bile canaliculi?

4. What is the role of serum binding proteins such as albumin and sex steroid binding protein in the elimination of toxicants?

TOXIC ACTION

Acute Toxicity

GERALD A. LEBLANC

10.1 INTRODUCTION

Acute toxicity of a chemical can be viewed from two perspectives. Acute toxicity may be viewed as a qualitative descriptor of an incident of poisoning. Consider the following statement: "methyl isocyanate gas, accidentally released from a chemical manufacturing facility in 1984, was *acutely toxic* to the residents of Bhopal, India." This statement implies that the residents of Bhopal were exposed to sufficiently high levels of methyl isocyanate over a relatively short time to result in immediate harm. High-level, short-term exposure resulting in immediate toxicity are all characteristics of acute toxicity. Alternatively, acute toxicity may represent a quantifiable characteristic of a material. For example, the statement: "the *acute toxicity* of methyl isocyanate, as measured by its LD_{50} in rats, is 140 mg/kg" defines the acute toxicity of the chemical. Again, the characterization of the quantified effects of methyl isocyanate as being acute toxicity implies that this quantification was derived from a short-term dosing experiment and that the response measured occurred within a short time period following dosing. Considering these qualitative and quantitative aspects, acute toxicity can be defined as *toxicity elicited immediately following short-term exposure to a chemical*. In accordance with this definition, two components comprise acute toxicity: acute exposure and acute effect.

10.2 ACUTE EXPOSURE AND EFFECT

In contrast to acute toxicity, chronic toxicity is characterized by prolonged exposure and sublethal effects elicited through mechanisms that are distinct from those that cause acute toxicity. Typically, acute and chronic toxicity of a chemical are easily distinguished. For example, mortality occurring within 2 days of a single dose of a chemical would be a prime example of acute toxicity (Figure 10.1a). Similarly, reduced litter size following continuous (i.e., daily) dosing of the parental organisms would be indicative of chronic toxicity (Figure 10.1b). However, defining toxicity as being acute or chronic is sometimes challenging. For example, chronic exposure

A Textbook of Modern Toxicology, Fourth Edition. Edited by Ernest Hodgson
Copyright © 2010 John Wiley & Sons, Inc.

(a) Short-term exposure resulting in immediate effects

(b) Continuous exposure resulting in sublethal effects

(c) Continuous exposure resulting in acute effects

(d) Short-term exposure resulting in later sublethal effects

Figure 10.1   Examples of exposure/effect scenarios that result in either acute toxicity (a), chronic toxicity (b), or mixed acute/chronic toxicity (c, d). Examples for each scenario are provided in the text.

to a persistent, lipophilic chemical may result in sequestration of significant levels of the chemical in adipose tissue of the organism with no resulting overt toxicity. Upon entering the reproductive phase, organisms may mobilize fatty stores, releasing the chemical into the bloodstream, resulting in overt toxicity including death (Figure 10.1c). One could argue under this scenario that chronic exposure ultimately resulted in acute effects. Lastly, acute exposure during a susceptible window of exposure (e.g., embryo development) may result in reproductive abnormalities and reduced fecundity once the organism has attained reproductive maturity (Figure 10.1d). Thus, acute exposure may result in chronic toxicity.

An additional consideration is noteworthy when comparing acute and chronic toxicity. All chemicals elicit acute toxicity at a sufficiently high dose, whereas, all chemicals do not elicit chronic toxicity. Paracelsus' often cited phrase "all things are poison … the dose determines … a poison" is clearly in reference to acute toxicity. Even the most benign substances will elicit acute toxicity if administered at a

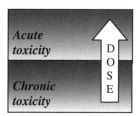

Figure 10.2 Relationships among chemical dose, acute toxicity, and chronic toxicity. All chemicals elicit acute toxicity at a sufficiently high dose. However, chronic toxicity may not occur since dosage elevation may simply lead to acute toxicity.

sufficiently high dose. However, raising the dose of a chemical does not ensure that chronic toxicity will ultimately be attained. Since chronic toxicity typically occurs at dosages below those that elicit acute toxicity, toxicity observed at the higher dosage may simply reflect acute, and not chronic, toxicity (Figure 10.2).

Effects encountered with acute toxicity commonly consist of mortality or morbidity. From a quantitative standpoint, these effects are measured as the lethal dose 50 (LD_{50}), effective dose 50 (ED_{50}), lethal concentration 50 (LC_{50}), or effective concentration 50 (EC_{50}). The LD_{50} and ED_{50} represent the dose of the material that causes mortality (LD_{50}) or some other defined effect (ED_{50}) in 50% of a treated population. The LC_{50} and EC_{50} represent the concentration of the material to which the organisms were exposed that causes mortality (LC_{50}) or some other defined effect (EC_{50}) in 50% of an exposed population. LD_{50} and ED_{50} are typically normalized to the weight of the animal (e.g., milligram chemical/kilogram body weight); whereas, LC_{50} and EC_{50} are typically normalized to the environment in which the organisms were exposed (e.g., milligram chemical/liter water for aquatic organisms). These measures of acute toxicity are used to assign a degree to toxicity to a chemical. For example, the following categories are used by the Organization for Economic Cooperation and Development (OECD):

LD_{50} (mg/kg)	Label
<5 mg/kg	Very toxic
>5 < 50	Toxic
>50 < 500	Harmful
>500	No label required

Results of acute toxicity tests are also used to identify dosages of chemical to be used in sublethal, chronic toxicity evaluations. Finally, results of acute toxicity tests can be used to provide insight to the mode of action of the toxicant as described further below.

10.3 DOSE–RESPONSE RELATIONSHIPS

Acute toxicity of a chemical is quantified by its dose–response curve. This relationship between dose of the chemical administered and the resulting response

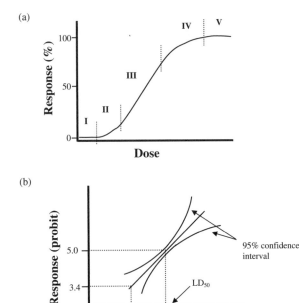

Figure 10.3 The dose–response relationship. (a) Five segments of the sigmoidal dose–response curve as described in the text. (b) Linearized dose–response relationship through log (dose)–probit (response) transformations. Locations of the LD_{50} and LD_{05} are depicted.

is established by treating groups of organisms with various amounts of the chemical. Dosing may be administered orally (e.g., gavage, food, water), through injection, from external application (dermal, ocular), or via environmental exposure (e.g., air). Ideally, doses are selected that will elicit >0% response but <100% response during the course of the experiment. At defined time periods following dosing, effects (e.g., mortality) are recorded. Results are plotted in order to define the dose–response curve (Figure 10.3a). A well-defined dose–response curve generated with a population of organisms whose susceptibility to the chemical is normally distributed will be sigmoidal in shape. The various segments (see Figure 10.3a) of the curve are represented as follows:

Segment I: This portion of the line has no slope and is represented by those doses of the toxicant that elicit no mortality to the treated population of organisms.

Segment II: This segment represents those doses of the toxicant that affects only the most susceptible members of the exposed population. Accordingly, these effects are elicited at low doses and only a small percentage of the dosed organisms are affected.

Segment III: This portion of the line encompasses those doses at which most of the groups of organisms elicit some response to the toxicant. Because most of the groups of exposed organisms respond to the toxicant within this range of dosages, segment III exhibits the steepest slope among the segments.

Segment IV: This portion of the line encompasses those doses of the toxicant that are toxic to even the most tolerant organisms in the populations. Accordingly, high doses of the toxicant are required to affect these organisms.

Segment V: Segment V has no slope and represents those doses at which 100% of the organisms exposed to the toxicant have been affected.

A well-defined dose–response curve can then be used to calculate the LD_{50} for the toxicant. However, in order to provide the best estimate of the LD_{50}, the curve is typically linearized through appropriate transformations of the data. A common transformation involves converting concentrations to logarithms and percentage effect to probit units (Figure 10.3b). Zero percent and 100% responses cannot be converted to probits; therefore, data within segments I and V are not used in the linearization. A 95% confidence interval also can be determined for the linearized dose–response relationship (Figure 10.3b). As depicted in Figure 10.3b, the greatest level of confidence (i.e., the smallest 95% confidence interval) exists at the 50% response level which is why LD_{50} values are favored over some other measure of acute toxicity (e.g., LD_{05}). This high level of confidence in the LD_{50} exists when ample data exist between the 51% and 99% response as well as between the 1% and 49% response.

Additional important information can be derived from a dose–response curve. The slope of the linearized data set provides information on the specificity of the toxicant. Steep slopes to the dose–response line are characteristic of toxicants that elicit toxicity by interacting with a specific target while shallow slopes to the dose–response line are characteristic of toxicant that elicit more nonspecific toxicity such as narcosis (see below). The dose–response line also can be used to estimate the threshold dose. The threshold dose is defined as the lowest dose of the chemical that would be expected to elicit a response under conditions at which the assay was performed. The threshold dose is often empirically estimated as being a dose less than the lowest dose at which an effect was measured but higher than the greatest dose at which no effect was detected. Conceptually, the threshold dose is defined as the intercept of segments I and II of the dose–response curve (Figure 10.3a). Statistically, the threshold dose can be estimated from the linearized dose–response curve as the LC_{05}. This value will closely approximate the threshold dose and can be statistically derived from the entire data set (i.e., the dose–response line). However, confidence in this value is greatly compromised since it is derived from one end of the line (Figure 10.3b).

10.4 NONCONVENTIONAL DOSE–RESPONSE RELATIONSHIPS

The low-level effects of chemicals have received attention among pharmacologists for over 100 years. A current resurgence in interest among pharmacologists in low-level effects stems from use of homeopathic approaches to treating disease. Proponents of homeopathy maintain that low levels of toxic materials stimulate physiological responses that can target disease without eliciting adverse effects in the individual undergoing treatment. Homeopathic principles may have application in toxicology based upon the premise that exposure to some chemicals at subthreshold levels, as defined by standard acute toxicity evaluations, can elicit toxicological

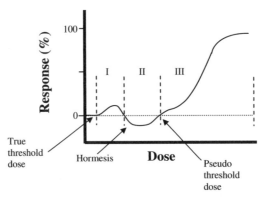

Figure 10.4 Nonconventional dose–response relationship involving low-dose effects and compensation. I: True initiation of the response followed by a compensatory response that returns the effects to the 0% level. II: A negative response due to overcompensation (hormesis) followed by recovery to the 0% response level. III: The standard sigmoidal dose–response relationship.

as well as pharmacological effects. Both pharmacological and toxicological homeopathy may be the consequence of hormesis.

Hormesis is defined as an overcompensatory response to some disruption in homeostasis resulting in a U or inverted U-shaped deflection at the low end of the dose–response curve. Accordingly, hormesis typically presents as an effect opposite to that elicited at higher doses of the chemical. For example, a chemical that stimulates corticosteroid secretion at high doses resulting in hyperadrenocorticism might elicit a hormetic response at low doses resulting in corticosteroid deficiency. A hypothetical nonconventional dose–response relationship resulting from such interactions is depicted in Figure 10.4. At the true threshold dose, the organisms begin to exhibit increased stimulation in corticosteroid secretion. However, at slightly higher doses, a compensatory response occurs whereby corticosteroid secretion is decreased in order to maintain homeostasis within the organism. Overcompensation may actually result in a decrease in corticosteroid secretion at certain toxicant dosages. Finally, the compensatory abilities of the organism are overcome by the high doses of the toxicant at the "pseudo" threshold dose, above which the standard dose–response relationship occurs. Nonconventional dose–response relationships have been observed with respect to both acute and chronic toxicity and are particularly relevant to the risk assessment process when establishing levels of exposure that are anticipated to pose no harm.

10.5 ALTERNATIVE METHODS

Conventional acute toxicity assessments performed with rodents can involve the use of a significant number of animals (e.g., 10 doses with10 animals per dose performed with males and females = 200 animals). Various approaches have been proposed as a means of reducing animal usage in the performance of acute toxicity tests.

10.5.1 Up-Down Method

Various derivations of the up-down method have been reported, but essentially, this method involves dosing a single animal with a starting dose of the chemical. This starting dose can be judged based upon toxicity tests performed with other species or with similar compounds. If the animal dies, then a second animal is dosed at a lower concentration. This progression of single dose levels administered to a single animal continues until several dosages are identified that do not elicit mortality. Similarly, if the starting dose is not lethal, then subsequent doses are administered at progressively higher levels until lethal dosages are identified. This method has proven valuable in bracketing the LD_{50} often with the use of less than 10 animals.

10.5.2 Fixed-Dose Method

The fixed-dose method has proven utility in assigning a chemical into a toxicity category. For example, fixed-doses of 5, 50, and 500 mg/kg might each be administered to five animals and responses used to classify a chemical accordingly the scheme present in Section 10.2. The acute toxicity of the chemical would be classified based upon the lowest dose that elicited no significant response among the treated animals.

10.5.3 *In Vitro* Methods

Cytotoxicity tests performed with human cell lines have been shown to be good predictors of blood concentrations of chemicals that are lethal to humans. The strength of the predictive value of cytotoxicity tests seems to be in the fact that for many chemicals, the cellular targets through which the chemicals cause toxicity are common to all cells (e.g., membranes, mitochondrial respiratory enzymes, etc.). The predictive value of cytotoxicity tests can be improved when used in conjunction with toxicokinetic modeling to consider factors related to adsorption, distribution, metabolism, and elimination that are relevant to *in vivo* toxicity but are inadequately represented in cultured cells.

10.6 MECHANISMS OF ACUTE TOXICITY

An exhaustive review of the mechanisms by which chemicals cause acute toxicity is beyond the scope of this chapter. However, certain mechanisms of toxicity are relevant since they are common to many important classes of toxicants. Some of these mechanisms of acute toxicity are discussed.

10.6.1 Narcosis

Narcosis in toxicology is defined as toxicity resulting from chemicals associating with and disrupting the lipid bilayer of membranes. Narcotics are classified as either nonpolar (Class 1) or polar (Class 2) compounds. Members of both classes of compounds are lipid soluble. However, Class 2 compounds possess constituents that confer some charge distribution to the compound (e.g., aliphatic and aromatic amines, nitroaromatics, alcohols). The aliphatic hydrocarbon (C5 through C8) are examples of powerful Class 1 narcotics, whereas ethanol is an example of a Class 2 narcotic. The affinity of narcotics to partition into the nonpolar core of membranes (Class 1 narcotics) or to distribute in both the polar and nonpolar components of membranes (Class 2 narcotics) alters the fluidity of the membrane. This effect compromises the ability of proteins and other constituents of the membranes to function properly, leading to various manifestation of narcosis. The central nervous system is the prime target of chemical narcosis and symptoms initially include disorientation, euphoria, giddiness, and progress to unconsciousness, convulsion, and death.

10.6.2   Acetylcholinesterase Inhibition

Acetylcholine is a neurotransmitter that functions in conveying nerve impulses across synaptic clefts within the central and autonomic nervous systems and at junctures of nerves and muscles. Following transmission of an impulse across the synapse by the release of acetylcholine, acetylcholinesterase is released into the synaptic cleft. This enzyme hydrolyzes acetylcholine to choline and acetate and transmission of the nerve impulse is terminated. The inhibition of acetylcholineasterase results in prolonged, uncoordinated nerve or muscle stimulation. Organophosphorus and carbamate pesticides (Chapter 4) along with some nerve gases (e.g., sarin) elicit toxicity via this mechanism.

Inhibitors of acetylcholinesterase function by binding to the substrate-binding site of the enzyme (Figure 10.5). Typically, the inhibitor or a biotransformation derivative of the inhibitor (e.g., the phosphodiester component of organophosphorus compounds) covalently binds to the enzyme resulting in its inhibition. Inhibition persists until the bound inhibitor is hydrolytically cleaved from the enzyme. This inhibition may be range from minutes in duration to permanent. Toxic effects of cholinesterase inhibition typically are evident when the enzyme activity is inhibited by ~50%. Symptoms include nausea and vomiting, increased salivation and sweating, blurred vision, weakness, chest pains. Convulsions typically occur between 50% and 80% enzyme inhibition with death at 80–90% inhibition. Death is most commonly due to respiratory failure.

10.6.3   Ion Channel Modulators

Ion transport is central to nerve impulse transmission both along the axon and at the synapse, and many neurotoxicants elicit effects by interfering with the normal transport of these ions (Figure 10.6). The action potential of an axon is maintained by the high concentration of sodium on the outside of the cell as compared to the low concentration inside. Active transporters of sodium (Na^+K^+ ATPases) that actively transport sodium out of the cell establish this action potential. One action of the insecticide dichlorodiphenyltrichloroethane (DDT) resulting in its acute toxicity is the inhibition of these Na^+K^+ ATPases resulting in the inability of the nerve to establish an action potential. Pyrethroid insecticides also elicit neurotoxicity through this mechanism. DDT also inhibits $Ca^{2+}Mg^{2+}$ ATPases that are important to neuronal repolarization and the cessation of impulse transmission across synapses.

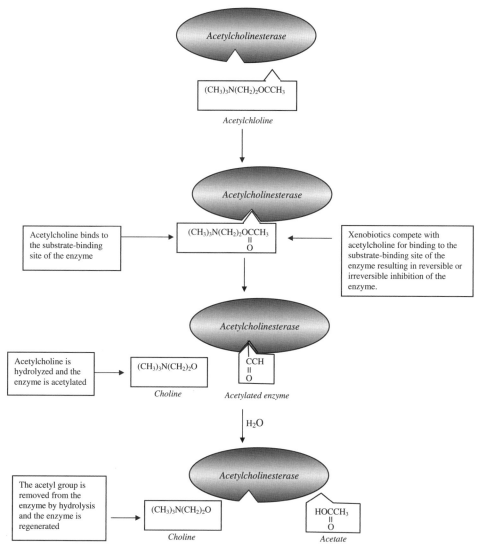

Figure 10.5 Hydrolysis of acetylcholine by the enzyme acetylcholinesterase and its inhibition by toxicants such as organophosphorus and carbamate insecticides.

The $GABA_A$ receptor is associated with chloride channels on the postsynaptic region of the neuron, and binding of gamma-aminobutyric acid (GABA) to the receptor causes opening of the chloride channel. This occurs after transmission of the nerve impulse across the synaptic cleft and postsynaptic depolarization. Thus, activation of $GABA_A$ serves to prevent excessive excitation of the postsynaptic neuron. Many neurotoxicants function by inhibiting the $GABA_A$ receptor resulting in prolonged closure of the chloride channel and excess nerve excitation. Cyclodiene insecticides (e.g., dieldrin), the organochlorine insecticide lindane, and some pyrethroid insecticides all elicit acute neurotoxicity, at least in part, through this mechanism. Symptoms of $GABA_A$ inhibition include dizziness, headache, nausea,

Figure 10.6 Ion channels that facilitate nerve impulse transmission and that are susceptible to perturbation by various toxicants and drugs. Ion transport inhibitors are indicated in parentheses.

vomiting, fatigue, tremors, convulsions, and death. Avermectins constitute a class of pesticides that are used extensively in veterinary medicine to treat a variety of parasitic conditions. While the mode of toxicity of these compounds is not precisely known, they appear to bind a distinct subset of chloride channels (GABA-insensitive chloride channels) resulting in disruptions in normal chloride transport across nerve cell membranes. Barbituates (e.g., phenobarbital) and ethanol elicit central nervous system effects, at least in part, by binding to $GABA_A$ receptors. However, unlike the previously discussed chemicals, these compounds enhance the ability of gamma-aminobutyric acid to bind the receptor and open the chloride channel. Accordingly, these compounds suppress nerve transmission which contributes to the sedative action of the chemicals.

10.6.4 Inhibitors of Cellular Respiration

Cellular respiration is the process whereby energy, in the form of ATP, is generated in the cell while molecular oxygen is consumed. The process occurs along respiratory assemblies that are located in the inner mitochondrial membrane. Electrons derived from NADH or $FADH_2$ are transferred along a chain of electron carrier proteins. This step-by-step transfer leads to the pumping of protons out of the mitochrondrial matrix resulting in the generation of a membrane potential across the inner mitochrondrial membrane (Figure 10.7). Protons are pumped out of the mitochrondrial matrix at three locations along the respiratory chain. Site 1 consists of the NADH-Q reductase complex; site 2 consists of the QH_2-cytochrome c reductase complex; and, site 3 is the cytochrome c-oxidase complex. Adenosine triphosphate (ATP) is generated from adenosine diphosphate (ADP) when protons flow back across the membrane through an ATP synthetase complex to the mitochrondrial matrix. The transfer of electrons culminates with the reduction of molecular oxygen to water.

Many chemicals can interfere with cellular respiration by binding to the cytochromes that constitute the electron transport chain and inhibiting the flow of electrons along this protein complex. The pesticide rotenone specifically inhibits electron transfer early in the chain with inhibition of proton transport beginning at site 1. Actimycin A inhibits electron transfer and proton pumping at site 2. Cyanide, hydrogen sulfide, and azide inhibit electron flow between the cytochrome oxidase

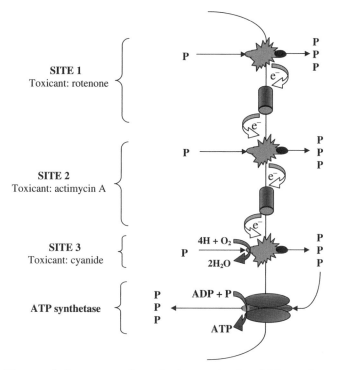

SITE 1
Toxicant: rotenone

SITE 2
Toxicant: actimycin A

SITE 3
Toxicant: cyanide

ATP synthetase

Figure 10.7 Electron (e^-) transport along the inner mitochondrial membrane resulting in the pumping of protons (P) out of the mitochondrial matrix. Protons are shuttled back into the matrix through the ATP synthetase complex where ATP is generated. Sites of toxicant action are indicated.

complex and O_2 preventing the generation of a proton gradient at site 3. Symptoms of toxicity from the inhibition of respiratory chain include excess salivation, giddiness, headache, palpitations, respiratory distress, and loss of consciousness. Potent inhibitors such as cyanide can cause death due to respiratory arrest immediately following poisoning.

Some chemicals do not interfere with electron transport leading to the consumption of molecular oxygen, but rather, interfere with the conversion of ADP to ATP. These uncouplers of oxidative phosphorylation function by leaking protons across the inner membrane back to the mitochrondrial matrix. As a result, a membrane potential is not generated, and energy required for the phosphorylation of ADP to ATP is lost. The uncoupling of oxidative phosphorylation results in increased electron transport, increased oxygen consumption, and heat production. The controlled uncoupling of oxidative phosphorylation is a physiologically relevant means of maintaining body temperature by hibernating animals, some newborn animals, and in some animals that inhabit cold environments. Chemicals known to cause uncoupling of oxidative phosphorylation include 2,4-dinitrophenol, pentachlorophenol, and dicumarol. Symptoms of intoxication include accelerated respiration and pulse, flushed skin, elevated temperature, sweating, nausea, coma, and death.

BIBLIOGRAPHY AND SUGGESTED READING

Calabrese, E. J. Hormesis: A revolution in toxicology, risk assessment and medicine. *EMBO Rep.* **5**:S37–S40, 2004.

Eaton, D. L. and C. D. Klaassen. Principles of toxicology. In *Casarrett & Doull's Toxicology*, 6th ed., ed. C. D. Klaassen, pp. 11–34. New York: McGraw-Hill, 2001.

Joy, R. M. Neurotoxicology: Central and peripheral. In *Encyclopedia of Toxicology*, Vol. 2, ed. P. Wexler, pp. 389–413. New York: Academic Press, 1998.

Walum, E. Acute oral toxicity. *Environ. Health Perspect.* **106** (Suppl. 2): 497–503, 1998.

SAMPLE QUESTIONS

1. The term "acute toxicity" can be used as a qualitative descriptor of an incidence of toxicity or a quantitative measure of toxicity. Describe how the term "acute toxicity" functions in both capacities.

2. To paraphrase Paracelsus: "all things are poison … the dose determines … a poison." Why is this old adage relevant primarily to acute (as opposed to chronic) toxicity?

3. Why is the LD_{50} value used as a measure of acute toxicity as opposed to the LD_{10}?

4. Which segment of a dose–response curve (I through V) is most important in defining the acute toxicity of a chemical? Why?

Chemical Carcinogenesis and Mutagenesis

ROBERT C. SMART

11.1 DNA DAMAGE AND MUTAGENESIS

DNA is chemically reactive and is subject to chemical modifications that can alter the coding properties of the bases and also cause single or double strand breaks in the DNA backbone. DNA can be modified by endogenous processes such as a base deamination (i.e., cytosine is spontaneously deaminated to form uracil), oxidative stress, lipid peroxidation, and spontaneous hydrolysis to produce apurinic/apyriminidinic base sites (AP site) in the DNA. DNA can also be modified by exogenous or environmental agents including ionizing radiation, UV radiation, chemotherapeutics, and chemical carcinogens (Figure 11.1). If there is an error in the repair of the DNA damage or if the damage is not repaired, an error could occur in the newly synthesized DNA resulting in a mutation in the daughter cell. A mutation is a permanent heritable alteration in the DNA which alters the base sequence. A germinal mutation occurs in ova or sperm cells and can be passed to future generations while a somatic mutation occurs in nongerm cells and cannot be passed to future generations. DNA damaging agents produce three general types of genetic alterations: (1) gene mutations which include point mutations involving single base pair substitutions that can result in amino acid substitutions in the encoded protein and frame shift mutations involving the loss or gain of one or two base pairs resulting in an altered reading frame and gross alterations in the encoded protein; (2) chromosome aberrations including gross chromosomal rearrangement such as deletions, duplications, inversions, and translocations; and (3) aneuploidy which involves the gain or loss of one or more chromosomes. Point mutations are further classified as missense or nonsense mutations. A missense mutation produces an altered protein in which an incorrect amino acid has been substituted for the correct amino acid. A nonsense mutation is an alteration that produces a stop codon and results in a truncated protein. A point mutation can also be characterized based on the mutagen-induced substitution of one base for another within the DNA. When a

A Textbook of Modern Toxicology, Fourth Edition. Edited by Ernest Hodgson
Copyright © 2010 John Wiley & Sons, Inc.

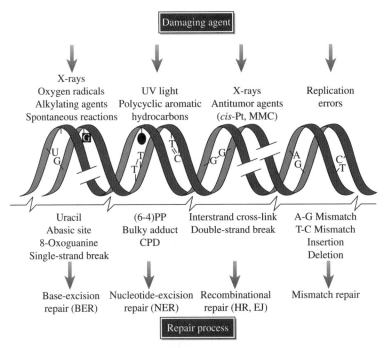

Figure 11.1   DNA damage and repair. DNA damaging agent produce specific type of DNA damage, and these are repaired by specific type DNA repair mechanisms. UVB produces 6-4 photoproducts (6-4)PP and cyclobutane pyrimidine dimers (CPD). *cis*-PT, cicplatinum; MMC, mitomycin C; HR, homologus recombinations; EJ, end joining. Adapted from Hoeijmakers, J. J. *Nature* **411**:366, 2001.

point mutation produces a substitution of a purine for another purine (i.e., guanine for adenine) or a pyrimidine for another pyrimidine (i.e., thymine for cytosine) the mutation is referred as a transition. If a purine is substituted for a pyrimidine or vice versa (i.e., thymine for adenine or guanine for cytosine), the mutation is referred to as a transversion.

Fortunately, cells of higher eukaryotes have four efficient repair systems that can repair specific types of DNA damage (Figure 11.1). Base excision repair (BER) repairs deaminated DNA, AP sites, alkylated DNA, oxidized bases, and single strand breaks. Nucleotide excision repair (NER) repairs DNA that contains large bulky adducts (i.e., polycyclic aromatic hydrocarbons [PAHs]) as well as UV-induced bulky cyclobutane pyrimidine dimers and 6-4 photoproducts. Recombinational repair includes homologous recombination repair and end joining repair, and these repair double strand breaks in DNA. The fourth system is mismatch repair, and it repairs base mismatches between bases on opposing strands of DNA. Cell of higher eukaryotes respond to DNA damage by engaging cell cycle checkpoints which pause the cell in the cell cycle to allow time for DNA repair, or if the damage is too extensive, damaged cells commit to apoptosis (programmed cell death).

11.2 GENERAL ASPECTS OF CANCER

Carcinogenesis is the process through which cancer develops. Chemical carcinogenesis is the study of the mechanisms through which chemical carcinogens induce cancer and also involves the development/utilization of experimental systems aimed at determining whether a substance is a potential human carcinogen. An important aspect of toxicology is the identification of potential human carcinogens.

Cancer is not a single disease but a large group of diseases, all of which can be characterized by the uncontrolled growth of an abnormal cell to produce a population of cells that have acquired the ability to multiply and invade surrounding and distant tissues. It is this invasive characteristic that imparts its lethality on the host.

Epidemiology studies have revealed that the incidence of most cancers increase exponentially with age (Figure 11.2). Epidemiologists have interpreted this exponential increase in cancer incidence to denote that three to seven critical mutations or "hits" within a single cell are required for cancer development. Molecular analyses of human tumors have confirmed the accumulation of mutations in critical genes in the development of cancer. Most cancers are monoclonal in origin (derived from a single cell) and do not arise from a single critical mutation but from the accumulation of multiple critical mutations in relevant target genes within a single cell (Figure 11.3). Initially, a somatic mutation occurs in a critical gene, and this mutation provides a proliferative advantage to the cell and results in the expansion or proliferation of the mutant clone. Over time, an additional mutation in a critical gene that occurs within this clone provides a further selective growth advantage. This process of mutation and selection is repeated over time and eventually results in clone of cells with mutations in multiple critical genes. It often requires decades

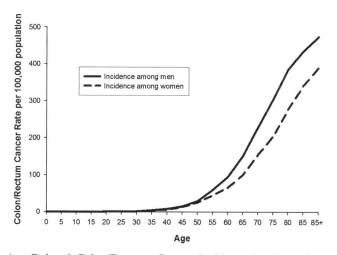

Figure 11.2 Age-Related Colon/Rectum Cancer Incidence in the United States (2000–2003). Epidemiological studies have revealed the incidence of most human cancers increase exponentially with age. NCI SEER. *Cancer Statistics Review, 1975–2003.* http://seer.cancer.gov.

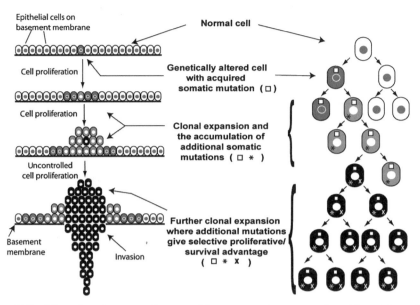

Figure 11.3   Monoclonal nature of cancer. Most cancers are monoclonal in origin and are derived from the accumulation of sequential mutations in an individual cell. Each additional mutation within the cell provides a further proliferative/survival advantage, and this process drives clonal expansion and tumor development.

for a cell clone to accumulate multiple critical mutations and for the progeny of this cell to clonally expand to produce a clinically detectable cancer. Thus, the time required for accumulation of mutations in critical genes within a cell is related to the observation that cancer incidence increases exponentially with age.

Specific genes found in normal cells, termed proto-oncogenes, are involved in the positive regulation of cell proliferation and cell survival and are frequently mutated in cancer. Mutational activation of these proto-oncogenes results in a gain of function in which the altered gene product continually stimulates cell proliferation or increases cell survival (block apoptosis). Such proto-oncogenes with gain-of-function mutations are now referred to as oncogenes. Another family of genes, known as tumor suppressor genes, can be mutationally inactivated during carcino-genesis, resulting in a loss of function. Tumor suppressor genes and the proteins they encode often function as negative regulators of cell proliferation or positive regulators of apoptosis. Tumor suppressor genes containing loss-of-function mutations encode proteins that are inactive so they are unable to inhibit cell proliferation or induce apoptosis in response to DNA damage or activated oncogenes. DNA stability genes responsible for genome maintenance are also mutated in some cancers, and their reduced activity contributes to genomic instability and the accumulation of mutations in oncogenes and tumor suppressor genes. The activation of oncogenes and inactivation of tumor suppressor genes within a cell as well as the alteration in genes responsible for genomic maintenance (repair, checkpoints, etc.) are important mutational events in carcinogenesis (Figure 11.4). A simple analogy can be made to the automobile; tumor suppressor genes are analogous to the brakes on the car while

Figure 11.4 General aspects of the cancer process. Somatic mutations involving gain-of-function mutations in proto-oncogenes and loss-of-function mutations in tumor suppressor genes provide for a selective proliferative/survival advantage and are critical events in tumorigenesis. Inactivating mutations in genes involved in genomic maintenance can result in genomic instability.

the proto-oncogenes are analogous to the accelerator pedal, and DNA stability genes are analogous to the automobile mechanic. Mutations within tumor suppressor genes inactivate the braking system while mutations in proto-oncogenes activate the acceleration system. Altering both the cellular brakes and cellular accelerator results in uncontrolled cell proliferation while mutations in genes responsible for DNA stability and genomic maintenance is akin to having an inept mechanic. Mutations in oncogenes, tumor suppressor genes, and DNA stability genes provide a selective growth advantage to the cancer cell through enhanced cell proliferation, decreased apoptosis, and increased genomic instability (Figure 11.4).

Cancer is a type of a neoplasm or tumor. While technically a tumor is defined as only a tissue swelling, it is now used as a synonym for a neoplasm. A neoplasm or tumor is an abnormal mass of tissue, the growth of which exceeds and is uncoordinated with the normal tissue, and persists after cessation of the stimuli that evoked it. There are two basic types of neoplasms, termed benign and malignant. The general characteristics of these tumors are defined in Table 11.1. Cancer is the general name for a malignant neoplasm. In terms of cancer nomenclature, most adult cancers are derived from epithelial cells (colon, lung, breast, skin, etc.) and are termed carcinomas. Sarcomas are derived from mesenchymal tissues, while leukemias and lymphomas are derived from blood-forming cells and lymphocytes, respectively. Melanoma is derived from melanocytes, and retinoblastoma, glioblastoma, and neuroblastoma are derived from the stem cells of the retina, glia, and neurons, respectively. According to the American Cancer Society, (1) the lifetime risk for developing cancer in the United States is one in three for women and one

TABLE 11.1   Some General Characteristics of Malignant and Benign Neoplasms

Benign	Malignant
Generally slow growing	May be slow to rapid growing
Few mitotic figures	Numerous mitotic figures
Well differentiated and architecture resembles that of parent tissue	Some lack differentiation, disorganized; loss of parent tissue architecture
Sharply demarcated mass that does not invade surrounding tissue	Locally invasive, infiltrating into surrounding normal tissue
No metastases	Metastases

in two for men; (2) in 2009 about 1.4 million new cancer cases are expected to be diagnosed, not including carcinoma *in situ* or basal or squamous cell skin cancer; and (3) cancer is a leading cause of death in the United States and approximately 23% of all deaths are due to cancer.

11.3   HUMAN CANCER

Although cancer is known to occur in many groups of animals, the primary interest of most cancer research is focused on human cancer. Nevertheless, much of the mechanistic research and carcinogen assessment is carried out in experimental models, usually rodent models. To begin to appreciate the complexity of this subject, it is important to first have some understanding of human cancer and its etiologies.

11.3.1   Causes, Incidence, and Mortality Rates of Human Cancer

Cancer cases and cancer deaths by sites and sex for the United States are shown in Figure 11.5. Breast, lung, and colon and rectum cancers are the major cancers in females while prostate, lung, and colon and rectum are the major cancer sites in males. A comparison of cancer deaths versus incidence for a given site reveals that prognosis for lung cancer cases is poor while that for breast or prostate cancer cases is much better. Age-adjusted cancer mortality rates (1930–2005) for selected sites in males are shown in Figure 11.6 and for females is shown in Figure 11.7. The increase in the mortality rate associated with lung cancer in both females and males is striking and is due to cigarette smoking. It is estimated that 87% of lung cancers are due to smoking. Lung cancer death rates in males and females began to increase in the mid 1930s and mid 1960s, respectively. These time differences are due to the fact that cigarette smoking among females did not become popular until the 1940s while smoking among males was popular in the early 1900s. Taking into account these differences along with a 20–25-year lag period for the cancer to develop explains the differences in the temporal increase in lung cancer death rates in males and female. Another disturbing statistic is that lung cancer, a theoretically preventable cancer, has recently surpassed breast cancer as the cancer responsible for the greatest number of cancer deaths in women. In addition to lung cancer, smoking also plays a significant role in cancer of the mouth, esophagus, pancreas, pharynx,

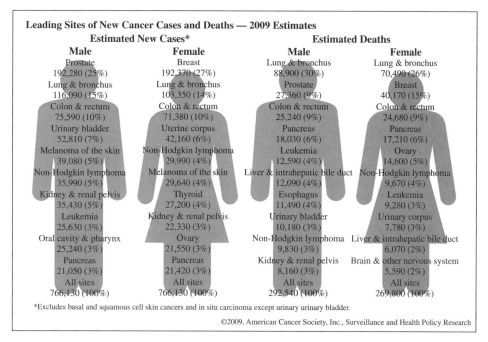

Leading Sites of New Cancer Cases and Deaths — 2009 Estimates

Estimated New Cases*		Estimated Deaths	
Male	**Female**	**Male**	**Female**
Prostate 192,280 (25%)	Breast 192,370 (27%)	Lung & bronchus 88,900 (30%)	Lung & bronchus 70,490 (26%)
Lung & bronchus 116,090 (15%)	Lung & bronchus 103,350 (14%)	Prostate 27,360 (9%)	Breast 40,170 (15%)
Colon & rectum 75,590 (10%)	Colon & rectum 71,380 (10%)	Colon & rectum 25,240 (9%)	Colon & rectum 24,680 (9%)
Urinary bladder 52,810 (7%)	Uterine corpus 42,160 (6%)	Pancreas 18,030 (6%)	Pancreas 17,210 (6%)
Melanoma of the skin 39,080 (5%)	Non-Hodgkin lymphoma 29,990 (4%)	Leukemia 12,590 (4%)	Ovary 14,600 (5%)
Non-Hodgkin lymphoma 35,990 (5%)	Melanoma of the skin 29,640 (4%)	Liver & intrahepatic bile duct 12,090 (4%)	Non-Hodgkin lymphoma 9,670 (4%)
Kidney & renal pelvis 35,430 (5%)	Thyroid 27,200 (4%)	Esophagus 11,490 (4%)	Leukemia 9,280 (3%)
Leukemia 25,630 (3%)	Kidney & renal pelvis 22,330 (3%)	Urinary bladder 10,180 (3%)	Uterine corpus 7,780 (3%)
Oral cavity & pharynx 25,240 (3%)	Ovary 21,550 (3%)	Non-Hodgkin lymphoma 9,830 (3%)	Liver & intrahepatic bile duct 6,070 (2%)
Pancreas 21,050 (3%)	Pancreas 21,420 (3%)	Kidney & renal pelvis 8,160 (3%)	Brain & other nervous system 5,590 (2%)
All sites 766,130 (100%)	All sites 766,130 (100%)	All sites 292,540 (100%)	All sites 269,800 (100%)

*Excludes basal and squamous cell skin cancers and in situ carcinoma except urinary urinary bladder.

©2009, American Cancer Society, Inc., Surveillance and Health Policy Research

Figure 11.5 Cancer Cases and Cancer Deaths by Sites and Sex—2009 Estimates. Reprinted with permission of the American Cancer Society. *Cancer Facts and Figures 2007*. Atlanta: American Cancer Society.

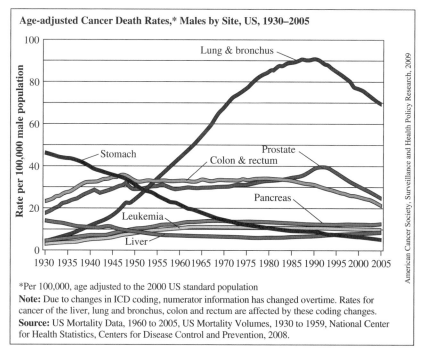

Age-adjusted Cancer Death Rates,* Males by Site, US, 1930–2005

American Cancer Society, Surveillance and Health Policy Research, 2009

*Per 100,000, age adjusted to the 2000 US standard population

Note: Due to changes in ICD coding, numerator information has changed overtime. Rates for cancer of the liver, lung and bronchus, colon and rectum are affected by these coding changes.

Source: US Mortality Data, 1960 to 2005, US Mortality Volumes, 1930 to 1959, National Center for Health Statistics, Centers for Disease Control and Prevention, 2008.

Figure 11.6 Age-Adjusted Cancer Death Rates in Males from 1930–2005. Reprinted with permission of the American Cancer Society. *Cancer Facts and Figures 2009*. Atlanta: American Cancer Society.

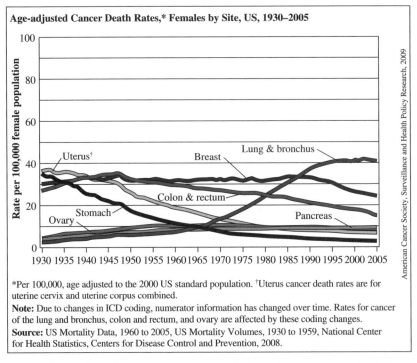

Figure 11.7 Age-Adjusted Cancer Death Rates in Females from 1930–2005. Reprinted with permission of the American Cancer Society. *Cancer Facts and Figures 2009.* Atlanta: American Cancer Society.

larynx, bladder, kidney, and uterine cervix. In general, it is important to realize that death and incidence rates for some types of cancers are increasing while the rates for others are decreasing or remaining constant.

Major insights into the etiologies of cancer have been attained through epidemiological studies that relate the role of hereditary, environmental, and cultural influences on cancer incidence as well as through laboratory studies using model rodent/cellular systems. Cancer susceptibility is determined by complex interactions between age, environment, and an individual's genetic makeup. It is estimated from epidemiological studies that 35–80% of all cancers are associated with the environment in which we live and work. The geographic migration of immigrant populations and differences in cancer incidence between communities has provided an important information regarding the role of the environment and specific cancer incidences. For example, Japanese immigrants and the sons of Japanese immigrants living in California begin to assume a cancer death rate similar to the California white population (Figure 11.8). These results implicate a role for the environment in the etiology of cancer. It should be noted that the term environment is not restricted to exposure to man-made chemicals in the environment but applies to all aspects of our lifestyle, including smoking, diet, cultural and sexual behavior, occupation, natural and medical radiation, and exposure to substances in air, water, and

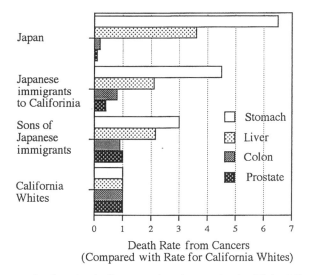

Figure 11.8 Cancer death rates in Japanese immigrants in the United States. The cancer death rate for each type of cancer is normalized to one for California White males and then compared to Japanese immigrants of a similar age. Adapted from Cairns, J. *Readings in Scientific American—Cancer Biology*, p. 13, 1986.

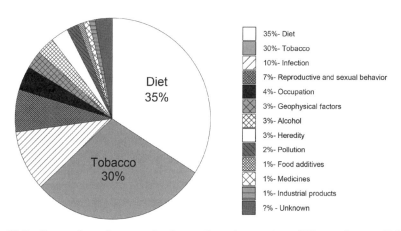

Figure 11.9 Proportion of cancer deaths attributed to various different factors. Values are a best estimate as determined by epidemiological studies. Adapted from Doll, R. and R. Peto. *The Causes of Cancer: Quantitative Estimates of Avoidable Risks of Cancer in the United States Today*. Oxford Medical Publications, 1981.

soil. The major factors associated with cancer and their estimated contributions to human cancer incidence are listed in Figure 11.9. Only a small percentage of total cancer occurs in individuals with a hereditary mutation/hereditary cancer syndrome (~5%). However, an individual's genetic background is the "stage" in which the cancer develops and susceptibility genes have been identified in humans. For example, genetic polymorphisms in enzymes responsible for the activation of chemical carcinogens may represent a risk factor as is the case for polymorphisms in the *N*-acetyltransferase gene and the risk of bladder cancer. These types of genetic risk factors are of low penetrance (low to moderate increased risk); however, increased risk is usually associated with environmental exposure. While the values presented in Figure 11.9 are a best estimate, it is clear that tobacco use and diet constitute the major factors associated with human cancer incidence. If one considers all of the categories that pertain to man-made chemicals, it is estimated that their contribution to human cancer incidence is approximately 10%. However, the factors listed in Figure 11.9 are not mutually exclusive since there is likely to be interaction between these factors in the multistep process of carcinogenesis.

11.3.2 Known Human Carcinogens

Two of the earliest observations that exposure of humans to certain chemicals or substances is related to an increased incidence of cancer were made independently by two English physicians, John Hill in 1771 and Sir Percival Pott in 1776. Hill observed an increased incidence of nasal cancer among snuff users while Pott observed that chimney sweeps had an increased incidence of skin cancer of the scrotum. Pott attributed this to topical exposure to soot and coal tar. It was not until a century and a half later in 1915 when two Japanese scientists, Yamagiwa and Ichikawa, substantiated Pott's observation by demonstrating that multiple topical applications of coal tar to rabbit skin produced skin carcinomas. This experiment is important for two major reasons: (1) it was the first demonstration that a chemical or substance could produce cancer in animals, and (2) it confirmed Pott's initial observation and established a relationship between human epidemiology studies and animal carcinogenicity. Because of these important findings, Yamagiwa and Ichikawa are considered the fathers of experimental chemical carcinogenesis. In the 1930s, Kennaway and coworkers isolated active carcinogenic chemicals from coal tar and identified one as benzo[a]pyrene, a PAH that results from the incomplete combustion of organic molecules. Benzo[a]pyrene has also been identified as one of the carcinogens in cigarette smoke. The p53 tumor suppressor gene can be mutationally inactivated by numerous carcinogens, including the carcinogenic metabolite of benzo[a]pyrene.

Epidemiological studies have provided sufficient evidence that exposure to a variety of chemicals, agents, or processes are associated with human cancer. For example, the following causal associations have emerged between exposure and the development of specific cancers; vinyl chloride and hepatic cancer, amine dyes and bladder cancer, benzene and leukemia, diethylstilbestrol and clear cell carcinoma of the vagina and cigarette smoking and lung cancer. Naturally occurring chemicals or agents such as asbestos, aflatoxins, nickel, and certain arsenic compounds are also associated with an increased incidence of certain human cancers. Both epidemiological studies and rodent carcinogenicity studies are important in the identification and classification of potential human carcinogens. The strongest evidence for establishing whether exposure to a given chemical is carcinogenic in humans comes from epidemiological studies. However, these studies are complicated by the fact that it often takes 20–30 years after carcinogen exposure for a clinically detectable cancer to develop. This delay is problematic and can result in inaccurate historical exposure information and additional complexity due to the interference of a large number of confounding variables. Most importantly, this lag period can also prevent the timely identification of a putative carcinogen and result in unnecessary human exposure. Therefore, methods to identify potential human carcinogens have been developed. The long-term rodent bioassay, also known as the 2-year rodent carcinogenesis bioassay (see Chapter 20), is currently used in an attempt to identify potential human carcinogens. It is clear that almost all, if not all, human carcinogens identified to date are rodent carcinogens; however, it is not known if all rodent carcinogens are human carcinogens. Indeed, identification of possible human carcinogens based on rodent carcinogenicity is complicated (see below). Table 11.2 contains the list of the known human carcinogens as reported by the National Toxicology Program's 11th Report on Carcinogens. The entire 11th Report on Carcinogens which succinctly summarizes the information for the classification of each listed agent (246 agents) is available at http://ntp.niehs.nih.gov. In addition, the

TABLE 11.2   Known Human Carcinogens Listed in 11th Report on Carcinogens (National Toxicology Program)

Aflatoxins
Alcoholic Beverage Consumption
4-Aminobiphenyl
Analgesic Mixtures Containing Phenacetin
Arsenic Compounds, Inorganic
Asbestos
Azathioprine
Benzene
Benzidine
Beryllium and Beryllium Compounds
1,3-Butadiene
1,4-Butanediol Dimethanesulfonate (Myleran)
Cadmium and Cadmium Compounds
Chorambucil
1-(2-Choroethyl)-3-(4-methylcyclohexyl)-1-nitrosourea (MeCCNU)
Bis(Chloromethyl) Ether and Technical Grade Chloromethyl Methyl Ether
Chromium Hexavalent Compounds
Coal Tar Pitches
Coal Tars
Coke Oven Emissions
Cyclophosphamide
Cyclosporin A
Diethylstilbestrol
Dyes Metabolized to Benzidine
Environmental Tobacco Smoke
Erionite
Estrogens, Steroidal
Ethylene Oxide
Hepatitis B Virus
Hepatitis C Virus
Human Papilloma Viruses: Some Genital-Mucosal Types
Melphalan

TABLE 11.2 *Continued*

Methoxsalen with Ultraviolet A Therapy (PUVA)
Mineral Oils (Untreated and Mildly Treated)
Mustard Gas
2-Naphthylamine
Neutrons
Nickel Compounds
Radon
Silica, Crystalline (Respirable Size)
Smokeless Tobacco
Solar Radiation
Soots
Strong Inorganic Acid Mists Containing Sulfuric Acid
Sunlamps/sunbeds
Tamoxifen
2,3,7,8-tetrachlorodibenzo-*p*-dioxin (TCDD)
Thiotepa
Thorium Dioxide
Tobacco Smoking
Vinyl Chloride
Ultraviolet Radiation, Broad Spectrum
Wood Dust
X-Radiation and Gamma Radiation

International Agency for Research on Cancer (IARC) maintains their own complete list of human carcinogens as well as carcinogenic complex mixtures and occupations associated with increased cancer incidence (http://monographs.iarc.fr/ENG/Classification/index.php). *In vitro* mutagenicity assays are also used to identify mutagenic agents that may have carcinogenic activity (see Section 11.3.4).

11.3.3 Classification of Human Carcinogens

Identification and classification of potential human carcinogens through the 2-year rodent carcinogenesis bioassay is complicated by species differences, use of high doses (MTD, the maximum tolerated dose), the short life span of the rodents, high background tumor incidence in some organs, sample size, and the need to extrapolate from high to low doses for human risk assessment. MTD is the highest dose used in the rodent bioassay and is operationally defined in toxicology as the highest daily dose of a chemical that does not cause overt toxicity in a 90-day study in laboratory mice or rats. Although these problems are by no means trivial, the rodent 2-year bioassay is still considered the "gold standard" assay for the identification of potential human carcinogens. The criteria used for the classification of carcinogens by the National Toxicology Program, 11th Report on Carcinogens 2005 is as follows: *known human carcinogen category* which is reserved for those substances for which sufficient evidence of carcinogenicity from studies in humans exists, indicating a cause and effect relationship between exposure to the substance and human cancer. The *reasonable anticipated to be a human carcinogen category* includes those substances for which there is limited evidence of carcinogenicity in humans and/or sufficient evidence of carcinogenicity in experimental animal, indicating a cause and effect relationship between exposure to the substance and cancer. Conclusions regarding carcinogenicity in humans or experimental animals are based on expert, scientific judgment with consideration given to all relevant information. The 11th Edition of the Report on Carcinogens contains 246 entries, 58 of which are listed as known to be human carcinogens and the remaining 188 listed as reasonably anticipated to be human carcinogens. The complexity of classifying agents as to their human carcinogenic potential is complex, and this is best demonstrated by examining the criteria and the classification system used by the IARC (Table 11.3). Carcinogens are generally classified by the weight of evidence for carcinogenicity referred to as sufficient, limited, or inadequate based on both epidemiological studies and animal data. In 2005, Environmental Protection Agency (EPA) revised their guidelines for carcinogen risk assessment and their carcinogen classification scheme. New guidelines emphasize the incorporation of biological mechanistic data in the analysis and do not rely solely on rodent tumor data. The six alphanumeric categories listed in Table 11.3 have been replaced by five descriptors for classifying

TABLE 11.3   International Agency for Research on Cancer (IARC) Classification of Carcinogens

Group 1: The agent is *carcinogenic to humans.*

This category is used when there is *sufficient evidence of carcinogenicity* in humans. Exceptionally, an agent may be placed in this category when evidence of carcinogenicity in humans is less than *sufficient* but there is *sufficient evidence of carcinogenicity* in experimental animals and strong evidence in exposed humans that the agent acts through a relevant mechanism of carcinogenicity.

Group 2

This category includes agents for which, at one extreme, the degree of evidence of carcinogenicity in humans is almost *sufficient,* as well as those for which, at the other extreme, there are no human data but for which there is evidence of carcinogenicity in experimental animals. Agents are assigned to either Group 2A (*probably carcinogenic to humans*) or Group 2B (*possibly carcinogenic to humans*) on the basis of epidemiological and experimental evidence of carcinogenicity and mechanistic and other relevant data. The terms *probably carcinogenic* and *possibly carcinogenic* have no quantitative significance and are used simply as descriptors of different levels of evidence of human carcinogenicity, with *probably carcinogenic* signifying a higher level of evidence than *possibly carcinogenic.*

Group 2A: The agent is *probably carcinogenic to humans.*

This category is used when there is *limited evidence of carcinogenicity* in humans and *sufficient evidence of carcinogenicity* in experimental animals. In some cases, an agent may be classified in this category when there is *inadequate evidence of carcinogenicity* in humans and *sufficient evidence of carcinogenicity* in experimental animals and strong evidence that the carcinogenesis is mediated by a mechanism that also operates in humans. Exceptionally, an agent may be classified in this category solely on the basis of *limited evidence of carcinogenicity* in humans. An agent may be assigned to this category if it clearly belongs, based on mechanistic considerations, to a class of agents for which one or more members have been classified in Group 1 or Group 2A.

TABLE 11.3 *Continued*

Group 2B: The agent is *possibly carcinogenic to humans.*

This category is used for agents for which there is *limited evidence of carcinogenicity* in humans and less than *sufficient evidence of carcinogenicity* in experimental animals. It may also be used when there is *inadequate evidence of carcinogenicity* in humans but there is *sufficient evidence of carcinogenicity* in experimental animals. In some instances, an agent for which there is *inadequate evidence of carcinogenicity* in humans and less than *sufficient evidence of carcinogenicity* in experimental animals together with supporting evidence from mechanistic and other relevant data may be placed in this group. An agent may be classified in this category solely on the basis of strong evidence from mechanistic and other relevant data.

Group 3: The agent is *not classifiable as to its carcinogenicity to humans.*

This category is used most commonly for agents for which the evidence of carcinogenicity is *inadequate* in humans and *inadequate* or *limited* in experimental animals. Exceptionally, agents for which the evidence of carcinogenicity is *inadequate* in humans but *sufficient* in experimental animals may be placed in this category when there is strong evidence that the mechanism of carcinogenicity in experimental animals does not operate in humans. Agents that do not fall into any other group are also placed in this category. An evaluation in Group 3 is not a determination of noncarcinogenicity or overall safety. It often means that further research is needed, especially when exposures are widespread or the cancer data are consistent with differing interpretations.

Group 4: The agent is probably not carcinogenic to humans.

This category is used for agents for which there is *evidence suggesting lack of carcinogenicity* in humans and in experimental animals. In some instances, agents for which there is *inadequate evidence of carcinogenicity* in humans but *evidence suggesting lack of carcinogenicity* in experimental animals, consistently and strongly supported by a broad range of mechanistic and other relevant data, may be classified in this group.

Source: IARC. http://monographs.iarc.fr/ENG/Preamble/currentb6evalrationale0706.php.

human carcinogenic potential. Carcinogens are classified by the EPA as: (1) carcinogenic to humans, (2) likely to be carcinogenic in humans, (3) suggestive evidence of carcinogenic potential, (4) inadequate information to assess carcinogenic potential, and (5) not likely to be a human carcinogen.

11.3.4 Usefulness and Limitations of Mutagenicity Assays for the Identification of Carcinogens

Identification and classification of potential human carcinogens through the 2-year rodent carcinogenesis bioassay is complicated by species differences, use of high doses (MTD), the short life span of the rodents, sample size, and the need to extrapolate from high to low doses for human risk assessment. In addition, the 2-year rodent bioassay to is costly to conduct (>2–4 million dollars) and can take 2–4 years before complete results can be obtained. Since many carcinogens are mutagens, short-term test systems to evaluate the mutagenicity or genetic toxicity of compounds were developed with the idea that these tests could be used to quickly and inexpensively detect/identify chemical carcinogens. Short-term genotoxicity/mutagenicity assays were developed in a variety of organisms including bacteria, yeast, *Drosophila*, and human and rodent cells. These mutagenic assays or short-term genotoxicity tests directly or indirectly measure point mutations, frame-shift mutations, chromosomal damage, DNA damage and repair, and cell transformation.

In the 1970s, it was reported that mutagenicity could predict rodent carcinogenicity 90% of the time; however, after extensive evaluation, it is now considered that mutagenicity can predict rodent carcinogenicity approximately 60% of the time. For certain classes of carcinogens, such as the PAHs, short-term mutagenicity tests are generally highly accurate at predicting rodent carcinogenicity. However, for other classes of carcinogens such as the halogenated hydrocarbons, short-term genotoxicity tests often fail to detect these rodent carcinogens. Many of these halogenated hydrocarbons probably function through an epigenetic mechanism/tumor promoting mechanism.

In a seminal study published in 1987 by Tennant et al., 73 chemicals previously tested in the rodent 2-year carcinogenesis bioassay were examined in four widely used short-term tests for genetic toxicity. The short-term assays measured mutagenesis/genotoxicity using the *Salmonella* assay (Ames Assay) and mouse lymphoma assay, and chromosome aberrations and sister chromatid exchanges in Chinese hamster ovary cells. The concordance (% agreement between short-term genotoxicity test and rodent bioassay results) of each assay with the rodent bioassay data was approximately 60%. Within the limits of the study, there was no evidence of complementarity among the four tests, and no battery of tests constructed from these assays improved substantially on the overall performance of the *Salmonella* assay. When interpreting the results of short-term test for genetic toxicity assays, it is important to consider (1) the structure and physical properties of the test compound; (2) there is only 60% concordance between short-term test for genetic toxicity and rodent carcinogenicity; (3) epigenetic versus genetic mechanisms of carcinogenesis; (4) the existence of noncarcinogenic mutagens; (5) a positive result in the *Salmonella* assay is a good predictor (83%) of rodent carcinogenicity; and (6) a negative result in the *Salmonella* assay only predicts a negative rodent bioassay 51% of the time. It is this latter statistic that lowers the concordance of the assay to 60%, which is largely due to agents that function through an epigenetic mechanism. It is also important to keep in mind that there is accumulating evidence that some compounds that are negative in short-term tests for mutagenicity can induce oxidative DNA damage *in vivo* through the direct or indirect production of reactive oxygen species. These compounds are *in vivo* mutagens but are negative in the short-term test of genetic toxicity. Several bacterial and mammalian short-term tests for mutagenicity and genotoxicity as well as their biochemical and genetic rationale are described in Chapter 20 on toxicity testing.

11.4 CLASSES OF AGENTS THAT ARE ASSOCIATED WITH CARCINOGENESIS

Chemical agents that influence cancer development can be divided into two major categories based on whether or not they are mutagenic in *in vitro* mutagenicity assay. DNA damaging agents (genotoxic) are mutagenic in *in vitro* mutagenicity assays and are considered to produce permanent alterations in the genetic material of the host *in vivo*, and epigenetic agents (nongenotoxic) are not mutagenic in *in vitro* assays. These agents are not believed to alter the primary sequence of DNA but are considered to alter the expression or repression of certain genes and/or produce perturbations in signal transduction pathways that influence cellular events related to proliferation, differentiation, or apoptosis. Epigenetic/nongenotoxic agents contribute to the clonal expansion of cells containing an altered genotype (DNA alterations) to form tumors; however, in the absence of such DNA alterations, these epigenetic agents have no effect on tumor formation.

11.4.1 DNA Damaging Agents

DNA damaging agents can be divided into four major categories: (1) Direct-acting carcinogens are intrinsically reactive compounds that do not require metabolic activation by cellular enzymes to covalently interact with DNA. Examples include *N*-methyl-*N*-nitrosourea and *N*-methyl-*N'*-nitro-*N*-nitrosoguanidine; the alkyl alkanesulfonates such as methyl methanesulfonate; the lactones such as beta propiolactone and the nitrogen and sulfur mustards. (2) Indirect-acting carcinogens require metabolic activation by cellular enzymes to form the ultimate carcinogenic species that covalently binds to DNA. Examples include dimethylnitrosamine, benzo[a]pyrene, 7,12-dimethylbenz[a]anthracene, aflatoxin B1 and 2-acetylaminofluorene (Figure 11.10). (3) Radiation and oxidative DNA damage can occur directly or indirectly. Ionizing radiation produces DNA damage through direct ionization of DNA to produce DNA strand breaks or indirectly via the ionization of water to reactive oxygen species that damage DNA bases. Ultraviolet radiation (UVR) from the sun is responsible for approximately 1 million new cases of human (nonmelanoma) skin cancer each year. Reactive oxygen species can also be produced by various chemicals and cellular processes including respiration and lipid peroxidation. (4) Inorganic agents such as arsenic, chromium, and nickel are considered DNA damaging agents, although in many cases, the definitive mechanism is unknown.

11.4.2 Epigenetic Agents

Epigenetic agents that influence carcinogenesis are not thought to alter the primary sequence of DNA, but rather they are considered to alter the expression or repression of certain genes and/or produce perturbations in signal transduction pathways that influence cellular events related to proliferation, differentiation, or apoptosis. Some mechanisms through which epigenetic agents produce their effects include: (1) alterations posttranslational modification of transcription factors, (2) activation of specific cellular kinases, (3) receptor–ligand interactions, (4) chromatin modifications involving promoter region methylation and histone modifications; and (5) immunosuppression. Many epigenetic agents favor the proliferation of cells with an altered genotype (cells containing a mutated oncogene(s) and/or tumor suppressor gene(s)) and allow the clonal expansion of these altered or "initiated" cells. Epigenetic agents can be divided into four major categories: (1) hormones such as conjugated estrogens and diethylstilbestrol, (2) immunosuppressive xenobiotics such as azathioprine and cyclosporin A; (3) solid state agents that include plastic implants and asbestos, and (4) tumor promoters in rodent models that include

Figure 11.10   Examples of DNA-damaging carcinogens.

12-*O*-tetradecanoylphorbol-13-acetate, peroxisome proliferators, TCDD, and phe-nobarbital (Figure 11.11). In humans, diet (including caloric, fat, and protein intake), excess alcohol, late age of pregnancy are considered to function through a promo-tion mechanism. While smoking and UVR have initiating activity, both are also considered to have tumor-promoting activity. By definition, tumor promoters are not classified as carcinogens since they are considered inactive in the absence of initiated cells. However, an altered genotype or an initiated cell can arise from spontaneous mutations resulting from imperfect DNA replication/repair, oxidative

Figure 11.11   Examples of tumor promoters.

DNA damage, or can result from environmental carcinogens. Theoretically, in the presence of a tumor promoter, these mutant cells would clonally expand to form a tumor. Therefore, the nomenclature becomes somewhat a matter of semantics as to whether the tumor promoter should or should not be classified as a carcinogen. Certain hormones and immunosuppressive agents are classified as human carcinogens although it is generally considered that these agents are not carcinogenic in the absence of initiated cells, but rather, like tumor promoters, may only allow for the clonal expansion of cells with an altered genotype.

11.5   GENERAL ASPECTS OF CHEMICAL CARCINOGENESIS

There is indisputable evidence to support the somatic mutation theory of carcinogenesis which states that cancer is caused by mutations within somatic cells. As stated earlier, cancer development (carcinogenesis) involves the accumulation of mutations in multiple critical genes, and these mutations can result from endogenous processes or from environmental carcinogens. Many chemical carcinogens can alter DNA through covalent interaction (DNA adducts or alkylation) or direct and/ or indirect oxidative DNA damage. Some chemical carcinogens are intrinsically reactive and can directly covalently bind to DNA while others require metabolic activation via cytochrome P450 to produce reactive electrophilic intermediates capable of covalently binding to DNA (Figure 11.12). In the 1950s, Elizabeth and James Miller observed that a diverse array of chemicals with divergent structures could produce cancer in rodents. In an attempt to explain this, they hypothesized that these diverse chemicals are metabolically activated to common electrophilic metabolites that are capable of interacting with nucleophilic sites in the DNA. The Millers termed this the electrophilic theory of chemical carcinogenesis. From this concept of metabolic activation of carcinogens, the terms parent, proximate, and ultimate carcinogen were developed (Figure 11.12). A parent carcinogen is a compound that must be metabolized in order to have carcinogenic activity; a proximate carcinogen is an intermediate metabolite requiring further metabolism resulting in the ultimate carcinogen which is the actual metabolite that covalently binds to the DNA and is responsible for producing mutations. The cell has many defense systems to detoxify the carcinogenic species, including cellular antioxidants (i.e., vitamin C

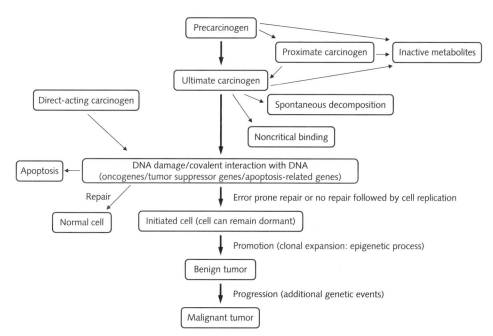

Figure 11.12   General aspects of multistage chemical carcinogenesis. Carcinogens can form reactive species that damage DNA. If damage occurs in a critical gene and either error prone repair or no repair is followed by cell replication, a mutation can result. This mutated cell is now referred to as an "initiated cell." Tumor promoters allow for the clonal expansion of an initiated cell to produce a benign tumor. This benign tumor can progress to a malignant tumor, and this involves additional genetic changes in critical genes.

and E) and nucleophiles (i.e., glutathione [GSH]) as well as a whole host of Phase I and Phase II enzymes. In addition, reactive carcinogenic species may bind to non-critical sites in the cell resulting in detoxification or they can undergo spontaneous decomposition. If the carcinogenic species binds to DNA, the adducted DNA can be repaired, producing a normal cell. If there is error in the repair of the DNA or the DNA adduct is not repaired before the cell replicates, an error in the newly synthesized DNA could occur and, if so, a mutation would occur in the DNA of the daughter cell. If this change has occurred in a critical gene, for example, in a proto-oncogene or tumor suppressor gene, it would represent an important mutagenic event(s) in carcinogenesis.

The mutationally altered cell or "initiated cell" has an altered genotype and may remain dormant (not undergo clonal expansion) for the lifetime of the animal; however, additional mutations or "hits" in critical genes followed by clonal expansion could lead to tumor development as described earlier in this chapter. In addition to this mechanism, chemical carcinogenesis in experimental models can be divided into at least three stages termed initiation, promotion, and progression (Figure 11.12); this model is often referred to as the initiation/promotion model of chemical carcinogenesis. As mentioned above, the "initiated cell" may remain dormant (not undergo clonal expansion) for the lifetime of the animal. However, if the animal is repeatedly exposed to a tumor promoter, it will provide a selective growth advantage to the "initiated cell" which will clonally expand to produce a benign tumor. This process is termed tumor promotion and is an epigenetic process favoring the growth of cells with an altered genotype. The development of a malignant tumor from a benign tumor encompasses a third step, termed progression, and involves additional genetic changes.

Higher doses of carcinogen or multiple doses of carcinogen can produce tumors without tumor promoter treatment; under these circumstances, the chemical agent is often referred to as a complete carcinogen and the model as the complete carcinogenesis model.

11.5.1 Initiation-Promotion Model

Experimentally, the initiation-promotion process has been demonstrated in several organs/tissues including skin, liver, lung, colon, mammary gland, prostate and bladder, as well as in variety of cells in culture. While tumor promoters have different mechanisms of action and many are organ specific, all have common operational features (Figure 11.13). These features include: (1) following a subthreshold dose of initiating carcinogen, chronic treatment with a tumor promoter will produce many tumors; (2) initiation at a subthreshold dose alone will produce very few if any tumors; (3) chronic treatment with a tumor promoter in the absence of initiation will produce very few if any tumors; (4) the order of treatment is critical; that is, you must first initiate and then promote; (5) initiation produces an irreversible change; and (6) promotion is reversible in the early stages; for example, if an equal number of promoting doses are administered but the doses are spaced further apart in time, tumors would not develop or would be greatly diminished in number. Many tumor promoters are organ specific. For example, 12-O-tetradecanoylphorbol-13-acetate (TPA), also known as phorbol 12-myristate 13-acetate (PMA), belongs to a family of compounds known as phorbol esters. Phorbol esters are isolated from

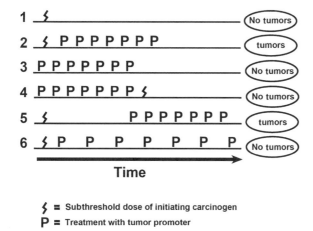

Figure 11.13 Diagrammatic scheme of the initiation promotion model. Topical application of a subthreshold dose of an initiating carcinogen to mouse skin results in no tumor formation; however, if this dose is followed by repetitive treatment with a tumor promoter, then tumors develop. Initiation is an irreversible genetic event leading to the development of an "initiated cell" that can remain dormant until exposed to a tumor promoter. Tumor promoters produce the clonal expansion of the initiated cell to form a tumor.

croton oil (derived from the seeds of the croton plant) and are almost exclusively active in skin. Phenobarbital, DDT, chlordane, TCDD, and peroxisome proliferators Wy 24,643, clofibrate, and nafenopin are hepatic tumor promoters. TCDD is also a promoter in lung and skin. Some bile acids are colonic tumor promoters while various estrogens are tumor promoters in the mammary gland and liver. There are multiple mechanisms of tumor promotion, and this may explain the organ-specific nature of the many promoters.

It is generally accepted that many tumor promoters allow for the clonal expansion of initiated cells by interfering with signal transduction pathways and/or altering expression of genes that are involved in the regulation of cell proliferation, differentiation, and/or apoptosis. While the precise mechanisms of many tumor promoters are not completely understood at the molecular/biochemical level, current research is providing new and promising mechanistic insights into how tumor promoters allow for the selective growth of initiated cells.

11.5.2 Metabolic Activation of Chemical Carcinogens and DNA Adduct Formation

Having described the general aspects of chemical carcinogenesis including the initiation-promotion model, we will now examine some aspects of chemical carcinogenesis in more detail. Metabolic activation of chemical carcinogens by cytochrome P450 is well documented. The metabolism of benzo[a]pyrene has been extensively studied, and at least 15 major Phase I metabolites have been identified. Many of these metabolites are further metabolized by Phase II enzymes to produce numerous different metabolites. Extensive research has elucidated which of these

Figure 11.14 Metabolic activation of benzo[a]pyrene to the ultimate carcinogenic species. Benzo[a]pyrene is metabolized by cytochrome P450 and epoxide hydrolase to form the ultimate carcinogen, (+)benzo[a]pyrene 7,8 diol-9,10 epoxide-2. Adapted from Conney, A. H. *Cancer Res.* **42**:4875, 1982.

metabolites and pathways are important in the carcinogenic process. As shown in Figure 11.14, benzo[a]pyrene is metabolized by cytochrome P450 to benzo[a]pyrene-7,8 epoxide which is then hydrated by epoxide hydrolase to form benzo[a]pyrene-7,8-diol. Benzo[a]pyrene-7,8-diol is considered the proximate carcinogen since it must be further metabolized by cytochrome P450 to form the ultimate carcinogen, the bay-region diol epoxide, (+)-benzo[a]pyrene-7,8-diol-9,10-epoxide-2. It is this reactive intermediate that binds covalently to DNA-forming DNA adducts. (+)-Benzo[a]pyrene-7,8-diol-9,10-epoxide-2 binds preferentially to deoxyguanine residues, forming N-2 adduct. (+)-Benzo[a]pyrene-7,8-diol-9,10-epoxide-2 is highly mutagenic in eukaryotic and prokaryotic cells and is carcinogenic in rodents. It is important to note that not only is the chemical configuration of the metabolites of many PAHs important for their carcinogenic activity, but so is their chemical conformation/stereospecificity (Figure 11.11). For example, four different stereoisomers of benzo[a]pyrene-7,8-diol-9,10 epoxide are formed, each one only differs with respect to whether the epoxide or hydroxyl groups are above or below the plane of the flat benzo[a]pyrene molecule and yet only one, (+)-benzo[a]pyrene-7,8-diol-9,10-epoxide-2, has significant carcinogenic potential. Many PAHs are metabolized to bay-region diol epoxides. The bay region theory suggests that the bay-region diol epoxides are the ultimate carcinogenic metabolites of PAHs.

DNA can be altered by strand breakage, oxidative damage, large bulky adducts, and alkylation. Carcinogens such as *N*-methyl-*N'*-nitro-*N*-nitrosoguanidine and methyl methanesulfonate alkylate DNA to produce N-alkylated and O-alkylated purines and pyrimidines. Ionizing radiation and reactive oxygen species commonly oxidize guanine to produce 8-oxoguanine. Formation of DNA adducts may involve any of the bases, although the N7 position of guanine is one the most nucleophilic sites in DNA. Of importance is how long the adduct is retained in the DNA. (+)-Benzo[a]pyrene-7,8-diol-9,10-epoxide-2 forms adducts mainly at guanine N-2 while aflatoxin B1 epoxide, another well-studied rodent and human carcinogen, binds preferentially to the N7 position of guanine. For some carcinogens, there is a strong correlation between the formation of very specific DNA adducts and tumorigenicity. Quantitation and identification of specific carcinogen adducts may be useful as biomarkers of exposure. Importantly, the identification of specific DNA adducts has allowed for the prediction of specific point mutations that would likely occur in the daughter cell providing there was no repair of the DNA adduct in the parent cell. As will be discussed in a later section, some of these expected mutations have been identified in specific oncogenes and tumor suppressor genes in chemically induced rodent tumors, providing support that the covalent carcinogen binding produced the observed mutation. In several cases, specific base pair changes in p53 tumor suppressor gene in human tumors are associated with a mutational spectrum that is consistent with exposure of the individual to a specific carcinogen. For example, the mutation spectrum identified in p53 in human tumors thought to result from the exposure of the individual to UVR, aflatoxin, and benzo[a]pyrene (from cigarette smoke) are consistent with the observed specific mutational damage in p53 induced by these agents in experimental cellular systems.

11.6 ONCOGENES

If the interaction of a chemical carcinogen with DNA leading to a permanent alteration in the DNA is a critical event in chemical carcinogenesis, then the identification of these altered genes and the function of their protein products is essential to our understanding of chemical carcinogenesis. It was not until the early to mid-1980s that the identification of specific genes that were mutationally altered by chemical carcinogens became known. Certain normal cellular genes, termed proto-oncogenes, can be mutated by chemical carcinogens providing a selective growth advantage to the cell. The mutational activation of proto-oncogenes is strongly associated with tumor formation, carcinogenesis, and cell transformation. Proto-oncogenes are highly conserved in evolution and their expression is tightly regulated. Their protein products function in the control of normal cellular proliferation, differentiation, and survival. However, when these genes are altered by a mutation, chromosome translocation, gene amplification, or promoter insertion, an abnormal protein product or an abnormal amount of product is produced. Under these circumstances these genes have the ability to transform cells *in vitro* and are termed oncogenes. Over 200 oncogenes have been identified with approximately 30 oncogenes having a major role in human cancer.

Most oncogene protein products appear to function in one way or another in cellular signal transduction pathways that are involved in regulating cell growth, differentiation, or apoptosis. Signal transduction pathways are used by the cells to receive and process information to ultimately produce a biological cellular response. These pathways are the cellular circuitry conveying specific information from the outside of the cell to the nucleus (Figure 11.15). In the nucleus, specific genes are expressed, and their encoded proteins produce the evoked biological response. Oncogenes encode proteins that are components of this cellular circuitry and can be classified with respect to their biological function (Table 11.4). If a component of the circuit is altered, then the entire cellular circuit of which the component is a part is altered. It is not difficult to imagine how an alteration in a pathway that regulates cellular proliferation could have very profound effects on cellular homeo-

Figure 11.15 Generic signal transduction pathway involving receptor tyrosine kinase (RTK). An extracellular growth factor signal is conveyed via receptors, GTPases (Ras), kinases and, ultimately, to transcription factors that alter gene expression and produce a cellular response.

TABLE 11.4 Human Oncogene Classification

Oncoprotein Families	Oncogenes
Growth factors	PDGF, HGF, TGFα, VEGF, WNT-1, IGF-2
Receptor tyrosine kinases (RTKs)	ERBB1, ERBB2, KIT, RET, MET
Nonreceptor tyrosine kinase	SRC, ABL, YES, LCK
Guanosine triphosphatases (GTPases)	H-RAS, K-RAS, N-RAS
Serine/threonine kinases	RAF-1, B-RAF, AKT, PIM-1, BCR
Transcription factors	MYC, FOS, JUN, ETS, REL, MYB, GLI
Survival proteins	BCL-2, AKT, E2F1, MDM2

ERBB1, EGF receptor (EGF-R); ERBB2, HER2 or NEU receptor; MET, HGF receptor (HGF-R).

stasis. Indeed, the alteration of pathways by oncogenes is the molecular basis through which oncogenes contribute to the cancer process.

11.6.1 Ras Oncogene

Ras genes are frequently mutated in chemically induced animal tumors and are among the most frequently detected mutated oncogenes in human tumors. Approximately 20–30% of all human tumors contain mutated *RAS*. The Ras sub-family includes H-Ras, K-Ras, N-Ras, and all have been found to be mutationally activated in numerous types of tumors from a large variety of species, including humans.

Activated Ras oncogenes have been detected in a large number of animal tumors induced by diverse agents including physical agents, such as radiation and a large number of chemical carcinogens. Some chemical carcinogens bind covalently to DNA, forming specific adducts which, upon DNA replication, yield characteristic alterations in the primary sequence of the H-Ras proto-oncogene. The study of the Ras oncogene as a target for chemical carcinogens has revealed a correlation between specific carcinogen-DNA adducts and specific activating mutations of Ras in chemically induced tumors. For example, 7,12-dimethylbenz[a]anthracene, a PAH carcinogen, is metabolically activated to a bay-region diol epoxide which binds preferentially to adenine residues in DNA. Skin tumors isolated from mice treated with 7,12 dimethylbenz[a]anthracene (DMBA) contain an activated H-Ras onco-gene with an A to T transversion of the middle base in the 61st codon of H-ras. Therefore, the identified mutation in Ras is consistent with the expected mutation produced by the DMBA-DNA adduct. Likewise, rat mammary carcinomas induced by nitrosomethylurea contain a G to A transition in the 12th codon of H-Ras, and this mutation is consistent with the modification of guanine residues by this carcino-gen. Based on these events, the alteration of Ras by specific chemical carcinogens appears to be an early event in carcinogenesis.

Ras proteins function as membrane-associated molecular switches operating downstream of a variety of membrane receptors. Ras is in the off position when it is bound to guanosine diphosphate (GDP); however, when a growth factor receptor is activated by the binding of its ligand, the activated receptor stimulates the guanine nucleotide exchange factor, SOS (son of sevenless), which causes Ras to exchange guanosine triphosphate (GTP) for GDP, and now Ras is bound to GTP and is in the "on" position. Ras communicates this "on" signal downstream to the other proteins in the signaling circuitry. The best-characterized pathways involve the acti-vation of a kinase cascade that results in the activation of various transcription factors. Theses transcription factors regulate the expression of genes involved in cell proliferation, and the cell is instructed to proliferate. As mentioned, Ras is a molecu-lar switch and once Ras has conveyed the "on" signal, Ras must turn itself "off." Ras has intrinisic GTPase activity which hydrolyzes GTP to form GDP and Ras is now "off" position. Another protein, termed GAPp120 (GTPase activating protein) aids Ras in GTP hydrolysis. When Ras is mutated by a gain of function mutation in certain codons, including the 12th, 13th, or 61st codon, the intrinsic GTPase activity of Ras is greatly diminished as is its ability to interact with GAP. The net effect is that mutated Ras is now an oncogene and is essentially stuck in the "on" position, continually sending a proliferative signal to the downstream circuitry.

11.7 TUMOR SUPPRESSOR GENES

Activation of oncogenes results in a gain of function while inactivation of tumor suppressor genes results in a loss of function. Tumor suppressor genes encode proteins that generally function as negative regulators of cell proliferation or positive regulators of apoptosis. The majority of tumor suppressor genes were first identified in rare familial cancer syndromes and later found be mutated in sporadic cancers through somatic mutation. Major tumor suppressor genes, their proposed function, as well as the cancer syndrome they are associated with, are shown in Table 11.5. When tumor suppressor genes that negatively regulate cell proliferation are inactivated by allelic loss, point mutation, or chromosome deletion, the result is uncontrolled cell proliferation. Generally, if one allele of a tumor suppressor gene is inactivated, the cell is normal (this gene is referred to as haplosufficient). However, when both alleles are inactivated, the ability to control cell proliferation is lost.

11.7.1 p53 Tumor Suppressor Gene

p53 aka TP53 encodes a 53 kDa protein. p53 is mutated in 50% of all human cancer and is the most frequently known mutated gene in human cancer. The majority (~80%) of p53 mutations are missense mutations and p53 is mutated in approximately 70% of colon cancers, 50% of breast and lung cancers, and 97% of primary melanomas. In addition to point mutations, allelic loss, rearrangements, and deletions of p53 occur in human tumors. p53 is a transcription factor and participates in many cellular functions including cell cycle regulation, DNA repair,

TABLE 11.5 Human Tumor Suppressor Genes

Gene Name	Familial Cancer Syndrome	Protein Function	Sites/Types of Commonly Associated Neoplasms
TP53	Li–Fraumeni syndrome	Transcription factor	Most human cancers
RB1	Hereditary retinoblastoma	Transcriptional modifier	Retinoblastoma, osteosarcoma
APC	Familial adenomatous Polyposis	B-catenin degradation	Colon, stomach, intestine
CDKN2A (p16^{INK4A})	Familial malignant melanoma	Cyclin-dependent kinase inhibitor	Melanoma, pancreas
CDKN2A (p14^{ARF})		p53 stabilizer	Melanoma
PTCH	Gorlin syndrome	Transmembrane receptor	Basal cell skin carcinoma, ovary, heart
PTEN	Cowden syndrome	PIP3 phosphatase	Hamartoma, glioma, uterus
TGFBR2		Transmembrane receptor	Colon, stomach, ovary

and apoptosis. The p53 protein is composed of 393 amino acids and single missense mutations can inactivate the p53. Unlike Ras genes which have a few mutational codons that result in its activation, the p53 protein can be inactivated by hundreds of different single point mutations in p53. It has been proposed that the mutation spectrum of p53 in human cancer can aid in the identification of the specific carcinogen that is responsible for the genetic damage; that is to say that different carcinogens cause different characteristic mutations in p53. Some of the mutations in p53 reflect endogenous oxidative damage, while others such as the mutational spectrum in p53 in hepatocellular carcinomas from individuals exposed to aflatoxin demonstrate a mutation spectrum characteristic aflatoxin. In sun-exposed areas where skin tumors develop, the mutations found in p53 in these tumors are characteristic of UV light-induced cyclobutane pyrimidine dimers and, finally, the mutation spectrum induced by (+)-benzo[a]pyrene-7,8-diol-9,10-epoxide-2 in cells in culture is similar to the mutational spectrum in p53 in lung tumors from cigarette smokers. Thus, certain carcinogens produce a molecular signature which may provide important information in understanding the etiology of tumor development.

p53 has been termed the "guardian of genome" because it controls a G1 checkpoint, regulates DNA repair and apoptosis. DNA damage results in the accumulation of p53 and the activation of p53 function. p53 prevents cells with damaged DNA from entering the S-phase of the cell cycle until the DNA damage is repaired. If the DNA damage is severe, p53 can cause the cell to undergo apoptosis (Figure 11.16). Mutation of p53 disrupts these functions, leading to the accumulation of mutations as cells enter S phase with damaged DNA (mutator phenotype; genetic instability) and further development of malignant clones.

Figure 11.16 *p53* regulates apoptosis and cell cycle progression. In response to DNA damage or oncogene activation, the *p53* protein undergoes posttranslational modifications that increase its stability and activity. *p53* accumulates in the cell and can regulate the expression of genes involved in apoptosis and cell cycle arrest. Oncogene activation is also believed to activate components of the DNA damage response pathway to further increase *p53*. MDM2 is a feedback inhibitor of *p53* and targets *p53* for proteasomal degradation.

BIBLIOGRAPHY AND SUGGESTED READING

11th Report on Carcinogens. US Department of Health and Human Services, Public Health Service, National Toxicology Program. 2005. http://ehp.niehs.nih.gov/roc.toc10.html.

American Cancer Society. *Cancer Facts and Figures 2009*. Atlanta, GA: American Cancer Society, 2009.

Cairns, J. *Readings in Scientific American—Cancer Biology*, p. 13. 1986.

Cancer Facts and Figures. American Cancer Society. 2009. www.cancer.org.

Doll, R. and R. Peto. *The Causes of Cancer: Quantitative Estimates of Avoidable Risks of Cancer in the United States Today*. New York: Oxford Medical Publications, 1981.

Hoeijmakers, J. J. Genome maintenance mechanisms for preventing cancer. *Nature* **411**:366, 2001.

IARC. http://monographs.iarc.fr/ENG/Preamble/currentb6evalrationale0706.php.

International Agency for Research on Cancer. www.iarc.fr.

Mellon, I. DNA repair. In *Molecular and Biochemical Toxicology*, 4th ed., eds. R. C. Smart and E. Hodgson. Hoboken, NJ: John Wiley and Sons, 2008.

NCI SEER. *Cancer Statistics Review, 1975–2003*. http://seer.cancer.gov.

Pitot, H. C. and Y. P. Dragan. Chemical carcinogenesis. In *Casarett and Doull's Toxicology*, ed. C. D. Klaassen, pp. 241–319. New York: McGraw Hill, 2001.

Smart, R. C. Carcinogenesis. In *Molecular and Biochemical Toxicology*, 4th ed., eds. R. C. Smart and E. Hodgson. Hoboken, NJ: Wiley, 2008.

Tennant, R. W., B. H. Margolin, M. D. Shelby, et al. Prediction of chemical carcinogenicity in rodents from in vitro genetic toxicity tests. *Science* **236**:933–941, 1987.

Wang, Z. DNA damage and mutagenesis. In *Molecular and Biochemical Toxicology*, 4th ed., eds. R. C. Smart and E. Hodgson. Hoboken, NJ: Wiley, 2008.

SAMPLE QUESTIONS

1. What does the somatic mutation theory state?

2. What are the three major categories of genes involved in cancer development?

3. Cancer susceptibility is determined by complex interactions between_____, _____and _____.

4. What is the major reason for the 60% concordance between rodent carcinogenicity and mutagenicity in short term tests?

5. Briefly describe how the study of cancer rates of groups of people that emigrate from one country to another has provided important information on the causes of cancer.

6. Describe the role of Ras in cancer. Be sure to describe its normal function and how this function is altered in carcinogenesis.

7. Describe the normal function of p53 protein in the cell and how a mutation in this gene contributes to the development of cancer.

Teratogenesis

JILL A. BARNES and IDA M. WASHINGTON

12.1   INTRODUCTION

Teratology is the study of abnormal development, and teratogenesis is the production of an abnormal organism. The term teratology is derived from the Greek word *teratos*, which means "monster." An agent is considered to be a teratogen if it increases the occurrence of structural or functional abnormalities in offspring when given to either parent before conception, to the mother during pregnancy, or to the developing embryo or fetus. Teratogens affect the developing embryo or fetus without significant toxicity in the mother; these agents may include chemicals, environmental factors, viruses, radiation, toxic plants, and metabolite deficiencies or excesses. The mechanisms by which teratogens disrupt development are still largely unknown. However, a number of general principles have emerged regarding the interaction of teratogens with the developing embryo.

The field of teratology had its origins in the early twentieth century with the observation in the 1920s that pregnant women exposed to ionizing radiation produced children with neural and skeletal defects. In the 1940s, a connection between maternal rubella infection and neonatal death and abnormalities was recognized. Experiments by Warkany and colleagues in the 1940s demonstrated abnormal growth and development of mammalian embryos after maternal exposure to dietary deficiency or irradiation. Interest in the field of teratology increased significantly in the 1950s and 1960s when human infants with severe limb defects were born to mothers dosed with the sedative thalidomide during pregnancy.

In order to understand the principles and mechanisms of teratogenesis, one must first understand how the embryo develops normally. Thus, this chapter will begin with an overview of normal embryonic development, followed by a review of basic principles of teratogenesis. Mechanisms of teratogenesis will be illustrated by describing specific teratogenic agents and current knowledge about how these factors disrupt normal embryogenesis. The chapter concludes with a discussion of future considerations in the field of teratology. Significant progress has been made in this field of study during the past half-century, but there is much yet to understand about molecular aspects of embryonic development and the mechanisms of teratogenesis.

A Textbook of Modern Toxicology, *Fourth Edition*. Edited by Ernest Hodgson
Copyright © 2010 John Wiley & Sons, Inc.

12.2 OVERVIEW OF EMBRYONIC DEVELOPMENT

12.2.1 Fertilization

Fertilization typically occurs in the ampulla of the uterine tube and represents the union of male and female germ cells to form a single-cell embryo, the zygote. Maternal and paternal chromosomes arrange on the mitotic spindle for the first mitosis, followed by a series of rapid mitotic divisions. The genetic sex of the mammalian embryo is determined at fertilization, when a spermatozoon carrying an X or Y sex chromosome combines with an oocyte carrying an X sex chromosome, to produce a female (XX) or male (XY) offspring.

12.2.2 Cleavage Stages

Morula The single-cell zygote undergoes a series of rapid mitotic divisions to produce a solid ball of cells, the morula, which is surrounded by an acellular layer, the zona pellucida. A group of large cells (inner cell mass) located centrally within the morula will form the embryo, while the smaller peripheral cells (outer cell mass) will form the extraembryonic membranes and placenta.

Blastula A fluid-filled cavity, the blastocele, begins to form between the cells of the morula as the embryo transitions to the blastula stage. During this stage, the blastocele enlarges to form a large central fluid-filled cavity. The cells of the embryoblast (future embryo) move to one pole of the blastula and form two layers, the epiblast and hypoblast. The outer cell mass becomes the trophoblast, which forms the wall of the blastula. The zona pellucida degenerates and disappears as the embryo "hatches" and then implants in the uterine mucosa. Implantation occurs in most species at approximately 5–8 days post fertilization (Table 12.1).

12.2.3 Determination

The zygote is capable of forming all cells of the body, a quality called totipotency. This capability persists through several cell divisions. As development proceeds, the potential of each cell becomes narrowed as its fate is progressively fixed. This process is called determination and is necessary for subsequent cellular differentiation.

TABLE 12.1 Comparison of Gestation in Several Species

Species	Number of Days after Conception		
	Implantation	Embryonic Period[a]	Fetal Period
Human	6–7	20–56	56–280
Rabbit	6–8	8–16	17–34
Rat	6–8	9–17	18–22
Mouse	5–7	7–16	17–20

[a] Period of organogenesis and greatest teratogenic risk.

12.2.4 Gastrulation

Gastrulation is the stage of development during which the three primary germ layers (ectoderm, mesoderm, and endoderm) are formed. Gastrulation begins with the appearance of the primitive streak on the surface of the embryonic disc. Cells on the surface of the embryo migrate to the primitive streak and invaginate to form two new layers, the endoderm and mesoderm. This process occurs in a cranial to caudal direction. When gastrulation is complete, the primitive streak disappears and the remaining surface layer forms the ectoderm.

12.2.5 Differentiation

After the three germ layers of the embryo are established, cells in different regions of these layers begin to differentiate into components of developing organs to serve specific functions. During differentiation, cells pass through several stages of increasing complexity to achieve a fully functional state. They develop characteristics specific to their cell type, which involves the proliferation or disappearance of certain organelles and the synthesis of certain intracellular or secreted proteins.

12.2.6 Organogenesis

Organogenesis is the stage during which organ systems are formed from the three primary germ layers (ectoderm, mesoderm, endoderm) that were established during gastrulation.

Ectoderm The original ectoderm layer is composed of neural ectoderm, nonneural ectoderm, and neural crest. Neural ectoderm forms the central nervous system, retina and olfactory epithelium, pineal gland, and posterior pituitary gland. Nonneural ectoderm forms surface structures and their derivatives, such as epidermis and associated hair, nails, and glands. The neural crest originates between the neural and nonneural ectoderm and migrates to form numerous derivatives, including most of the peripheral nervous system.

Mesoderm The original mesoderm layer becomes subdivided into paraxial, intermediate, and lateral plate regions. The paraxial mesoderm forms somitomeres in the head region and somites in the body region of the embryo. These temporary structures will further subdivide to form dermis, voluntary muscles, cartilage, bone, and connective tissue of the trunk and limbs, as well as voluntary muscles of the head and a few bones of the skull. Intermediate mesoderm forms the kidneys, gonads, ducts, and accessory glands of the urogenital system, as well as the adrenal cortex. Lateral plate mesoderm splits into somatic and splanchnic layers, which form the body wall of the embryo and wall of the gut tube, respectively.

Endoderm The endoderm layer forms the lining of the gut tube and derivatives of the embryonic gut, including respiratory tract and pancreas, liver, thyroid, parathyroid, tonsils, and thymus. Endoderm also lines the urinary bladder and urethra, and the auditory tube, middle ear, and tonsillar fossa.

12.2.7 Fetal Period

There is no distinct demarcation between the end of the embryonic period and the beginning of the fetal period. In general, organ primordia are established during the embryonic period, and rapid growth and differentiation of these organs occurs during the fetal period. In the fetus, organs and organ systems undergo structural and functional maturation. Species-dependent features start to become apparent in the early fetal period.

12.3 PRINCIPLES OF TERATOGENESIS

12.3.1 Wilson's Principles

In 1959, James Wilson proposed six basic principles of teratology. Fifty years later, these principles remain important basic tenets in the field of teratology. These principles include the following:

1. Susceptibility to teratogenesis depends on the genotype of the conceptus and the manner in which it interacts with environmental factors.
2. Susceptibility to teratogens varies with the developmental stage at the time of exposure.
3. Teratogenic agents act in specific ways on developing cells and tissues to initiate abnormal developmental processes.
4. The access of adverse environmental influences to developing tissues depends on the nature of the influences.
5. The final manifestations of altered development are death, malformation, growth retardation, and functional disorder.
6. Manifestations of altered development increase in frequency and in degree as dosage increases from no effect to 100% lethality.

12.3.2 Critical Period

The result of exposing an embryo to a teratogenic compound or condition depends on its developmental stage at the time of exposure (Principle #2 above). During the zygote to blastula stages, teratogens may affect numerous cells and cause embryonic death. Alternatively, few cells of the early embryo may be affected by a teratogenic compound, resulting in embryonic compensation and recovery. Malformations are most likely to result from teratogenic exposure during the stage of organogenesis, when organ systems are formed. Each organ system has a different critical period, during which it is most susceptible to the effects of teratogenic agents. In general, there is a decline in teratogenic susceptibility during the fetal period, which is the stage of organ growth. Exposure to teratogens during the fetal period may result in growth retardation or functional impairment (Figure 12.1).

12.4 MECHANISMS OF TERATOGENESIS

In general, factors that cause congenital abnormalities can be either genetic or environmental. In humans, it is estimated that approximately 20% of malformations

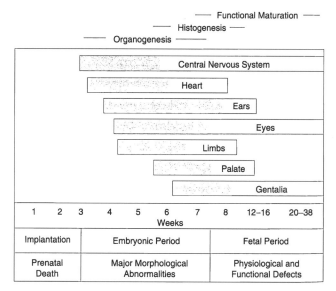

Figure 12.1   Sensitivity to teratogenic exposure at different stages of embryonic and fetal development.

are due to genetic factors and approximately 10% are due to environmental exposure to teratogens such as drugs, chemicals, or infectious agents. This leaves the vast majority of malformations, approximately 70%, for which the etiology is unknown.

12.4.1 Genetic Factors

Mutations   Mutations are alterations in the DNA sequence of an organism. In general, mutations can be classified as spontaneous or induced and will cause a structural change to the DNA, which may then lead to altered function of a gene. Spontaneous mutations typically occur at rates of one per million. Induced mutations are generally a result of exposure to chemical or physical agents (e.g., radiation), which alter DNA.

Some examples of known mutations include X-linked muscular dystrophy (in cats and dogs) which leads to an abnormal dystrophin gene or gangliosidosis, resulting in a deficiency of β-galactosidase. In humans, Marfan syndrome is an example of a mutation where a defective glycoprotein product of the fibrillin gene (FBN1) antagonizes the product of the normal allele.

Chromosomal Abnormalities   Large-scale alterations to DNA segments can lead to chromosomal abnormalities. When the chromosome number of a cell is altered by either the addition or loss of a chromosome, the condition is called aneuploidy. Monosomy and trisomy refer to the condition where a pair of chromosomes either loses or adds to its pair, respectively. Examples in humans include Down syndrome, which is trisomy of chromosome 13, and Klinefleter syndrome, which is characterized by the addition of an X chromosome.

12.4.2 Teratogens

According to the principles of teratogenesis, a teratogen must cause a specific mal-formation through a specific mechanism during a period in which the conceptus is susceptible to that mechanism (Karnofsky, 1965). Clearly, there are multiple mecha-nisms known to cause malformations that are in agreement with these principles. It is difficult, if not impossible, to discuss all of the known or potential mechanisms responsible for inducing malformations. These include DNA attack, enzyme inhibi-tion, interference with hormonal action, alterations of gene signaling pathways, reactive oxygen species, and insult to membranes, proteins, and mitochondria. Examples of agents and/or mechanisms known to cause malformations are described below.

Drugs and Other Xenobiotics

Ethyl Alcohol   Fetal alcohol syndrome occurs in infants of women with severe alcoholism during pregnancy. Since ethyl alcohol can readily cross the placenta, this agent is exceptionally dangerous to the developing embryo and fetus. Children who are affected are developmentally and mentally retarded. Studies in mice show that ethyl alcohol interferes with neural crest cell migration, causes apoptosis (cell death) of neurons in the developing forebrain, and detrimentally alters the activity of cell adhesion molecules.

Dioxin   Dioxins are halogenated hydrocarbons which are used in many industrial processes and have been linked to congenital defects in humans who have been exposed to the compound as an herbicide. Exposure of pregnant mice to dioxin leads to cleft palate as well as kidney, brain, and other defects in the offspring. *In vitro* studies of palate cells demonstrated that exposure to dioxins altered cell pro-liferation and differentiation of the palate epithelial cells which have high-affinity receptors for the compound.

Diethylstilbestrol (DES)   DES is a synthetic estrogen that was used for nearly 30 years in the prevention of miscarriage or other complications of pregnancy. Unfortunately, female offspring of women treated with DES in the early stages of pregnancy showed an increased risk of reproductive tract abnormalities. After decades of experiments, the complex genetic messenger mechanisms responsible for DES-induced defects are better understood. Studies have shown that pregnant mice exposed to DES have repressed expression of HOX-a-10 gene in the parame-sonephric duct. DES, acting primarily through the estrogen receptor, represses Wnt 7a gene expression, which in turn prevents Hox expression. Lack of Hox expression prevents activation of the gene Wnt 5a which codes for a protein required for cel-lular division of the developing uterus.

Thalidomide   Thalidomide was chiefly sold and prescribed during the late 1950s and early 1960s to pregnant women as an antiemetic and as an aid to help them sleep. However, this drug turned out to be a potent teratogen in rabbits and pri-mates, including humans. Thalidomide has severe teratogenic effects from 20 to 36 days of gestation in humans. Lack of long bone development in the limbs, defects of the gastrointestinal (GI) tract, heart, eye, ear, and renal defects have all been documented as a result of thalidomide exposure. The teratogenic effects of thalidomide have been attributed to its ability to detrimentally affect the production of angiogenesis factors in the developing limb buds and other target tissues by causing the downregulation of specific genes.

Plants

Numerous poisonous plants have been identified to cause congenital defects in animals with considerable species variations.

Veratrum Californicum **(Skunk Cabbage/False Hellebore)**   Ewes that consume this particular plant on the fourteenth day of gestation produce offspring with congenital cyclopean deformities of the head, cleft palate, limb deformities, and tracheal stenosis. Teratogeneic compounds present in this plant include cyclopamine, cycloposine, and jervine. These toxic alkaloids have been shown to interfere with Sonic Hedgehog signaling pathways.

Lupinus Species   There are more than 100 species of lupins and some of these have been shown to be teratogenic. Pregnant cows that ingest these particular plants produce calves with malformations of the forelimbs. This condition is often referred to as "crooked calf disease." Limb abnormalities consist of contracture of the flexure muscles, arthrogryposis associated with disproportionate growth of joints, and shortening and rotation of bones. A quinolizidine alkaloid is considered to be the teratogenic agent.

Infectious Agents

Several infectious agents that can cross the placenta and infect the developing fetus are significant causes of defects in humans as well as domestic animals. These can include bacteria, protozoa, fungi, or viruses.

Rubella Virus (German Measles)   Infants born to women infected with the rubella virus during the first 3 months of pregnancy are at a significant risk of developing congenital defects. Abnormalities include cardiac malformations, microcephaly, deafness, ocular defects, and mental retardation. As the fetus matures, the risk of defects is reduced, and defects are infrequently seen after the twentieth week of gestation in humans. Maternal immunity, either from immunization or following infection, will prevent congenital infection.

Feline Panleukopenia Virus   Transplacental infection with this particular parvovirus in cats can have significant effects on fetal development which relate directly to the stage of gestation at the time of infection. Early infection may result in fetal resorption or death. Cerebellar hypoplasia and retinal dysplasia occurs in cats infected during late pregnancy. If the dam is infected during the last 2 weeks of pregnancy, kittens will have severe cerebellar hypoplasia, which is characterized by ataxia, tremors, and hypermetria.

12.5 FUTURE CONSIDERATIONS

The discovery of environmental agents that cause congenital malformations is extremely important to the health of human as well as animal populations. The question remains: how do we determine which agents are teratogens? Recent reports show that experimental data from 11 groups of known human teratogens across 12 species showed huge amounts of variation in positive predictability (Bailey et al., 2005). Thus, it appears that animal studies are reasonably predictive for animals but, to date, the best human data may, in fact, be epidemiological. This is not particularly surprising, given the amount of variables one must consider. In summary: susceptibility to teratogenesis varies between different species, different strains, and among individuals; affected individuals frequently show different phenotypes, and all these aspects are influenced by genetic makeup, environmental factors, and metabolic and placental differences. Results are further affected by anatomical differences, differences in routes of administration, dose levels, and strategies, differences in absorption, distribution, metabolic activation, sensitivity and excretion, and by typically stressful laboratory handling and housing conditions which can impair health (Bailey et al., 2005). The future dictates that we should employ any and all experimental strategies, including *in vitro* embryonic stem cell tests and whole embryo culture as well as animal studies, to best determine which, if any, of the thousands of chemicals that humans and animals are continuously exposed to may be dangerous to the developing offspring.

BIBLIOGRAPHY AND SUGGESTED READING

Bailey, J., A. Knight, and J. Balcombe. The future of teratology research is *in vitro*. *Biog. Amines* **19** (2): 97–145, 2005.

Karnofsky, D. A. Mechanisms of actions of certain growth inhibiting drugs. In *Teratology, Principles and Techniques*, ed. J. G. Wilson, pp. 185–213. Chicago, IL: University of Chicago Press, 1965.

Warkany, J. Congenital Malformations Induced by Maternal Dietary Deficiency-Experiments and Their Interpretation, The Harvey Lectures, series 48; New York, Academic Press, Inc, 1952–53.

Wilson J. G. Experimental studies on congenital malformations. *J Chronic. Dis.* **10**: 111–130, 1959.

SAMPLE QUESTIONS

1. An agent that specifically disrupts cells in the developing endoderm layer could produce congenital defects in the
   a. Adrenal cortex
   b. Retina
   c. Vertebrae
   d. Pancreas
   e. Skin

2. Exposure to a teratogenic agent during organogenesis would most likely cause

   a. Fetal growth retardation but no congenital defects

   b. Structural or functional congenital defects

   c. Embryonic death

   d. Maternal toxicity

   e. Delayed implantation

3. Which of the following best describes the critical period?

   a. The critical period is the same for all organs.

   b. The critical period always occurs prior to implantation.

   c. The critical period usually occurs during organogenesis.

   d. The critical period is the stage at which teratogenic agents cause embryonic death.

   e. The critical period is the primary stage of organ growth and functional maturation.

ORGAN TOXICITY

Hepatotoxicity

ANDREW D. WALLACE and SHARON A. MEYER

13.1 INTRODUCTION

Hepatotoxicity is a consequence of exposure to natural toxins and many man-made chemicals including industrial compounds, pesticides, and pharmaceutical drugs. Mechanisms of hepatotoxicity are well understood for several chemicals such as halogenated solvent CCl_4 and analgesic acetaminophen. Drug induced liver injury (DILI) remains one the major reasons for new drugs to fail to meet regulatory approval. The progressive injury to the liver due to repeated exposure to toxic doses of ethanol remains a leading human health concern. The liver has many critical functions in the body, and the unique structures and functions of the liver are important reasons for the liver's susceptibility to chemical toxicity.

13.1.1 Liver Structure

The liver consists of a variety of cell types, but the basic architecture of the hepatic parenchyma consists of rows of functionally diverse hepatocytes separated by spaces called sinusoids (see Chapter 9, Figure 9.2). Blood flows into the sinusoidal spaces via the hepatic portal vein blood from the gastrointestinal (GI) tract, which is the main blood supply, and oxygenated blood also enters from the hepatic artery. Blood subdivides and drains into the sinusoids then exits via the terminal hepatic venule (THV) or central vein. The blood that perfuses the liver exits by these hepatic veins, which merge into the inferior vena cava and return blood to the heart. The hepatocytes located near the THV are referred to as centrilobular, while those near the portal vein are periportal hepatocytes, and these hepatocytes differ in size and functions.

Although hepatocytes comprise the majority of liver cells, other nonparenchymal cells are present in sizable numbers at specific locations (Figure 13.1). Bile duct epithelial cells are located in portal triads and endothelial cells line the sinusoids. Kupffer cells are macrophages, which engulf and destroy materials such as solid particles, bacteria, and dead blood cells, and are attached to the intralumenal side of the sinusoidal wall, while hepatic stellate cells (HSCs) (also known as fat-storing

A Textbook of Modern Toxicology, Fourth Edition. Edited by Ernest Hodgson

Figure 13.1 Diagram illustrating different types of liver cells and their spatial relationship. HC, hepatocytes; Ku, Kupffer cells; En, vascular endothelial cells; St, Stellate (Ito) cells; NK, lymphocytes.

or Ito cells) are in the perisinusoidal space of Disse, a region between the sinusoidal endothelium and hepatocytes. In chemically injured liver, the periportal region can become populated with a morphologically distinct cell, the "oval" cell, which is thought to be a stem cell capable of differentiating into either hepatocytes or bile duct epithelia.

Other materials, such as bile acids and many xenobiotics, move from the hepatocytes into the bile from their sites of synthesis at the hepatocyte canalicular membrane, which merge into larger ducts that follow the portal vein branches. The ducts merge into the hepatic duct from which bile drains into the upper part of the small intestine, the duodenum. The gall bladder, in all species but rat, serves to hold bile until it is emptied into the intestine.

13.1.2 Liver Function

The liver has many important physiological functions that impact the body, but the liver's three main functions include storage, metabolism, and biosynthesis, and the heterogeneity of hepatocytes in the conduct of these functions occurs largely differentiated by position along the sinusoid. Glucose is converted to glycogen and stored as needed for energy, and is converted back to glucose as the need arises by periportal hepatocytes due to their enrichment in gluconeogenic enzymes. Fat-soluble vitamins and other nutrients are also stored in the liver. Fatty acids are metabolized and converted to lipids, which are then conjugated with proteins synthesized in the liver and released into the bloodstream as lipoproteins. The liver also synthesizes numerous functional proteins, such as enzymes and plasma proteins including blood-coagulating factors. In addition, the liver, which contains numerous xenobiotic metabolizing enzymes, is the main site of xenobiotic metabolism, which predominates in the centrilobular hepatocytes. Liver metabolism of xenobiotics absorbed from the gut can greatly reduce the xenobiotic blood levels reaching systemic circulation and is known as the first-pass effect.

13.2 SUSCEPTIBILITY OF THE LIVER

The liver, the largest organ in the body, is often the target organ for chemically induced injuries. Several important factors are known to contribute to the liver's susceptibility. First, most xenobiotics enter the body through the GI tract and, after absorption, are transported by the hepatic portal vein to the liver. Thus, the liver is the first organ perfused by chemicals that are absorbed in the gut and is exposed to the highest concentrations of xenobiotics. A second factor is the high concentration in the liver of xenobiotic metabolizing enzymes, primarily the cytochrome P450-dependent monooxygenase system. Although most biotransformations of xenobiotics act as detoxification reactions, many oxidative reactions produce reactive metabolites (Chapters 7 and 8) that can induce lesions within the liver. Often, areas of damage are in the centrilobular region, as hepatocytes in this localization have the highest concentration of cytochrome P450s (CYPs), and therefore, the greatest amount of reactive metabolites are produced in this region. Third, the process of bile formation and movement of bile to the GI tract can concentrate xenobiotics that are transported with the bile. Xenobiotics and most of the bile released into the intestines are reabsorbed and transported back to the liver by the hepatic portal circulation, which can increase the concentration of xenobiotics in hepatocytes.

13.3 TYPES OF LIVER INJURY

The classification of hepatotoxicity is primarily based on the pattern of incidence and the histopathological morphology. *Intrinsic* hepatotoxicants demonstrate broad incidence, dose-dependent relationship, and usually similar toxicities are seen in humans and animal models. *Idiosyncratic* hepatotoxicants demonstrate limited toxicity seen in susceptible individuals and results from hypersensitivity or unusual metabolic conversions that may occur due to polymorphisms in drug metabolizing genes. The types of injury to the liver depend on the type of toxic agent, the severity of intoxication, and whether the type of exposure is acute or chronic. The main types of liver damage are discussed briefly in this section. The hallmarks of hepatotoxicity are impaired hepatocyte function and viability that are observed histopathologically as steatosis (fatty liver), cholestasis, fibrosis, and necrosis, or apoptosis. Whereas some types of damage—for example, cholestasis—are liver specific, others such as necrosis and carcinogenesis are a more general phenomena. Damaged liver cells release liver-specific enzymes such as alanine aminotransferase (ALT), aspartate aminotransferase (AST), and alkaline phosphatase into the blood.

The enzymes ALT and AST are used as biomarkers of injured hepatocytes, while alkaline phosphatase indicates bile duct epithelial damage. These enzymes are commonly monitored clinically and in animal studies to detect hepatotoxicity.

13.3.1 Fatty Liver

Fatty liver or steatosis refers to the abnormal accumulation of lipid in hepatocytes, primarily as triglycerides, due to an imbalance between the uptake of extrahepatic triglycercides and the hepatic secretion of triglyceride-containing lipoproteins and fatty acid catabolism. Although many toxicants may cause lipid accumulation in the liver (Table 13.1), the mechanisms may be different. Basically, lipid accumulation is related to disturbances in either the synthesis or the secretion of lipoproteins. Excess lipid can result from an oversupply of free fatty acids from adipose tissues or, more commonly, from impaired release of triglycerides from the liver into the plasma. Triglycerides are secreted from the liver as lipoproteins, such as very low density lipoprotein (VLDL). As might be expected, there are a number of points

TABLE 13.1 Examples of Hepatotoxic Agents and Associated Liver Injury

Necrosis and Fatty Liver		
Carbon tetrachloride	Dimethylnitrosamine	Phosphorous
Chloroform	Cyclohexamide	Beryllium
Trichloroethylene	Tetracycline	Allyl alcohol
Tetrachloroethylene	Acetaminophen	Galactosamine
Bromobenzene	Mitomycin	Azaserine
Thioacetamide	Puromycin	Aflatoxin
Ethionine	Tannic acid	Pyrrolizidine alkaloids
Troglitazone	Zidovudine (AZT)	
Cholestasis (Drug Induced)		
Chlorpromazine	Imipramine	Carbarsone
Promazine	Diazepam	Chlorthiazide
Thioridazine	Methandrolone	Methimazole
Mepazine	Mestranol	Sulfanilamide
Amitriptyline	Estradiol	Phenindione
Phenytoin		
Hepatitis (Drug Induced)		
Iproniazid	Methoxyflurane	Halothane
Isoniazid	Papaverine	Zoxazolamine
Imipramine	Phenyl butazone	Indomethacin
6-Mercaptopurine	Colchicine	Methyldopa
Carcinogenesis (Experimental Animals)		
Aflatoxin B1	Dimethylbenzanthracene	Acetylaminofluorene
Pyrrolizidine alkaloids	Dialkyl nitrosamines	Urethane
Cycasin	Polychlorinated biphenyls	
Safrole	Vinyl chloride	

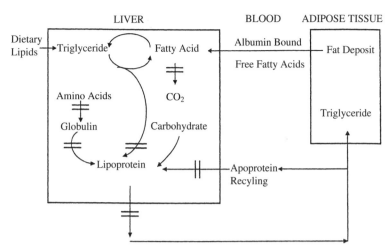

Figure 13.2 Triglyceride cycle in the pathogenesis of fatty liver. "=" are metabolic blocks. From Wallace, A. D. and S. A. Meyer. *Molecular and Biochemical Toxicology*, 4th ed. Wiley, 2008.

at which this process can be disrupted. Some of the more important ones are as follows (Figure 13.2):

- Interference with synthesis of proteins
- Impaired conjugation of triglyceride with lipoprotein
- Interference with transfer of VLDL across cell membranes
- Decreased synthesis of phospholipids
- Impaired β-oxidation of lipids by mitochondria
- Inadequate energy (adenosine triphosphate [ATP] for lipid and protein synthesis)

The role that fatty liver plays in liver injury is not clearly understood, and fatty liver in itself does not necessarily mean liver dysfunction. The onset of lipid accumulation in the liver is accompanied by changes in blood biochemistry, as indicated by changes in ALT and AST, and for this reason, blood chemistry analysis can be a useful diagnostic tool.

13.3.2 Cholestasis

Cholestasis is the suppression or stoppage of bile flow, and may have either intrahepatic or extrahepatic causes. Inflammation or blockage of the bile ducts results in retention of bile salts as well as bilirubin accumulation, an event that leads to jaundice. Other mechanisms causing cholestasis include changes in membranes permeability of either hepatocytes or biliary canaliculi. The formation of bile depends on ATP-dependent transport of bile into the canalicular lumen. Chemicals that have effects on membrane permeability and disrupt cellular Na^+ and K^+ gradients can cause cholestatis by their impact on the ATP-dependent movement of bile.

Cholestasis is usually drug induced (Table 13.1) and is difficult to produce in experimental animals. Again, changes in blood chemistry can be a useful diagnostic tool.

13.3.3 Fibrosis and Cirrhosis

Chemicals that are hepatotoxicants cause damage to hepatocytes that results in hepatic fibrosis as part of the wound-healing response. Fibrosis is characterized by the deposition of collagen, proteoglycans, and glycoproteins, and chronic fibrosis results in formation of an extracellular matrix (ECM) that can be observed histopathologically. After a toxicant exposure, hepatic stellate cells (HSC) proliferate and differentiate into fibroblast-like cells that secrete the components of the ECM. Extensive fibrosis can disrupt the liver architecture and blood flow resulting in irreversible liver damage. Reversibility of fibrosis is possible upon HSC becoming quiescent or undergoing apoptosis, breakdown of ECM, and hepatocyte regeneration.

Cirrhosis is a result of hepatotoxicant exposure that is characterized by fibrosis to the extent that deposition of collagen is found throughout the liver and results in the formation of scar tissue. In most cases, cirrhosis results from chronic chemical injury, which results in the accumulation of ECM that causes severe restriction in blood flow and also inhibits the liver's normal metabolic and detoxication processes. This situation can in turn cause further damage and eventually lead to liver failure. In humans, chronic use of ethanol is the single most important cause of cirrhosis, although there is some dispute as to whether the effect is due to ethanol alone or is also related to the nutritional deficiencies that usually accompany alcoholism.

13.3.4 Necrosis

Necrosis refers to an irreversible loss of cell viability that occurs due to loss of normal cellular function. Necrosis, usually an acute injury, may be localized and affect only a few hepatocytes (focal necrosis), or it may involve an entire lobe (massive necrosis). Cell death is "unordered" and occurs along with rupture of the plasma membrane, and is preceded by a number of morphologic changes such as cellular swelling, dilation of the endoplasmic reticulum, accumulation of triglycerides, swelling of mitochondria with disruption of cristae, and dissolution of organelles and a shrunken nucleus. In areas of necrosis, increased eosinophilic staining of the cytoplasm and an immune response is seen as neutrophils infiltrate the damaged area. Biochemical events that may lead to these changes include binding of reactive metabolites to proteins and unsaturated lipids (inducing lipid peroxidation and subsequent membrane destruction, disturbance of cellular Ca^{+2} homeostasis, inference with metabolic pathways, shifts in Na^+ and K^+ balance, and inhibition of protein synthesis. Changes in blood chemistry resemble those seen with fatty liver, except they are quantitatively larger. Because of the regenerating capability of the liver, necrotic lesions are not necessarily critical. Massive areas of necrosis, however, can lead to severe liver damage and failure.

13.3.5 Apoptosis

Apoptosis is a controlled form of cell death that serves as a regulation point for biologic processes and can be thought of as the counterpoint of cell division by mitosis. This "ordered" mechanism of cell death, unlike necrosis, is particularly active during development and senescence. Although apoptosis is a normal physiological process, it can also be induced by a number of exogenous factors such as xenobiotic chemicals, oxidative stress, anoxia, and radiation. (A stimulus that induces a cell to undergo apoptosis is known as an apogen.) If, however, apoptosis is suppressed in some cell types, it can lead to accumulation of these cells. For example, in some instances, clonal expansion of malignant cells and subsequent tumor growth results primarily from inhibition of apoptosis.

Apoptosis can be distinguished from necrosis by morphologic criteria, using either light or electron microscopy. A hallmark of apoptosis is the absence of inflammatory infiltrate. Toxicants, however, do not always act in a clear-cut fashion, and some toxicants can induce both apoptosis and necrosis either concurrently or sequentially.

13.3.6 Hepatitis

Hepatitis is an inflammation of the liver and is usually viral in origin; however, certain chemicals, usually drugs, can induce a hepatitis that closely resembles that produced by viral infections (Table 13.1). It is characterized by the increase in immune cells and this type of liver injury is sometimes associated with idiosyncratic hepatotoxicants, such as diclofenac. This type of idiosyncratic response is not usually demonstrable in laboratory animals and is often manifest only in susceptible individuals. Fortunately, the incidence of this type of disease is very low.

13.3.7 Carcinogenesis

The most common type of primary liver cancer is hepatocellular carcinoma; other types include cholangiocarcinoma, biliary cystadenocarcinoma, and undifferentiated liver cell carcinoma. Although a wide variety of chemicals are known to induce liver cancer in laboratory animals (Table 13.1), the incidence of primary liver cancer in humans in the United States is very low.

Some naturally occurring liver carcinogens are aflatoxin, cycasin, and safrole. A number of synthetic chemicals have been shown to cause liver cancer in animals, including the dialkylnitrosamines, dimethylbenzanthracene, aromatic amines such as 2-naphthylamine and acetylaminofluorene, and vinyl chloride. The structure and activation of these compounds can be found in Chapters 6 and 7. In humans, the most noted case of occupation-related liver cancer is the development of angiosarcoma, a rare malignancy of blood vessels, among workers exposed to high levels of vinyl chloride in manufacturing plants. For a discussion of chemical carcinogenesis, see Chapter 11.

13.4 MECHANISMS OF HEPATOTOXICITY

Chemically induced cell injury can be thought of as involving a series of events occurring in the affected animal and often in the target organ itself:

- The chemical agent is activated to form the initiating toxic agent.
- The initiating toxic agent is either detoxified or causes molecular changes in the cell.

- The cell recovers or there are irreversible changes.
- Irreversible changes may culminate in cell death.

Cell injury can be initiated by a number of mechanisms, such as inhibition of enzymes, depletion of cofactors or metabolites, depletion of energy (ATP) stores, interaction with receptors, elevated intracellular free calcium, formation of a reactive metabolite, and alteration of cell membranes. In recent years, attention has focused on the role of biotransformation of chemicals to highly reactive metabolites that initiate cellular toxicity. Many compounds, including clinically useful drugs, can cause cellular damage through metabolic activation of the chemical to highly reactive compounds such as free radicals, carbenes, and nitrenes causing oxidative stress (Chapters 6 and 7).

These reactive metabolites can bind covalently to cellular macromolecules such as nucleic acids, proteins, cofactors, lipids, and polysaccharides, thereby changing their biologic properties. The liver is particularly vulnerable to toxicity produced by reactive metabolites because it is the major site of xenobiotic metabolism. Most activation reactions are catalyzed by CYP enzymes, and agents that induce these enzymes, such as phenobarbital and 3-methylcholanthrene, often increase toxicity. Conversely, inhibitors of CYPs, such as SKF-525A and piperonyl butoxide, frequently decrease toxicity.

Formation of reactive metabolites can result in oxidative stress, which has been defined as an imbalance between the pro-oxidant/antioxidant steady state in the cell, with the excess of pro-oxidants being available to interact with cellular macromolecules to cause damage to the cell, often resulting in cell death. To date, a number of liver diseases, including alcoholic liver disease, metal storage diseases, and cholestatic liver disease, have been shown to have an oxidative stress component.

Reactive oxygen and reactive nitrogen radicals can be formed in a number of ways (Figure 13.3), the former primarily as a by-product of mitochondrial electron transport. Superoxide, hydrogen peroxide, singlet oxygen, and hydroxyl can all arise

Figure 13.3   Origin of reactive oxygen and nitrogen species and sites of blocking their oxidant challenges by antioxidant defenses. From Reed, D. J. *Molecular and Biochemical Toxicology*, 4th ed. Wiley, 2008.

from this source. Other sources include monooxygenases and peroxisomes. If not detoxified, reactive oxygen species can interact with biological macromolecules such as DNA and protein or with lipids. Once lipid peroxidation of unsaturated fatty acids in phospholipids is initiated, it is propagated in such a way as to have a major damaging effect on cellular membranes. The formation, detoxication by superoxide dismutase and by glutathione-dependent mechanisms, and interaction at sites of toxic action are illustrated in Figure 13.3.

Mechanisms such as conjugation of the reactive chemical with glutathione are protective mechanisms that exist within the cell for the rapid removal and inactivation of many potentially toxic compounds. Because of these interactions, cellular toxicity is a function of the balance between the rate of formation of reactive metabolites and the rate of their removal. Examples of these interactions are presented in the following discussions of specific hepatotoxicants.

13.5 EXAMPLES OF HEPATOTOXICANTS

13.5.1 Carbon Tetrachloride

Carbon tetrachloride has probably been studied more extensively, both biochemically and pathologically, than any other hepatotoxicant. It is a classic example of a chemical activated by CYPs to form a highly reactive free radical (Figure 13.4). First, CCl_4 is converted to the trichloromethyl radical ($CCl_3\cdot$) and then to the trichloromethylperoxy radical ($CCl_3O_2\cdot$). Such radicals are highly reactive and generally have a small radius of action. For this reason, the necrosis induced by CCl_4 is most severe in the centrilobular liver cells that contain the highest concentration of the CYP isozyme responsible for CCl_4 activation.

Typically free radicals may participate in a number of events (Figure 13.5), such as covalent binding to lipids, proteins, or nucleotides as well as lipid peroxidation. It is now thought that $CCl_3\cdot$, which forms relatively stable adducts, is responsible for covalent binding to macromolecules, and the more reactive $CCl_3O_2\cdot$, which is formed when $CCl_3\cdot$ reacts with oxygen, is the prime initiator of lipid peroxidation. Lipid peroxidation (Figure 13.6) is the initiating reaction in a cascade of events, starting with the oxidation of unsaturated fatty acids to form lipid hydroperoxides, which then break down to yield a variety of end products, mainly aldehydes, which can go on to produce toxicity in distal tissues. For this reason, cellular damage

Figure 13.4 Metabolism of carbon tetrachloride and formation of reactive metabolites. From Hodgson, E. and Levi, P. E. *A Textbook of Modern Toxicology*, 3rd ed., Wiley, 2004.

Figure 13.5 Summary of targets for free radicals. From Hodgson, E. and Levi, P. E. *A Textbook of Modern Toxicology*, 3rd ed., Wiley, 2004.

Figure 13.6 Metabolism of bromobenzene. From Hodgson, E. and Levi, P. E. *A Textbook of Modern Toxicology*, 3rd ed., Wiley, 2004.

results not only from the breakdown of membranes such as those of the endoplasmic reticulum, mitochondria, and lysosomes but also from the production of reactive aldehydes that can travel to other tissues. It is now thought that many types of tissue injury, including inflammation, may involve lipid peroxidation.

13.5.2 Ethanol

Alcohol-related liver diseases are complex, and ethanol has been shown to interact with a large number of molecular targets. Ethanol can interfere with hepatic lipid metabolism in a number of ways and is known to induce both inflammation and necrosis in the liver. Ethanol increases the formation of superoxide by Kupffer cells thus implicating oxidative stress in ethanol-induced liver disease. Similarly, pro-oxidants (reactive oxygen species) are produced in the hepatocytes by partial reactions in the catalytic cycle of CYP2E1, an ethanol-induced CYP isoform. The formation of protein adducts in the microtubules by acetaldehyde, the metabolic product formed from ethanol by alcohol dehydrogenase, plays a role in the impairment of VLDL secretion associated with ethanol.

13.5.3 Bromobenzene

Bromobenzene is a toxic industrial solvent that is known to produce centrilobular hepatic necrosis through the formation of reactive epoxides. Figure 13.6 summarizes the major pathways of bromobenzene metabolism. Both bromobenzene 2,3-epoxide and bromobenzene 3,4-epoxide are produced by CYP oxidations. The 2,3-epoxide, however, is the less toxic of the two species, reacting readily with cellular water to form the nontoxic 2-bromophenol. The more stable 3,4-epoxide is the form most responsible for covalent binding to cellular proteins. A number of pathways exist for detoxication of the 3,4-epoxide: rearrangement to the 4-bromophenol, hydration to the 3,4-dihydrodiol catalyzed by epoxide hydrolase, or conjugation with glutathione. When more 3,4-epoxide is produced than can readily be detoxified, cell injury increases. Pretreatment of animals with inhibitors of CYPs is known to decrease tissue necrosis by slowing down the rate of formation of the reactive metabolite, whereas pretreatment of animals with certain CYP inducers can increase the toxicity of bromobenzene, as the CYP inducer phenobarbital increases hepatotoxicity by inducting a P450 isozyme that preferentially forms the 3,4-epoxide. However, pretreatment with another CYP inducer, 3-methylcholanthrene, decreases bromobenzene hepatotoxicity by inducing a form of CYP that produces primarily the less toxic 2,3-epoxide.

13.5.4 Acetaminophen

Acetaminophen is a widely used analgesic that is normally safe when taken at therapeutic doses. Overdoses, however, may cause an acute centrilobular hepatic necrosis that can be fatal. Although acetaminophen is eliminated primarily by formation of glucuronide and sulfate conjugates, a small proportion is metabolized by CYPs to a reactive electrophilic intermediate believed to be a quinoneimine (see Chapter 8). This reactive intermediate is usually inactivated by conjugation with reduced glutathione and excreted. Higher doses of acetaminophen will progressively deplete hepatic glutathione levels, however, resulting in extensive covalent binding of the reactive metabolite to liver macromolecules with subsequent hepatic necrosis. The early administration of sulfhydryl compounds such as cysteamine, methionine, and N-acetylcysteine is very effective in preventing liver damage, renal failure, and death that would otherwise follow an acetaminophen overdose. These agents are thought to act primarily by stimulating glutathione synthesis.

In laboratory animals, the formation of the acetaminophen-reactive metabolite, the extent of covalent binding, and the severity of hepatotoxicity can be influenced by altering the activity of various CYP isozymes. Induction of CYP isozymes with phenobarbital, 3-methylcholanthrene, or ethanol increases toxicity, whereas inhibition of CYPs with piperonyl butoxide, cobalt chloride, or metyrapone decreases toxicity. Consistent with these effects in animals, it appears that the severity of liver damage after acetaminophen overdose is greater in chronic alcoholics and patients taking drugs that induce the levels of the CYP isozymes responsible for the activation of acetaminophen.

13.5.5 Troglitazone

Troglitazone (Rezulin®; Pfizer, Inc., New York, NY, USA) was a type II diabetes drug approved for use in 1997 and subsequently withdrawn from the market due to hepatotoxicity, which was seen in susceptible patients, but was not observed in preclinical animal studies. Troglitazone represented a new type of drug treatment for diabetes and acted as a peroxisome proliferator-activated receptor (PPAR) gamma agonist. In a small number of cases, complete liver failure was seen resulting in liver transplant or death. During therapy elevations of blood liver enzymes indicating hepatic injury were not seen until months after the initiation of treatment. The spectrum of liver injury in patients was broad with a heterogenous pattern of injury that included steatosis, cholestasis, fibrosis, cirrhosis, inflammation, and necrosis. Much effort has been made to elucidate if the mechanism(s) of toxicity involve genetic differences of metabolic enzymes in susceptible patients, formation of toxic metabolites, mitochondrial toxicity, oxidative stress, apoptosis, or a combination of these mechanisms. While it remains unclear the exact mechanisms responsible for troglitazone hepatotoxicity, evidence suggests a combination of unknown genetic and/or environmental factors lead to mitochondrial dysfunction. Fortunately, this type of idiosyncratic hepatotoxicity is rare, but much research still needs to be done to understand the mechanisms responsible.

13.6 METABOLIC ACTIVATION OF HEPATOTOXICANTS

Studies of liver toxicity caused by bromobenzene, acetaminophen, and other compounds have led to some important observations concerning tissue damage:

- Toxicity may be correlated with the formation of a minor but highly reactive intermediate.
- A threshold tissue concentration of the reactive metabolite must be attained before tissue injury occurs.
- Endogenous substances, such as glutathione, play an essential role in protecting the cell from injury by removing chemically reactive intermediates and by keeping the sulfhydryl groups of proteins in the reduced state.
- Pathways such as those catalyzed by glutathione transferase and epoxide hydrolases play an important role in protecting the cell.
- Agents that selectively induce or inhibit the xenobiotic metabolizing enzymes may alter the toxicity of xenobiotic chemicals.

These same principles are applicable to the toxicity caused by reactive metabolites in other organs, such as kidney and lung as will be illustrated in the following sections.

BIBLIOGRAPHY AND SUGGESTED READING

Hodgson, E. and P. E. Levi. *A Textbook of Modern Toxicology*, 3rd ed. Hoboken, NJ: Wiley, 2004.

Hodgson, E. and S. A. Meyer. Pesticides. In *Comprehensive Toxicology: Hepatic and Gastrointestinal Toxicology*, Vol. 9, 2nd ed., ed. C. A. McQueen. New York: Elsevier Science, 2010, in press.

Jaeschke, H. Toxic responses of the liver. In *Casarett and Doull's Toxicology: The Basic Sciences of Poisons*, 7th ed., ed. C. D. Klaassen, p. 557. New York: McGraw-Hill, 2008.

Reed, D. J. Glutathione-dependent mechanisms in chemically induced cell injury and cellular protection mechanisms. In *Molecular and Biochemical Toxicology*, 4th ed., eds. E. Hodgson and R. C. Smart, p. 333. Hoboken, NJ: Wiley, 2008.

Wallace, A. D. and S. A. Meyer. Hepatotoxicity. In *Molecular and Biochemical Toxicology*, 4th ed., eds. E. Hodgson and R. C. Smart, p. 671. Hoboken, NJ: Wiley, 2008.

SAMPLE QUESTIONS

1. Sinusoidal endothelial cells form loose connections such that the sinusoids are relatively leaky. When intra-sinusoidal Kupffer cells encounter certain agents, such as bacterial lipopolysaccharide (endotoxin), they become activated and secrete various small protein molecules, the cytokines. These cytokines can cause toxic responses in hepatocytes. Discuss how these spatially separated liver cells interact to mediate endotoxin-mediated hepatocellular toxicity.

2. How would you determine in an experimental animal study whether a hepatotoxicant required metabolic activation by cytochrome P450?

3. Hepatotoxicants can be classified into two different groups based on the pattern of injury. Name these two groups and describe them.

4. What are the hallmarks of hepatotoxicity and what tests can be done to detect hepatotoxicity?

Nephrotoxicity

JOAN B. TARLOFF and ANDREW D. WALLACE

14.1 INTRODUCTION

Nephrotoxicity can be a potentially serious complication of drug therapy or chemical exposure. Although in most instances the mechanisms mediating nephrotoxicity are unclear, susceptibility of the kidney to toxic injury appears to be related, at least in part, to the complexities of renal anatomy and physiology.

14.1.1 Structural Organization of the Kidney

Upon gross examination, three major anatomical areas of the kidney are apparent: cortex, medulla, and papilla. The cortex is the outermost portion of the kidney and contains proximal and distal tubules, glomeruli, and peritubular capillaries. Cortical blood flow is high relative to cortical volume and oxygen consumption; the cortex receives about 90% of total renal blood flow. A blood-borne toxicant will be delivered preferentially to the renal cortex and therefore have a greater potential to influence cortical, rather than medullary or papillary, functions.

The renal medulla is the middle portion of the kidney and consists of the loops of Henle, vasa recta, and collecting ducts. Medullary blood flow (about 6% of total renal blood flow) is considerably lower than cortical flow. However, by virtue of its countercurrent arrangement between tubular and vascular components, the medulla may be exposed to high concentrations of toxicants within tubular and interstitial structures.

The papilla is the smallest anatomical portion of the kidney. Papillary tissue consists primarily of terminal portions of the collecting duct system and the vasa recta. Papillary blood flow is low relative to cortex and medulla; less than 1% of total renal blood flow reaches the papilla. However, tubular fluid is maximally concentrated, and the volume of luminal fluid is maximally reduced within the papilla. Potential toxicants trapped in tubular lumens may attain extremely high concentrations within the papilla during the process of urinary concentration. High intraluminal concentrations of potential toxicants may result in diffusion of these chemicals into papillary tubular epithelial and/or interstitial cells, leading to cellular injury.

The nephron is the functional unit of the kidney and consists of vascular and tubular elements. Both elements have multiple specific functions, which may be

---

A Textbook of Modern Toxicology, Fourth Edition. Edited by Ernest Hodgson
Copyright © 2010 John Wiley & Sons, Inc.

influenced by toxicants. The glomerulus is the portion of the nephron where ultrafiltrate of the plasma is formed and is governed by physical processes across capillaries. The renal tubule begins as a blind pouch surrounding the glomerulus, and consists of multiple segments which modify the composition of the ultrafiltrate. The segments of the renal tubule include the proximal tubule, loop of Henle, distal tubule, and collecting duct. The unique properties and functions of the cells that compose these segments can lead to susceptibility to toxicants.

14.1.2 Function of the Renal System

The kidneys participate in regulation of extracellular fluid volume, blood pressure, acid–base balance, and electrolyte balance. Blood-borne substances are exposed to kidney cells through the processes of filtration and reabsorption. A primary function of the kidneys is to eliminate waste products. During the process of reabsorption, potentially toxic chemicals may achieve higher concentrations than present in plasma, which may predispose the kidney to injury.

Renal tubules consist of multiple segments. These tubular elements selectively modify the composition of glomerular filtrate, enabling conservation of electrolytes and metabolic substrates while allowing elimination of waste products. For example, renal tubules reabsorb 98–99% of filtered electrolytes and water, and virtually 100% of filtered glucose and amino acids. Additionally, renal tubules participate in the reabsorption of bicarbonate and secretion of protons, thereby participating in acid–base balance.

Other functions of the kidney include synthesis of hormones. For example, 25-hydroxy-vitamin D_3 requires metabolism by the kidneys to the active 1,25-hydroxy-vitamin D_3. The kidney also secretes erythropoietin, which is involved in differentiation and development of red blood cells. Renin is an important enzyme released by the kidney in response to low blood pressure and catalyzes a step in the formation of angiotensin II, a powerful vasoconstrictor hormone.

Kidney toxicity is usually diagnosed by changes in excretory function, such as increases in urinary glucose, amino acid, or protein excretion, changes in urine volume, osmolarity, or pH. Changes in blood urea nitrogen (BUN) or serum creatinine concentrations are also indicators of altered renal function. Recently, several biomarkers have been approved by the Food and Drug Administration (FDA) as reliable indicators of kidney toxicity. A biomarker is a biochemical feature that can be used to diagnose a disease or monitor the effects of treatment. The biomarkers approved by the FDA are all proteins that appear in the urine when kidney damage has occurred and include proteins such as kidney injury molecule-1 (KIM-1), β_2-microglobulin, and albumin. Excretion of higher molecular weight proteins in the urine such as albumin is suggestive of injury to the glomerulus, while the presence of low molecular weight proteins, such as β_2-microglobulin is more suggestive of proximal tubule injury.

14.2 FACTORS CONTRIBUTING TO NEPHROTOXICITY

Several factors contribute to the unique susceptibility of the kidney to toxicants (Table 14.1). First, renal blood flow is high relative to organ weight. For an organ constituting less than 1% of body weight, the kidneys receive about 25% of the

TABLE 14.1   **Factors Influencing Susceptibility of the Kidney to Toxicants**

High renal blood flow
Concentration of chemicals in intraluminal fluid
Reabsorption and/or secretion of chemicals through tubular cells
Biotransformation of protoxicants to reactive intermediates

resting cardiac output. Thus, the kidneys will receive higher concentrations of toxicants (per gram of tissue) than poorly perfused tissue such as skeletal muscle, skin, and fat. Renal blood flow is unequally distributed, with cortex receiving a disproportionately high flow compared to medulla and papilla. Therefore, a blood-borne toxicant will be delivered preferentially to the renal cortex and thereby have a greater potential to influence cortical, rather than medullary or papillary, functions.

Second, the processes involved in forming concentrated urine also will serve to concentrate potential toxicants present in the glomerular filtrate. Reabsorptive processes along the nephron may raise the intraluminal concentration of a toxicant from 10 mM to 50 mM by the end of the proximal tubule, 66 mM at the hairpin turn of the loop of Henle, 200 mM at the end of the distal tubule, and as high as 2000 mM in the collecting duct. Progressive concentration of toxicants may result in intraluminal precipitation of poorly soluble compounds, causing acute renal failure secondary to obstruction. The potentially tremendous concentration gradient for passive diffusion between lumen and cell may drive even a relatively nondiffusible toxicant into tubular cells.

Third, active transport processes within the proximal tubule may further raise the intracellular concentration of an actively transported toxicant. During active secretion and/or reabsorption, substrates generally accumulate in proximal tubular cells in much higher concentrations than present in either luminal fluid or peritubular blood.

Fourth, certain segments of the nephron have a capacity for metabolic bioactivation. For example, the proximal and distal tubules contain isozymes of the cytochrome P450 monooxygenase system that may mediate intrarenal bioactivation of several protoxicants. Additionally, prostaglandin synthase activity in medullary and papillary interstitial cells may be involved in co-oxidation of protoxicants, resulting in selective papillary injury.

14.3   EXAMPLES OF NEPHROTOXICANTS

Many compounds have been implicated as nephrotoxicants (Table 14.2). Only rarely have specific receptors for specific nephrotoxicants been identified. Rather, in many cases it appears that toxicants exert multiple effects on intracellular systems. This is not to say, however, that there are not specific targets for certain nephrotoxicants in the kidney. For example, the proximal convoluted tubule seems to be more susceptible than other nephron segments to certain metals, such as chromium. The straight portion of the proximal tubule seems to be more susceptible to damage due to halogenated hydrocarbons (i.e., hexachlorobutadiene and dichlorovinyl-L-cysteine). Some agents, such as analgesic mixtures (usually aspirin, phenacetin, and caffeine) taken over long periods can produce a unique toxicity characterized

TABLE 14.2 Segments of the Nephron Affected by Selected Toxicants

Glomerulus
Immune complexes
Aminoglycoside antibiotics
Puromycin aminonucleoside
Adriamycin
Penicillamine

Proximal Tubule
Antibiotics
Cephalosporins
Aminoglycosides
Antineoplastic agents
Nitrosoureas
Cisplatin and analogs
Radiographic contrast agents
Halogenated hydrocarbons
Chlorotrifluoroethylene
Hexafluropropene
Hexachlorobutadiene
Trichloroethylene
Chloroform
Carbon tetrachloride
Maleic acid
Citrinin
Metals
Mercury
Uranyl nitrate
Cadmium
Chromium

Distal Tubule/Collecting Duct
Lithium
Tetracyclines
Amphotericin
Fluoride
Methoxyflurane

Papilla
Aspirin
Phenacetin
Acetaminophen
Nonsteroidal anti-inflammatory agents
2-bromoethylamine

by renal medullary and papillary necrosis. Histological evaluation following intoxication with analgesic mixtures reveals damage to the ascending limbs of the loop of Henle. Likewise, fluoride ion and outdated tetracyclines produce damage in this area.

14.3.1 Metals

Many heavy metals are potent toxicants. Exposure to relatively low amounts of metal can produce renal toxicity characterized by functional changes such as glucosuria, aminoaciduria, and polyuria. As the exposure level increases, renal necrosis, anuria, increased BUN, and overt renal failure may occur. Several mechanisms operate to protect the kidney from heavy metal toxicity. After low level exposure and often before detectable signs or symptoms of nephrotoxicity, metals may be found in renal lysosomes. Accumulation in lysosomes occurs following uptake of metal–protein complexes, digestion of metal-damaged organelles such as mitochondria, or interactions of metals with lipoproteins within lysosomes.

Cadmium Human exposure to cadmium is through food or industrial processes. Cadmium is excreted in urine complexed with metallothionein (MT), a low molecular weight protein synthesized in liver. MT contains free sulfhydryl groups that bind metals such as cadmium. Binding of cadmium to MT may protect some organs, such as testis and brain, from toxicity. However, the cadmium–MT complex may be taken up by kidney cells more readily than unbound cadmium. Thus, complexing of cadmium with MT may contribute to selective renal toxicity of cadmium. Cadmium–MT probably accumulates in lysosomes following uptake into kidney cells. Once in lysosomes, the cadmium may dissociate and persist in cells as free metal. The half-life of cadmium is extremely long in humans, 10–12 years, so that low level exposure to cadmium over time may result in renal accumulation and toxicity.

In Japan, Itai-itai Byo (literally, ouch-ouch disease) occurred among women who consumed rice grown in cadmium-contaminated areas. The disease is characterized by anemia, bone and joint pains, and kidney failure, and the severity of disease is correlated with the extent of cadmium contamination.

Mercury Mercury is found in the environment and many industrial settings, and exposure may occur from dietary sources such as contaminated water or food items such as large predator fish. Mercury can exist as elemental (Hg), mercury salts ($HgCl_2$), or organic mercury (R-Hg). In the body, elemental mercury is a cation (Hg^{2+}) that binds to sulfhydryl-containing molecules including glutathione, cysteine, homocysteine, and metallothionein. Within the kidney, inorganic and organic mercury accumulate rapidly.

The nephrotoxicity of mercury is characterized by increasing excretion levels of enzymes, such as alkaline phosphatase and γ-glutamyltransferase, amino acids, and albumin in the urine. Intracellular toxicity of mercury occurs due to its high affinity for thiol-containing proteins that can lead to oxidative stress involving mitochondrial dysfunction. Thiol-containing metal chelating agents such as *meso*-2,3-dimercaptosuccinic acid or 2 2,3-dimercapto-1-propanesulfonic acid are often utilized as antidotes for mercury poisoning and allow excretion of mercuric conjugates in the urine.

14.3.2 Antimicrobial Agents

Aminoglycosides Aminoglycoside antibiotics, such as gentamicin, amikacin, and netilmicin, are powerful drugs for the treatment of serious gram-negative infections. However, about 10% of patients treated with aminoglycosides will develop moderate but significant signs and symptoms of renal toxicity. Aminoglycoside nephrotoxicity is characterized by proximal tubular necrosis, proteinuria, and a profound decline in glomerular filtration rate.

Aminoglycoside antibiotics are organic polycations and carry net positive charges. The primary route of elimination of aminoglycosides is by renal excretion. Gentamicin, a typical nephrotoxic aminoglycoside, is filtered at the glomerulus and appears to be reabsorbed via active transport processes at the proximal tubular brush border. Intracellular accumulation of gentamicin appears to occur following binding to plasma luminal membrane sites and incorporation of bound drug into apical vesicles such as lysosomes. Lysosomal alterations and the presence of myelin bodies and cytosegresomes are characteristic of aminoglycoside nephrotoxicity (Figure 14.1).

The sequence of biochemical events leading to gentamicin-induced proximal tubular dysfunction is unknown. Perhaps owing to its polycationic structure, gentamicin interferes with a number of intracellular proteins and macromolecules, producing a variety of biochemical effects. Several mechanisms have been proposed to account for gentamicin cytotoxicity, including (1) lysosomal damage, (2) altered phospholipid metabolism, (3) inhibition of critical intracellular enzymes, (4) inhibition of mitochondrial respiration, (5) lipid peroxidation, and (6) misreading of mRNA.

Cephalosporins   Cephalosporins are broad-spectrum antibiotics similar in structure to penicillin. For several cephalosporins, therapy is limited by the development of nephrotoxicity.

Cephaloridine is zwitterionic and the principal route of elimination is by the kidneys. Cephaloridine clearance approximates inulin clearance, indicating absence

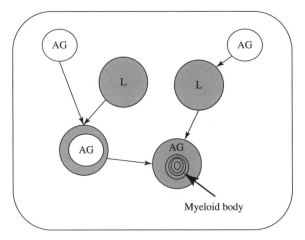

Figure 14.1   Interaction of aminoglycosides with lysosomes. Aminoglycosides (AG) enter the cell by pinocytosis and endocytosis, subsequently fusing with a primary lysosome (L). Aminoglycosides may interfere with normal lysosomal function, forming myeloid bodies (arrow).

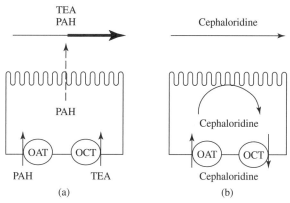

Figure 14.2 Schematic representation of proximal tubular transport and urinary excretion of *para*-aminohippurate (PAH), tetraethylammonium (TEA), and cephaloridine in the kidney. (a) PAH and TEA are excreted following both filtration and active secretion by the proximal tubule. PAH is transported across the basolateral membrane by organic anion transporter(s) (OAT) and TEA is secreted by organic cation transporter(s) (OCT). Intracellular concentrations of PAH and TEA may become great enough to drive passive diffusion from intracellular fluid to tubular fluid. Alternately, anion and cation exchangers may facilitate movement across the luminal membrane. (b) Cephaloridine is excreted primarily following filtration. Active cortical uptake of cephaloridine, inhibited by probenecid and PAH, indicates a secretory component for cephaloridine transport. However, diffusion of cephaloridine from proximal tubular cell to lumen is restricted, leading to high intracellular concentrations of cephaloridine. Some efflux of cephaloridine from proximal tubular cells appears to be mediated by organic cation transporter(s) since inhibitors of this transport system potentiate cephaloridine nephrotoxicity.

of net secretion for cephaloridine. However, inhibitors of organic anion transport, such as penicillin and probenecid, attenuate nephrotoxicity of cephaloridine while inhibitors of organic cation transport, such as cyanine 863 and mepiperphenidol, exacerbate toxicity. Taken together, these data suggest that, owing to its zwitterionic charge, cephaloridine is actively accumulated into proximal tubular cells via the organic anion transport system (inhibited by probenecid, PAH) and that a portion of cephaloridine efflux occurs via the organic cation transport system (inhibited by mepiperphenidol, cyanine). Once cephaloridine is transported into proximal tubular cells, it diffuses across the luminal membrane into tubular fluid only to a limited extent. Thus, active transport of cephaloridine into proximal tubular cells results in extremely high intracellular cephaloridine concentrations compared to other organs, which, in turn, contributes to selective nephrotoxicity (Figure 14.2).

Although the role of renal tubular transport in cephaloridine nephrotoxicity has been well defined, the exact molecular mechanisms mediating cephaloridine nephrotoxicity are less well understood. Several mechanisms have been postulated to mediate cephaloridine nephrotoxicity, including: (1) production of a highly reactive acylating metabolite(s) by cytochrome P450-dependent monooxygenases, (2) production of mitochondrial respiratory toxicity, and (3) production of lipid peroxidation.

Amphotericin B Amphotericin B is a polyene antifungal agent used in the treatment of systemic mycoses caused by opportunistic fungi. Clinical utility of amphotericin B is limited by its nephrotoxicity, characterized functionally by polyuria resistant to antidiuretic hormone administration, hyposthenuria, hypokalemia, and mild renal tubular acidosis.

Amphotericin B is highly lipophilic and interacts with membrane lipid sterols, such as cholesterol, to disrupt membrane permeability. Since amphotericin is freely filtered, it achieves high concentrations in distal tubular fluid and easily forms complexes with cholesterol and other lipids present in distal tubular luminal membranes. Amphotericin effectively transforms the "tight" distal tubular epithelium into an epithelium leaky to water, H^+ and K^+. Functional abnormalities observed with amphotericin B are attenuated when the antifungal agent is administered as an emulsion formulation whereby amphotericin is incorporated into lipid micelles. Antifungal activity of emulsion-formulated amphotericin B is equivalent to the standard non-emulsion formulation, whereas polyuria and hyposthenuria are significantly reduced by emulsion formulation.

14.3.3 Agents that Precipitate in Renal Tubules

The kidneys are responsible for producing small volumes of waste products and are involved in maintenance of water balance by antidiuretic hormone-dependent water reabsorption. However, this function may lead to relatively high concentrations of poorly soluble substances and in some cases, these poorly soluble substances may precipitate and obstruct urine outflow. Kidney stones represent a form of precipitate formation. The most common type of kidney stones contains calcium in combination with either oxalate or phosphate.

Ethylene Glycol Ethylene glycol is commonly found in antifreeze and hydraulic brake fluids. The cause of toxicity is not ethylene glycol but its metabolites, particularly oxalic acid. Ethylene glycol is metabolized initially to glycolic acid and ultimately to oxalic acid (Figure 14.3). Oxalic acid binds with calcium to form a poorly soluble product that precipitates and blocks urine flow. In addition, oxalic acid may be directly toxic to kidney cells.

Melamine Recalls of pet food in 2007 and infant formula in 2008 focused interest on the toxicity of melamine. Melamine is a nitrogen-containing compound used in

Figure 14.3 Metabolism of ethylene glycol, showing solubility of parent compound and metabolites.

Figure 14.4 Chemical structures of melamine and cyanuric acid.

Figure 14.5 A hematoxylin and eosin stained paraffin embedded kidney tissue depicts the renal parenchyma from an adult cat, which was presented with acute renal failure during the recent outbreak of commercial pet food-associated melamine toxicosis. Characteristic melamine-containing crystals (arrow) occluding the lumen of a renal tubule and necrosis of tubular epithelial cells (arrowheads) are visible. Image courtesy of Drs. Ronald Baynes and Keith Linder (North Carolina State University).

commercial applications as a component of plastics. Because of its high nitrogen content, it is incorporated into animal feed as a nonprotein source of nitrogen. Cyanuric acid is also a nitrogen-containing compound used in the manufacture of bleaches. It can serve as a nonprotein nitrogen source in animal feed (Figure 14.4).

While both melamine and cyanuric acid have been reported as relatively safe, necropsies of animals fed contaminated pet food revealed yellowish-brown crystals present in kidney tissue (Figure 14.5). Subsequently, investigators found that a combination of melamine and cyanuric acid produced renal toxicity in rats and observed crystals containing both components in kidney tissue and urine from rats.

14.3.4 Halogenated Hydrocarbons

Chloroform Chloroform is a nephrotoxicant that most likely undergoes metabolic bioactivation within the kidney. Chloroform ($CHCl_3$), a common organic solvent

Figure 14.6   Proposed mechanism of chloroform biotransformation. Chloroform undergoes cytochrome P450-catalyzed conversion to trichloromethanol (CCl_3-OH), which spontaneously decomposes to form phosgene. Phosgene is highly reactive and may be detoxified by reacting with sulfhydryl-containing chemicals (cysteine, glutathione [GSH]). Alternately, phosgene can react with sulfhydryl groups on protein, leading to covalent binding and possibly to toxicity.

widely used in the chemical industry, produces hepatic and renal injury in humans and experimental animals. Tissue injury by chloroform is probably not due to chloroform per se, but is mediated by a chloroform metabolite. The initial step leading to chloroform-induced tissue injury is believed to be the biotransformation of chloroform to a reactive intermediate, phosgene ($COCl_2$). Phosgene is a highly reactive intermediate and may react with intracellular macromolecules to induce cell damage (Figure 14.6).

Hexachlorobutadiene   Hexachlorobutadiene is an industrial solvent used in various applications. It is a widespread environmental contaminant and a relatively specific nephrotoxicant. The nephrotoxicity of hexachlorobutadiene is of interest because it is an example of formation of a more toxic compound due to glutathione conjugation. Glutathione is a major intracellular antioxidant and conjugation with glutathione is thought to represent a detoxification or protective pathway. However, hexachlorobutadiene–glutathione conjugates are further processed into species that can be accumulated by kidney cells. Once inside cells, the conjugate is metabolized by a specific renal enzyme, cysteine conjugate β-lyase, into a reactive intermediate.

14.3.5 Analgesics

Chronic consumption of large dosages of combination analgesics, typically phenacetin and/or caffeine-containing preparations, may be associated with renal papillary necrosis.

Renal function may be compromised modestly by a loss of concentrating ability or, in severe cases, anuria, sepsis, and rapid deterioration of renal function may occur. Morphologically, there is loss of renal papilla (containing terminal collecting ducts), medullary inflammation, and interstitial fibrosis, and loss of renomedullary interstitial cells. A variety of nonnarcotic analgesics have been implicated in the etiology of renal papillary necrosis, including acetaminophen, aspirin, acetanilid, and nonsteroidal anti-inflammatory agents such as ibuprofen, phenylbutazone, and indomethacin.

The mechanism of renal injury of these compounds is unclear. Chronic consumption over a period of many years is required to demonstrate loss of concentrating ability. Although these agents are dissimilar structurally and chemically, they share a common mechanism of action, acting as analgesics by inhibiting prostaglandin synthesis. In the kidney, prostaglandin H synthase activity is distributed asymmetrically, with highest activity in renal medulla and lowest activity in renal cortex. The renal papilla may be injured selectively by nonnarcotic analgesic agents due to the combination of high concentrations of potential toxicants present in tubular fluid and specialized enzymes capable of biotransforming protoxicants to active intermediates.

14.4 SUMMARY

Susceptibility of the kidney to chemically induced toxicity is related, at least in part, to several unique aspects of renal anatomy and physiology. By virtue of high renal blood flow, active transport processes for secretion and reabsorption, and progressive concentration of the glomerular filtrate following water removal during the formation of urine, renal tubular cells may be exposed to higher concentrations of potential toxicants than are cells in other organs. Additionally, intrarenal metabolism, via cytochrome P450 or prostaglandin H synthase, may contribute to the generation of toxic metabolites within the kidney.

The precise biochemical mechanisms leading to irreversible cell injury and nephrotoxicity are not well defined. Many diverse biochemical activities occur within the kidney, and interference with one or more of these functions may lead to irreversible cell injury. Rather than any one single mechanism mediating chemically induced nephrotoxicity, it is likely that a toxicant alters a number of critical intracellular functions, ultimately leading to cytotoxicity and cellular necrosis.

BIBLIOGRAPHY AND SUGGESTED READING

Coca, S. G. and C. R. Parikh. Urinary biomarkers for acute kidney injury: Perspectives on translation. *Clin. J. Am. Soc. Nephrol.* **3**:481–490, 2008.

Dekant, W. Chemical-induced nephrotoxicity mediated by glutathione S-conjugate formation. *Toxicol. Lett.* **124**:21–36, 2001.

Schnellmann, R. G. Toxic responses of the kidney. In *Casarett and Doull's Toxicology: The Basic Science of Poisons*, 6th ed., ed. C. D. Klaassen, pp. 491–514. New York: McGraw-Hill, 2001.

Van Vleet, T. R. and R. G. Schnellmann. Toxic nephropathy: Environmental chemicals. *Semin. Nephrol.* **23**:500–508, 2003.

Zhou, Y., V. S. Vaidya, R. P. Brown, et al. Comparison of kidney injury molecule-1 and other nephrotoxicity biomarkers in urine and kidney following acute exposure to gentamicin, mercury, and chromium. *Toxicol. Sci.* **101**:159–170, 2008.

SAMPLE QUESTIONS

1. Discuss how aspects of renal physiology, particularly water reabsorption, can contribute to selective kidney toxicity.

2. Discuss the role of renal transporters in exposure of kidney cells to potentially toxic concentrations of chemicals.

3. Compare and contrast the toxicity of chloroform and hexachlorobutadiene.

Toxicology of the Nervous System

BONITA L. BLAKE

15.1 INTRODUCTION

Many substances alter the normal activity of the nervous system. Sometimes these effects are immediate and transient, like the stimulatory effect of a cup of coffee, or a headache from smelling fresh paint. Other effects can be much more insidious, like the movement disorders suffered by miners after years of breathing toxic manganese dust. Many agents are safe and even therapeutic at lower doses, but become neurotoxic at higher levels. Trace metals and pyridoxine (vitamin B6) fall into this category of dose-dependent neurotoxicants. Since these agents affirm the maxim, "the dose makes the poison," it becomes necessary to have a meaningful definition of nervous system poisoning, or neurotoxicity. *Neurotoxicity refers to the ability of an agent to adversely affect the structural or functional integrity of the nervous system.* It is often easier to identify changes in the structure or function of the nervous system than it is to say whether or not these events are adverse. For example, while some individuals need the stimulant effect of a morning cup of coffee, the same amount of coffee might provoke anxiety in others. Certainly, the function of the central nervous system (CNS) is altered (albeit temporarily) in both cases, but only those people who became jittery or nervous would characterize the effect as adverse.

In this chapter, a brief introduction to the nervous system and how it functions is described. A discussion of some of the mechanisms of structural and functional neurotoxicant effects follows. These descriptions are not exhaustive, but are meant to illustrate the concepts of toxicant interaction with the nervous system. Finally, some methods for testing toxicant effects in the nervous system are explored.

15.2 THE NERVOUS SYSTEM

Most multicellular animals possess a nervous system. In each case, the function of the nervous system is to receive signals about the external and internal environment,

---

integrate this information, and then to coordinate a response that is appropriate to the environmental stimulus. All of the other organ systems of the body are subject to control by the nervous system, thus damage to this "master system" by toxicants can have far-reaching and even devastating effects. In addition to these basic, vital functions, the nervous systems of higher organisms are responsible for thinking and learning.

In vertebrates, there are two major components of the nervous system. The brain and spinal cord comprise the CNS, and the nervous tissue (ganglia and peripheral nerves) outside the brain and spinal cord comprise the peripheral nervous system (PNS). Although these two systems are thought of as separate anatomical divisions, they are contiguous and function interactively. The PNS can be further divided into the somatic nervous system (SNS) and the autonomic nervous system (ANS). The somatic division consists of neurons that carry sensory information from the skin, muscle, and joints to the CNS, and motor nerves that originate in the CNS and innervate skeletal muscle to cause contractive movement. The ANS is often thought of as an involuntary motor system for visceral organs, since it innervates and controls the function of smooth muscle, cardiac muscle, and endocrine and exocrine glands. The ANS consists of sympathetic and parasympathetic subdivisions that control functions that are needed in preparation for expending energy ("fight or flight," sympathetic) or conserving energy ("rest and digest," parasympathetic). For example, stimulation of sympathetic nerves increases heart rate, while stimulation of the vagus nerve, the primary parasympathetic innervation of the heart, slows the rate of cardiac contraction. Nearly all glands and organs are innervated by both sympathetic and parasympathetic nerves, and their influences generally oppose one another.

15.2.1 The Neuron

The basic unit of the nervous system is the neuron, a type of cell that is structurally and functionally specialized to receive, integrate, conduct, and transmit information. Although neurons are a far more diverse group than any other cell type in the body, some common features can be found. Neurons are polarized cells, meaning that they have different characteristics on one end of the cell compared to the other (Figure 15.1). A typical neuron has a receiving end and a transmitting end. The end of the neuron that receives information from other neurons, usually in the form of neurotransmitter stimulation, is highly branched and is known as the dendritic tree. The branches are sometimes studded with tiny projections, known as spines, which contain clusters of neurotransmitter receptors on the surface. In such areas of high receptor density, the neuron is in close contact with other neurons via specialized structures called synapses. Synapses are areas of close apposition where one neuron (called the presynaptic neuron) releases neurotransmitter into the gap between the two neurons. The receptors on the dendritic spine of the receiving neuron (called the postsynaptic neuron) are selective for certain types of neurotransmitters. Receptor stimulation by neurotransmitter is translated into intracellular and electrochemical signals, and these signals from multiple regions of the dendritic tree are integrated together intracellularly. Neurotransmitters and their receptors are discussed in more detail below. In the typical neuron, the arborizations of the dendritic tree converge on the soma, or cell body, where the nucleus and most of

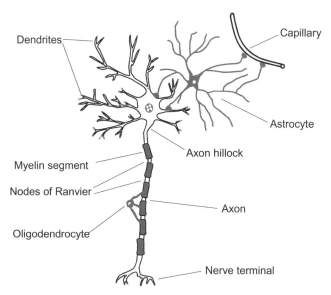

Dendrites

Capillary

Astrocyte

Axon hillock

Myelin segment

Nodes of Ranvier

Axon

Oligodendrocyte

Nerve terminal

Figure 15.1   A typical neuron with myelinated axon. The neuron is shown with two types of glial cells, the myelinating oligodendrocyte and an astrocyte that is also interacting with a capillary.

RNA- and protein-synthesizing machinery exist. Integrated signals that reach the nucleus modulate the expression of a multitude of molecules within the neuron, many of which help to fashion the neuron's responsiveness to further neurotransmitter stimulation.

The area of the neuron designed to transmit information is the axon, and most neurons have only a single axon. The initial segment of the axon as it leaves the cell body is called the axon hillock. This area is particularly sensitive to the summation of signals from dendritic regions that arrive at the cell body. If enough signals arrive over a short period of time to reach a certain threshold, an action potential will be formed in the hillock. Here it is thus determined whether the neuron will transmit its information (or "fire"), causing the release of neurotransmitter at its terminal (see below).

In the resting state, the interior of the neuronal membrane is negatively charged compared to the exterior surface and with this difference in charge, the resting membrane is said to be "polarized." The charge difference, or potential, across the membrane in the resting state is approximately $-70\,\text{mV}$, due primarily to an excess of sodium ions on the exterior which have been actively pumped out of the neuron by the energy-dependent Na^+/K^+ ATPase pump. Sodium, however, can be transferred back across the membrane through selective channels on the membrane surface. These channels are normally closed, but are sensitive to changes in the charge difference across the membrane, as well as to intracellular signaling pathways. Signals arriving from the dendritic regions of the neuron stimulate the opening of these channels and sodium moves inward down its own concentration gradient. The incoming sodium brings its positive charges with it, and this alters the resting state potential. The net charge difference across the membrane is thus reduced as positive ions pour inward, and the membrane is said to be "depolarized." When the summation of these depolarization signals over a short time period reaches a certain threshold at the axon hillock (generally about +50 mV), the axon will generate an action potential. Once this occurs, all of the sodium channels in the nearby vicinity are stimulated to open, allowing a massive influx of sodium. Sodium channels stay open for only a short period of time, and once they close, they cannot reopen for a while, so the amount of time sodium can flow inward through a single channel is limited. However, as sodium channels a little further down the axon sense the voltage change across the membrane, they also open and thus, a feed-forward effect is created (Figure 15.2). The membrane is repolarized by the opening of potassium channels, which respond to the very same signals that stimulated the sodium channels but are slightly delayed in time. Therefore, as sodium channels begin to close after being stimulated, the potassium channels open, and potassium rushes out of

Figure 15.2   An action potential is a wave of electrical impulse that is propagated down an axon only in one direction. The ATP-dependent sodium/potassium pump (a) exchanges three sodium ions (transported to the outside of the cell) for two potassium ions (transported inward), maintaining a polarized membrane. When an action potential is initiated at the axon hillock, nearby voltage-gated sodium channels temporarily open and allow sodium to enter the neuron, depolarizing more of the membrane. This stimulates other channels to open, propagating the depolarization (b and c). Potassium channels open more slowly than sodium channels, and these allow potassium to exit, restoring the polarized state of the membrane (c and d). The ATPase pump reestablishes the sodium and potassium gradient needed to drive the next impulse (d).

the cell down its own concentration gradient. This produces a net outflow of positive charge, restoring the resting state condition of more positive charges on the outside and repolarizing the membrane. This process of depolarization/repolarization continues propagating itself down the length of the axon. Behind the action potential, the resting state sodium and potassium ion concentrations are restored by the ongoing activity of the Na^+/K^+ ATPase pump, pumping sodium back out and potassium back into the cell.

A segmented sheath of myelin (see Section 15.2.3) is found around the trunk of some axons. In myelinated axons, the ion channels that mediate action potential are clustered in regions between the segments of myelin. These regions are known as nodes of Ranvier. Myelin protects and insulates the axon, preventing any leakage of charge across the membrane and allowing the current to flow from one node to the next. The action potential is in effect regenerated at each node. This process of action potential jumping from node to node is called saltatory conduction (Figure 15.3) and results in much faster conduction velocity down the length of the axon.

Axons terminate at synapses with other neurons, at neuromuscular junctions, or in effector organs such as a gland or the heart. When the action potential reaches the terminal of the axon, the depolarizing impulse stimulates the release of neurotransmitter from the terminal into the cleft between the presynaptic membrane and its effector or receiving neuron (Figure 15.4). Neurotransmitter is packaged into vesicles docked at the presynaptic membrane. Upon stimulation by an incoming action potential, these vesicles fuse with the membrane to release their contents

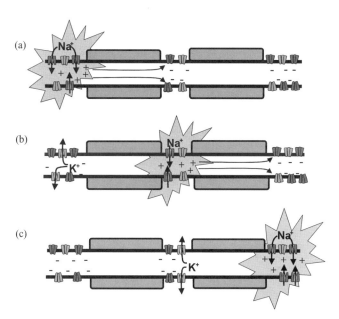

Figure 15.3 Saltatory conduction. (a) Myelin acts as an insulator to prevent current loss as the action potential travels down the axon. (b) Sodium and potassium channels are clustered at the Nodes of Ranvier, where there is no myelin. (c) Action potentials jump from one node to the next, reducing the overall membrane area involved in conduction, and speeding up electrical transmission.

Figure 15.4 Neurotransmitter release. (a) Presynaptic nerve terminal is shown containing vesicles and other organelles. (b) Neurotransmitter-containing vesicles are made of lipid bilayers and contain membrane-associated proteins that participate in the release process. (c) These proteins form a complex with proteins on the presynaptic membrane to dock the vessels as they wait for the signal to release neurotransmitter. (d) The protein complexes alter their conformation when stimulated by calcium promoting fusion of the vesicle with the presynaptic membrane. Neurotransmitter within the vesicle is then free to diffuse into the synapse (d).

into the synaptic cleft. The actual primary signal to fuse is an influx of calcium, mediated by calcium channels on the presynaptic membrane that, like the sodium channels described above, are sensitive to changes in voltage across the membrane. Specific proteins on the vesicle membrane and on the presynaptic membrane form complexes with one another, and when stimulated by the localized increase in calcium ion concentration, mediate the fusion and pulling apart of the two membranes to release neurotransmitter. Electrical signals transferring information within the neuron are thus converted to chemical signals that transfer information between neurons in the form of neurotransmitters.

15.2.2 Neurotransmitters and Their Receptors

Neurotransmitters are recognized by receptors that lie on the postsynaptic membrane of receiving neurons, at neuromuscular junctions, or on end effector organs. Receptors are generally selective for the neurotransmitter that they bind, just like the lock-and-key mechanism of an enzyme/substrate interaction. Often, more than one selective receptor is associated with a specific neurotransmitter. An example of this is acetylcholine, which binds to two very different subclasses of selective receptors, the nicotinic and muscarinic acetylcholine receptors. The acetylcholine receptor found in neuromuscular junctions belongs to the nicotinic subclass, and these receptors are ion channels that are permeable to sodium. Stimulation of nicotinic receptors by acetylcholine results in the opening of the channel, and the influx of sodium serves to rapidly depolarize the muscle membrane that receives acetylcholinergic innervation. Neurotransmitter receptors that are ion channels thus mediate very fast and short-lived neurotransmission. This is particularly evident when one compares its signaling to that of the other major type of neurotransmitter receptor, the G protein-coupled receptor. Unlike nicotinic receptors, muscarinic acetylcholine receptors are coupled intracellularly to G proteins, which then activate a variety of intracellular signaling pathways. G protein-coupled receptors thus produce a more slow and sustained response to neurotransmitter stimulation. G protein-coupled receptors can modulate ion channel neurotransmission by stimulating kinase and phosphatase pathways, altering the phosphorylation state, and thus the activity, of ion channels. G protein-coupled receptors also signal to the nucleus to maintain and mediate changes in RNA and protein expression, and promote cellular survival.

Neurotransmitters stimulate receptors on postsynaptic membranes, but the message mediated by the receptor may be either excitatory or inhibitory to the receiving neuron. For example, the neurotransmitter glutamate binds to selective ion channel receptors and G protein-coupled receptors, and both of these receptor types transmit a signal that enhances the excitability of the receiving neuron. On the other hand, the neurotransmitter GABA (for gamma-amino butyric acid), while also binding both ion channel GABA receptors and G protein-coupled GABA receptors, is known for its ability to decrease the excitability of the postsynaptic neuron. Its message is therefore inhibitory to the propagation of signaling within a group of neurons. The nervous system works on a balance of excitatory and inhibitory neurotransmission, primarily mediated in the brain by glutamate and GABA, respectively.

15.2.3 Glial Cells

While neurons constitute the definitive unit of the nervous system, their function is critically dependent on the presence of glial cells. In fact, glial cells make up about 90% of cells in the nervous system. Glial cells perform many functions, including nutritive and protective support, electrical insulation, modulation of synaptic function, and guidance of migration during development.

Astrocytes are the most numerous of all glial cells, and their roles in the nervous system are probably the most diverse. Of critical importance to toxicology, astrocytes make up the interface between the bloodstream and neurons. They help comprise part of the blood–brain barrier by extending processes that enwrap and interact with blood vessels, prohibiting some substances from reaching neurons while actively transporting glucose and other substances to neurons. Astrocytes also signal changes in neuronal activity to blood vessels, resulting in changes in regional blood flow. This allows more glucose and oxygen to be delivered to neurons when they are highly active, a mechanism that is the basis for the study of brain activity by functional magnetic resonance imaging or fMRI.

Astrocytes are also intimately associated with synapses, where they take up excess neurotransmitter and ions, and serve as a physical barrier to isolate synaptic connections between neighboring neurons. In this manner, private signals are transmitted between two communicating neurons while diffusion of neurotransmitter into the extrasynaptic space (where it could interact with other neurons) is limited. Astrocytes express many of the same neurotransmitter receptors that neurons do.

Upon stimulation, glutamate that has been taken up by astrocytes can be released to interact with neurons, thus these cells are active participants in synaptic signaling. Other molecules released by astrocytes include growth factors and neuromodulators (usually peptides or small molecules like ATP) that inhibit or enhance overall levels of neuronal activity.

Metabolic enzymes expressed within and on the surface of astrocytes regulate neuronal signaling by catabolizing excessive amounts of neurotransmitter. Monoamine oxidases, for example, catalyze the biotransformation of dopamine, norepinephrine, and serotonin into oxidation products that are substrates for further enzymatic reactions *en route* to excretion. Several drugs and neurotoxicants are also substrates of these enzymes.

Astrocytes are very sensitive to the homeostatic status of the tissue in which they reside. In response to a toxic insult or other injury, astrocytes are activated to multiply and undergo morphological changes. Activated astrocytes have greatly enlarged cytoplasmic processes, and produce increased amounts of a protein known as glial fibrillary acidic protein (GFAP). GFAP is often used as a quantitative histochemical marker for toxicant-mediated injury in the nervous system.

Another class of glial cell performs the important function of insulating axons with myelin. The myelinating cells in the CNS are oligodendrocytes, while Schwann cells myelinate axons in the PNS. These cells wrap layer upon layer of their plasma membrane around an axon with very little cytoplasm between layers; thus, myelin is composed chiefly of lipids. The white matter areas of the brain appear white because they are dense in myelinated axons. Myelin aids in speeding electrical transmission by insulating axons from leakage of current. The loss of myelin can disrupt neurotransmission between different areas of the brain, or between the brain and the body. Several neurotoxicants that target myelin or myelinating glial cells are discussed in the following sections.

A third class of glial cell is called microglia. Unlike neurons and other glial cells that are derived from neuroectoderm, microglial cells are derived from hematopoietic precursor cells that migrate to the nervous system during development. Microglial cells are the resident immune cells of the nervous system, monitoring neural tissue for signs of injury or infection. When they encounter signals of injury, such as changes in ionic balance or inflammatory factors, microglia can migrate toward the source of such signals. At the same time, they change their morphology in a process known as activation and begin secreting inflammatory factors that attract other microglia and stimulate astrocyte involvement. At the final stage of activation, microglial cells transform into macrophages capable of engulfing cellular debris. While many of the functions of microglia are beneficial, they can also release factors that are cytotoxic to neural tissue, such as damaging inflammatory cytokines and reactive oxygen species. Often, the most devastating consequences of toxicant action in the nervous system arise indirectly due to inflammatory responses that have spiraled out of control.

15.2.4   The Blood–Brain Barrier

The blood–brain barrier was conceptualized when it was noted that dyes injected into the bloodstream of animals stained nearly all tissues except the brain. This barrier and its PNS equivalent, the blood–nerve barrier, prevent all but a select few molecules from entering the nervous system. The barrier itself is not a single unitary structure but is a combination of unique anatomical and biochemical features that prevent the translocation of blood-borne agents from brain capillaries into the surrounding tissue. As mentioned above, astrocytes help form the barrier, surrounding capillary endothelial cells with extensions of their cytoplasm known as endfeet. There are also pericytes, the function of which is not well-known, that associate with the capillaries and may participate in blood flow regulation and inflammation. Another component of the barrier is the relatively impermeable nature of the endothelial cells that line the interior of capillaries in the nervous system. For example, capillary endothelial cells in the brain are different from those in the periphery in at least three ways. First, brain capillaries form tight junctions of very high resistance between cells. In contrast, peripheral capillaries have low resistance tight junctions, and even openings, or fenestrations, which allow compounds to pass between cells. Second, compared to peripheral endothelial cells, brain endothelial cells are deficient in their ability to transport agents by pinocytosis, and only small lipophilic molecules are transported transcellularly by this mechanism. For larger molecules, carrier-mediated transport mechanisms are highly selective, and allow only one-way transport. Third, there is an enzymatic barrier that metabolizes nutrients and other compounds. Enzymes such as gamma-glutamyl transpeptidase, alkaline phosphatase, and aromatic acid decarboxylase are more prevalent in cerebral microvessels than in non-neuronal capillaries. Most of these enzymes are present at the lumenal side of the endothelium. Additionally, the P-glycoprotein (P-gp) multidrug efflux transporter is presently thought to exist at the interior surface of the capillary, although some scientists argue that P-gp is actually associated with astrocytes. Finally, the CNS endothelial cell displays a net negative charge at its luminal side and at the basement membrane. This provides an additional selective mechanism by impeding passage of anionic molecules across the membrane.

Most of the toxicants that enter the nervous system do so by exploiting mechanisms designed to allow entry of essential molecules, such as nutrients, ions, neurotransmitter precursors, and the like. Small, lipophilic molecules are able to cross the blood–brain barrier relatively easily. Some agents can be recognized by active transport systems and thereby traverse the blood–brain barrier along with endogenous ligands. For example, the neurotoxicant methylmercury forms a complex with cysteine and enters the brain through amino acid transporters due to its structural similarity to methionine. In some cases, the blood–brain barrier is itself subject to damage by neurotoxicants. Metals such as lead, cadmium, mercury, and manganese accumulate in endothelial cells and damage their membranes, leading to brain hemorrhage and edema.

15.2.5 The Energy-Dependent Nervous System

Nervous tissue has a high demand for energy, yet nerve cells can only synthesize ATP through glucose metabolism in the presence of oxygen. Critical ATP-dependent processes in the nervous system include regulation of ion gradients, release and uptake of neurotransmitters, anterograde and retrograde axonal transport, active transport of nutrients across the blood–brain barrier, P-gp function, phosphorylation reactions, assembly of mitochondria, and many others. The highest demand for energy (up to 70%) is created by the maintenance of resting potential in the form of sodium and potassium concentration gradients across the nerve cell membrane. As discussed earlier, these gradients are maintained primarily by the activity of the Na^+/K^+ ATPase pump. The pump uses the energy of hydrolyzing each ATP molecule to transport three sodium ions out of the cell and two potassium ions into the cell. Maintenance of the resting potential is not the only benefit of this pump's activity, however. The gradients created by the pump are also important for maintaining osmotic balance, and for the activity of indirect pumps that make use of the sodium gradient to transport other molecules against their own concentration gradient. Neurotransmission is thus heavily dependent on the proper functioning of the Na^+/K^+ ATPase pump.

Another process dependent on energy metabolism is axonal transport. Axonal transport carries organelles, vesicles, viruses, and neurotrophins between the nerve nucleus and the terminal. This distance can be quite long when one considers that the length of the sciatic nerve, for example, can be up to one meter. Anterograde transport (from cell body to terminal) is accomplished by two mechanisms defined by their rate: fast axonal transport and slow axonal transport. Fast axonal transport proceeds at rates of approximately 400 mm/day, and is mediated by the ATP-dependent motor protein kinesin. Kinesin forms cross-bridges between vesicles or organelles and microtubules, and dual projections of these cross-bridges shift back-to-front in a coordinated, ATP-dependent manner, such that the entire molecule "walks" along the microtubule. Slow axonal transport is used to carry cytoskeletal elements such as tubulin and neurofilaments to the far ends of the axon, and it proceeds at approximately 0.2–5 mm/day. Traditionally, slow transport has been regarded as passively dependent on axoplasmic flow; however, recent evidence suggests that the cytoskeletal elements actually move rather quickly, but frequently stall in a stop-and-go fashion. Fast axonal transport also proceeds retrogradely, mediated by the ATP-dependent motor protein dynein. The rate of retrograde transport is about 200 mm/day. Neurons use retrograde transport for recycling membranes, vesicles, and their associated proteins. Neurotrophic factors, and some viruses and toxins (e.g., tetanus toxin) are also transported by this mechanism.

15.3   TOXICANT EFFECTS ON THE NERVOUS SYSTEM

Neurotoxicants affect the nervous system in a number of different ways. Some neurotoxicants damage the distal portions of axons without much effect on the remainder of the cell, while others produce outright cell death. Still others affect signaling processes in the nervous system without causing structural damage. Neurons may also be secondarily affected by neurotoxicants that target other cells in the nervous system, disrupting normal homeostatic function and causing structural or functional damage.

15.3.1   Structural Effects of Toxicants on Neurons

Demyelination   The role of myelin in the nervous system is to aid in signal transduction. Myelin acts like an electrical insulator by preventing loss of ionic currents, and intact myelin is critical for the rapid saltatory nerve conduction as discussed above. Neurotoxicants that target the synthesis or integrity of PNS myelin may cause numbness and tingling, muscle weakness, poor coordination, and paralysis. The nerve disorder associated with the loss of myelin from peripheral nerves is called myelinopathy. In the brain, white matter tracts that connect neurons within and between hemispheres may be destroyed, in a syndrome known as toxic leukoencephalopathy. Clinical manifestations of toxic leukoencephalopathy are extremely varied; some of these include headaches, to mild to severe cognitive dysfunction, paralysis, and death.

Neurotoxicants that produce primary demyelination are uncommon but may be divided into those that affect the integrity of the myelin sheath without or prior to damage to the myelinating cells, and those that directly injure myelin-producing cells. The former include agents like hexachlorophene, isoniazid, and the organotins. These compounds cause reversible edema between the layers of myelin by a mechanism that is yet unclear. The optic nerve is particularly susceptible to demyelination by hexachlorophene and organic solvents, whereas other cranial nerves, such as the trigeminal and vestibulocochlear, are vulnerable to styrene, xylene, and to tricholoroethylene, an agent used in dry-cleaning. The metalloid tellurium damages myelin by inhibiting an enzyme involved in the synthesis of cholesterol, a major component of myelin. In many cases, complete recovery from the effects of these agents is possible once the source of exposure is removed.

In contrast to agents that target the integrity of the myelin sheath, chronic exposure to cyanide and carbon monoxide is thought to directly injure myelin-producing Schwann cell bodies in the PNS and oligodendrocytes in the CNS. Inorganic lead also causes direct damage myelinating cells. Oligodendrocytes appear more sensitive to lead toxicity than astrocytes or neurons. One mechanism for the devastating developmental effects of lead exposure may be the preferential inhibition of oligodendrocyte precursor cell differentiation.

Axonopathy   Axonopathy is a specialized form of neuronal damage, involving selective degeneration of the axon while leaving the cell body intact. In many cases, the most distal portions of the longest and largest diameter axons are most vulnerable to this type of toxicity, and these areas degenerate first. With continued exposure to the toxicant, however, the degeneration progresses proximally and may eventually affect the entire neuron. This distal-to-proximal degeneration is called "dying back neuropathy." As the axon degenerates, the myelin associated with it breaks down as well; yet Schwann cells may survive and guide regeneration of the axon in some cases. If exposure to the toxicant is discontinued before death of the entire proximal axon and cell body, axons in the PNS will often regenerate, but axonal regeneration does not occur within the CNS.

It has been speculated that the reason for the enhanced vulnerability of distal axons to toxic effects is because these regions are the most heavily dependent on intact axonal transport mechanisms. Since axonal transport is energy-dependent, toxicants that interfere with ATP production, such as the nicotinamide analog Vacor, may cause distal regions to degenerate initially. Agents that target tubulin, like the vinca alkaloids, also cause this type of injury, because the tubulin-derived microtubules are critically important for axonal transport.

In the 1850s Augustus Waller described the sequence of degenerative events that occurred following transection, or slicing in half, of a nerve fiber. These events have subsequently become known as Wallerian degeneration. The essential features of this type of degeneration include swelling of the axon in the proximal segment at the, dissolution and phagocytosis by inflammatory cells of the axon segment distal to the transection, and dissolution of myelin, with preservation and proliferation of Schwann cells along the length of the former axon. Certain neurotoxicants are capable of chemically transecting an axon, producing Wallerian degeneration similar to that occurring after slicing the nerve in half. Hexane, for example, forms covalent adducts with neurofilament proteins resulting in secondary cross-linking of neurofilaments. This cross-linking is thought be the source of axonal swellings that contain high levels of neurofilament. These swellings essentially block transport to regions of the axon distal to the swelling, performing in effect a chemical transection. The distal regions then die due to lack of communication with the neuron cell body, undergoing Wallerian degeneration.

Axonopathy can manifest as defects in sensory or motor functions, or a combination of the two. For most neurotoxicants, sensory changes are noticed first, followed by progressive involvement of motor neurons. One historically important case that illustrates these effects is that of the epidemic poisoning resulting from the consumption of "Ginger Jake" during Prohibition. Tonics containing extracts of ginger were legally required to contain 5 g of ginger per milliliter of alcohol. To check for compliance with this requirement, the Department of Agriculture sampled the tonics by boiling off the ethanol and weighing the solid content. Bootleggers soon discovered that money could be saved by cutting back on the ginger and substituting it with adulterating agents like castor oil and molasses that would give the tonics the same amount of solid content. It was such an attempt at adulterating Ginger Jake that led to the addition of Lyndol, a triorthocresyl phosphate (TOCP)-containing oil used in lacquers and varnishes, to tonic that was consumed by hundreds of thousands of people. Days to weeks after consuming the product, people developed problems beginning with tingling and numbness in the hands and feet. In many, this progressed to leg cramps, weakness of the legs and arms, and loss of coordination and balance. Those with minor symptoms improved, but perhaps thousands of people were left permanently paralyzed by the incident. Today, TOCP is used to study the syndrome of delayed effects caused by some organophosphate compounds, commonly known as organophosphate-induced delayed neuropathy (OPIDN). The nature of OPIDN is still poorly understood. It appears not to be associated with organophosphate inhibition of acetylcholinesterase, but rather with another neuronal enzyme, the neuropathy target esterase (NTE). Recently, a physiological role for NTE in phospholipid homeostasis has been proposed.

Neuronopathy Neuronopathy refers to generalized damage to nerve cells, with the primary damage occurring at the nerve cell body. Many neurotoxicants produce their effects by promoting cell death in neurons. One area of intense research focus has been the toxic effects of excessive signaling by glutamate and other excitatory amino acids (EAAs), and the role that EAAs may play in neurodegenerative disorders. Glutamate activates ion channel receptors that open to allow influx of calcium and other ions into the neuron. This influx of ions, combined with other second messenger events that promote further intracellular release of calcium, contribute to calcium overload. Signaling cascades are then activated in response to the intracellular calcium, and these pathways eventually lead to oxidative stress and cell death. This type of injury, known as excitotoxicity, has been extensively studied for its role in ischemic and seizure-induced brain damage. Domoic acid, a toxin produced by algae, binds to glutamate receptors and produces excitotoxic cell death. In 1987, several people died and dozens became ill with dizziness, seizures, and memory loss after consuming shellfish that were contaminated with domoic acid. The domoic acid had been produced at high levels following an algae bloom near Prince Edward Island, Canada, contaminating the shellfish. More recently, domoic acid produced by algae blooms has been blamed for episodes of abnormal behavior and deaths of pelicans, cormorants, and sea lions on the California coast.

15.3.2 Toxicant-Mediated Alterations in Synaptic Function

Nervous system function may be adversely affected by neurotoxicants without necessarily causing structural damage to tissue. In many cases, neurotoxicants interfere with signaling processes within the nervous system by activating or inhibiting receptors, or altering the amount of neurotransmitter available to activate receptors. This type of neurotoxicity is illustrated by the well-characterized actions of the organophosphates and carbamates on acetylcholine signaling.

Organophosphates inhibit acetylcholinesterase, the enzyme responsible for breaking down acetylcholine into acetic acid and choline. After acetylcholine has been released into the synapse or the neuromuscular junction, acetylcholinesterase terminates receptor-stimulating activity by binding acetylcholine in its active site. Separate sites within the binding pocket of acetylcholinesterase bind the quaternary nitrogen of the choline group, and the carbonyl of the ester group. A hydrolytic reaction results in the loss of choline, leaving an acylated serine residue, which is then rapidly hydrolyzed. The biologically active oxon forms of organophosphates also bind to the active site of acetylcholinesterase, covalently phosphorylating the serine residue in the catalytic site of the enzyme. The phosphorylation of acetylcholinesterase creates a relatively stable inactive enzyme that persists for hours to days before hydrolysis of the phosphate moiety occurs spontaneously, restoring acetylcholinesterase activity. The rate of spontaneous hydrolysis is increased with larger alkyl groups attached to the phosphate moiety. When one or more of these alkyl groups is lost, in a process known as "aging," spontaneous reactivation of acetylcholinesterase by hydrolysis of the phosphate moiety is impossible, and the enzyme is permanently inactivated. Carbamates similarly inhibit acetylcholinesterase by carbamylating the enzyme active site. The stability of carbamylation is much less than phosphorylation, however, and spontaneous reactivation thus occurs faster than with organophosphates.

The effects of acetylcholinesterase inhibition can be seen throughout the nervous system. Acetylcholine and its receptors mediate neurotransmission in sympathetic and parasympathetic autonomic ganglia, in the effector organs where autonomic nerves terminate, in neuromuscular junctions, and in the brain and spinal cord. The signs of hypercholinergic activity are thus very diverse, and include effects mediated by both nicotinic and muscarinic types of acetylcholine receptor. Hyperstimulation of nicotinic receptors in neuromuscular junctions results in muscle weakness, in rapid, localized contractions called fasciculations, and in paralysis. Nicotinic receptors are also found in sympathetic and parasympathetic ganglia, and so stimulation of both divisions of the autonomic system is apparent as hypertension, increased heart rate, and dilation of the pupils. Muscarinic receptors in the PNS mediate postganglionic parasympathetic effects on the smooth muscle present in the end organs such as the lung, gastrointestinal tract, eye, bladder, and secretory glands. Hyperstimulation of these receptors results in a pattern of toxicity known by the mnemonic SLUDGE (salivation, lacrimation, urination, defecation, g.i. upset, emesis). Bronchospasm and bradycardia are also muscarinic effects. In the CNS, confusion, anxiety, restlessness, ataxia, seizures, and coma are effects of both muscarinic and nicotinic receptor overstimulation. Death generally occurs from respiratory paralysis.

Anticholinergic toxicity by organophosphates and carbamates is directed at counteracting hyperstimulation and regenerating acetylcholinesterase enzymatic activity. Atropine is a muscarinic receptor antagonist (it blocks acetylcholine from binding to the muscarinic receptor), and is used to counteract the effects of cholinergic overactivity. Since muscarinic receptors are found primarily at parasympathetic sites, atropine blocks parasympathetic symptoms of organophosphate toxicity. Atropine has no effect at the nicotinic receptor, however, so the effects on skeletal muscle and some of the sympathetic responses to cholinergic hyperstimulation will remain after administration of atropine. Inhibition of acetylcholinesterase activity by organophosphates can be reversed by administration of oxime compounds (such as pralidoxime and 2-PAM). These compounds contain a quaternary nitrogen that binds to the choline binding site of acetylcholinesterase, positioning the oxime portion of the molecule near the esteratic site. Oximes are themselves reversible inhibitors of acetylcholinesterase, but their mechanism of organophosphate reversal is by attack of the covalent phosphoserine bond, releasing the phosphate group. Oximes are not effective on dealkylated or "aged" enzymes, so they must be administered soon after organophosphate intoxication in order to be effective. They are also ineffective against carbamate-mediated toxicity, and some researchers believe they actually worsen carbamate effects by stabilizing the carbamylation of the enzyme.

Whereas organophosphates enhance neurotransmitter activity by inhibiting the breakdown of acetylcholine, many biological toxins produce hyperstimulation of receptors by directly binding and activating them (agonism). Others reduce receptor stimulation by prohibiting the neurotransmitter from activating them (antagonism). A number of natural toxins such as snake and spider venoms, mushroom and plant alkaloids affect nervous system function by these mechanisms. As the binding of receptors by these agents is usually reversible, their effects are reversible as well, although some may still cause death by massively altering neuronal signaling.

On the other hand, the *Clostridium* bacterial toxins, botulinum (causing botulism) and tetanospasmin (causing tetanus), block neurotransmission by inhibiting release of neurotransmitter into synapses and at motor end plates in muscle. Both of these agents are structurally similar proteases, but the effects they cause are vastly different. Botulinum toxin enters presynaptic motor neurons in the PNS, where it cleaves proteins that are involved in the fusion of synaptic vesicles with membranes. This cleavage results in the inhibition of acetylcholine release from the presynaptic terminal, and thus muscles cannot be stimulated to contract. The clinical result of botulinum intoxication (usually by ingestion) is a flaccid paralysis. Since the release of acetylcholine onto muscles is blocked, the muscles are unable to contract and are thus flaccid or limp. Recovery occurs when the presynaptic neuron sprouts new nerve endings that contact the muscle and create new motor end plates. Tetanospasmin causes a completely different clinical picture, even though its substrate specificity for the cleavage of proteins is very similar. Once taken up into the presynaptic nerve endings, tetanospasmin is transported retrogradely toward the neuron cell body and then to the dendritic regions of neurons. There it is released into synapses within the spinal cord. In the spinal cord, tetanospasmin prevents the release of the inhibitory neurotransmitter GABA onto the motor neurons. GABA normally acts as an inhibitory "brake" to keep motor neurons from becoming hyperactive, but when tetanospasmin blocks GABA release, the neurons fire readily in response to the many excitatory signals they receive. This firing of motor neurons results in overstimulation of the muscles with acetylcholine, resulting in spasms, stiffness, and whole-body paralysis. Thus, the clinical effect of tetanus toxin is a spastic paralysis that is quite the opposite of the effects of botulinum toxin. Similar to botulinum toxin, however, the interneurons themselves do not die, but they must form new synapses with the motor neurons to regain their ability to inhibit them. Fortunately, in all but the most severe cases, recovery is complete. The reformation of new synapses by neurons, even in the CNS, is an example of the remarkable plasticity of the nervous system. The continual formation and reformation of synaptic connections allows the organism to change and adapt to an inconstant environment.

15.4 NEUROTOXICITY TESTING

A large number of the chemicals used in industry today remain poorly characterized with respect to their toxic effects on the nervous system. In order to determine potential risks to human and environmental well-being, existing neurotoxicants must be identified, and the approximately 2000 new chemicals introduced each year must be screened for their potential neurotoxic effects. A tiered approach is recommended, with the first tier consisting of general screening tests to assess neurotoxic exposures. These include a functional observational battery (FOB, see below) to evaluate sensory, motor, and autonomic effects, as well as quantitative tests that identify changes in motor activity, and neuropathological or postmortem assessment. More selective testing and examining the effects of repeated exposures are used to characterize effects in the second tier. Specialized tests for behavioral effects, developmental neurotoxicity, or delayed organophosphate effects may be required. If necessary, a third tier of testing characterizes dose–response effects and identifies mechanisms of neurotoxicant-induced injury. Complete and comprehensive evaluation of potential neurotoxicant effects requires that data from different types of sources be considered; this can range from molecular interactions to whole animal and human exposure analysis. Some examples of techniques commonly used for testing neurotoxic effects are described below.

15.4.1 *In Vivo* Tests of Animal Exposure

The primary approach currently used to detect and characterize potential neurotoxicants involves the use of animal models, particularly rodents. Behavioral and neurophysiological tests, often similar to the ones used in humans, are typically administered. The sensitivity of these measures to neurotoxicant exposure is widely accepted. Although it is often not possible to test toxicant effects on some higher behavioral functions in animals (e.g., verbal ability, cognitive flexibility), there are other neurobehavioral outcomes such as memory loss, motivational defects, somato-sensory deficits, and motor dysfunction that can be successfully modeled in rodents. These behaviors are based on the ability of the nervous system to integrate multiple inputs and outputs, factors that are difficult to model adequately *in vitro*. Although the bulk of neurotoxicity data has been collected in rodents, birds and primates are also used to model human behavioral outcomes.

As mentioned above, a FOB is a useful screening tool for the effects of drugs and chemicals that are potentially neurotoxic. Designed to assess autonomic, neuromuscular, sensorimotor, and behavioral status of animals, FOBs are a noninvasive method of detecting overt changes in behavior and physiology of animals that have been exposed to neurotoxicants. In the typical exam, an observer documents cageside observations regarding the appearance and activity of the animal, such as whether the animal is sitting, running, lying on its side, having seizures, etc. Then, the animal is handled and examined for obvious signs such as lacrimation, salivation, or piloerection. The ease of handling the animal is noted at this time. The animal is then placed in an open field, such as the top of a laboratory cart, and observed for a set period of time, during which the observer records exploratory behaviors, excretion rate, mobility, and level of arousal. Following the open field measurements, the animal's response to various types of stimulation is tested. The latter tests are designed to assess hearing, sensitivity to touch and noise, righting reflex, coordination, and grip strength. Pupillary light responses, weight, and temperature are recorded. A more thorough test of locomotor activity can be administered along with the FOB, consisting of quantitative evaluations of the animal's movement in either an open field or a maze. A number of agents, such as toluene, triadimefon, and chlorinated hydrocarbons increase or decrease motor activity in a toxicant-specific manner, unrelated to their general effects on the health of the animal.

More in-depth behavioral tests are required if dose-related toxicant effects are noted in screening tests. These tests may also be required as part of more selective toxicological screening, such as for developmental neurotoxicity. Focused tests of neuromotor function and activity, sensory functions, memory, attention, and motivation help to identify sites of toxicant-mediated lesioning, aid in the classification of neurotoxicants, and may suggest mechanisms of action. Some of these tests, like the schedule-controlled operant behavior tests for cognitive function, require animal training and extensive operator interaction with the animals.

15.4.2   *In Vivo* Tests of Human Exposure

Historically, the first signs indicating neurotoxic potential by a chemical have often followed accidental human exposure in the workplace. Case reports of incidents involving individuals, or clusters of individuals, are useful for documentation, but generally provide a limited amount of information about the specific details of an exposure. Procedures included in most case reports include a patient's medical history and clinical neurological exam, sometimes supplemented with psychiatric or neurophysiological tests, and/or neuroimaging. Although the specific tests involved vary depending on the clinician, most basic clinical neurological exams rely heavily on evaluation of mental status (level of consciousness, orientation, mood, etc.) and sensorimotor function (gait, coordination, muscle tone, sensitivity to touch, reflexes).

Human epidemiological studies generally represent a deeper investigation into the causal relationship between an exposure and neurotoxicological effects. Some of the methods used to identify neurotoxic effects in epidemiological studies include behavioral assessments, neurophysiological evaluations, and neuroimaging techniques. Neurobehavioral assessments examine a variety of psychological and cognitive functions such as mood, attention, memory, perceptual and visuospatial ability, and psychomotor performance. In an effort to standardize neurotoxicological testing of human behavioral effects, particularly for studies involving worksite exposure, the World Health Organization (WHO) and the U.S. National Institute for Occupational Safety and Health (NIOSH) devised a the Neurobehavioral Core Test Battery (NCBT). The NCBT (Table 15.1) consists of seven tests that were shown previously to be sensitive indicators of neurotoxicant exposure. The battery is designed to be administered one-on-one by an examiner. Although this battery has a relatively narrow focus (primarily on the effects most commonly seen in CNS toxicity), it also provides suggestions for the selection of further testing depending on the exposure setting. The NCBT has been widely used because of its ease of administration, relatively low cost, and its large base of control data. A broader battery of cognitive and psychomotor tests that is often used is the Neurobehavioral Evaluation System (NES). The NES consists of a combination of automated

TABLE 15.1   The WHO Neurobehavioral Core Test Battery (NBCT)

Domain	Analysis	Test	Task
Psychomotor performance	Motor speed, motor steadiness	Pursuit aiming	Follow a pattern of small circles, placing a dot in each circle around a pattern; subject's score is number of taps in circle within one minute.
	Manual dexterity, hand–eye coordination	Santa Ana Dexterity test	Perform skillful movements with hands and arms.
Perceptual coding and perceptual motor speed		Wechsler digit Symbol test	Each number in a list is associated with a simple symbol. On a list of random digits with blank spaces below them, write the correct symbols in blank spaces as fast as possible.
Attention and short-term memory	Attention and response speed	Simple reaction time	Reactions of hands or feet from visual and auditory signals.
	Visual perception and memory	Benton visual retention test	Recall and reproduce figures.
	Auditory memory	Wechsler digit span test	Recall digits in series forward and backward immediately after hearing them.
Mood and affect		Profile of mood states	Questionnaire to evaluate anger, tension, confusion, depression, etc.

(computerized) and hand-administered tests. The sensitivity of the NES to effects caused by neurotoxicants in industrial settings has been validated internationally.

Neurobehavioral examinations are useful for identifying neurotoxicant-mediated deficits, but it is often difficult to localize the site of toxic action from such tests. For example, sensorimotor tests of reaction time, manual dexterity, hand–eye coordination, and finger tapping can indicate either neuromuscular or psychomotor damage. The results of these tests should thus be interpreted in the context of other experiments. For example, electrophysiological techniques can help to focus an investigation to the site of the lesion and characterize electrical dysfunction within the damaged nerves. Electrophysiological nerve conduction studies can distinguish between proximal and distal axonal lesions in peripheral nerves and can be performed noninvasively (i.e., with skin surface electrodes). Characteristic changes in the velocity, duration, amplitude, waveform, or refractory period of peripheral nerves may be detected, depending on the agent. Evoked potentials represent another useful electrophysiological end point. These procedures measure the function of an entire system, such as the visual, auditory, or motor systems. The specific pathway is stimulated by an evoking stimulus, such as a flash of light or electrical nerve stimulation. In response to the stimulation, evoked potentials are read as changes in ongoing electroencephalograms (EEGs) measuring electrical brain activity or electromyograms measuring electrical muscle activity. Evoked potentials can be very sensitive indicators of changes in neural activity when performed in a carefully controlled environment, and when interpreted in light of behavioral or other physiological findings.

An increasingly popular method of documenting brain pathology is the use of neuroimaging methods. Computerized axial tomography (CAT) and magnetic resonance imaging (MRI) can produce images of the brain that can show structural changes in the volume or density of a specific region or ventricle. Other techniques, such as positron emission tomography (PET) and single photon emission computerized tomography (SPECT), use radioactive tracer molecules to determine functional biochemical changes in processes like glucose utilization or receptor binding. The number of cases so far analyzed with neuroimaging techniques is still relatively small and thus, specific toxicant-mediated effects are not well characterized. Nevertheless, this growing field promises to contribute significantly to neurotoxicity studies in the future.

15.4.3   *In Vitro* Neurochemical and Histopathological End points

In vitro methods for studying neurotoxicant effects are a valuable part of whole animal and human testing, allowing the researcher to supplement findings, test hypotheses, and reduce the number of animals used for toxicity testing. Much of the neurochemical and histopathological data on neurotoxicant effects in humans and animals is gathered concomitantly with, or immediately after, performing behavioral tests. This may involve collection of bodily fluids or samples from living subjects for the purpose of analyzing acetylcholinesterase or NTE activity in blood, determining hormone or neurotransmitter concentration, or detecting the presence of toxicant or metabolite in the cerebrospinal fluid.

Postmortem tissues can provide a wealth of information about the location, timing, extent, and mechanism of neurotoxicant-induced damage. For example, changes in the gross morphology and weight of brain or nerves may be seen at higher levels of toxicant exposure. Microscopically, fixed and stained tissues reveal characteristics regarding the type of damage to target cells, such as axonopathic or demyelinating lesions. Degenerative changes in cells may be indicative of the processes leading to injury, and may indicate whether cells are dying by necrosis or apoptosis. Typical stains such as Nissl stain (cresyl violet) and Golgi impregnation (with potassium dichromate and silver nitrate) are useful for cell morphology and counting. Other stains are selective for damaged cells, like the specialized silver degeneration staining techniques that are frequently used to identify neurotoxicant-mediated injury to neurons.

Tissue sections may also be processed for immunohistochemical staining. A frequently used immunochemical marker for neuropathologic insult is GFAP. GFAP is produced in large amounts by reactive astrocytes that proliferate in response to tissue injury. Stress proteins, apoptotic signals, and immediate early genes are also utilized as markers of neuronal activity and injury. Other protein markers can be used to quantitatively identify specific types of neurons, which may be reduced in numbers after selective neurotoxicant-induced cell death. For example, tyrosine hydroxylase (TH) is an enzyme involved in dopamine synthesis, and as such, is selectively expressed in dopamine-containing neurons. Loss of TH immunoreactivity is used to identify dopaminergic cell death.

In homogenized tissue preparations, mechanistic information can be obtained from analyzing tissue levels of neurotransmitter and metabolites, signaling proteins, and receptor binding affinities. Protein and lipid peroxidation and oxygen radical formation are commonly seen with toxicants that target mitochondrial function. Neurotoxicants may alter the levels or activation state of many proteins, including kinases, phosphatases, and proteases, quantifiable with activity or immunological techniques.

Cell culture protocols are a useful adjunct to neurotoxicity testing. Individual cell lines are particularly well suited for identifying selective cellular and molecular toxicity and for studying the mechanistic aspects of neurotoxicant injury. Clonal cell lines, as well as primary cultures of neurons or glial cells may be used, and the choice of cell type or particular clonal line depends on the particular end points under study. For example, if a researcher wished to study the effects of a given neurotoxicant on neurotransmitter release, he or she might choose the rat pheochromocytoma PC12 cell, which releases catecholamine neurotransmitter upon stimulation with a variety of agents. The relative inexpense and ease of manipulating exposure make cellular techniques an attractive alternative for many types of studies. Cultured cell studies cannot, however, reproduce systemic metabolic and kinetic effects, or mimic the complex neuronal circuitry that is present *in vivo*. Thus, while cell studies provide a vehicle for in-depth examination of the nature of toxicant–cellular interactions, extrapolation to *in vivo* conditions is often not possible.

15.5 SUMMARY

The nervous system is at once unique in structure and staggeringly complex, exquisitely sensitive, yet capable of amazing adaptability. Because of these attributes, the neurotoxic potential of many agents, to say little of their underlying mechanisms, remains unknown. Particularly concerning are the possibilities that chronic low levels of chemical exposure are having an effect on the behavioral development of children, and contributing subtly to neurodegenerative diseases in the elderly. The huge task of testing natural and synthetic chemicals for neurotoxic effects has been facilitated in recent years with the development of behavioral testing batteries, advances in pathological and biochemical techniques, and a more focused attention of regulatory agencies on issues relating to neurotoxicology.

BIBLIOGRAPHY AND SUGGESTED READING

Harry, J., B. Kulig, M. Lotti, D. Ray, H. Tilson, and G. Winneke, eds. Serial: Environmental Health Criteria Series, no. 223 (WHO, 0250-863x). In *Neurotoxicity Risk Assessment for Human Health: principles and approaches*, Geneva, Switzerland: World Heath Organization, 223 p, 2001.

Massaro, E., ed. *Handbook of Neurotoxicology*. Totowa, NJ: Humana Press, 2002.

Tilson, H. and J. Harry, eds. *Neurotoxicology*. Philadelphia: Taylor and Francis, 1999.

US Congress, Office of Technology Assessment, *Neurotoxicity: Identifying and Controlling Poisons of the Nervous System*, OTA-BA-436 (Washington, DC: US Government Printing Office, April 1990).

SAMPLE QUESTIONS

1. Which of the following components of a neuron is specialized for receiving chemical signals from other neurons?

    a. Nodes of Ranvier

    b. Dendrites

    c. Somas

    d. Axons

2. Which cell type is considered the resident immune cells of the nervous system? Why?

3. Describe four components of the blood–brain barrier.

4. Atropine administration is unable to reverse all of the neurotoxic effects of organophosphates because it has no effect on ...

    a. Nicotinic receptors in the brain.

    b. Nicotinic receptors in skeletal muscle.

    c. Nicotinic receptors in the sympathetic nervous system.

    d. Muscarinic receptors in the parasympathetic nervous system.

    e. a, b and c.

5. Compare and contrast botulinum and tetanospasmin.

6. What is the purpose of a rodent functional observational battery?

Reproductive System

HEATHER PATISAUL

16.1 INTRODUCTION

In her 1985 dystopian novel, *The Handmaid's Tale*, Margret Atwood described a human population rendered largely sterile by overwhelming nuclear, biological, and chemical pollution. Science fiction or foreshadowing? In 1992, Carlsen and colleagues conducted a comprehensive review of the literature on human semen quality. Their systematic analysis of 61 published papers, incorporating data collected from nearly 15,000 men, revealed a statistically significant decline in mean seminal volume and sperm concentration over the last 50 years (Carlsen et al., 1992). This finding was recently confirmed by a different group of investigators (Swan et al., 2000). In Demark, it is now estimated that more than 10% of men have sperm counts in the infertile range and up to 30% are in the subfertile range (Joensen et al., 2008). The rapidity of the decline suggests an environmental, rather than a genetic, etiology. It is now widely hypothesized that exposure to endocrine disrupting compounds, both naturally occurring as well as synthetic, are at least partially responsible for this decline and may also be contributing to a decline in female fecundity. Is this a plausible hypothesis? If so, did this decline in fecundity result from exposure in the womb, when the gonads were forming, puberty, when the reproductive system was maturing, or adulthood, when conception is desired? Can such a hypothesis be tested? Can decreased fecundity which results from environmental factors be improved or corrected? Could it affect subsequent generations? This chapter aims to illustrate both the major principles and complexity of reproductive toxicity, a field which spans conception through adulthood and includes the unborn.

16.1.1 Defining Reproductive Toxicity

What is reproductive toxicity? Reproductive toxicity is the occurrence of adverse effects on male or female sexual anatomy, function, maturation, or behavior resulting from exposure to exogenous chemical agents. This definition includes lactation, sexual maturation, the ability to produce viable, fertile offspring, sex-specific

A Textbook of Modern Toxicology, *Fourth Edition*. Edited by Ernest Hodgson
Copyright © 2010 John Wiley & Sons, Inc.

behavior, and premature reproductive senescence. There is now speculation that gender preference and gender identity may also be influenced by toxic insult. Because the potential for successful reproduction begins in the womb, it can be difficult to distinguish reproductive toxicity from developmental toxicity. This chapter will first provide an overview of normal male and female reproductive physiology and discuss how the ontogeny and function of this system might be affected by toxicants, particularly endocrine disrupting compounds.

16.1.2 Defining Endocrine Disruption

Endocrine disruption is thought to be one of the primary ways in which the developing and adult reproductive system is altered by exposure to toxicants. The U.S. Environmental Protection Agency defines an endocrine disrupting compounds (EDC) as an exogenous chemical substance or mixture that alters the structure or function(s) of the endocrine system and causes adverse effects. Hundreds of compounds have now been identified as possible endocrine disruptors. There are six basic mechanisms of endocrine disruption.

1. Alteration of the production and/or secretion of hormones.
2. Altered sensitivity of the target cell to the hormone. For example, this could result from interference with the ability to upregulate or downregulate a hormone receptor.
3. Agonism of the hormone receptor, effectively mimicking the effect of the hormone.
4. Antagonism of the hormone receptor, effectively blocking the effect of the hormone.
5. Alteration of hormone metabolism and/or clearance resulting in abnormally high or low levels of the hormone.
6. Displacement of a hormone from its binding proteins.

Think about these mechanisms as you read the rest of the chapter and how a compound, through any of these five mechanisms, could disrupt either the development or function of the reproductive system across the lifespan (Li et al., 2008).

16.2 THE HYPOTHALAMIC-PITUITARY-GONADAL AXIS

In mammals, reproductive maturation and function is coordinated by the hypothalamic-pituitary-gonadal (HPG) axis (Figure 16.1). The hypothalamus is a sexually dimorphic brain region that lies just below the thalamus. It forms the ventral part of the forebrain (diencephalon) and is responsible for coordinating the majority of neuroendocrine functions including hunger, thirst, circadian cycles, emotion, body temperature, stress, and reproduction. It is responsive to a number of external signals including day length, hormones, olfactory cues, and glucose levels among others. Mammalian reproduction is regulated by the temporal release of gonadotropin releasing hormone (GnRH) from the hypothalamus to the anterior portion of the pituitary gland (adenohypophysis). GnRH is released from the hypothalamus in

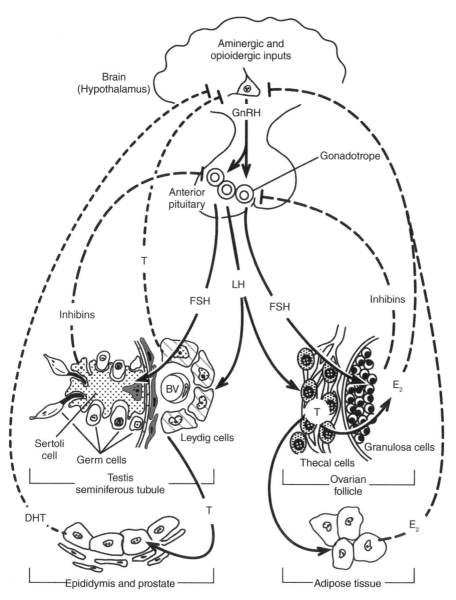

Figure 16.1 Steroid positive and negative feedback loops of the mammalian hypothalamic-pituitary-gonadal (HPG) axis. (E_2, estradiol; T, testosterone, DHT, dihydrotestosterone; FSH, follicle-stimulating hormone; LH, luteinizing hormone.) Reprinted with permission from Couse, J. F. Reproductive toxicology. In *Molecular and Biochemical Toxicology,* 4th ed., eds. R. C. Smart and E. Hodgson, p. 810. Hoboken, NJ: Wiley, 2008.

pulses, approximately once per hour in humans, and stimulates the synthesis and secretion of the gonadotropins follicle-stimulating hormone (FSH) and luteinizing hormone (LH) from the anterior pituitary into the bloodstream. FSH and LH then stimulate the production of steroid hormones, primarily estrogens in females and androgens (such as testosterone) in males, by the gonads. Elevated steroid hormone levels then feed back on the hypothalamus to suppress the production of GnRH in a process called steroid negative feedback. In females, ovulation is triggered by steroid positive feedback, the specifics of which will be discussed later in this chapter.

GnRH neurons also coordinate the timing of pubertal onset. Early in fetal development, GnRH neurons arise in the olfactory portion of the brain, migrate to their ultimate position in the hypothalamus, and begin secreting GnRH in low amplitude pulses. At puberty, the amplitude of GnRH secretion increases dramatically which, in turn, signals the HPG axis to undergo maturation. Pubertal transformation includes the development of secondary sex characteristics, the ability to produce and release gametes, and, in females, support a full-term pregnancy. Disruption of GnRH release can impair HPG axis function. Effects may be permanent if GnRH neurons, their axonal projections to the anterior pituitary gland, or synaptic inputs on GnRH neurons are damaged. In contrast, effects may be reversible if exposure occurs in adulthood because the neural circuitry surrounding GnRH neurons has already developed. For example, opiate drugs such as morphine and other narcotics, are tonic inhibitors of GnRH secretion and men who abuse narcotics often develop impotence as a result. Cessation of drug taking eliminates the suppression and normal, pulsatile release of GnRH resumes. Many industrial solvents, such as toluene and other aromatic hydrocarbons, and some pesticides can also suppress GnRH secretion. Polychlorinated biphenyl (PCB) mixtures (such as Aroclor 1221 and Aroclor 1254), organochlorine pesticides (such as methoxychlor and chlorpyrifos), and the plastic component bisphenol A have also been found to affect hypothalamic GnRH gene expression, cell survival, and neurite outgrowth (Gore, 2001). Exposure to the phytoestrogen genistein, which is found in soy and soy-based products, during early development can result in impaired steroid positive feedback in adult female rats and thus compromised reproductive capacity.

16.3   MALE REPRODUCTIVE PHYSIOLOGY

The presence of the SRY gene on the Y chromosome initiates a gene cascade that ultimately triggers the development of the testes from the mammalian fetal urogenital tract. In humans, the testes begin to develop around the eight week of pregnancy. Once formed, they begin secreting androgens and anti-Müllerian hormone. Androgens produce their effects by binding to androgen receptors (ARs) on target tissues throughout the reproductive tract and the brain. Androgens promote the maturation of the Wolffian ducts, which ultimately develop into the epididymis, vas deferens, and seminal vesicles (Figure 16.2). Anti-Müllerian hormone suppresses the development of the Müllerian ducts, which ultimately form

Figure 16.2   A cross-sectional diagram of the human testis depicting the overall structure of the seminiferous tubules and interstitial tissue. Leydig cells, along with connective tissue and blood vessels lie in the interstitial tissue. Sertoli cells and the germ cells lie within the seminiferous tubles. Maturation of the germ cells progresses from the basement membrane toward the lumen. Each stage is depicted with the requisite number of chromosomes indicated. Reprinted with permission from Jones, R. E. *Human Reproductive Biology*. San Diego, CA: Academic Press, 1991, p. 75.

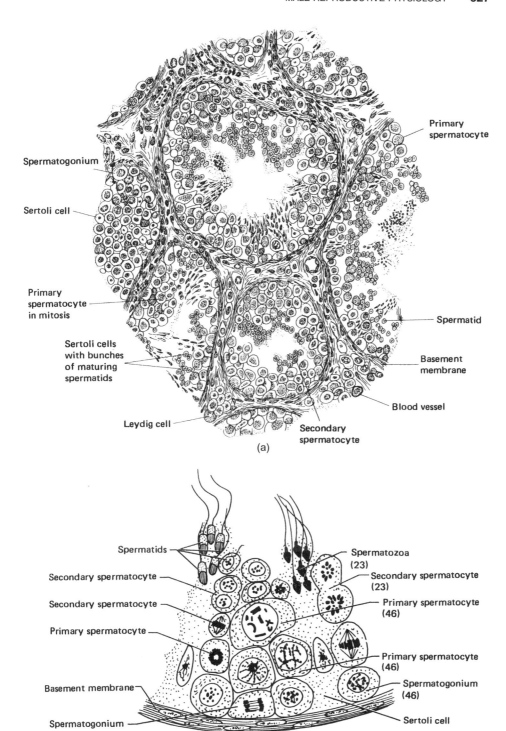

Primary spermatocyte

Spermatogonium

Sertoli cell

Primary spermatocyte in mitosis

Sertoli cells with bunches of maturing spermatids

Leydig cell

Spermatid

Basement membrane

Blood vessel

Secondary spermatocyte

(a)

Spermatids

Secondary spermatocyte

Secondary spermatocyte

Primary spermatocyte

Basement membrane

Spermatogonium

Spermatozoa (23)

Secondary spermatocyte (23)

Primary spermatocyte (46)

Primary spermatocyte (46)

Spermatogonium (46)

Sertoli cell

(b)

the uterus and fallopian tubes in females. Differentiation of the prostate and external genitalia requires 5α-dihydrotestosterone (DHT), which is the most potent androgen produced by the testis and is a metabolite of testosterone. Failure to produce DHT, or sufficient levels of DHT, can lead to poorly developed or malformed external genitalia. The enzyme 5α-reductase catalyzes the conversion of testosterone to DHT; therefore, toxicants that interfere with the action of this enzyme, or the interaction of DHT with ARs during development can impair proper development of male genitalia and the prostate.

Androgen levels are relatively high during early fetal development, drop toward the end of pregnancy, and then rise again in infancy. The functional significance of these androgens in infancy is not known but has been shown to be critical for the masculinization of the brain and sex-specific behavior in other mammals. In late infancy, androgen levels fall precipitously and remain low through adolescence, then rise at puberty as GnRH pulse amplitude increases. Elevated androgen levels following pubertal onset ultimately initiate the development of secondary sex characteristics, the maturation of the external genitalia, and sperm production.

The testis contains three major cell types: Leydig cells, Sertoli cells, and germ cells. Leydig cells reside in the interstitial space of the testis, outside of the seminiferous tubules (Figure 16.2). The principal function of Leydig cells is androgen production. Androgen production is stimulated by LH which binds to its receptor within Leydig cells. Cholesterol, the basic building block of all steroid hormones, is delivered to the mitochondria by the steroidogenic acute regulatory (StAR) protein. Cholesterol is then cleaved to form pregnenolone and ultimately, after many steps, testosterone (Figure 16.3). Most of the enzymes needed for this conversion process belong to the cytochrome P450 family of enzymes. Interference with enzyme action within this pathway can result in insufficient androgen production. For example, some therapeutics for prostate cancer suppress the activity of the enzymes needed to convert androstenedione to testosterone or testosterone to dihydrotestosterone, therefore reducing androgen levels. Ketoconazole, an antifungal agent, is a potent inhibitor of the enzyme CYP17A1, which is needed for the production of androgens (and estrogens).

Sertoli cells are sometimes called the somatic nurse cells of the testis because their primary function is to support the germ cells as they differentiate into sperm. Sperm count is directly proportional to the number of Sertoli cells contained within the testis and this population is nonrenewable in the adult. Sertoli cells reside within the seminiferous tubule (Figure 16.2), are irregular in shape, and envelop the maturing germ cells. Tight junctions, called occluding junctional complexes, bind neighboring Sertoli cells together and ring the seminiferous tubule, forming the blood–testis barrier. This barrier is selectively permeable, much like the blood–brain barrier, and protects the developing germ cells from the immune system. In addition to providing a physical scaffold for the seminiferous tubules, Sertoli cells play a vital role in the development of the male reproductive tract and germ cell differentiation. During development, anti-Müllerian hormone is produced and secreted by the Sertoli cells. In adulthood, Sertoli cells synthesize a number of critical hormones and growth factors, most notably inhibin, which plays an important role in steroid negative feedback in the HPG axis (Figure 16.1). Sertoli cells express ARs, and are thus the primary targets for androgens in the testes. They also express the receptor for FSH, which is a critical regulator of Sertoli cell function.

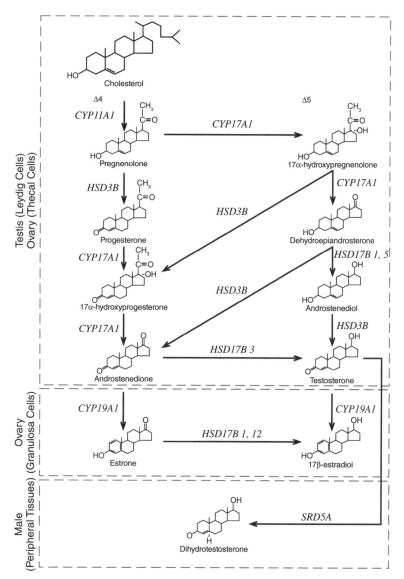

Figure 16.3   Biosynthetic pathway for sex steroid hormones indicating the enzyme required for each conversion.

Germ cells make up approximately 95% of the seminiferous epithelium and are especially sensitive to toxicological damage because, in the adult, they are undergoing a high rate of proliferation. Early in testicular development, primordial germ cells and Sertoli cell precursors aggregate and initiate the development of the seminiferous tubules. As the testis differentiates, the germ cells undergo mitotic division to form the primary spermatogonia. Division then slows considerably until pubertal onset. Spermatogonia then differentiate into sperm (spermatozoa) through a multistep process called spermatogenesis. In the mature testis, there are two major types of spermatogonia, Type A and Type B (Figure 16.2) which lie along the basement membrane of the seminiferous tubules. Maturation of the germ cells then proceeds inward, toward the lumen, and concludes with the production of spermatozoa. Type A spermatogonia are the precursors of Type B spermatogonia which ultimately undergo meiosis and develop into primary spermatocytes. Type A spermatogonia undergo mitosis to produce Type B spermatogonia and replenish the supply of Type A spermatogonia. Type B spermatogonia then divide by mitosis to produce primary spermatocytes. These cells then undergo meiosis to produce haploid, secondary spermatocytes. The progression from Type A spermatogonia through secondary spermatocytes is called spermatocytogenesis. The haploid secondary spermatocytes then differentiate into round spermatids and then, ultimately, spermatozoan through a process called spermiogenesis. In human males, the entire process takes 70 days.

The immature spermatids then move into the epididymis where they mature, become more motile, and ultimately become viable. The prostate and seminal vesicles (secretory organs of the male reproductive tract) contribute to the maturation of the sperm and the chemical composition of the semen. Sperm are only viable after they mature in the epididymis.

16.4   DISRUPTION OF MALE REPRODUCTION BY TOXICANTS

Male reproductive physiology is vulnerable to disruption by toxicants during all stages of development and function, including gestation. Toxic insult to the testes can result in numerous effects including reduced sperm count and blood androgen levels that are either above or below the normal range. Damage to the epididymis can result in poor sperm motility or inadequate semen quality. Because sperm production is directly proportional to Sertoli cell number and once lost, cannot be replenished, failure to properly differentiate this critical cell population can irreversibly limit the capacity to produce viable sperm. Disruption to the testis or accessory sex glands can occur via a number of different mechanisms including DNA-damage, inhibition of cell division, interference with GnRH secretion, or alteration of testicular vasculature. The most widely studied and best understood mechanism of action is interference with ARs and/or androgen production.

16.4.1   Pesticides

One of the earliest animal studies designed to test the hypothesis that chemical agents could interfere with androgen action was conducted in 1950 using chickens. Injection of the pesticide dichlorodiphenyltrichloroethane (DDT) resulted in markedly undersized testes and inhibited the development of the comb and wattle. It was later determined that DDT and its metabolites function as anti-androgens and compete with endogenous androgens for access to the ARs. Because androgens, especially DHT, are essential for the development of male genitalia, it is widely speculated that gestational exposure to compounds like DDT with anti-androgen properties may be responsible for the documented increase in the frequency of developmental defects in male genitalia including hypospadias (misplaced urethral opening on the penis) and cryptorchidism (undescended testes) in boys.

Impaired fertility from pesticide exposure has also been documented in humans. For example, one of the earliest known examples of reduced male fertility in humans following pesticide exposure emerged in the 1970s in a group of 25 Californian men exposed occupationally to the pesticide 1,2-dibromo-3-chloropropane (DBCP). DBCP was a potent nematocide that went into commercial production in 1956 and was later banned in 1977. The exact mechanism by which DBCP impairs male reproduction has still not been elucidated, but it is hypothesized that one of its metabolites acts directly on germ cells, thereby impairing spermatogenesis. Of the 25 occupationally exposed men, most had reduced sperm counts and 9 had no viable sperm. Subsequent studies, which examined the long-term effects of exposure on these and other occupationally exposed men, found that sperm production ultimately improved in most cases, but not in individuals whose sperm counts were lowest. These studies illustrate two important concepts. First, effects on the male reproductive system from exposure to pesticides and other toxicants during adulthood are largely temporary and reversible because the system is already developed and, in most cases, can recover. Second, recovery is unlikely if primary spermatogonia or Sertoli cells are killed. In the case of DBCP, long-term exposure depleted the pool of healthy spermatogonia, ultimately resulting in irreversible infertility.

Other pesticides, herbicides, and fungicides are also known to affect male reproductive development and function. Parathion, an organophosphate, used on nine commercial crops in the United States, most notably corn, and vinclozolin, a fungicide commonly used on vineyards, are two examples. Parathion is a potent inhibitor of cholinesterase, an enzyme that facilitates the transmission of nerve impulses. Therefore, acute exposure to high doses can cause neurotoxicity and result in death. At lower doses, it is suspected of being an endocrine disruptor but the specific mechanism(s) by which it acts is not yet known. Vinclozolin is an androgen antagonist. Animal studies have shown that exposure to vinclozolin in the womb, when the male reproductive organs are forming, can result in the impaired development of the male reproductive organs. This process is called demasculinization.

16.4.2 Metals

Of the metals, the effects of lead on the male reproductive system are the best documented and understood. Lead exposure can reduce sperm counts and increase the frequency of malformed sperm, resulting in reduced fertility. Lead can affect the testis in several ways. For example, it can directly interfere with the HPG axis, resulting in hypogonadism or low testosterone levels. It can also disrupt the formation and function of the vascular system within the testis, ultimately causing damage to the seminiferous tubules. Lead exposure most frequently occurs in industrialized settings but is also common in older homes by inhalation or ingestion of lead paint residues. Children are most likely to be exposed through this route or through unacceptably high levels in drinking water. There is some concern that exposure to lead levels currently considered to be "safe" can affect brain development and impair male fertility but, to date, the data remain inconclusive.

Cadmium is also well-known to impair male reproduction but only at near fatal doses. It causes Sertoli cell death, testicular necrosis, and germinal cell damage through a direct effect on the vascular system within the testis. Exposures of this magnitude likely only occur in industrial settings. Other metals suspected of having reproductive effects in men are mercury and boron.

16.4.3 Plastics

Considerable attention is now being placed on the potential for chemicals with endocrine disrupting properties to leach from plastics into food. Of these, the most widely studied group that pertains to male reproductive health is the phthalates. There are many different kinds of phthalates, and the two considered to have the greatest potential to impact male reproduction are dibutyl phthalate (DBP) and diethylhexyl phthalate (DEHP). DBP is used in many personal care products such as lotions, cosmetics, nail polish, and perfume. It is also found in the coatings of many time release pharmaceuticals. DEHP is primarily used as a plasticizer in the production of flexible products including vinyl, medical tubing, and other medical devices, baby toys, and flooring.

A series of studies conducted in rats in the late 1990s was the first to demonstrate that phthalates could interfere with the ability of testosterone to masculinize the male reproductive tract. Exposure *in utero*, when the genitals are being formed and the action of DHT is critical, resulted in a number of genital malformations in laboratory animals including hypospadias and hemorrhagic testes (Gray et al., 1999; Wolf et al., 2000). Interestingly, the phthalates do not produce their effects by antagonizing ARs, but rather by interfering with the production of testosterone in the fetal testis. Exposure to phthalates during pregnancy has now been associated with a greater risk of genital malformations in infant boys (Swan et al., 2005). Epidemiological evidence has also positively correlated higher urinary phthalate levels with lower sperm counts and an increased likelihood of sperm with damaged DNA in adult men. Phthalate exposure is also linked with lower circulating levels of estrogens and androgens, indicating that the HPG axis may be impaired. Although it is important to keep in mind that correlation does not prove causation, these newly emerging epidemiology studies are the best evidence to data that phthalates have the potential to affect male reproductive health.

Studies of phthalate exposure in animals and humans emphasize the significance of when exposure occurs. In late 2008, the United States banned the use of phthalates in toys. Although this will help remove a source of exposure for children, it fails to eliminate the potential for exposure when the developing male is most vulnerable—in the womb (Lottrup et al., 2006). Exposure during development is more likely to result in permanent effects because disrupted organization of the male reproductive system is largely irreparable. Therefore, it is important to always be mindful of the mechanism by which a compound is suspected of impacting the male reproductive system. Understanding the mechanism is critical for making regulatory decisions that could protect human health.

16.5 FEMALE REPRODUCTIVE PHYSIOLOGY

Absence of the SRY gene allows the bipotential gonad, contained within the fetal urogenital tract, to develop into an ovary. As in the male, both the Wolffian and Müllerian ducts are present during early fetal life. In the female, the Wolffian ducts regress because the androgens needed to promote their differentiation are not present (Figure 16.2). Similarly, absence of anti-Müllerian hormone preserves this structure which ultimately differentiates into the oviducts, uterus, cervix, and upper vagina. The absence of androgens also results in the formation of the female genitalia. Because androgens are required for the masculinization of the genitalia, exposure to androgenic compounds can cause masculinization, also called virilization, of the female genitalia. Generally, this virilization is incomplete and can lead to the formation of ambiguous genitalia, including an enlarged clitoris, enlarged labia, or partial fusion of the labia.

The ovary contains three major cell types: germ cells, granulosa cells, and thecal cells. Early in gestation, primordial germ cells, termed gonocytes, migrate to the genital ridge and induce the formation of the ovary. As the ovary forms, the gonocytes then undergo rapid mitotic divisions producing oogonia. In humans, by the second trimester of gestation, the fetal ovary contains several million oogonia, the most the fetus will ever have in her lifetime. At this point in development, the mitotic divisions end and most oogonia enter the prophase of meiosis I, becoming primary oocytes. The primary oocytes remain in the diplotene stage prophase (meiosis I) until they are recruited for maturation during the ovulatory cycle in adulthood. Over time, the number of germ cells decline through a process called atresia, decreasing to about a million at birth, approximately 200,000 at puberty, and only a few thousand in early adulthood. It has been estimated that, of the approximately 7 million oogonia produced in human fetal life, only 400 will ever be ovulated. This phenomenon appears to be universal, since all vertebrate species examined to date are born with far fewer oocytes than their peak number generated during early gestation.

The primary oocytes lie within ovarian follicles, each of which comprises a single oocyte surrounded by follicular cells (Figure 16.4). Thecal cells form the outer layer of follicular cells. They are similar to the Leydig cells of the testes in that they constitutively express the receptor for LH and their primary role is to produce androgens. Granulosa cells lie inwardly adjacent to the thecal cells and line the interior of the ovarian follicle. The functional role of these cells is similar to the Sertoli cells. They are tightly linked together by gap junctions, express the receptor for FSH, and convert the androgens, produced by the thecal cells, into estrogens, most significantly estradiol. The enzyme need to complete this conversion is cytochrome P450 (or CYP19), otherwise know as aromatase. Aromatase inhibitors are now commonly used to treat hormone-dependent breast cancers.

Estrogens produce their effects by binding to estrogen receptors (ERs) of which there are two major subtypes in mammals, ERα and ERβ. ERβ was discovered relatively recently, and its functional role in the reproductive system is still being described. Thecal cells express ERα while granulosa cells express ERβ. Mice lacking ERα (ERα knockout mice) are infertile and have numerous ovarian malformations. In contrast, mice lacking ERβ have relatively normal ovulatory cycles and are capable of getting pregnant, although they appear to be subfertile compared to wild-type control mice.

16.5.1 The Ovulatory Cycle

The HPG axis signals the ovary to initiate follicular growth by the secretion of FSH. Several follicles are recruited, enlarge, and develop into primary follicles. Granulosa

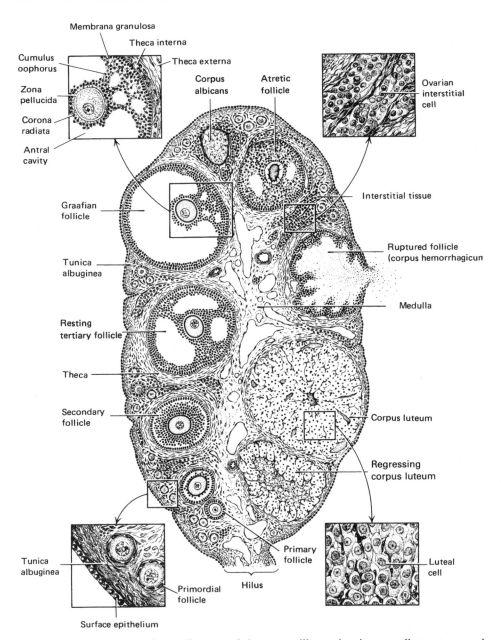

Figure 16.4 A cross-sectional diagram of the ovary illustrating its overall structure and component parts. The different phases of follicular development are also depicted along with the cellular components of the mature follicle. Reprinted with permission from Jones, R. E. *Human Reproductive Biology*. San Diego, CA: Academic Press, 1991, p. 42.

cells proliferate, increase their estrogen production, and support the growing oocyte. The thecal cells surrounding each developing primary follicle become heavily vascularized and help coordinate the flow of nutrients to the developing follicle. As the follicular phase progresses, for reasons that still remain largely unknown, most of these primary follicles ultimately begin to degenerate, as does the oocyte within them. In humans, only one a month will ultimately reach maturity. A cavity, called the antrum, begins to form within the remaining primary follicles at which point these follicles are called secondary follicles. The granulosa cells also promote the development of a clear membrane, called the zona pellucida, around the oocyte. When the follicle is ready to be ovulated, it is called a mature or graafian follicle. Elevated estradiol levels, produced by the granulosa cells, stimulate GnRH neurons to release a surge of gonadotropins, which in turn generates the surge of pituitary LH needed to trigger ovulation (Figure 16.1). This process is called steroid positive feedback.

At ovulation, the secondary oocyte, surrounded by the zona pelludica, erupts from the graafian follicle and begins its migration down the oviduct. The remaining follicular cells enlarge and transform into a corpus luteum, a glandular structure that secretes progesterone along with some estrogens. If the oocyte is ultimately fertilized, the progesterone produced by the corpus luteum will sustain the first several days of conception and be stimulated by the fetus to remain through the early part of pregnancy until the placenta forms and takes over the production of progesterone. If fertilization does not take place, the corpus luteum will regress. In humans, this process takes approximately 10 days.

In humans, the menstrual cycle is approximately 28 days but in rodents, it is only about 4 days. Hormone profiles across the ovulatory cycle vary somewhat across different species, and these differences should be kept in mind when considering the effects of toxicant exposure on ovarian function. Hormone profiles also differ during development. For example, the role of estrogen in the development of the female reproductive tract is species specific. In rodents, estrogen is not required, and exposure to estrogens during development, including the neonatal period, can in fact masculinize the female reproductive system, particularly the brain. This may not be the case in humans, however, as androgens, rather than estrogens, appear to be more important for masculinizing the primate brain. Therefore, caution must be taken when extrapolating data from female rodents to women.

16.6 DISRUPTION OF FEMALE REPRODUCTION BY TOXICANTS

Data on female reproductive health trends are limited, but recent data suggest that conception rates are declining in the developing world and the prevalence of reproductive disorders such as advanced pubertal onset, premature ovarian failure, and polycystic ovary syndrome are becoming more common (Hamilton and Ventura, 2006). Delayed childbearing resulting from cultural changes arguably has the biggest impact on fertility trends in Western countries. Obesity, dietary habits, and chronic stress may also be important factors affecting female reproductive health. These characteristics do not adequately explain, however, why the sharpest increase in reported infertility is among women under the age of 25. It is now widely hypothesized that environmental exposure, either to the mother or her developing fetus, may also contribute.

Female reproductive physiology is vulnerable to disruption by toxicants during all stages of development and function, including gestation. Toxic insult to the ovaries can result in structural abnormalities or damage to the follicles. Compounds that act as androgen agonists could masculinize aspects of the female reproductive tract or brain. For example, exposure to low levels of androgens during fetal life is thought to increase the risk of developing polycystic ovary syndrome. Disruption of uterine development can occur via a number of different mechanisms including endocrine disruption, inhibition of cell division, or improper maturation of the Müllerian ducts. Improper ductal differentiation could result in malformations within the fallopian tubes, potentially compromising the ability to sustain pregnancy. The most widely studied and best-understood mechanism of action is interference with estrogen receptors and/or estrogen production.

16.6.1 Cigarette Smoke

Smoking has long been known to induce premature menopause and reduce fertility. Smokers have fewer eggs, fewer quality eggs, an impaired capacity to recruit quality eggs when administered fertility drugs, and decreased rates of successful *in vitro* fertilization. Cigarette smoke contains several hundred known toxicants, all of which could, either individually or in combination, produce these effects. In animal models, nicotine has been shown to have disruptive effects on ovulation rates, oocyte maturation, and implantation success. Chromosomal abnormalities within oocytes are also more prevalent in smokers compared to the general population and may contribute to lower conception rates among smokers. Fertility in the offspring of smokers may also be compromised. Studies in mice have found that *in utero* exposure to polycyclic aromatic hydrocarbons, a group of more than 100 different chemicals found in cigarette smoke, results in significantly fewer ovarian follicles and therefore compromised reproductive capacity (Jurisicova et al., 2007).

16.6.2 Diethylstilbestrol (DES)

The most well understood example of how the administration of a synthetic estrogen can affect female reproduction is the tragic case of diethylstilbestrol (DES) used by pregnant women in the middle of the twentieth century. DES, a potent synthetic estrogen, was first manufactured in 1938 and prescribed to women for the prevention of miscarriage until 1971. It is estimated that 4–6 million women were prescribed DES in the United States alone. The consequences of *in utero* DES exposure were first identified in 1971 by a group of keen-eyed physicians who noticed that DES daughters were more likely to develop an extremely rare type of cervicovaginal cancer (Herbst et al., 1970, 1971). Since that initial discovery, other abnormalities in DES daughters have been described, including uterine malformations, increased risk of ectopic pregnancy, premature menopause, increased risk of developing uterine fibroids and endometriosis, increased risk of breast cancer, and decreased fertility. There is also newly emerging evidence that the children of the DES daughters (referred to as DES granddaughters) are also experiencing reproductive problems. For these women, their exposure occurred when they were germ cells in their mothers' developing ovary in the womb of their grandmother. It is one of the first instances in humans which demonstrate that persistent, generational effects can result from an *in utero* exposure to a potent estrogen (Newbold et al., 1998, 2006). Nearly all of these outcomes were replicated in or predicted by laboratory animal studies. This unfortunate event in human medical history illustrates both the vulnerability of the developing fetus to estrogenic endocrine disruptors and the importance of animal models for predicting these vulnerabilities. The tragedy of DES introduced the concept of "fetal origins of adult disease," a principle that is still used in reproductive toxicology.

What remains to be determined is if chronic, low-dose exposures to far weaker estrogens can also impair female reproduction. DES has a higher binding affinity for ERα and ERβ than estradiol and is therefore a potent estrogen agonist. Other compounds, some of which will be discussed in detail below, have binding affinities 100–10,000 times lower and are thus considered "weak" estrogens. In some cases, however, blood levels of these compounds can be severalfold higher than endogenous estrogen levels. Therefore, understanding the potential impact of these compounds on reproductive health is paramount.

16.6.3 Pesticides

A variety of pesticides and herbicides have been found to interfere with hormone action, especially estrogen. The most famous of these is DDT (still sold under many names including Anofex, Dicophane, Dinocide, Neocid, and Neocidol among others). DDT is usually sold and used as a mixture of several, closely related compounds of which dichlorodiphenyldichloroethylene (DDE) is a component and the principle breakdown product. Created in 1874 by Othmar Zeidler, the insecticidal properties were not revealed until 1939 by Hermann Müller, for which he was awarded the 1948 Nobel Prize in Physiology and Medicine. This insecticide, which is structurally similar to DES, was widely used all over the globe until the early 1970s to eradicate lice and mosquitoes. It was, and continues to be, a critical weapon in the fight against malaria and other mosquito-borne illnesses which, collectively, cause the greatest number of preventable deaths in the world. More people die from malaria than HIV, heart disease, or cancer. For that reason, it is still used in many places where the risk of contracting malaria remains high although some mosquito populations have developed resistance to the insecticide. The effects of DDT on wildlife populations was popularized by the landmark 1962 book *Silent Spring* by biologist Rachel Carson. The book became an instant best seller and launched the environmental movement in the United States as well as a governmental investigation into its claims that persistent pesticides, including DDT, were poisoning wildlife and, potentially, humans. DDT and its metabolites are toxic to a wide range of animals and insects. It thins eggshells in birds, and is believed to be one of the primary culprits for the decline of the bald eagle and other large birds of prey.

DDT has multiple modes of action. It is moderately toxic to mammals with a rat oral lethal dose 50 (LD_{50}) of 113 mg/kg and is similar to the pyrethroids in that it is a nervous system stimulant which acts by opening neuronal sodium channels, causing spasms, seizure, and death. At far lower doses, it is also thought to be both an estrogen agonist and androgen antagonist. Whether or not DDT can impair reproductive development in women is still the subject of investigation. Exposure *in utero* or early in life is associated with an increased risk of breast cancer (Cohn et al., 2007). There is also emerging evidence from areas where DDT is still used, that DDT exposure is associated with preterm birth, early pregnancy loss, reduced semen quality, disrupted menstruation, and problems with lactation (Rogan and Chen, 2005; Venners et al., 2005). Infants from these regions are frequently exposed to levels which exceed the acceptable daily intake established by the World Health Organization via breast milk. Therefore, the potential reproductive risk of DDT exposure is ongoing for many populations outside the United States.

Synthetic pyrethroids also appear to have endocrine disrupting properties in female vertebrates. Natural pyrethroids are derived from chrysanthemums, and their synthetic brethren are similar in structure but more lipophilic and environmentally persistent, making them potent weapons against ticks, mites, lice, and mosquitoes. Synthetic pyrethroids can act either as estrogen agonists or estrogen antagonists, depending on the level of exposure and the endogenous hormone levels of the exposed individual.

16.6.4 Plastics

A large number of chemicals are used in plastics manufacturing to make them more useful. For example, compounds may be added to reduce microbial growth, or to improve the stability, clarity, durability, or pliability of the product. One additive that has been the subject of intense scientific and political attention is bisphenol A (BPA). BPA was developed in the 1930s as a synthetic estrogen but was abandoned as a pharmaceutical with the invention of DES. It is now used in a variety of products, most notably polycarbonate plastics, to increase their strength and durability. It is also present in dental sealants, epoxy resins, and the linings of food cans and drink cartons, including soda cans. Annual worldwide production of BPA is estimated to exceed 6 billion pounds and in 2008, the U.S. Center for Disease Control (CDC) estimated that BPA was present in approximately 95% of the U.S. population. A study published in early 2009 found that exposure to BPA, along with the phthalates and other endocrine disruptors that leach from plastics, is particularly high among premature infants being cared for in neonatal intensive care units (Weuve et al., 2006). BPA does not persist in the environment like DDT or the PCBs, but daily human exposure is estimated to be much higher because it readily leaches from plastic containers, especially if heated. Laboratory rodents exposed to BPA *in utero* or just after birth develop a variety of reproductive abnormalities including accelerated puberty, abnormal ovarian follicles, premature loss of the ovulatory cycle, and alterations in mammary gland formation that suggest an elevated risk of developing breast cancer (Crain et al., 2008). In 2008, the U.S. Food and Drug Administration concluded that the reference dose of 50 μg/kg body weight was adequate, a conclusion that was widely criticized and revised in January, 2010. The FDA now states that it has "some concern" for neuroendocrine effects in infants and children. Public attention has resulted in considerable pressure upon manufacturers and as a result, many products advertised to be "BPA free" are now widely available. The history of BPA is an informative example of how scientific discovery, both by academic and industry scientists, policymakers, and public pressure interact to affect how a compound is used (Box 16.1).

16.6.5 Phytoestrogens

Not all endocrine disruptors are produced by humans. Phytoestrogens are a class of compounds produced by plants, primarily legumes. Legumes, such as soy, use these

BOX 16.1   THE HISTORY OF BISPHENOL-A (BPA): AN EXAMPLE OF HOW SCIENCE, THE MEDIA, AND PUBLIC SCRUTINY SHAPE PUBLIC POLICY

1891	BPA synthesized by Aleksandr Dianin, a chemist in Saint Petersburg, Russia.
1953	Polycarbonate plastic developed by Bayer and General Electric.
1957	BPA enters commercial production and is incorporated in epoxy resins.
1982	The National Toxicology Program (NTP) establishes a lowest observed adverse effect level (LOAEL) of 50 mg/kg body weight or 1000 parts per million.
1988	The Environmental Protection Agency (EPA) sets the safe dose (reference dose) at 50 μg/kg body weight. This standard remains today.
1996	The Food and Drug Administration (FDA) concludes that adults are exposed to approximately 11 μg per day and infants are exposed to 7 μg per day.
1997	First laboratory animal study showing adverse reproductive effects at doses equal or lower than those seen in humans is published by Fred vom Saal at the University of Missouri-Columbia.
1997	FDA finds BPA contamination in infant formula, demonstrating for the first time that BPA can leach into food.
1999	The Consumers Union, which publishes Consumer Reports, finds that BPA can leach from baby bottles, especially when heated.
1999	FDA asserts that BPA is safe and human doses are too low to cause adverse effects.
2003	An advisory panel for the NTP is convened to evaluate the safety of BPA. A company called Sciences International is hired to help organize the evaluation.
2003	Worldwide production of BPA exceeds 6 million pounds.
2006	A draft report from the NTP advisory panel is published and generally concludes that BPA is safe. Many scientists are critical of the report noting that some low-dose animal studies were excluded from the evaluation.
February–March 2007	The advocacy group Environmental Working Group discovers that the largest BPA manufacturers are clients of Sciences International, a revelation that launches a Congressional investigation into how the NTP advisory panel evaluation was conducted. A series of investigational reports by the *LA Times* heightens public and media awareness of the issue.

April 2007	The NTP panel continues working on the report, but the contract with Sciences International is canceled. Scrutiny of the panel by scientists, private citizens, the media, and advocacy groups intensifies.
August 2007	A separate, NIH-funded, panel of 38 scientists considered to be experts in BPA research (known as the "Chapel Hill Panel") are convened to conduct their own evaluation and publish a consensus statement. This panel produced five peer-reviewed articles published in the journal *Reproductive Toxicology*.
November 2007	The NTP advisory panel issues its final report and concludes that there is "some concern" about the neural and behavioral impacts of fetal exposure to BPA. This is considered a substantial change from the initial draft report.
February 2008	Congress requests that the FDA "clarify" its position on BPA.
April 2008	The NTP, weighing the conclusions of both the advisory panel and Chapel Hill Panel, concludes that BPA may pose a threat to humans. Canada announces its intention to ban the chemical from baby bottles and ultimately declares it a "hazardous substance." 10 States now have legislation pending to minimize use of or ban BPA. Major manufacturers including Playtex and Nalgene announce that they will stop using BPA and Wal-Mart announces that it will phase out BPA-containing bottles.
January 2010	the FDA reverses it's prior decision and declares that it has "some concern" for developmental toxicity in fetuses, infants and children.

compounds to recruit and store nitrogen-fixing bacteria. They generally have a higher binding affinity for both forms of the estrogen receptor than most man-made products (with the exception of DES) and, like most endocrine disruptors, readily cross the placenta. Consumption of a soy-rich diet is associated with a lower risk of heart disease and osteoporosis, but the reproductive health impacts of soy phytoestrogen exposure either *in utero* or after birth (through the use of soy infant formula) is generally unknown in humans. Evidence from rodents suggests that the phytoestrogen genistein can inhibit oocyte development resulting in multinucleated follicles and premature loss of the ovulatory cycle. In 2001, a team of epidemiologists examined the reproductive health of females that, as infants, had taken part in a controlled feeding study at the University of Iowa between 1965 and 1978 (Strom et al., 2001). The 2001 survey took place when the women were between 20 and 34 years of age. Women fed soy formula were more likely to have a longer period of menstrual bleeding each month, abnormally painful menstrual periods, or to have visited a gynecologist for pain or discomfort during menstruation. It was not possible to examine any potential effects on pregnancy outcomes because at the time of the study, most of the women had not yet attempted a first pregnancy. These findings are generally consistent with studies done in animals. In rodents, the premature cessation of ovulatory cycles is commonly observed. It has yet to be determined if phytoestrogen consumption in infancy results in premature menopause in women.

16.6.6   Others

PCBs were banned in the 1970s because of their toxicity and environmental persistence, but they remain in the environment and human tissues. Once used as coolants and within insulation for electrical equipment, they are present at alarmingly high levels in places thought to be pristine and free from environmental contamination, such as the arctic. A group of Inuit women and their children in Arctic Quebec is currently under study because they are known to have significantly elevated blood levels of PCBs and other contaminants compared to the general population. Dioxins, produced from incineration processes, many flame retardants, phthalates, and some fungicides are also thought to interfere with female reproduction, especially when exposure occurs *in utero*.

16.7   SUMMARY

This chapter has provided an overview of reproductive toxicology. However, as a whole, reproductive toxicology overlaps with neurotoxicology, endocrine toxicology, and developmental toxicology, all of which are covered in depth in prior or subsequent chapters. The next chapter will explore the endocrine system in more detail and build upon the concepts presented in this chapter.

BIBLIOGRAPHY AND SUGGESTED READING

Carlsen, E., A. Giwercman, et al. Evidence for decreasing quality of semen during past 50 years. *BMJ* **305**(6854):609–613, 1992.

Cohn, B. A., M. S. Wolff, et al. DDT and breast cancer in young women: New data on the significance of age at exposure. *Environ. Health Perspect.* **115**(10):1406–1414, 2007.

Couse, J. F. Reproductive toxicology. In *Molecular and Biochemical Toxicology*, 4th ed., eds. R. C. Smart and E. Hodgson. Hoboken, NJ: Wiley, 2008.

Crain, D. A., S. J. Janssen, et al. Female reproductive disorders: The roles of endocrine-disrupting compounds and developmental timing. *Fertil. Steril.* **90**(4):911–940, 2008.

Gore, A. C. Environmental toxicant effects on neuroendocrine function. *Endocrine* **14**(2): 235–246, 2001.

Gray, L. E. Jr., C. Wolf, et al. Administration of potentially antiandrogenic pesticides (procymidone, linuron, iprodione, chlozolinate, p,p′-DDE, and ketoconazole) and toxic substances (dibutyl- and diethylhexyl phthalate, PCB 169, and ethane dimethane sulphonate) during sexual differentiation produces diverse profiles of reproductive malformations in the male rat. *Toxicol. Ind. Health.* **15**(1–2):94–118, 1999.

Hamilton, B. E. and S. J. Ventura. Fertility and abortion rates in the United States, 1960–2002. *Int. J. Androl.* **29**(1):34–45, 2006.

Herbst, A. L., T. H. Green Jr., et al. Primary carcinoma of the vagina. An analysis of 68 cases. *Am. J. Obstet. Gynecol.* **106**(2):210–218, 1970.

Herbst, A. L., H. Ulfelder, et al. Adenocarcinoma of the vagina. Association of maternal stilbestrol therapy with tumor appearance in young women. *N. Engl. J. Med.* **284**(15): 878–881, 1971.

Joensen, U. N., N. Jorgensen, et al. Testicular dysgenesis syndrome and Leydig cell function. *Basic Clin. Pharmacol. Toxicol.* **102**(2):155–161, 2008.

Jones, R. E. *Human Reproductive Biology*. San Diego, CA: Academic Press, 1991.

Jurisicova, A., A. Taniuchi, et al. Maternal exposure to polycyclic aromatic hydrocarbons diminishes murine ovarian reserve via induction of Harakiri. *J. Clin. Invest.* **117**(12): 3971–3978, 2007.

Li, A. A., M. J. Baum, et al. Building a scientific framework for studying hormonal effects on behavior and on the development of the sexually dimorphic nervous system. *Neurotoxicology* **29**(3):504–519, 2008.

Lottrup, G., A. M. Andersson, et al. Possible impact of phthalates on infant reproductive health. *Int. J. Androl.* **29**(1):172–180, 2006; discussion 181–185.

Newbold, R. R., R. B. Hanson, et al. Increased tumors but uncompromised fertility in the female descendants of mice exposed developmentally to diethylstilbestrol. *Carcinogenesis* **19**(9):1655–1663, 1998.

Newbold, R. R., E. Padilla-Banks, et al. Adverse effects of the model environmental estrogen diethylstilbestrol are transmitted to subsequent generations. *Endocrinology* **147** (Suppl. 6):S11–S17, 2006.

Rogan, W. J. and A. Chen. Health risks and benefits of bis(4-chlorophenyl)-1,1,1-trichloroethane (DDT). *Lancet* **366**(9487):763–773, 2005.

Strom, B. L., R. Schinnar, et al. Exposure to soy-based formula in infancy and endocrinological and reproductive outcomes in young adulthood. *JAMA* **286**(7):807–814, 2001.

Swan, S. H., E. P. Elkin, et al. The question of declining sperm density revisited: An analysis of 101 studies published 1934–1996. *Environ. Health Perspect.* **108**(10):961–966, 2000.

Swan, S. H., K. M. Main, et al. Decrease in anogenital distance among male infants with prenatal phthalate exposure. *Environ. Health Perspect.* **113**(8):1056–1061, 2005.

Venners, S. A., S. Korrick, et al. Preconception serum DDT and pregnancy loss: A prospective study using a biomarker of pregnancy. *Am. J. Epidemiol.* **162**(8): 709–716, 2005.

Weuve, J., B. N. Sanchez, et al. Exposure to phthalates in neonatal intensive care unit infants: Urinary concentrations of monoesters and oxidative metabolites. *Environ. Health Perspect.* **114**(9):1424–1431, 2006.

Wolf, C. J., G. A. LeBlanc, et al. Characterization of the period of sensitivity of fetal male sexual development to vinclozolin. *Toxicol. Sci.* **55**(1):152–161, 2000.

SAMPLE QUESTIONS

1. What is meant by the terms "steroid positive" and "steroid negative" feedback?

2. List three different mechanisms by which endocrine disruptors can impact the reproductive system and give an example of each.

3. Leydig cells are most similar to which cell type in the ovary? Why?

4. You are asked to test the toxicity of a compound on male sperm production. In your experiments you find that the compound damages Sertoli cells. Will this likely affect sperm production?

5. True or false: all oocytes are made prior to birth.

6. What is Bisphenol-A and why is there so much public pressure to discontinue using it?

Endocrine Toxicology

GERALD A. LEBLANC

17.1 INTRODUCTION

Among the various organ systems of the body, the endocrine system is somewhat unique. While most systems are associated with a specific physiological task (i.e., respiration, reproduction, excretion) the endocrine system functions to regulate many of the activities associated with these other systems. Accordingly, the endocrine system is integral to the maintenance of total normal bodily function (homeostasis), and disruption of normal endocrine function by exogenous chemicals can result in multiple, diverse, and dire consequences. Toxicity to the endocrine system is most commonly associated with altered development, growth, maturation, and reproduction (Table 17.1). However, endocrine toxicity also can present as gastrointestinal dysfunction, malaise, neurological disorders, etc. (Table 17.1). Accordingly, endocrine toxicity often can be misconstrued as toxicity to some other endocrine-regulated system of the body.

The endocrine system, as an authentic target of chemical toxicity, tragically entered the limelight as a consequence of the widespread use of the drug diethylstilbestrol (DES). DES, a nonsteroidal synthetic estrogen, was prescribed to pregnant women from the 1940s to the 1960s as a prophylactic against miscarriage (see Section 17.4.1). Following the discovery of the endocrine toxicity of this drug, many additional drugs and environmental chemicals have been shown to mimic the action of hormones or interfere with their normal function. These activities often have been clearly shown, in laboratory studies, to result in endocrine-related toxicity. In some instances, drug use or exposure to ambient environmental chemicals also have been shown to result in endocrine toxicity. Such examples will be presented at the end of this chapter.

17.2 ENDOCRINE SYSTEM

The endocrine system can be broadly described as *an assemblage of organs (glands) that produce chemical messengers (hormones) that regulate various bodily*

A Textbook of Modern Toxicology, Fourth Edition. Edited by Ernest Hodgson
Copyright © 2010 John Wiley & Sons, Inc.

TABLE 17.1 Processes Regulated by Some Hormones of the Endocrine System and, Accordingly, That Are Susceptible to Disruption by Endocrine Toxicants

Hormone Group	Example	Origin	Regulated Process
Androgens	Testosterone	Testes, adrenals	Sexual differentiation, fertility, secondary sex characteristics, sexual function, libido
Estrogens	17β-estradiol	Ovaries, testes	Sexual differentiation, fertility, secondary sex characteristics, bone density maintenance, blood coagulation
Glucocorticoids	Cortisol	Adrenals	Bone formation, wound healing growth, development
Thyroid hormones	Thyroxine	Thyroid gland	Fetal brain and bone development oxygen consumption, gut motility

functions. The bodily functions regulated by the endocrine system can be categorized as those involved in the maintenance of homeostasis and those involved in physiological progression. Functions regulated by the endocrine system resulting in homeostasis include maintenance of the reproductive system, energy production, and metabolism. Functions regulated by the endocrine system resulting in physiological progression include fetal development, growth, and maturation. Endocrine processes related to physiological progression historically have received the greatest attention in endocrine toxicology and will be emphasized in this chapter.

Both the maintenance of homeostasis and the regulation of physiological progression require that the endocrine system detect signals, either external or internal, and transduce these signals to the appropriate target sites within the body. These target sites then respond in the appropriate manner to maintain homeostasis or institute change related to development, maturation, etc. In many species, these initial signals are of external origin. For example, many species initiate reproductive maturation in response to changes in environmental temperature, day length, etc. Reproductively mature organisms often respond to external visual or olfactory stimuli produced by sexually receptive individual to initiate sexual behavior.

The signal to be transduced by the endocrine system initiates in the central nervous system. In mammals, the hypothalamus commonly initiates the endocrine signaling pathway by secreting peptide hormones. These neuroendocrine hormones can be rapidly synthesized, secreted, and degraded to allow near instantaneous, short-lived responses to the stimulatory signal. Accordingly, they can be present in the body in pulses and secretory rhythms that often contribute to their signaling function. For example, the hypothalamic peptide hormones "growth hormone releasing hormone" (GHRH) and somatostatin are secreted in an alternating pulsatile fashion. Both hormones target the pituitary gland, though GHRH stimulates and somatostatin inhibits growth hormone secretion by the pituitary. As a result, the secretory pattern of the secondary hormone messenger in this cascade,

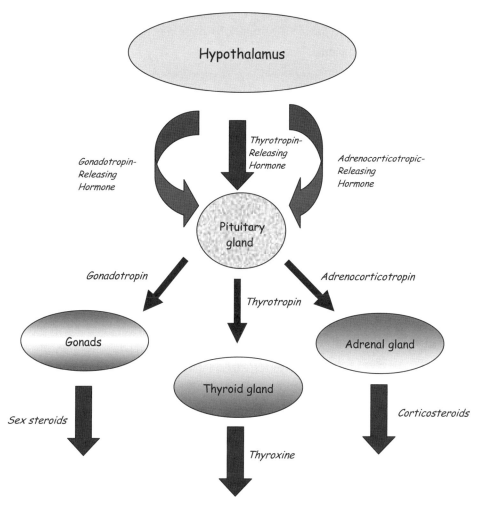

Figure 17.1 Some major neuroendocrine axes that transduce endocrine signals to target organs. Neuroendocrine signaling is initiated by the secretion of releasing hormones or, in some instances, inhibiting hormones that regulate secretion of the secondary hormone signal by the pituitary. Pituitary hormones then regulate secretion of the tertiary hormone, often a steroid hormone, by the appropriate endocrine gland. The tertiary hormones then stimulate gene transcription at target organs.

growth hormone, is highly controlled. Disruption of this rhythm in rodent models can alter hepatic enzyme expression and other dynamic processes. Disruption of the growth hormone secretory rhythm associated with sleep has been shown to interfere with normal growth in children. Hormone secretory rhythms have been associated with other physiological processes including sleep, sexual behavior, and ovulation.

Endocrine signaling pathways from the central nervous system to the target organ typically occur along axes (Figure 17.1). An axis is defined by the endocrine glands that produce signaling hormones along the cascade (e.g.,

Figure 17.2   The hypothalamic-pituitary-gonadal axis. Endocrine signaling cascades provide multiple sites for regulation and to ensure optimum signaling.

hypothalamic-pituitary-gonadal axis), and sometimes, a terminal target organ of the signaling pathway (e.g., hypothalamic-pituitary-gonadal-hepatic axis).

Endocrine signaling cascades offer several advantages over a single hormone signaling strategy. Cascades provide several sites at which the signal can be regulated thus ensuring maintenance of the appropriate endocrine signal (Figure 17.2). For example, testosterone is secreted by the testis, but regulates its own secretion by acting upstream in the axis at the pituitary gland and hypothalamic gland. Signaling cascades also utilize multiple hormones with differing properties to contribute to the process. Peptide hormones are commonly the intermediate messengers along a signaling cascade while the terminal hormone is often of non-peptide origin (e.g., steroids). Peptide hormones offer advantages as intermediate messengers in that they can be rapidly synthesized and degraded (i.e., turned "on" and "off"). Peptide hormones also do not require cell entry to elicit activity, but rather bind to cell surface receptors. This facilitates a rapid physiological response to the hormone. Steroid and other non-peptide hormones are typically more stable; they are maintained in circulation at a relatively constant, physiologically appropriate level; they can be stored as precursor molecules or apolar conjugates; they can be mobilized as polar conjugates; and most often, they require cell entry to interact with its receptor and elicit a response. Accordingly, the non-peptide terminal hormones offer the advantages of constant availability but lack the advantages of rapid modulation.

17.2.1 Nuclear Receptors

Toxicologically, the function of the terminal hormones of endocrine cascades (i.e., steroid, retinoid, thyroid hormones) appear to be most susceptible to disruption by chemicals. This is because many foreign molecules share sufficient characteristics with these hormone molecules to allow binding to the nuclear receptors of these hormones in either an agonistic or antagonistic fashion. The binding of the xenobiotic to the receptor results in aberrant receptor function with associated toxicological outcome. The nuclear receptors are so-called since these receptors initiate their classical physiological responses within the cell nucleus. Cell surface receptors to peptide hormones, on the other hand, can likely discriminate between peptide molecules and non-peptide xenobiotics thus minimizing the likelihood of interaction and associated disruption of function.

Nuclear receptors comprise a superfamily of transcription factors divided among seven distinct subfamilies. Humans express at least 48 different nuclear receptors, many of which are considered orphan receptors because no hormone is known to activate the receptor. Many of these orphan receptors may indeed be ligand-independent transcription factors. Others may have, as of yet, unrecognized ligands. Androgen, estrogen, progesterone, glucocorticoid, and mineralocorticoid receptors are all members of the nuclear receptor subfamily NR3. The retinoic acid, vitamin D, and others are members of the NR1 subfamily. The NR3 subfamily receptors typically homodimerize to form active transcription factors while the NR1 family members typically heterodimerize with RXR (an NR2 subfamily member) to form the active transcription factor (Figure 17.3a).

Nuclear receptors typically share a common five-domain structure (Figure 17.3b). The DNA-binding (domain C) and ligand-binding (domain E) are most highly conserved among receptors. The remaining domains are highly variable among receptors and, in the case of domain F, is absent from some receptors. Nuclear receptors modulate transcription of target genes by binding to hormone response elements in the promoter region of the target gene. Hormone response elements typically consist of two half-sites each consisting of a core sequence of AGGTCA or some derivative thereof (Figure 17.3c). Half-sites are separated by nucleotide spacers, and the number of spacers contributes to the receptor specificity of the response element. Each partner receptor in a dimeric complex binds to one of the half-sites to properly stabilize the complex on the DNA. Some orphan receptors bind DNA as a monomer. Response elements for these receptors consist of a single half-site with an adjacent A/T-rich region that further promotes stabilization of the complex. Binding of hormone-activated receptor on the response element initiates the recruitment of coactivators and other factors required to initiation gene transcription.

The NR3 subfamily members typically exist in the extranuclear matrix of the cell in association with various accessory proteins. These accessory proteins stabilize the receptor molecule and help maintain the molecule's integrity. Binding of hormone ligand to the receptor protein stimulates dissociation with the accessory

(a)

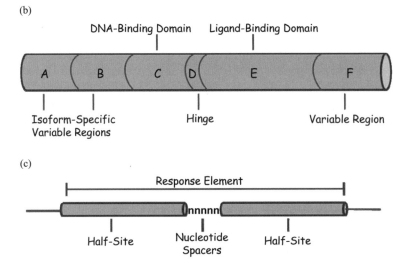

Receptor	Group	Major partnership	Example ligand
Thyroid hormone (THR)	NR1A	THR:RXR	Thyroxine
Retinoic acid (RAR)	NR1B	RAR:RXR	*all trans*-retinoic acid
Vitamine D (VDR)	NR1I	VDR:RXR	Vitamin D$_3$
Retinoid X (RXR)	NR2B	RXR:RXR	9-*cis*-retinoic acid
Estrogen (ER)	NR3A	ER:ER	17b-estradiol
Androgen (AR)	NR3C	AR:AR	Dihydrotestosterone
Progesterone (PR)	NR3C	PR:PR	Progesterone
Glucocorticoid (GR)	NR3C	GR:GR	Cortisol
Mineralcorticoid (MR)	NR3C	MR:MR	Aldosterone

(b)

DNA-Binding Domain   Ligand-Binding Domain

A | B | C | D | E | F

Isoform-Specific   Hinge   Variable Region
Variable Regions

(c)

Response Element

Half-Site   Nucleotide   Half-Site
             Spacers

Figure 17.3   (a) Some hormone-activated nuclear receptors of vertebrates. (b) Basic common structure of nuclear receptors. (c) General characteristics of hormone response elements to which receptors bind.

proteins, homodimerization of two receptor molecules, and nuclear localization (Figure 17.4). The NR1 group members typically form heterodimeric combinations with the retinoid X receptor (RXR). RXR also is capable of homodimerization in association with its ligand.

17.2.2 Membrane-Bound Steroid Hormone Receptors

Some cellular responses occur too rapidly following steroid hormone exposure to involve the multistep process of nuclear receptor activation. For example, 17β-estradiol can rapidly stimulate adenylate cyclase and cause a near instantaneous increase in intracellular cyclic adenosine monophosphate (cAMP) in cultured prostate cells. These effects are mediated by the interaction of steroid hormones with cell surface proteins.

Due to their lipophilic nature, steroid hormones are mobilized in the circulatory system by transfer proteins. Sex hormone-binding globulin (SHBG) is one such

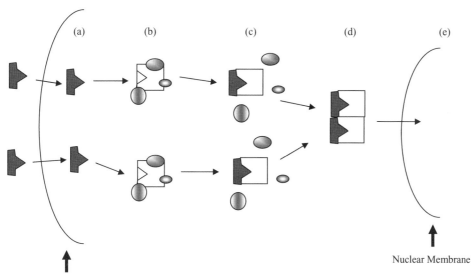

Cell Surface Membrane

Nuclear Membrane

Figure 17.4 Intracellular steroid receptor activation by hormone ligands. (a) Steroid hormones diffuse across the cell membrane into the cell. (b) Steroid hormone receptors in the basal state bound to accessory proteins. (c) Steroid hormones bind to receptors and accessory proteins are dissociated from the receptors. (d) Hormone–receptor complexes dimerize. (e) Dimer complexes enter the nucleus and initiate transcription of responsive genes.

transfer protein that binds testosterone, 17β-estradiol, and other sex steroids. Receptors exist on the surface of some cells that are capable of binding unliganded SHBG (Figure 17.5). Unliganded SHBG, that is bound to the cell surface receptor, can subsequently bind steroid hormone. Binding of an appropriate hormone to the SHBG then stimulates a signal-transduction pathway within the cell. Some steroid hormones (17β-estradiol, 5α-androstan-3α,17β-diol) function as SHBG:SHBG-receptor agonists while others (testosterone, 5α-dihydrotestosterone) function as antagonists. Interestingly, 5α-androstan-3α,17β-diol had previously been considered an inactivation product of the potent androgen 5α-dihydrotestosterone (DHT). Studies in human prostate cells have shown that activation of this SHBG-dependent pathway stimulates DNA synthesis and cell growth. These observations, in combination with studies in dogs that have shown 5α-androstan-3α,17β-diol to stimulate benign prostatic hyperplasia, have led to suggestions that the SHBG-receptor pathway is involved in this disease condition. The susceptibility of these membrane-bound receptor pathways in endocrine toxicology has received little attention, although conceivably, toxicants could perturb these pathways by competing with endogenous hormone for binding to surface membrane receptors resulting in the loss of stimulatory activity (antagonists) or inappropriate stimulation of activity (agonists). Several xeno-estrogens have been shown to stimulate calcium influx in cultured cells through interaction with a surface membrane-bound estrogen receptor at exposure concentrations much lower than is required to activate the nuclear estrogen receptor.

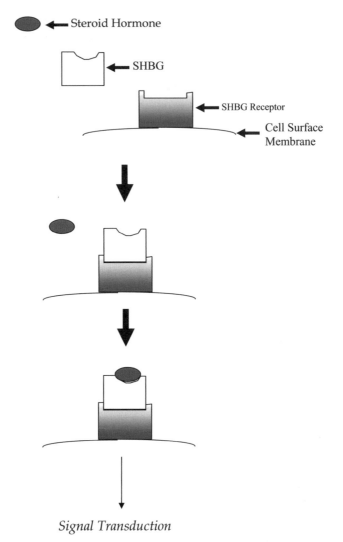

Figure 17.5 Endocrine signaling pathway involving steroid hormone, sex hormone-binding globulin (SHBG), and the SHBG receptor.

17.3 ENDOCRINE DISRUPTION

Xenobiotics have the ability to disrupt hormone activity through a variety of mechanisms though the predominant mechanisms appear to involve binding to the hormone receptor, either as an agonist or antagonist, or by modulating endogenous steroid hormone levels.

17.3.1 Hormone Receptor Agonists

A hormone receptor agonist is defined as *a compound that binds to and activates a hormone receptor*. Endogenous hormones function as agonist to their respective receptors. Xenobiotics can act as receptor agonist and stimulate receptor-dependent physiological processes in the absence of the endogenous receptor ligand (hormone). Such inappropriate stimulation can result in the errant expression of hormone-dependent processes such as breast development in males (gynecomastia).

Estrogen Receptor   Among the steroid hormone receptors, the estrogen receptor appears most susceptible to the agonistic action of xenobiotics. Estrogen receptor agonists are quite diverse in molecular structure (Figure 17.6). Several steric considerations, associated with the steroid structure, in conjunction with electrostatic (charge) properties of the outer surface of the molecule, seem to dictate whether a xenobiotic can fit into the binding pocket of the receptor and function as a receptor agonist. It is not clear why the estrogen receptor would be more susceptible to the agonistic action of xenobiotics as compared to other steroid hormone receptors. The estrogen receptor is often referred to as a promiscuous receptor because of this susceptibility to agonistic interactions with xenobiotics.

Some drugs are rather potent estrogens (e.g., DES); however, environmental chemicals with estrogenic activity are typically weak agonists with activity several

Figure 17.6   Diverse structures of estrogen receptor agonists.

TABLE 17.2   Potency of Some Xeno-Estrogens Relative to 17β-Estradiol

Chemical	Potency
17β-Estradiol	100
Diethylstilbestrol	74
4-Nonylphenol	0.005
4-Octylphenol	0.003
4-*tert*-Octylphenol	0.00036
o′,p′-DDT	0.00011
o′,p′-DDE	0.00004
2′,5′-Dichloro-4-biphenylol	0.62
2′,4′,6-Trichloro-4-biphenylol	1.0
2′,3′,4′,5′-Tetrachloro-4-biphenylol	0.82
Bisphenol A	0.005
Butylbenzylphthalate	0.0004

Note: Estrogenic potency of the compounds was measured using a recombinant yeast cell bioassay (Coldham et al. *Environ. Health Perspect.* **105** (7): 734–742, 1997).

orders of magnitude less than that of 17β-estradiol (Table 17.2). Because of this weak activity, xeno-estrogens are typically not associated with endocrine toxicity to adult females owing to the large amount of 17β-estradiol in these individuals. However, adult males, immature individuals, and embryos all have been shown to exhibit endocrine toxicity resulting from xeno-estrogen exposure. For example, *in utero* exposure of male or female rodents and humans to DES causes proliferation of epithelial cells associated with the reproductive system, resulting in abnormalities of this system. Gynecomastia is a common side effect of estrogenic drugs such as fosfestrol when administered to adult males. The physiological consequences of xeno-estrogenic activity is typically characteristic of feminization, that is, the acquisition of female characteristics.

Ecdysteroid Receptor   Ecdysteroids are a class of steroid hormones that regulate a variety of processes related to development, growth, and reproduction in insects and other arthropods but are not utilized by vertebrates. Many compounds of plant origin, or derivations thereof, have been identified that are ecdysteroid receptor agonists (i.e., cucurbitacins, withasteroids). The ecdysteroid agonists are presumed to have evolved in plants as a means of protection against insect predation. Some environmental chemicals of anthropogenic origin also have been shown to exhibit ecdysteroid receptor agonistic activity (e.g., tebufenozide) and have been exploited as insecticides due to their ability to interfere with insect development and growth.

Retinoic Acid Receptor   Most of the biological effects of retinoids are mediated through the retinoic acid receptor (RAR) and the retinoid X receptor (RXR). Both all-*trans*-retinoic acid and 9-*cis*-retinoic acid serve as agonists of RAR while only 9-*cis*-retinoic acid functions as an agonist of RXR. The functional RAR exists as a heterodimer with RXR, while functional RXR exists as a homodimer. Methoprene is a juvenile hormone III analog that mimics the activity of this insect hormone. Exposure of juvenile insects to methoprene results in various abnormalities associated with development and ultimately death. An environmental degradation product of methoprene, methoprenic acid, was found to serve as an RXR agonist and specifically activate genes responsive to RXR homodimers. In addition, exposure of frog larvae to methoprenic acid caused developmental deformities consistent with those that have been observed in recent years in wild populations and consistent with those caused by exposure to retinoic acid under laboratory conditions. The biocide tributyltin is a high-affinity agonist of the RXR. This activity has been causally associated with a condition, documented worldwide, where females of some marine snail species have acquired male sex characteristics. These observations indicate that activation of the RXR by xeno-agonist may contribute to the occurrence of various deformities documented in wildlife species.

17.3.2 Hormone Receptor Antagonists

While the estrogen receptor appears somewhat unique among vertebrate nuclear hormone receptors in its promiscuity toward receptor agonists, many nuclear hormone receptors have been shown to be susceptible to chemical antagonism. Receptor antagonists are defined as *chemicals that bind to a hormone receptor but do not activate the receptor*. Rather, these chemicals inhibit receptor activity by preventing the endogenous hormone from binding to and activating the receptor.

Estrogen Receptor   Chemicals often bind to the estrogen receptor and function as mixed agonists/antagonists (discussed below). Drugs that bind to the estrogen receptor as an antagonist or mixed agonist/antagonist include tamoxifen, raloxifene, ICI 164,384, and toremifene. Environmental estrogen receptor antagonists include some phytochemicals (e.g., flavonoids) and industrial chemicals such as some polychlorinated biphenyls (PCBs). Consequences of estrogen receptor antagonism are typically considered defeminization (loss of female traits). In laboratory animal studies, estrogen receptor antagonists have been shown in females to disrupt estrous cycles, impair fertility, increase preimplantation loss, and cause embryo lethality.

Androgen Receptor   Chemicals that bind to the androgen receptor in an antagonistic fashion include the pharmaceuticals spironolactone, cimetidine, cyproterone acetate, and hydroxyflutamide. Environmental chemicals that have been shown to act as androgen receptor antagonists include the metabolites of the agricultural fungicide vinclozolin, the DDT metabolite p,p'-DDE, some hydroxylated PCBs, and the organophosphate insecticide fenitrothion. The consequence of androgen receptor antagonism is typically considered demasculinization (loss of male traits). Demasculinizing effects of anti-androgens in laboratory animal studies have included reductions in the size of the ventral prostate and seminal vesicle weights along with deformities of the penis.

Glucocorticoid Receptor   Some drugs (e.g., mifepristone) elicit antagonistic activity toward the glucocorticoid receptor. This property has been associated with adverse side effects of the drugs and also has been capitalized upon therapeutically for the modulation of the glucocorticoid receptor. Anti-glucocorticoids typically are steroidal compounds that are capable of binding to the receptors but are relatively ineffective in activating the receptor. As such, these compounds are typically mixed agonists/antagonists (see below). Glucocorticoid receptor antagonists can adversely affect growth, development, and other glucocorticoid-regulated processes (Table 17.1). Little is known of the ability of environmental chemicals to function as glucocorticoid receptor antagonists.

Mixed Agonists/Antagonists Chemicals often can function as either a receptor agonist or antagonist depending upon the level of endogenous hormone. A weak agonist may bind to a receptor and stimulate some low-level receptor-mediated activity in the absence of the endogenous hormone. However, in the presence of the hormone, binding of the xenobiotic to the receptor may prevent binding of the endogenous hormone and, if the xenobiotic is a much weaker activator of receptor-mediated activity, then the net effect is loss of activity. Thus, in the presence of the endogenous hormone, the xenobiotic functions as a receptor antagonist. Whether a weak xeno-agonist functions as an agonist or antagonist typically depends upon (1) the concentration of the xeno-agonist, (2) the potency of the xeno-agonist, and (3) the concentration of the endogenous hormone. These compounds are classified as mixed agonists/antagonists.

17.3.3 Organizational Versus Activational Effects of Endocrine Toxicants

Effects of receptor agonists or antagonists on endocrine-related processes are often described as being either organizational or activational. An organizational effect of an endocrine toxicant is one that typically results from neonatal or prenatal exposure during which time hormones are directing various irreversible aspects of development. Accordingly, the disrupting effect of the toxicant also is irreversible. These organizational effects may be evident only later in life during maturation or reproduction. Neonatal exposure to DES resulting in proliferation of epithelial cells of the reproductive tract at reproductive maturity is an example of an organizational effect of an endocrine toxicant. Organizational effects of endocrine toxicants have been of great concern to toxicologists and are the most difficult type of toxicity to diagnose, owing to the temporal separation between exposure and effect.

An activational effect of an endocrine toxicant occurs in the same time frame as the exposure and is the consequence of the toxicant disrupting the immediate role of a hormone in some physiological process. Activational effects are typically reversible following cessation of exposure to the toxicant. For example, androgens contribute to maintenance of the prostate gland in the adult male. Exposure of adult males to an anti-androgen can result in a decrease in prostate size. Cessation of exposure to the anti-androgen then results in restoration of the prostate gland to its normal size.

17.3.4 Inhibitors of Hormone Synthesis

Endocrine toxicants can elicit antihormone activity by lowering levels of endogenous hormone in the body. With steroid hormones, chemicals typically elicit this effect by inhibiting enzymes necessary for synthesis of the hormone. For example, the cytochrome P450 enzyme CYP19 is responsible for the aromatization of testosterone to form 17β-estradiol. CYP19 inhibitors such as fadrozol, anastrozole, and letrozole, can lower endogenous 17β-estradiol levels resulting in defeminization. Cytochrome P450 enzymes also are critical to various hydroxylation reactions that contribute to the synthesis of androgens and other steroid hormones, and inhibition of these enzymes can result in a variety of anti-steroid hormone effects. For example, the agricultural and medicinal fungicides propiconazole, ketoconazole, and fenarimol are capable of inhibiting P450 enzymes and reducing synthesis and circulating levels of testosterone and other steroid hormones. Toxicological consequences of the lowering of endogenous steroid hormone levels are typically comparable to those effects elicited by antagonists of the hormone's receptor.

17.3.5 Inducers of Hormone Clearance

In most species, steroid and thyroid hormones are inactivated and cleared from the body by the same biotransformation processes that are involved in chemical detoxification (see Chapter 6). Predominant among the hormone biotransformation processes in vertebrates are hydroxylation, glucuronic acid conjugation, and sulfate conjugation. The thyroid hormones T_3 and T_4 are inactivated and cleared following sulfate and glucuronic acid conjugation, respectively. The glucuronosyl transferase enzymes that are responsible for the elimination of T_4 are induced following exposure to phenobarbital-type inducers and Ah receptor ligands (see Chapter 8). Thus, exposure to chemicals such as some dioxins and PCBs can result in enhanced clearance of thyroid hormone resulting in low circulating thyroid hormone levels. The resulting hypothyroid state can result in a variety of pathological conditions (Table 17.3). In newborn infants, hypothyroidism is associated with cretinism. This organizational syndrome is characterized by mental retardation, short stature, and various neurological abnormalities. In children, hypothyroidism can cause delayed growth and mental development while advancing the onset of puberty in adolescents. Hypothyroidism in adults results in various activational abnormalities including impaired cardiovascular, pulmonary, intestinal, and renal function. Chronic fatigue, lethargy, and difficulty in concentration are also associated with hypothyroidism in adults.

TABLE 17.3 Clinical Manifestations of Hypothyroidism

Organ System	Manifestation
Skin	Puffy appearance, dry, course, yellow tinted skin
	Brittle nails, wound healing slowed, hair loss
Cardiovascular	Enlarged heart, changes in electrocardiographs
Respiratory	Maximal breathing capacity reduced, obstructive sleep apnea, fluid accumulation in the pleural cavity
Digestive system	Reduced appetite with modest weight gain
Muscle	Stiffness, aching
Nervous	Slowing of intellectual functions, lethargy, headaches

Increased clearance of steroid hormones due to induction of hepatic biotransformation enzymes following chemical exposure often has been cited as a possible mechanism by which toxicants could lower circulating testosterone or 17β-estradiol levels. While enhanced clearance of sex steroids has been demonstrated following chemical exposure and induction of hepatic biotransformation enzymes, elegant feedback control mechanisms tend to ensure that more hormone is produced and homeostasis is maintained (Figure 17.2). Enhanced clearance of sex steroids can contribute to endocrine disruption if the toxicity also results in impaired hormone synthesis (i.e., gonadal toxicity or interference with the feedback control of hormone synthesis). 2,3,7,8-Tetrachlorodibenzodioxin appears to lower circulating sex steroid levels via this dual effect.

17.3.6 Hormone Displacement from Binding Proteins

Steroid and thyroid hormones are distributed throughout the body while bound to serum-binding proteins such as sex hormone-binding globulin, corticosteroid-binding globulin, thyroxine-binding globulin (transthyretin), and albumin. Most steroid and thyroid hormones (>95%) are present in the blood reversibly bound to proteins. This bound hormone is not available for cell entry where it may interact with nuclear receptors or undergo inactivation/elimination reactions. Rather, the bound hormone serves as a reservoir from which hormone can be liberated (free hormone) for cell entry.

Some xenobiotics can compete with hormones for binding to the blood proteins. As a result, the circulating hormone reservoir can be depleted and free hormone becomes limited. A variety of phenolic compounds including hydroxylated metabolites of PCBs, chlorophenols, chlorophenoxy acids, and nitrophenols have been shown to interfere with thyroxine binding to thyroxine-binding globulin during *in vitro* experiments. In some instances, compounds that displace thyroxine from the binding protein also have been shown to decrease circulating thyroxine levels in exposed animal models or in humans. *In vitro* experiments also have revealed that testosterone and 17β-estradiol can be displaced from sex hormone-binding globulin by some chemicals such as 4-nonylphenol, 4-tert-octylphenol, bisphenol A, O-hydroxybiphenyl, and pyrethroid insecticides. However, it is not clear whether these chemicals would significantly displace sex steroids from the binding globulin at concentrations typically measured in human blood.

17.4 INCIDENTS OF ENDOCRINE TOXICITY

17.4.1 Organizational Toxicity

In utero exposure to estrogens or anti-androgens has been shown, in animal models, to elicit a variety of organizational effects associated with development of the reproductive system. The best-described example of the organizational effects of a drug administered to humans involves the synthetic estrogen DES. DES was prescribed to over two million pregnant women in the United States between the 1940s and 1960s to prevent miscarriage. Offspring exposed to DES during fetal development experienced a variety of problems upon attainment of sexual maturity.

DES daughters experience a significantly increased risk of clear cell adenocarcinoma of the vagina and cervix. DES daughters had increased risk of a variety of reproductive disorders including structural abnormalities of the reproductive tract, infertility, ectopic pregnancy, miscarriage, and preterm delivery.

Less is known of the risks faced by males exposed to DES during fetal development. Animal studies have revealed that male rodents exposed to DES have increased incidence of prostatic metaplasia. Epidemiological studies of DES sons have suggested increased risk of various testicular abnormalities including epididymal cysts, testicular varicoceles, and undescended testis. Hyperplasia and metaplasia of the prostatic ducts in DES sons also have been reported.

The effects elicited by fetal exposure to DES appear to be largely the consequence of the estrogenic activity of this drug. Estrogens orchestrate organizational events during fetal development that promote female reproductive tract development. Excess estrogen exposure resulting from DES treatment of either female or male fetuses resulted in permanent alterations, many of which became evident only upon attainment of reproductive maturity.

Organizational effects on reproductive development resulting from perinatal exposure to endocrine toxicants of environmental origin also have been reported to occur. In 1973, a fire retardant containing polybrominated biphenyls (PBBs) was mistakenly added to cattle feed in Michigan. An estimated 4000 people subsequently were exposed to the PBBs by consuming dairy products derived from these cattle. PBBs are long-lived chemicals that are stored in the fat of exposed individuals. PBBs have been reported to elicit endocrine toxicity-like symptoms in animal models consistent with hypothyroidism. For example, offspring from maternal rats provided PBBs during gestation and lactation showed signs of neurological deficit and growth retardation. Daughters of mothers that were exposed to PBBs during the Michigan incident exhibited an earlier initiation of menarche (menstruation) that correlated with the likely severity of *in utero* PBB exposure. The most highly exposed daughters began menstruating approximately 1 year ahead of females that were less severely exposed. Early initiation of menarche is consistent with precocious puberty associated with hypothyroidism.

The distance between the anus and the genitals is a sexually dimorphic trait that is hormonally regulated during fetal development. Males have a greater anogenital distance than do females. A study published in 2005 revealed that anogenital distance among male babies was inversely correlated to the level of phthalate ester metabolites in the urine of their mothers. Babies produced by mothers with the highest phthalate ester metabolite concentrations were seven times more likely to have a shortened anogenital distance. Phthalate esters are widely used in plastics to confer flexibility to the plastic. Studies in rats have shown that some phthalate esters interfere with fetal testosterone production and thus have anti-androgenic effects in the developing fetus. Results with human babies suggest that levels of phthalate esters associated with the human population may be sufficient to disrupt male reproductive development.

17.4.2 Activational Toxicity

Estrogenic Pharmaceuticals Administration of estrogenic pharmaceuticals to children or adults can result in a variety of abnormalities associated largely with secondary sex characteristics that are reversible upon cessation of drug treatment.

Gynecomastia, the development of breast tissue in males, is often the consequence of perturbations in the normal androgen/estrogen ratio. Prolonged administration of drugs with estrogenic or anti-androgenic activity can cause gynecomastia. Gynecomastia had been reported in the medical literature to occur as a result of frequent intercourse when an estrogen-containing cream was used as a vaginal lubricant and among morticians who applied estrogen-containing skin creams to corpses without the use of gloves.

Similar to gynecomastia in adult males, activational toxicity from estrogenic drugs has been reported to cause pseudoprecocious puberty in children. Pseudoprecocious puberty is characterized by the development of some indicators of puberty (e.g., pubic or facial hair, morphological changes in sex organs, breast development) in preadolescent individuals. An outbreak of pseudoprecocious puberty was reported among a group of children ranging in age from 4 months to 2 years of age following application of a skin cream to treat dermatitis. Symptoms included pigmentation of the nipples, breast development, the presence of pubic hair, and vaginal discharge and bleeding among the females. Breast development also was reported in prepubertal boys following use of an estrogen-containing hair cream. These reports highlight the fact that dermal exposure can be adequate to attain a sufficient dose of endocrine-active compound to elicit adverse responses. In all of these cases, the symptoms of endocrine toxicity resolved following cessation of exposure to the causative agent.

Environmental Estrogens   Thelarche is defined as the development of breast tissue in preadolescent females (typically <8 years of age). An epidemic level of thelarche occurred on the island of Puerto Rico during the 1980s and 1990s. The cause of the outbreak was never discerned; however, evidence strongly implicated exposure to endocrine-disrupting agents. Analyses of blood samples from thelarche and non-thelarche children for environmental chemicals with known estrogenic activity revealed that 68% of the thelarche children contained significantly high levels of several types of phthalate esters. Only a single non-thelarche child contained a significant amount of phthalate ester and only one type of phthalate ester was found in this individual. The association between phthalate ester exposure and the high incidence of thelarche in Puerto Rico does not establish causality but has generated concern that environmental agents were responsible for this condition.

Kepone (chlordecone) is an organochlorine insecticide (Figure 17.5) that was manufactured in Hopewell, Virginia from the mid 1960s to 1975. In 1975, the Center for Disease Control determined that employees of the manufacturing facility and other residents of Hopewell, totaling over 200 individuals, had been significantly contaminated with this insecticide. Exposed individuals reported a variety of symptoms. Foremost, among the symptoms of "Kepone sickness" were neurological disorders presenting as tremors, weight loss, and nervousness. However, subsequent evaluations revealed that males exposed to Kepone also experienced testicular dysfunction that was characteristic of estrogen exposure. Later laboratory studies demonstrated that Kepone was an estrogen receptor agonist, which could explain its adverse effects on the male reproductive system.

17.4.3 Hypothyroidism

Hypothyroidism describes the clinical state arising from a deficiency in thyroid hormone. Toxicity resulting in hypothyroidism is manifested at several organ systems as described in Table 17.3, and individual effects may be misdiagnosed as organ-specific toxicity. Hypothyroidism can result from various causes other than chemical toxicity including diseases of the hypothalamic-pituitary-thyroidal axis, iodine deficiency, and heritable defects in thyroid hormone production. Chemical agents that have historically been recognized for their ability to cause hypothyroidism include phenylbutazone, resorcinol, lithium, and para-aminosalicylic acid.

Disruptions in thyroid hormone levels can occur through chemical-induced increases in the metabolic inactivation and elimination of the hormone. Chemicals that are capable of increasing the metabolic clearance of thryroid hormone include the polycyclic halogenated hydrocarbons (i.e., dioxins, furans, PCBs, polybrominated biphenyls). A study reported in *The New England Journal of Medicine* suggested that environmental or occupational exposure to such chemicals can result in hypothyroidism in humans. The study consisted of a comparison of thyroid status in workers who were occupationally exposed to polybrominated biphenyls as compared to workers who were not exposed to any polyhalogenated hydrocarbons. Four of 35 exposed workers and none of 89 unexposed workers exhibited signs of hypothyroidism that included increased plasma levels of thyrotropin and decreased plasma levels of thyroxine. Thyrotropin is secreted by the pituitary gland and stimulates the thyroid gland to produce thyroxine (see Figure 17.1). The increase in thyrotropin and decrease in thyroxine is consistent with hypothyroidism caused by increased clearance of the thyroxine. As discussed earlier in this chapter, perinatal exposure to PBBs during the Michigan milk contamination also produced symptoms characteristic of hypothyroidism.

17.5 CONCLUSION

The endocrine system possesses many targets at which toxicants can elicit either reversible or permanent effects on an individual. Effects of chemicals on endocrine-regulated processes such as development, maturation, growth, and reproduction have been well documented in both laboratory and epidemiological studies. Less is known of the potential effects of endocrine toxicants on more generalized endocrine-regulated processes such as bone maintenance, general organ function, and metabolism. The U.S. Environmental Protection Agency (EPA) has been mandated by the U.S. Congress to develop and implement a program for the screening and testing of chemicals for endocrine-disrupting toxicity. The EPA has developed a program that focuses upon the effects of chemicals on androgen, estrogen, and thyroid hormone-regulated processes. This required testing will greatly expand our knowledge of the extent to which humans are exposed to chemicals that interfere with processes regulated by these hormones. However, it is important to recognize that chemicals have the potential ability to interfere with other hormone cascades including those involving mineralocorticoids, glucocorticoids, retinoids, and perhaps some peptide hormones. Research is needed to increase our understanding of the susceptibility of endocrine signaling pathways involving these hormones to chemical toxicity and, ultimately, establishing chemical exposure limits that include these considerations.

BIBLIOGRAPHY AND SUGGESTED READING

Bahn, A. K., J. L. Mills, P. J. Snyder, et al. Hypothyroidism in workers exposed to polybrominated biphenyls. *N. Engl. J. Med.* **302**:31–33, 1980.

Beas, F., L. Vargas, R. P. Spada, et al. Pseudoprecocious puberty in infants caused by dermal ointment containing estrogens. *J. Pediatr.* **75**:127–130, 1962.

Blanck, H. M., M. Marcus, P. E. Tolbert, et al. Age at menarche and tanner stage in girls exposed in utero and postnatally to polybrominated biphenyls. *Epidemiology* **11**:641–647, 2000.

Colborn, T. and C. Clement, eds. Chemically induced alterations in sexual and functional development: The wildlife/human connection. In *Advances in Modern Environmental Toxicology*, ed. M. A. Mehlman. Princeton, NJ: Princeton Scientific Publishing, 1992.

DiRaimondo, C. V., A. L. Roach, and C. K. Meador. Gynecomastia from exposure to vagina estrogen cream. *N. Engl. J. Med.* **302**:1089–1090, 1980.

Guillette, L. Jr. and D. A. Crain, eds. *Environmental Endocrine Disrupters*. New York: Taylor & Francis, 2000.

Larsen, P. R., H. M. Kronenberg, S. Melmed, et al. *Textbook of Endocrinology*. Philadelphia: Saunders, 2003.

Laudet, V. and H. Gronemeyer. *The Nuclear Receptor Facts Book*. New York: Academic Press, 2002.

McLachlan, J. A. Environmental signaling: What embryos and evolution teach us about endocrine disrupting chemicals. *Endocr. Rev.* **22**:319–341, 2001.

SAMPLE QUESTIONS

1. Endocrine signaling typically occurs through neuroendocrine cascades that involve both peptide and non-peptide (e.g., steroid) hormones. What is the value of having both peptide and steroid hormones in the same endocrine cascade?

2. Why are steroid hormone receptors often the most susceptible link in an endocrine cascade to disruption by toxicants?

3. What are two mechanisms through which a chemical might act as an anti-androgen?

4. Define "organizational" and "activational" actions of endocrine toxicants.

Respiratory Toxicology

JAMES C. BONNER

18.1 INTRODUCTION

The respiratory system represents a unique target for the potential toxicity of toxicants due to the fact that the lungs are the primary portal of entry for inhaled gases and particles. In addition, the lungs receive the entire cardiac output, and therefore, toxicants that enter the bloodstream also have the potential to cause lung injury. Because of the essential function of the respiratory system in transporting atmospheric oxygen to the bloodstream, the alveolar region of the lung is the primary interface for exposure to a vast array of toxins. The mammalian lung has evolved exquisite cellular defense mechanisms to cope with the clearance of inhaled agents and finely tuned repair mechanisms to restore the delicate architecture of the lung after injury. Despite these protective mechanisms, toxic exposures to both natural and man-made toxicants result in lung diseases such as asthma, fibrosis, chronic obstructive pulmonary disease (COPD), and cancer.

18.2 ANATOMY AND FUNCTION OF THE RESPIRATORY TRACT

The general features of upper and lower regions of the respiratory tract are illustrated in Figure 18.1. The upper respiratory tract consists of the mouth, nose, and pharyngeal region. Air enters the respiratory systems of mammals through the nose or mouth in humans, while some species such as rodents are obligate nasal breathers. The primary functions of the upper respiratory tract are olfaction, temperature equilibration and humidification of inspired air, and uptake of inhaled particles and irritant gases. The lower respiratory tract begins distal to the pharyngeal region and consists of the tracheobronchial region and the pulmonary parenchyma or alveolar region. A primary function of the lower respiratory tract is harvesting oxygen from the atmosphere and the transfer of the metabolic gas CO_2 from the blood to the exhaled air. A second important function of the lower respiratory tract is defense against inhaled toxicants.

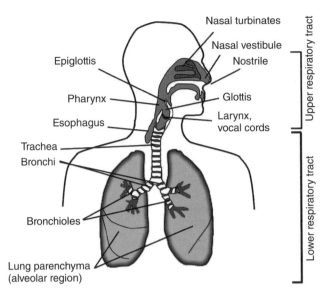

Figure 18.1 Anatomy of the human respiratory system. Adapted from *LifeART Illustration Series*. Hagerstown, MD: Lippincott Williams & Wilkins, 1994.

As the goal of the lung is to provide oxygen to the tissues and remove CO_2, the inflow and outflow of air between the atmosphere and the alveoli, termed *ventilation*, is a critical function. A simple method for studying pulmonary ventilation is to record the volume movement of air into and out of the lungs, a process called *spirometry*. Four different pulmonary lung volumes are defined as follows: (1) the *tidal volume* is the volume of air inspired or expired with each normal breath; (2) the *inspiratory reserve volume* is the maximum extra volume of air that can be inspired over and above the normal tidal volume; (3) the *expiratory reserve volume* is the maximum extra volume of air that can be expired by forceful expiration after the end of a normal tidal expiration; and (4) The *residual volume* is the volume of air remaining in the lungs after the most forceful expiration. Two or more volumes together are called *pulmonary capacities*, which are described as follows: (1) the *inspiratory capacity* equals the *tidal volume* plus the *inspiratory reserve volume*; (2) the *functional residual capacity* equals the *expiratory reserve volume* plus the *residual volume*; (3) the *vital capacity* equals the *inspiratory reserve volume* plus the *tidal volume* plus the *expiratory reserve volume*; and (4) the *total lung capacity* is the maximum volume to which the lungs can be expanded, equal to the *vital capacity* plus the *residual volume*.

Measures of the lung to expand and fill with air during inspiration and to deflate during exhalation are termed lung mechanics. These properties depend on the elasticity of the lung and the caliber of the airways. Lung mechanics are commonly reported as compliance and resistance. *Compliance* is the volume change per unit pressure change. *Resistance* is the pressure difference per change in airflow. Compliance and resistance are generally measured during steady tidal breathing by recording transpulmonary pressure, tidal volume, and airflow rate. Because the direct measurement of transpulmonary pressure require invasive procedures such as an intrapleural catheter, indirect measures approximating compliance and resistance are typically applied to humans using *plethysmography*.

Toxicity to the respiratory tract can interfere with this gas-exchange function by (1) altering the tone of the airways resulting in decreased airflow; (2) damaging the delicate architecture of the alveolar/capillary barrier of the deep lung resulting in impaired gas exchange; or (3) causing tissue damage that leads to chronic structural changes and decreased lung volumes or lung mechanics. The total lung capacity and vital capacity are reduced in fibrotic lung disease in which the lung becomes smaller and stiffer. This type of change in the lung is called *restrictive lung disease* and is marked by smaller lung volumes and little change in airflow. In an emphysematous lung, on the other hand, the total lung capacity may increase as a result of the breakdown of alveolar walls and loss of elastin fibers that allow the lung to deflate on exhalation, but vital capacity is reduced due to airway collapse during exhalation. This type of change in the lung is called *obstructive lung disease* and is marked by reduced airflow.

18.2.1 Upper Respiratory Tract as a Site of Toxicity

The upper respiratory tract, particularly the nose, has a unique anatomy that performs normal physiologic functions as well as innate defense against inhaled toxins. The nose extends from the nostrils to the pharynx. The nasal cavity is divided longitudinally by a septum into two nasal compartments. In most mammalian species, each nasal cavity is divided into a dorsal, ventral, and middle (lateral) meatus by two turbinate bones, the nasoturbinate and maxilloturbinate. These turbinates project from the dorsolateral and ventrolateral wall of the cavity, respectively. In the posterior portion of the nose, the ethmoid recess contains the ethmoturbinate. The nasal cavity is lined by a vascular mucosa that consists of four distinct types of epithelia. In rodents, these epithelia are (1) the stratified squamous epithelium that line the nasal vestibule and the floor of the ventral meatus in the anterior portion of the nose; (2) the nonciliated, pseudostratified, transitional epithelium that lies between the squamous epithelium and the respiratory epithelium and lines the lateral meatus; (3) the ciliated respiratory epithelium that lines the remainder of the nasal cavity anterior and ventral to the olfactory epithelium; and (4) the olfactory epithelium (neuroepithelium) that lines the dorsal meatus and ethmoturbinates in the caudal portion of the nose.

The olfactory epithelium is composed of basal, neuronal (olfactory), and sustentacular (support) cells (Figure 18.2). The portion of each olfactory cell that responds to the olfactory chemical stimuli is the cilia. The odorant substance first diffuses into the mucus that covers the cilia and then binds to specific receptor proteins in the membrane of each cilium. Next, receptor activation by the odorant activates multiple molecules of the G-protein complex in the olfactory epithelial cell. This in turn activates adenylyl cyclase inside the olfactory cell membrane, which in turn causes formation of a greater multitude of cyclic adenosine monophosphate (cAMP) molecules. Finally, the cAMP molecules trigger the opening of yet an even greater multitude of sodium ion channels. This amplification mechanism accounts for the exquisite sensitivity of the olfactory neurons to extremely small amounts of odorant. The olfactory epithelium is an important target of certain inhaled toxicants. Certain

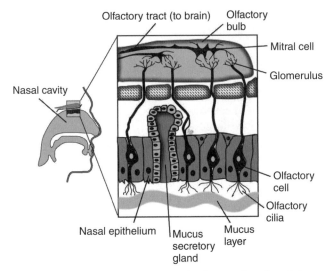

Figure 18.2   Anatomy of the olfactory apparatus in the nasal cavity of the upper respiratory tract. Adapted from *LifeART Illustration Series*. Hagerstown, MD: Lippincott Williams & Wilkins, 1994.

metals, solvents, proteins, and viruses are transported to the brain via transport from the olfactory epithelium to the olfactory tract and exert neurotoxicity.

As air passes through the nose, essential normal respiratory functions are performed by the nasal cavities. These functions are collectively referred to as the air conditioning function of the upper respiratory tract. First, the air is warmed and is almost completely humidified by the extensive mucosal surfaces of the nasal conchae and septum. Second, the nose plays a critical role in trapping the majority of inhaled particles before they reach the lower lung. Large particles (>5 μm in diameter) are stopped from entering the lower respiratory tract by *impaction* of particles on the surfaces of the nasal turbinates (also referred to as nasal conchae), which are folds of osseous tissue lined with ciliated and mucous-producing epithelial cells. Impaction means that particles in the inspired air passing through the nasal passageways collide with many obstructions, including the nasal turbinates and septum, and stick in the mucous lining of the nasal epithelium. The term "turbinates" was derived from the fact that these structures cause turbulence of the inspired air. As air encounters the turbinates, it changes direction of movement. However, particles that are suspended in the air have greater mass than the air itself and continue forward, impacting the surface of the obstruction. Impacted particles are transported by the beating cilia of the nasal epithelium in a unidirectional manner to the pharynx where they are swallowed.

18.2.2   Lower Respiratory Tract as a Site of Toxicity

The lower respiratory tract is composed of the conducting airways (trachea, bronchi, and bronchioles) and the lung parenchyma which consists primarily of gas exchange units (alveoli). The trachea, bronchi, and bronchioles conduct air to the pulmonary

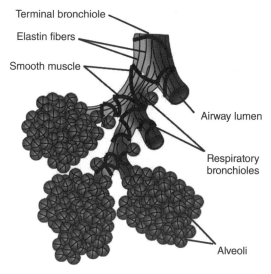

Figure 18.3 Anatomy of the distal airways and functional units (alveoli) in the lower respiratory tract. Adapted from *LifeART Illustration Series*. Hagerstown, MD: Lippincott Williams & Wilkins, 1994.

parenchyma. The trachea extends from the larynx distally where it divides to form the two main bronchi, which enter the right and left lungs. The bronchi bifurcate to form bronchioles and continue to progressively bifurcate in a tree-like fashion to form bronchioles of decreasing diameter. The most distal conducting segment of the tracheobronchial tree is called the terminal bronchiole, which bifurcates to form respiratory bronchioles that contain some alveolar ducts and terminate in clusters of alveolar sacs. This distal region of the lung where airways transition to alveoli is illustrated in Figure 18.3.

18.2.3 Airways of the Lower Respiratory Tract

The trachea and bronchi contain bands of cartilage in the airway wall that prevent collapse. In contrast, the walls of the intrapulmonary airways do not contain cartilage but are supported by flexible elastic fibers and bands of smooth muscle. These smaller airways are susceptible to bronchoconstriction in diseases such as asthma, where allergens provoke a neurogenic response that results in airway smooth muscle contraction. The airways of the lungs are lined with pseudostratified, columnar cells that are predominantly ciliated or mucus-producing serous cells (Figure 18.4). Together, these epithelial cell types contribute to the clearance of particles from the airways through a mechanism termed the "mucociliary escalator." Throughout the tracheobronchial region, mucociliary clearance is an important clearance and defense mechanism for moving inhaled particles up the airway tree where they are expelled from the trachea and swallowed. In humans, submucosal glands in the bronchi also contribute to mucus production, especially during chronic irritation such as cigarette smoking or during chronic respiratory viral infections. In the

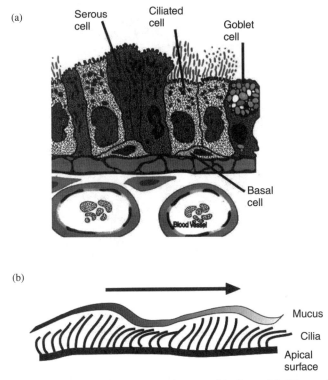

Figure 18.4 Structure and dynamics of the airway epithelium. (a) Illustration of tracheal and bronchial epithelial cell types. (b) The mucociliary escalator wherein epithelial cell cilia move in a low viscosity periciliary layer to propel mucus with their tips.

terminal bronchioles, the population of mucus and serous cells gradually transition to nonciliated cuboidal (Clara) cells. Clara cells have relatively high levels of cytochrome P450 enzymes and may be selectively damaged by toxicants that require metabolic activation by P450s. The epithelium of large and small airways contain triangular-shaped basal cells, which are thought to give rise to ciliated and mucous cells following toxicant injury. All airway epithelial cells are attached basally to a basement membrane or lamina composed of extracellular matrix.

The airway epithelium forms a continuous lining for the conducting airways. The varied composition of the epithelium allows it to perform a variety of functions. First, the epithelium, along with its apical mucus layer and its basal lamina, comprise an important barrier against inhaled toxicants and xenobiotics. The apical surfaces of the airway epithelial cells are connected by tight junctions and effectively provide a barrier that isolates the airway lumen. Second, the various airway epithelial cells produce a mixture of secretions composed of (a) an aqueous "sol" phase containing proteins, lipids, and ions and (b) a gel phase containing mucus. Third, ciliated cells comprise the largest proportion of exposed cells in the normal airway and as discussed above, they propel the mucus within the airway lumen proximally, thereby mediating clearance of inhaled particles and debris (Figure 18.4). Fourth, the airway epithelium exhibits repair following injury, thereby establishing normal airway architecture. Fifth, the airway epithelium can produce a variety of soluble mediators (cytokines, growth factors, proteinase, and lipid mediators) that modulate the responses of other lung cells including airway smooth muscle cells, fibroblasts, immune cells and phagocytes.

18.2.4 Parenchyma of the Lower Respiratory Tract

The primary function of the lung parenchymal region is gas exchange. The major structures are the respiratory bronchioles, alveolar ducts, alveolar sacs, and alveoli. The respiratory bronchioles are lined with cuboidal ciliated and Clara cells, and have alveoli opening into their lumina. Therefore, the respiratory bronchioles function both as conducting passages and as a gas exchange region. A number of mammalian species, including humans, have respiratory bronchioles whereas other species, including rats, have no respiratory bronchioles. In the latter case, the terminal bronchioles end in alveolar ducts. Alveolar ducts are tubular structures whose walls are covered by alveoli. The alveoli open polyhedral chambers lined with thin type I epithelial cells interspersed with cuboidal type II cells. Type I cells comprise 8–11% of the structural cells found in the alveolar region and yet cover 90–95% of the alveolar surface. Their major function is to allow gases to equilibrate across the air–blood barrier and to prevent leakage of fluids across the alveolar wall into the lumen. The type I epithelium is particularly sensitive to damage from a variety of inhaled toxicants due to their large surface area. Moreover, their repair capacity is limited because they have few organelles associated with energy production and macromolecular synthesis.

Type II cells comprise 12–16% of the structural cells in the alveolar region but cover only about 7% of the alveolar surface. They are cuboidal cells with a microvillus surface and unique organelles called lamellar bodies that store surfactant. The major function of type II cells is to secrete surfactant to lower the surface tension in the alveoli, thereby reducing the filling of the alveolar compartment with fluid and alveolar collapse. Type II cells also serve as a progenitor cell for type I cells, which cannot replicate. Therefore, type II cells are critical to alveolar epithelial repair after injury. An interesting third pneumocyte, called the brush cell, is sparsely distributed and appears at alveolar duct bifurcations. Even though this cell type was identified decades ago, essentially nothing is known about its function.

The wall of the alveolus is composed of the alveolar epithelium, a thin layer of collagenous and elastic connective tissue interspersed with fibroblasts (termed the pulmonary interstitium) and a network of capillaries lined by endothelial cells. This distance between the alveolar space and the capillary lumen is known as the air–blood barrier. The air–blood barrier is a multilayered structure approximately $0.4\,\mu m$ in thickness that consists of an alveolar type I cell, alveolar basement membrane, interstitial space, endothelial basement membrane, and a capillary endothelial cell (Figure 18.5). CO_2 and O_2 are exchanged between air and blood by diffusion across the air–blood barrier. A number of factors determine how rapidly a gas will pass through the respiratory membrane. First, the rate of gas diffusion through the membrane is inversely proportional to the thickness of the membrane. In situations where the thickness of the membrane increases, gas exchange between the air and blood is decreased. For example, edema fluid in the alveolar space results in gases requiring passage not only through the cellular membrane but also through a fluid layer. Another example is thickening of the lung interstitial space between the

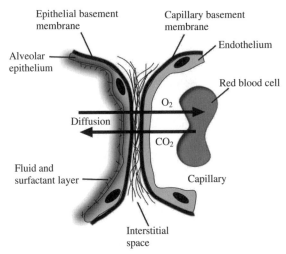

Figure 18.5 Ultrastructure of the alveolar respiratory membrane shown in cross section. Adapted from Guyton, A. C. and J. E. Hall. *Textbook of Medical Physiology*, 10th ed., ed. Philadelphia: Saunders, 2000.

alveolar membrane and the blood capillary membrane during pulmonary fibrogenesis. In general, any factor that increases the thickness of the interstitial space more than two to three times normal can significantly interfere with normal respiratory exchange of gases.

The pulmonary interstitium consists of extracellluar matrix (collagen and elastin) and resident interstitial cells. Although the amount of collagen and elastin in the pulmonary parenchyma is small, these structural proteins are key to normal pulmonary mechanics. Increases or decreases in these proteins lead to impairment, as in pulmonary fibrosis or emphysema. The major resident cell type are fibroblasts, although interstitial macrophages, lymphocytes, plasma cells, and mast cells are also present in the interstitium. Resident interstitial cells comprise about 35% of the structural cells in the alveolar region. However, during an inflammatory response, the relative abundance of these interstitial cells, as well as neutrophils, monocytes, and lymphocytes infiltrating from the blood greatly increase.

Capillary endothelial cells comprise 30–42% of cells in the alveolar region and comprise the walls of the extensive network of blood capillaries in the lung parenchyma. The endothelium forms a continuous, attenuated cell layer that transports respiratory gases, water, and solutes. However, it also forms a barrier to the leakage of excess water and macromolecules into the pulmonary interstitial space. Pulmonary endothelial cells, like type I cells, are vulnerable to injury from inhaled substances and substances in the systemic circulation. Injury to the endothelium results in fluid and protein leakage into the pulmonary interstitium and alveolar spaces, resulting in pulmonary edema.

The lung parenchyma is constantly surveyed by mobile phagocytes, which provide an essential defense against inhaled foreign materials. The most common of these is the alveolar macrophage, which patrol the surface of the alveolar spaces. Their major defense roles are phagocytosis, killing, and clearing of microorganisms, such as bacteria, as well as phagocytosis and clearance of a wide variety of inhaled

Figure 18.6 Macrophages migrate to a site of injury in the wall of a distal airway. Alveolar macrophages normally residing in the alveolar spaces migrate to chemoattractants released by injured epithelial cells. Other leukocytes such as monocytes and neutrophils also respond to chemotactic molecules and migrate from the blood across the pulmonary interstitium.

particulate matter. After engulfing a foreign particle, clearance may be accomplished via the *mucociliary escalator*. Particle-laden macrophages migrate from the alveolar spaces to the distal airways and are taken up the airway by the unidirectional beating of cilia on the airway epithelium, which moves the macrophages and their cargo up and out of the lungs where they are expelled into the pharynx and swallowed. In addition to the mucociliary escalator, macrophages may also clear engulfed foreign material by migrating into the pulmonary interstitium and into the lymphatics. Phagocytosis triggers the release of cytokines and chemokines, growth factors, proteolytic enzymes, and reactive oxygen species. These mediators recruit and activate other cells that participate either in the resolution of an inflammatory response or structural alterations in lung tissue that lead to a pathological disease outcome. The macrophage population in the lung is replenished by recruitment of bone marrow-derived monocytes via the bloodstream (Figure 18.6). Once in the lung, monocytes proliferate in response to specific growth factors and mature into alveolar macrophages under the direction of other specific cytokines and differentiation factors. Other subpopulations of macrophages that are less recognized are the pulmonary interstitial macrophage that resides beneath the epithelial lining and the intravascular macrophage that is present in humans and some other mammals, but not in rodents.

18.2.5 Circulatory, Lymphatic, and Nervous System of the Lung

The respiratory system communicates with other organ system primarily through direct connections with the circulatory, lymphatic, and nervous systems. The *circulatory system* bridges the closely adjacent heart and lungs and the entire cardiac output of the heart enters the lung to be replenished with oxygen. Oxygen-poor blood from the right ventricle travels through the pulmonary arteries to supply capillary beds of the respiratory bronchioles, alveolar ducts, and alveoli where gas is exchanged. Oxygenated blood then returns to the heart via pulmonary venules in the lung parenchyma that merge into pulmonary veins that feed the left atrium. A second arterial system, the bronchial system, supplies oxygenated blood from the left ventricle of the heart through the aorta and the bronchial arteries to the large airways of the lung, the pleura, and large pulmonary vessels.

The *pulmonary lymphatic system* is a vascular network that serves to remove excess fluid from the connective tissue spaces of the lung parenchyma. The lymphatic system is also important in clearing particulate material from the lung to the lymph nodes. The lymphatic system in the lung is divided into superficial and deep portions, but these two portions are connected. The superficial portion is located in the connective tissue of the pleural lining of the lung. The deep portion is in the connective tissue surrounding the bronchovascular tree. The two portions connect in the interlobular septa. The lymphatic vessels are structurally similar to thin-walled veins. The presence of valves in the lymphatic vessels and the movement of the lung during respiration promote the flow of lymph from the periphery and pleura toward the hilus. Afferent lymphatics from the lung drain into the tracheobronchial lymph nodes. Lymph from tracheobronchial and hilar nodes drain into the thoracic, right, and left lymphatic ducts and from these ducts drain into the systemic venous system.

The respiratory tract contains both sensory (afferent) and motor (efferent) innervation. Both the parasympathetic and sympathetic portions of the autonomic *nervous system* provide the motor innervation. Preganglionic parasympathetic fibers descend in the vagus nerves to ganglia located around airways and blood vessels. The postganglionic fibers innervate the smooth muscle of the airways and blood vessels, bronchial glands, and epithelial mucous cells. In general, the same structures are also innervated by postganglionic fibers from the sympathetic ganglia. Vagal stimulation causes airway constriction, dilation of the pulmonary circulation, and increased glandular secretion. Conversely, sympathetic nerve stimulation causes bronchial relaxation, constriction of pulmonary blood vessels, and inhibition of glandular secretion. Sensory receptors that respond to irritants or mechanical stress are located throughout the respiratory tract. Stimulation of these receptors leads to reflex responses such as stimulation of nasal receptors to cause sneezing. Three principal vagal sensory reflexes and their corresponding receptors are known as (1) bronchopulmonary stretch receptors, (2) irritant receptors, and (3) C-fiber receptors. Stretch receptors are associated with smooth muscle of the trachea and bronchi, are stimulated by lung inflation, and normally function to terminate inspiration. Rapidly adapting irritant receptors are located in the epithelium of extrapulmonary and, to a lesser extent, intrapulmonary bronchi. They respond to a variety of stimuli, including inhalation of irritant gases and mechanical stimulation of the airways, to cause bronchoconstriction, cough, and increased mucus secretion. C-fiber receptors are located both in the parenchyma and along conducting airways. Bronchial C-fibers along conducting airways respond to stimuli near the bronchial arterial system and when stimulated cause airway constriction. Pulmonary C-fibers may contribute to the sensation of dyspnea that accompanies pulmonary edema, pneumonia, and inhalation of noxious gases. Stimulation of both pulmonary and bronchial C-fibers causes a reflex increase in airway secretion.

18.3 TOXICANT-INDUCED LUNG INJURY, REMODELING, AND REPAIR

The respiratory tract is exposed to many environmental factors (particles, gases, infectious microbial agents) and has evolved sophisticated defense and repair systems. The response to injury by foreign agents that enter the respiratory tract involves *host recognition* of the toxic insult followed by *acute inflammation* and then *tissue remodeling* that can result in one of two general outcomes. First, an inflammatory response may resolve and lead to a tissue repair process where normal respiratory architecture and function are restored. Alternatively, an inflammatory response may not resolve but instead may progress to an abnormal tissue remodeling response leading to diseases such as fibrosis and emphysema. This general concept is illustrated in Figure 18.7.

A number of factors determine whether tissue in the lung parenchyma is successfully repaired after injury or whether an inflammatory response progresses to a pathologic outcome. As mentioned previously, the alveolar region is especially vulnerable to damage due to the delicate architecture of the type I epithelium and blood capillary endothelial membranes. An appropriate balance of catabolic and anabolic activity involving cytokines, growth factors, lipid mediators, and proteinases is required for tissue repair. Following injury, damaged cells are triggered to undergo apoptosis by specific cytokine signals (e.g., tumor necrosis factor (TNF)-α) released

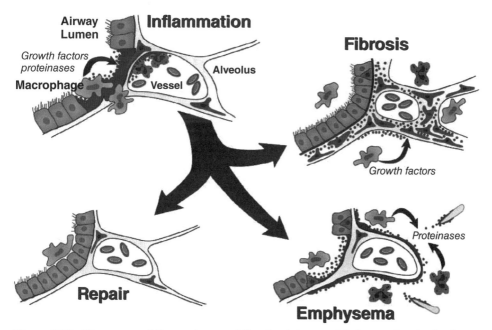

Figure 18.7 Tissue remodeling outcomes following injury and inflammation in the lower respiratory tract that result in repair or disease. (See text for details.)

from phagocytes or the damaged cells themselves release apoptotic factors in an autocrine manner. At the same time, extracellular debris is degraded by specific proteinases (e.g., elastases, collagenases) released by infiltrating neutrophils and mononuclear cells. Resident macrophages then mediate the clearance of the resulting cellular and extracellular debris. The rebuilding process involves a precise balance of growth and differentiation factors to stimulate repopulation of the epithelium and restore connective tissue, vascular tissue, and nerves. An over-exuberant production of growth factors (e.g., transforming growth factor beta (TGF-β), connective tissue growth factor (CTGF), platelet-derived growth factor (PDGF)) may lead to a fibrotic response characterized by increased fibroblasts and collagen. Alternatively, an imbalance in proteinase/anti-proteinase systems may lead to a progressive degradation of structural proteins (e.g., elastin) in the alveolar wall and may cause emphysema.

18.3.1   Oxidative Stress and Lung Injury

There is considerable evidence that links oxidants to the development of a number of human lung diseases. Oxidants can be generated by endogenous mechanisms involving nicotinamide adenine dinucleotide phosphate (NADPH) oxidase systems in phagocytic cells, as well as xanthine oxidase. Lung oxidants can also be increased from exogenous sources, such as inhaled air pollution (gases or particles) and cigarette smoke. The intake of oxygen through the lungs is required for aerobic life, and yet conversion of oxygen to reactive oxygen species (ROS) can have profound detrimental effects on tissues of the respiratory tract. Oxygen is converted to the oxidizing agent superoxide anion (O_2^-) by cellular NADPH oxidase systems (principally in phagocytes) and by xanthine oxidase (Figure 18.8). Hydrogen peroxide (H_2O_2) is formed by the further oxidation of O_2^- by superoxide dismutase (SOD):

$$2O_2^- + 2H^+ \rightarrow H_2O_2 + O_2$$

Three types of SODs occur: (1) extracellular (EC)-SOD, (2) manganese SOD found in the mitochondria, and (3) copper-zinc SOD found in the cytosol and nucleus. H_2O_2 can be converted to the highly toxic hydroxyl radical (•OH) via the iron-catalyzed Fenton reaction:

$$Fe^{2+} + H_2O_2 \rightarrow Fe^{3+} + \cdot OH + OH^-$$

Figure 18.8   Conversion of superoxide anion (O_2^-) to hydroxyl radical (•OH) in the presence of iron, or to hydrogen peroxide (H_2O_2) by superoxide dismutase (SOD), which is converted to hypochlorous acid by myeloperoxidase (MPO).

Moreover, H_2O_2 in the presence of O_2^- and a divalent metal can also produce $\cdot$OH via the iron-catalyzed Haber–Weiss reaction:

$$Fe^{2+} + H_2O_2 \rightarrow Fe^{3+} + \cdot OH + OH^-$$

$$Fe^{3+} + O_2^- \rightarrow Fe^{2+} + O_2$$

$$O_2^- + H_2O_2 \rightarrow \cdot OH + OH^-$$

At least three different sources of ROS contribute to oxidative stress in the respiratory system. First, macrophage-mediated phagocytosis involves the release of O_2^-, which is primarily generated by the NADPH oxidase system in these cells. This "respiratory burst" likely evolved as a microbial killing mechanism. However, the phagocytosis of a variety of particles and fibers also activate macrophages to undergo a respiratory burst. Some of these agents (e.g., asbestos fibers) are not easily cleared by macrophages and therefore stimulate chronic activation of phagocytes. Second, particles containing metal oxides generate an additional burden of reactive oxygen species via the Fenton and Haber–Weiss reactions shown above. Third, ozone and nitrogen dioxide gases in the environment are reactive species that further contribute to oxidative stress. Finally, cigarette smoke contains a multitude of oxidizing compounds, including nitrogen oxides, quinones, semiquinone radical, and hydroquinone moieties. The combined burden of ROS from these sources results in pulmonary cell damage by lipid peroxidation of the polyunsaturated fatty acids in cell membranes.

While the majority of toxic agents that generate ROS enter via inhalation, some chemicals cause injury via entry through the circulation. Ingestion of the herbicide paraquat causes pulmonary fibrosis by accumulating in the epithelium of the lung and causes oxidant formation as a result of redox cycling (Figure 18.9). The reduction of paraquat (PQ^{2+}) by NADPH is dependent on cytochrome P450 reductase present in the endoplasmic reticulum and mitochondria. The intravenous administration of the chemotherapeutic drug bleomycin also causes lung fibrosis via redox cycling, although it is strictly Fe-dependent. Because bleomycin binds avidly to DNA, much of the site-directed $\cdot$OH formation leads to DNA damage. Finally, ionizing radiation directly produces $\cdot$OH without the necessity for a transition metal.

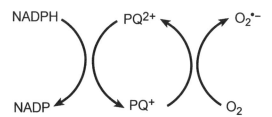

Figure 18.9 Formation of superoxide anion (O_2^-) by the herbicide paraquat (PQ^{2+}) via redox cycling.

18.3.2   Antioxidant Mechanisms in the Lung

There are a number of enzymatic and nonenzymatic antioxidant defense mechanisms that counterbalance the deleterious effects of oxidative stress in the lung. Several general concepts have emerged regarding the action of antioxidants. First, there is specificity in the scavenging ability of various antioxidants. Nearly all cells contain a number of enzymatic scavengers, including *superoxide dismutase* (SOD), *catalase*, and *glutathione redox systems* which degrade specific oxidants in specific ways. For example, SOD converts superoxide anion to H_2O_2, whereas catalase specifically degrades H_2O_2. Second, there is usually a selective localization of antioxidants within cells. Manganese SOD is localized in mitochondria, whereas copper-zinc SOD is primarily located in the cytoplasm. Third, antioxidant levels and activities are dynamic. Antioxidants can be inactivated by oxidants. For example, H_2O_2 can inactivate SOD and superoxide anion can inactivate catalase. Moreover, oxidative stress can induce the transcription of genes that encode oxidant-generating systems such as NADPH oxidase or alternatively induce antioxidant enzymes as a protective feedback mechanism. Catalase, SOD, and enzymes of the glutathione redox cycle are primary intracellular antioxidant defense mechanisms that eliminate oxygen radicals and hydroperoxides that may pose a threat to the cell by oxidizing cellular structures. Catalase is located primarily in peroxisomes which contain many of the enzymes that generate H_2O_2 in aerobic cells. Catalase reduces H_2O_2 to water and oxygen. In the lung, catalase is present mainly in type II cells, Clara cells, and macrophages.

The glutathione redox cycle is a central mechanism for reduction of intracellular fatty acid hydroperoxides and complements catalase as a reducing system for H_2O_2. Glutathione metabolism also degrades large molecule lipid peroxides formed by free radical action on polyunsaturated lipid membranes. The key enzyme in the glutathione redox cycle responsible for the reduction of hydroperoxides is glutathione peroxidase. Nonstressed cells contain a high intracellular ratio of reduced glutathione (GSH) to oxidized glutathione (GSSG). This high GSH/GSSG ratio ensures availability of GSH for reduction of hydroperoxides via the glutathione redox cycle. Regeneration of GSH requires nicotinamide-adenine dinucleotide phosphate (NADP)-reducing equivalents that are supplied through the glucose-6-phosphate dehydrogenase (G6PD) activity in the hexose monophosphate shunt. NADPH is an important antioxidant molecule as it is the cofactor for the regeneration of GSH from its oxidized form, GSSG. The NADPH, in turn, is regenerated by the enzyme, G6PD as part of the hexose monophosphate shunt pathway of energy metabolism.

18.3.3   Respiratory Tract Injury from Inhaled Particles and Fibers

A variety of naturally occurring and man-made particles and fibers pose a threat to the respiratory tract, including air pollution particulates, allergens such as pollen, soot from the industrial burning of oil, diesel exhaust particles, metal oxides, particles in cigarette smoke, and asbestos fibers. Many of these, including air pollution particulates and cigarette smoke, are complex mixtures of organic and inorganic chemical substances. Several different types of particles and fibers will be discussed below to highlight the heterogeneous nature of particle exposure in environmental or occupational settings.

The inhalation of urban air particles has been associated with increased morbidity and mortality principally due to the physiologic impact on the pulmonary and cardiovascular systems. Adverse respiratory effects in exposed human populations include increased asthmatic episodes and an increase in the prevalence of chronic bronchitis. Moreover, chronic exposure to air pollution particles could contribute to the increasing prevalence of COPD. A mixture of organic and inorganic agents contribute to the composition of air pollution particles, including transition metals released during the burning of petrochemicals, polycyclic aromatic hydrocarbons derived from diesel exhaust, and endotoxins from bacterial sources. Many of these constituents are known to stimulate a variety of intracellular signaling pathways that mediate cellular stress responses leading to the pathologic phenotypes that characterize airway remodeling.

Many of the pathophysiologic effects of inhaled particles are due to oxidative stress. The oxidative potential of air pollution particles is generally attributed to transition metals (e.g., zinc, copper, vanadium, iron) which can induce generation of ROS either via Fenton-like reactions, or by stimulating an oxidative burst in leukocytes that engulf particles via phagocytosis. Particles from a number of sources have been shown to induce oxidant generation that is associated with metal content of the sample.

18.3.4 Particle and Fiber Deposition and Clearance

While chemical composition is important in determining the toxicity of particles and fibers, it is equally or more important to determine where a particle or fiber will deposit in the respiratory tract and how long it will stay there. The mechanism of deposition is determined by the physical (size, shape, and density) and chemical (hygroscopicity and charge) characteristics of the inhaled particles. Particle deposition is also affected by biological factors inherent to the exposed individual such as breathing pattern (volume and rate), route of breathing (mouth vs. nose), and the anatomy of the airways.

Particle deposition in the respiratory tract occurs primarily by three mechanisms: *impaction*, *sedimentation*, and *diffusion*. Impaction and sedimentation depend on the particle's *aerodynamic diameter*. The aerodynamic diameter of a particle is the size parameter of greatest importance for deposition considerations. It is equal to the diameter of a unit-density sphere having the same terminal settling velocity as the particle in question. *Impaction* is the collision of a moving particle with a static structure. It occurs when an inhaled particle has too much momentum to change course with the directional change in airflow and as a result impacts against the airway surface. Most inhaled particles greater than 5 μm in aerodynamic diameter impact on the surfaces of the pharyngeal tracheobronchial regions of the respiratory tract and do not reach the distal lung. *Sedimentation* occurs by the gravitational settling of particles on a respiratory tract surface. Diffusion takes place when a particle reaches an airway surface by random *Brownian movement*. This is an important mechanism for particles with diameters in the nanometer range that reach the distal lung (terminal bronchioles and alveoli) where there is almost no airflow. The combined processes of diffusive and sedimentary deposition are important for particles in the range 0.1–1 μm. Impaction and sedimentation predominate above and diffusion predominates below this range (Figure 18.10). Air pollution particulate

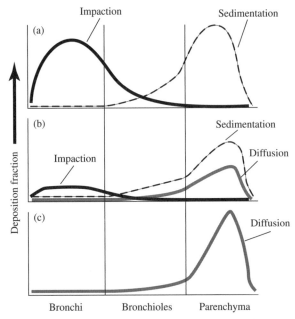

Figure 18.10 The deposition site of particles in the lung depends on particle size. Shown are predicted deposition patterns of (a) 5 μm particles, (b) 1 μm particles, and (c) 0.1 μm particles. Adapted from Bennett, W. D. and J. S Brown. Particulate dosimetry in the respiratory tract. In *Air Pollutants and the Respiratory Tract*, 2nd ed., eds. W. M Foster and D. L. Costa, pp. 21–73. New York: Taylor & Francis, 2005.

matter has been characterized into three categories: (1) ultrafine (<0.1 μm diameter), also referred to as nanosized particulates, (2) fine particles (ranging from 0.1 to 2.5 μm diameter), and course particles (>2.5 μm diameter). Epidemiological studies suggest that ultrafine and fine particles are better correlated with adverse health effects when compared to course particles.

Other mechanisms of deposition are *interception* and *electrostatic charge*. *Interception* is most important for fibers and occurs when an inhaled fiber contacts an airway wall or when long, thin fibers intercept the airway bifurcations. The likelihood of interception is enhanced with increasing fiber length. Such is the case for chrysotile asbestos fibers, which primarily deposit at alveolar duct bifurcations in the distal lung. While spherical particles with a diameter greater than 5 μm do not reach this region of the lung due to impaction in the upper respiratory tract, asbestos fibers with a length exceeding 20 μm can reach the alveolar region of the lung and persist for months or even years due to the inability of macrophages to effectively clear such long, thin structures. Therefore, for particles, decreasing diameter correlates with increasing toxicity because (1) smaller particles in the submicron or nanometer range reach the alveolar region (distal lung), and (2) an equivalent mass of smaller particles have a greater surface area per unit mass and therefore a greater potential to generate ROS and oxidative stress (Figure 18.11). For fibers, increasing

Figure 18.11   Size and shape characteristics of particles that determine deposition, clearance and toxicity in the respiratory tract. In general, smaller particles are more toxic whereas longer fibers are more toxic. A lung macrophage is shown as a size reference. (See text for details.)

length corresponds to greater toxicity given that the fiber is respirable (i.e., one that can be inhaled and be deposited in the alveolar region of the lung) because longer fibers are more difficult to clear from the lung and persist in the interstitial tissues to cause damage.

18.3.5   Respiratory Tract Injury from Gases and Vapors

The respiratory toxicity of gases and vapors is determined by several different physical and *chemical* properties. These factors include (1) chemical *dose*, (2) *water solubility* (hydrophilicity vs. lipophilicity), and (3) *chemical reactivity*.

The *dose* of the toxic substance to which an individual is exposed generally determines severity of injury. For example, the occupational exposure to gases (e.g., chlorine, ammonia, HCl) can be divided into three levels of dose exposure that directly correlate to the degree of airway injury and severity of symptoms. Low exposures to these gases cause sneezing, rhinitis, and sore throat. Repeated low exposures or mild exposure causes persistent cough and bronchitis. Accidental exposure to high concentrations of these gases cause laryngeal edema, acute respiratory distress syndrome, and possibly death. Moreover, high dose exposure to toxic gases and vapors can cause substantial airway epithelial cell damage in the respiratory tract, resulting in a reactive airways dysfunction syndrome (RADS) that is characterized by denuded epithelium and airway luminal fibroproliferative lesions.

The *solubility* of an inhaled substance influences the deposition pattern in the respiratory tract and the site of injury. Gases such as formaldehyde, hydrochloric acid, and sulfur dioxide are taken up by the mucosal surfaces of the upper

TABLE 18.1 Water Solubility, Irritant Classification, and Site of Injury for Selected Highly Reactive Gases

Gas/Vapor	Water Solubility	Irritant Class	Site of Injury
Chlorine	High	Sensory	Nasal, trachea
Formaldehyde	High	Sensory	Nasal, trachea
HCL	High	Sensory	Nasal, trachea
Ammonia	High	Sensory, bronchoconstrictor	Nasal, trachea
Phosgene	Low	Pulmonary	Terminal bronchioles
Ozone	Low	Pulmonary	Terminal bronchioles
Nitrogen dioxide	Low	Pulmonary	Terminal bronchioles
Sulfur dioxide	High	Sensory, bronchoconstrictor	Nasal, trachea

respiratory tract and will exert most of their toxicity in the nasal region. This is because a highly water-soluble chemical will tend to leave the air in the respiratory tract and enter the mucus lining. The relative amount of a chemical in two compartments (e.g., air and mucus lining) at equilibrium is called the *partition coefficient* for that chemical in those two compartments. Gases and vapors with low water solubility such as nitrogen dioxide and ozone tend to deposit farther down the respiratory tract and cause damage to terminal and respiratory bronchioles. Selected water-soluble and insoluble gases are listed in Table 18.1.

The *reactivity* of an inhaled chemical refers to a more unstable conformation (high-energy state) such as formaldehyde that can easily bond with other molecules. Also, chemical reactivity often means that the reactive substance has the ability to generate reactive oxygen or reactive nitrogen species as a consequence of its reaction with other molecules.

A number of gases also cause *irritant responses* in the respiratory tract. These can be classified as *sensory irritants*, *pulmonary irritants*, or *bronchoconstrictors*. Sensory irritants stimulate the trigeminal nerve endings in the upper respiratory tract, leading to a burning sensation in the nose. Examples of sensory irritants are ammonia, acrolein, formaldehyde, and sulfur dioxide. A *pulmonary irritant* is one that stimulates sensory receptors (C-fiber receptors) in the airways, thereby increasing respiratory rate and causing dyspnea (difficulty in breathing) or rapid, shallow breathing. Examples of pulmonary irritants are phosgene, nitrogen dioxide, ozone, and sulfuric acid mist. Bronchoconstrictors stimulate nerve endings in airways to cause contraction of airway smooth muscle, thereby narrowing the airway lumen and increasing resistance to airflow in conducting airways. Examples of bronchoconstrictors are sulfur dioxide and ammonia. In addition to gases and vapors, a number of particles and allergens stimulate bronchoconstriction in individuals with asthma.

18.4 OCCUPATIONAL AND ENVIRONMENTAL LUNG DISEASES

While the respiratory system is well-equipped to defend against exposure to a vast array of toxic substances, the intricate cellular and molecular mechanisms designed to repair injured lung tissues often fail, resulting in a number of chronic lung diseases, including cancer, fibrosis, asthma, hypersensitivity pneumonitis (HP), and COPD, which is a combination of bronchitis and emphysema.

18.4.1 Pulmonary Fibrosis

Fibrotic interstitial lung diseases are restrictive disorders that involve the proliferation of myofibroblasts in the interstitium, including the alveolar walls and perivascular and peribronchial tissues. These include the chronic progressive disorders such as idiopathic pulmonary fibrosis (fibrotic disorders of unknown etiology) and a number of fibrotic diseases that occur as a result of environmental or occupational exposures, including asbestosis and silicosis. A chronic fibrotic disease caused by exposure to a specific airborne inorganic dust (e.g., asbestos) is often referred to as a *pneumoconiosis*. Fibrosis can also occur following acute lung injury, as in the adult respiratory distress syndrome, where destruction of the epithelium and endothelium destroys the permeability barrier and permits flooding of the airspaces with proteinaceous edema and the infiltration of myofibroblasts. Myofibroblasts may also infiltrate into small airways of the lung, as occurs during the progression of obliterative bronchiolitis following tissue rejection after lung transplantation. Finally, it is becoming increasingly clear that more subtle fibrotic reactions are associated with airway remodeling in diseases such as asthma, chronic bronchitis, and COPD.

18.4.2 Asthma

Asthma is an *obstructive* allergic airways disease that is hallmarked by *acute bronchospasm* of airways called an "asthma attack," but also features *chronic airway inflammation and remodeling* that occurs over a period of years. Both the acute bronchospasm and the chronic airway remodeling process contribute to the obstructive nature of this disease. Asthma is common. Over the past few decades, asthma prevalence has increased and affects up to 10% of the population in developed countries. The reason for the increase in asthma is not entirely clear, but children are primarily affected. Some cases of asthma in children resolve, while some individuals with asthma develop irreversible changes in lung function due to the chronic airway remodeling process. A variety of allergens cause asthma and act as sensitizing agents (Table 18.2). High-molecular weight allergens from plants, animals, bacteria, or mold trigger an IgE-mediated immune response. Most cases of asthma are caused by indoor allergens, particularly components of chitin exoskeleton from house dust mite or cockroaches. Low-molecular weight allergens such as metals, anhydrides, and penicillin generally act as "haptens" (incomplete allergen) and must combine with serum proteins to elicit an allergic response.

Asthma features a chronic airway remodeling response that is characterized by: (1) eosinophilic inflammation, (2) airway smooth muscle thickening, (3) mucus cell hyperplasia and mucus hypersecretion, and (4) subepithelial fibrosis. Allergic diseases, including asthma, are thought to result from a dysregulated immune response to commonly encountered antigens in genetically predisposed individuals. Immunological research into the mechanisms of allergy has identified cytokine production by T-helper 2 (Th2) effector lymphocytes as being critical for

TABLE 18.2   Agents that Cause or Exacerbate Asthma

Agent
Allergens causing asthma
Cockroach
House dust mite
Cat
Plant debris
Endotoxin
Molds
Exacerbation of asthma
Environmental tobacco smoke
Sulfur dioxide
Nitrogen dioxide
Ozone
Particulates

orchestrating allergic inflammation rich in eosinophils. Upon recognition of their cognate antigen, Th2 lymphocytes produce cytokines that regulate IgE synthesis, growth and activation of eosinophils and mast cells, and expression of endothelial cell adhesion molecules. The first step in the allergic immune response is the uptake and presentation of allergen by antigen-presenting cells called dendritic cells. Macrophages and B lymphocytes may also serve as antigen-presenting cells. Following recognition and uptake, dendritic cells migrate to the T-cell-rich area of draining lymph nodes, display an array of antigen-derived peptides on the surface of major histocompatibility complex (MHC) molecules, and acquire the cellular specialization to select and activate naïve antigen-specific T cells. Allergen targeting to the dendritic cells occurs via membrane-bound IgE. Dendritic cells interact with many cell types, including mast cells, epithelial cells, and fibroblasts. Mediators released by these cells can activate the dendritic cells so that it is induced to mature and attract memory Th2 cells through release of Th2-selective chemokines. Mature effector Th2 cells play a central role in asthma pathogenesis by releasing cytokines (e.g., IL-13) that stimulate eosinophil recruitment, smooth muscle cell cytokine and chemokine production, and goblet cell hyperplasia.

18.4.3   Hypersensitivity Pneumonitis (HP)

Hypersensitivity pneumonitis (HP) is an exaggerated adaptive immune response that occurs in susceptible individuals. Unlike asthma, which affects the airways of the lung, HP is a disease of the lower lung (terminal bronchioles and alveoli) caused by the inhalation of allergens that elicit lymphocytic inflammation. The pathogenesis of HP involves three major steps. First, there is an *acute phase* that features lymphocyte and macrophage accumulation and activation. CD8+ cytotoxic lymphocytes and monocytes accumulate widely in the alveolar spaces. After inhalation, soluble antigens bind IgG antibody, and immune complex initiate the complement cascade, and the resulting C5 activates macrophages. Activated resident macrophages in the lung then secrete chemokines that attract circulating

TABLE 18.3 Causative Agents of Hypersensitivity Pneumonitis

Agent	Source	Disease
Microbes		
Thermophilic actinomycetes	Moldy plant materials	Farmer's lung
Saccharopolyspora rectivirgula	Moldy hay	Farmer's lung
Thermoactinomycetes vulgaris	Compost	Mushroom worker's lung
Aspergillus species	Tobacco mold	Tobacco worker's lung
Penicillium chrysogenum	Moldy wood dust	Woodworker's lung
Bacteria and fungi	Metalworking fluids	Machine operator's lung
Animals		
Avian proteins	Bird excreta, feathers	Bird fancier's lung
Animal fur protein	Animal fur	Furrier's lung
Plants		
Soybean	Soybean hulls	Soybean worker's lung
Coffee	Coffee bean dust	Coffee worker's lung
Chemicals		
Isocyanates	Paints, plastics	Paint refinisher's lung
Anhydrides	Plastics	Chemical worker's lung
Pyrethrum	Insecticides	Insecticide lung
Metals		
Cobalt	—	Hard metal lung disease
Beryllium	—	Chronic beryllium disease

Source: Patel et al. *J. Allergy Clin. Immunol.* **108**:661–670, 2001.

T lymphocytes and monocytes. In contrast to IgE-mediated allergic responses (i.e., asthma), the IgG-mediated response in HP does not feature eosinophilic inflammation. Second, a *subacute phase* occurs when the monocytes mature into foamy macrophages and develop into *granulomas* widely dispersed throughout the lung. Third, a *chronic phase* begins as the inflammatory cells produce growth factors such as TGF-β that, over time, stimulate fibroblasts and myofibroblasts to proliferate and deposit collagen to form scar tissue. The end result is interstitial fibrosis. A variety of agents cause HP and many of these agents are encountered occupationally (Table 18.3). Most of the causative allergens have been recognized in a wide variety of occupations. Consequently, once the problem has been identified, the exposure can be controlled and the disease prevented. As a result of this reduction in occupational exposures, the disease is now less common than it was 20 years ago. It is also important to recognize that many existing exposures also occur at home. They are especially associated with pet birds, contaminated humidifiers, and heavy concentrations of indoor molds.

18.4.4 COPD

COPD is a progressive disease that encompasses both *chronic bronchitis* and *emphysema*. The major cause of COPD is cigarette smoking. Chronic bronchitis is defined clinically as the presence of chronic productive cough for 3 months in each of 2 successive years in a person in whom other causes of cough have been excluded. Emphysema is defined anatomically by enlargement of the airspaces distal to the terminal bronchioles, accompanied by destruction of alveolar walls. Reduced airflow in COPD is caused by increased resistance to airflow due to inflammation, fibrosis, goblet cell metaplasia, and smooth muscle cell hypertrophy in small airways. A major factor in reducing airway function is loss of elastic recoil due to inflammation and the progressive loss of elastin-dependent attachment of alveoli to bronchioles. The progressive destruction of elastin in the distal lung is the major cause for loss of alveolar walls in emphysema. The loss of elastin is thought to be due to an imbalance between proteinases released from inflammatory cells and anti-proteinases in the lung that normally serve to protect against excessive proteolytic degradation. This is referred to as the *proteinase-anti-proteinase hypothesis* for the development of emphysema. This hypothesis is supported by the following: (1) individuals with a genetic deficiency in α1-antitrypsin (the major inhibitor of neutrophil elastase) are predisposed to the emphysema whether they smoke or not, and (2) emphysema can be induced experimentally in animals by the intratracheal instillation of neutrophil elastase. Moreover, ROS released by cigarette smoke inactivate the inhibitor of elastase (α1-antitrypsin) by oxidizing amino acids within the active site of the enzyme. While elastase appears to be a major proteinase involved in emphysema, other proteinases may be contributory, including cathepsins, matrix metalloproteinases, and collagenases. In addition to α1-antitrypsin, other anti-proteinases include secretory leukoproteinase inhibitor (SLPI), tissue inhibitors of metalloproteinases (TIMPs), and α2-macroglobulin.

18.4.5 Lung Cancer

Lung cancer is the leading cause of death from cancer in the United States. The major risk factor for lung cancer is cigarette smoke. A higher incidence of lung cancer in men versus women is directly correlated with a higher rate of smoking in men versus women. However, an increase in smoking women has led to a doubling in lung cancer incidence in women over the past 20 years, whereas a decline in smoking among men has led to a slight decline in lung cancer incidence in men over the same period. In addition to cigarette smoke, occupational exposures to a variety of agents, including arsenic, asbestos, polycyclic aromatic hydrocarbons, chromium, and nickel are also associated with increased incidence of lung cancer. The major histopathologic types of human lung cancer are squamous cell carcinoma (29%), adenocarcinoma (32%), small cell carcinoma (18%), and large cell carcinoma (9%). Adenocarcinomas and large cell carcinomas tend to occur more in the peripheral lung, whereas squamous cell carcinomas and small cell carcinomas are more likely to occur in the central lung adjacent to large airways. Mesothelioma is an unusual type of lung cancer that develops along the pleural lining of the lung as a result of asbestos exposure. The major histopathologic types of lung cancer found in rats and mice include adenomas, adenocarcinomas, and squamous cell carcinomas. The mouse has proven to be a suitable model for the study of lung cancer progression and most agents that cause cancer in humans also cause cancer in mice (Table 18.4).

TABLE 18.4   Pulmonary Tumor Response of Laboratory Rodents to Inhalation of Known Human Pulmonary Carcinogens

Agent	Human	Rat	Mouse
Chemicals			
Arsenic	+[a]	ND	ND
Asbestos	+	+	±
Beryllium	+	+	±
Chromium	+	±[b]	ND
Coal tar	+	+	+
Mustard gas	+	ND[c]	±
Nickel	+	+	±
Soots	+	+	ND
Vinyl chloride	+	+	+
Environmental agents			
Tobacco smoke	+	+	±
Radon	+	+	—[d]

Source: From Hahn, F. F. Lung carcinogenesis. In *Carcinogenicity*, ed. K. T. Kitchum. New York: Marcel Dekker, 1998.

[a]+, positive; [b]±, limited data; [c]ND, no data; [d]−, negative.

BIBLIOGRAPHY AND SUGGESTED READING

Bennett, W. D. and J. S. Brown. Particulate dosimetry in the respiratory tract. In *Air Pollutants and the Respiratory Tract*, 2nd ed., eds. W. M. Foster and D. L. Costa, pp. 21–73. New York: Taylor & Francis, 2005.

Crystal, R. G. and J. B. West, eds. *The Lung: Scientific Foundations*. New York: Raven Press, 1991.

Gardner, D. E., J. D. Crapo, and R. O. McClellan, eds. *Toxicology of the Lung*, 3rd ed. New York: Raven Press, 1999.

Guyton, A. C. and J. E. Hall, eds. *Textbook of Medical Physiology*, 9th ed. Philadelphia: Saunders, 2000.

Hahn, F. F. Lung carcinogenesis. In *Carcinogenicity*, ed. K. T. Kitchum. New York: Marcel Dekker, 1998.

LifeART Illustration Series. Hagerstown, MD: Lippincott Williams & Wilkins, 1994.

Patel, A. M., Ryu J. H. and C. E. Reed. *J. Allergy Clin. Immunol.* **108**(5): 661–670, 2001.

SAMPLE QUESTIONS

1. The lung is equipped with unique cellular defenses that operate to remove foreign inhaled materials such as particles and fibers. What are two mechanisms that act coordinately to remove inhaled particles from the lungs?

2. Give two reasons why the lung is a major site of toxicity from environmental exposures.

3. What are the three major factors that determine the toxicity of gases and vapors?

4. An antioxidant enzyme that converts hydrogen peroxide to water and oxygen is: (a) superoxide dismutase, (b) catalase, (c) elastase, (d) all of the above.

5. Which mechanism is most important for the deposition of inhaled particles in the nanometer size range? (a) Impaction, (b) Sedimentation, (c) Diffusion, (d) all of the above.

6. Which of the following lung diseases is caused by repeated exposures to agents such as *thermophilic actinomycetes* or beryllium that result in proliferation of Th1 lymphocytes? (a) Asthma, (b) Hypersensitivity pneumonitis, (c) Cystic fibrosis, (d) Chronic obstructive pulmonary disease.

Immune System*

MARYJANE K. SELGRADE

19.1 INTRODUCTION

A properly functioning immune system is essential to good health. It defends the body against infectious agents and, in some cases, tumor cells. Individuals with immune deficiencies resulting from genetic defects, diseases (e.g., AIDS, leukemia), or drug therapies are more susceptible to infections and certain types of cancer, the consequences of which can be life-threatening. On the other hand, the immune system may react to foreign substances that would otherwise be relatively inno-cuous, such as certain chemicals, pollens, or house dust. The resulting allergic reactions can produce an array of pathologies ranging from skin rashes and rhinitis to more life-threatening asthmatic and anaphylactic reactions. A crucial part of immune function is the ability to distinguish endogenous components ("self") from potentially harmful exogenous components ("nonself"). Failure to make this distinction results in autoimmune disease.

Immunotoxicology is the study of undesired effects resulting from the interac-tions of xenobiotics with the immune system (Figure 19.1). There is evidence that some xenobiotics can cause immune suppression. Xenobiotics can also interact with the immune system to either cause or exacerbate allergic disease. Finally, there is also evidence that xenobiotics have some involvement in autoimmune disease. This chapter provides a brief overview of the immune system, chemicals associated with immune suppression and immune pathologies, and approaches to testing for these effects.

*Disclaimer: This chapter has been reviewed by the National Health and Environmental Effects Research Laboratory and the U.S. Environmental Protection Agency, and has been approved for publication. Approval does not signify that the contents necessarily reflects the views and policies of the Agency, nor does mention of trade names or commercial products constitute endorsement or recom-mendation for use.

Figure 19.1   Potential consequences of immunotoxicity.

TABLE 19.1   Leukocytes

Granulocytes (Polymorphonuclear Leukocytes)
Neutrophils Eosinophils Basophils/mast cells[a]
Mononuclear Leukocytes
Lymphocytes Monocytes/macrophages[a] Natural killer cells

[a] Found in blood/more activated form found in tissues.

19.2   THE IMMUNE SYSTEM

Cells of the immune system include several types of leukocytes (white blood cells) (Table 19.1), which are derived from bone marrow. T lymphocytes, a subset of immune cells, undergo differentiation and maturation in the thymus. Leukocytes circulate throughout the body in blood and lymph and populate other lymphoid tissues including the spleen, lymph nodes (scattered throughout the body), tonsils, and adenoids, as well as aggregates of lymphoid tissue in the lung, gut, and skin referred to as bronchus-, gut- and skin-associated lymphoid tissue (BALT, GALT, and SALT). Also, immune cells can be recruited to almost any tissue in the body where there is injury or infection. Accumulation of leukocytes in tissues in response to injury is known as inflammation. Cytokines (e.g., interleukins, interferons, and chemokines), soluble mediators produced by immune cells, as well as cells outside the immune system, control the maturation, differentiation, and mobilization of immune cells. Immune responses are divided into innate responses directed non-specifically against foreign substances, and acquired responses directed against specific antigens. There is considerable interaction between these two types of immunity.

Innate immunity provides a rapid, although usually incomplete, antimicrobial defense and is characterized by inflammation. Granulocytes, natural killer cells, and macrophages are important mediators of innate immunity. Neutrophils (and to a limited extent eosinophils) have the capacity to phagocytose (engulf) infectious agents or other types of particles and destroy or remove them from the tissue. Granulocytes release a variety of soluble mediators that can kill invading organisms, increase vascular permeability, and recruit more leukocytes to the tissue. Natural killer cells are large granular lymphocytes that nonspecifically kill tumor and virus infected cells. Macrophages are also phagocytic, can release chemotactic and cytotoxic cytokines, and, when activated, can kill tumor or virus-infected cells. Mediators released from all of these cells during the acute inflammatory response influence the development of acquired immune responses.

Although the innate immune system does not recognize specific antigens, proteins known as pattern recognition receptors occur on macrophages, neutrophils, and dendritic cells and recognize features common to many pathogens. When the receptors bind to these highly conserved microbial constituents, a cascade of events is triggered that culminates in phagocytosis, chemotaxis, and production of molecules that influence the initiation and nature of subsequent adaptive immune responses. Toll-like receptors (TLR) are the most well studied of the pattern recognition receptors. Different TLRs recognize pathogen-associated molecular patterns (PAMPs) that are present in many pathogens. For example, TLR4 recognizes bacterial lipopolysaccharide, a component of the outer membrane of gram-negative bacteria, and TLR3 recognizes double-stranded RNA, a component of some viruses. Physical tissue damage, including damage that may occur as a result of exposure to a toxicant, also causes inflammatory responses. The release of nuclear components from cells that have died as a result of tissue trauma also appears to be recognized by certain TLRs.

Acquired immunity *specifically* recognizes foreign substances (called antigens) and *selectively* eliminates them. On re-encountering the same antigen, there is an enhanced response providing protection against reinfection. Vaccination against infectious agents is based on this principle. T lymphocytes and B lymphocytes (T cells and B cells) are the major players in acquired immunity (Figure 19.2). In both cases, there are millions of different clones, groups of immune cells that have specific receptors for a particular antigen. When a cell encounters that specific antigen, clonal expansion occurs; that is, B and T cells with that particular specificity divide and differentiate and are thus activated to respond to the current crisis (e.g., infection). Memory cells also develop that represent an enlarged clone of long-lived cells that are committed to respond rapidly, by clonal expansion, upon reexposure to the same antigen.

B cells recognize native or denatured forms of proteins or carbohydrates in soluble, particulate, or cell-bound form. Activated B cells differentiate into plasma cells and produce antibodies—soluble proteins known as immunoglobulins (Ig)—that circulate freely and react specifically with the invoking antigen. There are several classes (called isotypes) of Ig molecules—IgM, IgG, IgA, IgE, and IgD. IgM is the predominant antibody in the primary immune response (following initial exposure to an antigen). IgG usually appears later following a primary infection but is the predominant antibody in the response to subsequent exposures. IgE acts as a mediator of allergy and parasitic immunity. IgA is found in secretions such as mucous, tears, saliva, and milk, as well as serum, and acts locally to block entrance of pathogens through mucous membranes. IgD is mainly membrane bound on B cells. Little is known about the function of this isotype. It does not appear to have a unique role that affects host immunity.

Acquired Immune Response

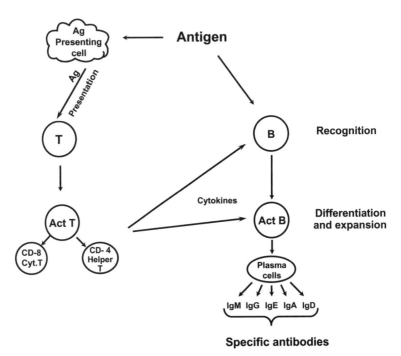

Specific antibodies

Figure 19.2   The acquired immune response: In response to a specific antigen, there is clonal expansion of B cells and subsequent production of antibodies (Ig) specific for that antigen. Antigen presenting cells process and present antigen to T cells. Again, there is clonal expansion of cells specific for that antigen.

A given B cell will form antibody against just one single antigen; however, during the lifetime of this cell, it can switch to make a different class of antibody. Isotype switching is mediated by T-helper (Th) cells. B cells recognize two types of antigen: T-independent antigens, which activate the cell without T cell help (predominantly an IgM response), and T-dependent antigens, which require T cell help in order to activate B cells. Most antigens belong to this latter category. Antibodies that specifically recognize microbial antigens can, in combination with plasma proteins known as complement, lyse bacterial cells or neutralize virus. Also, microbes complexed with antibody are more readily phagocytized.

T cells recognize antigen that is presented via an antigen-presenting cell (APC), usually a dendritic cell. APCs process and present short peptide fragments complexed with major histocompatibility complex (MHC) molecules on the surface of the APC. This processing and presentation is required for T cell activation. There are two major divisions of T cells that are distinguished by expression of different cell surface markers (CD4 and CD8) and recognize different MHC molecules. CD4 cells, also known as Th cells because they provide help for B cell activation, recognize antigen presented by MHC II. CD8 cells, also known as cytotoxic T cells because they lyse cells expressing specific viral or tumor antigens, recognize antigen presented by MHC I. There are two major subpopulations of CD4 cells, Th1 and Th2 that are distinguished by the cytokines they produce and the functions they carry out. There are also a smaller number to CD4 cells collectively known as regulatory T cells that have suppressive effects that control or turn off the effector cells when the crisis has passed.

As indicated above, the thymus plays a key role in T cell differentiation. Pre-T cells migrate from the bone marrow to the thymus. As relatively immature cells, T cells express both CD4 and CD8 molecules. As maturation progresses in the thymus, these cells undergo both positive and negative selection. During positive selection, only cells that bind to MHC with a certain affinity survive. As a result of this process, T cells become MHC restricted; that is, they will only respond to antigen presented in association with MHC. Cells that survive positive selection are potentially able to respond to self proteins. However, before T cells leave the thymus, negative selection occurs during which self-reactive cells are removed or are functionally inactivated. During the course of positive and negative selection, CD4+ CD8+ cells downregulate the expression of one of these molecules such that mature T cells express only CD4 or CD8. Mature T cells leave the thymus and populate secondary lymphoid organs.

19.3 IMMUNE SUPPRESSION

Experimental studies in laboratory rodents have demonstrated that a diverse array of chemical exposures suppress immune function (Table 19.2). In addition, a limited

TABLE 19.2 Selected Examples of Immunosuppressive Agents

Drugs
  Cyclosporin A, cyclophosphamide, glucocorticoids (Dexamethazone), azothioprine
Metals
  Arsenic, lead, cadmium, methylmercury, organotins[a]
Pesticides
  Chlorodane[a], DDT[a], dieldrin[a]
Industrial compounds
  2,3,7,8-tetrachlorodibenzo-p-dioxin (TCDD), polychlorinated and polybrominated
    biphenyls (PCBs and PBBs), benzene, polyaromatic hydrocarbons[a]
Addictive substances
  Cocaine, ethanol, opiates, cannabinoids, nicotine
Air pollutants
  Environmental tobacco smoke, ozone, nitrogen dioxide
Microbial toxins
  Aflatoxin[b], ochratoxin A[b], trichothecenes T-2 toxin[b]
Radiation
  Ionizing, UV
Other
  Asbestos, diethylstilbestrol (DES), dimethylnitrosamine

[a] Effects in humans are unknown; for all other compound without superscripts, changes have been demonstrated in both rodents and humans.
[b] Effects in humans unknown, but veterinary clinicians have noted immunosuppression in livestock ingesting mycotoxins at levels below those that cause overt toxicity.

TABLE 19.3 Tier I Tests (Screen) for Immune Suppression[a]

Immunopathology	Hematology: complete blood count and differential
	Weights: body, spleen, thymus
	Histology: spleen, thymus, lymph node
Antibody-mediated immunity	IgM plaque-forming cell (PFC) response to T cell-immune dependent antigen, e.g., SRBC, (TDAR)
Cell-mediated immunity	Lymphoproliferative response: T cell mitogens (Concanavalin A and phytohemagglutinin)
	Allogeneic mixed leukocyte response (MLR)
Nonspecific immunity	Natural killer (NK) cell activity

[a]For details on specific assays, see Luster et al. *Fund. Appl. Toxicol.* **10**:2–19, 1988.

number of clinical and epidemiologic studies have reported suppression of immune function (including responses to vaccines) and/or increased frequency of infectious and/or neoplastic disease following exposure of humans to some of these agents. From the above description, it is clear there are a number of cellular and molecular targets for chemicals that act as immunosuppressants. Clearly, a chemical that disrupts cell proliferation would impact clonal expansion. Disruption of T cell maturation in the thymus is another potential mechanism for immune suppression. Chemicals may also interfere with receptor ligand binding at the cell surface and/ or the cascade of signals that lead to transcription of genes responsible for generating and regulating the appropriate immune responses.

Because of the complexity of the immune system, tiered approaches to testing chemicals for immunosuppressive potential have been developed. Like other types of toxicity testing, the first level of the tier (Table 19.3) frequently relies solely on structural end points, including changes in the weight of thymus and other lymphoid organs, histopathology of these organs, or differential blood cell counts. This type of evaluation is convenient because it can be carried out along with an evaluation for other organ systems during routine toxicity testing using one set of animals. However, although these nonfunctional end points may be effective in identifying gross (high dose) immunotoxic effects, they are not very accurate in predicting changes in immune function or alterations in susceptibility to challenge with infectious agents or tumor cells at lower chemical doses. Hence, the first testing tier (Table 19.3) often includes functional end points designed to assess (1) antibody-mediated responses, (2) T cell-mediated responses, and (3) NK cell activity. The most commonly used immune function assay in laboratory animals assesses the ability of a mouse or rat to respond to challenge with an antigen, usually sheep red blood cells (SRBC) (Figure 19.3). The response is assessed by determining the number of antigen-specific antibody (IgM)-forming cells (AFC) in the spleen (Jerne assay) or by assessing antigen-specific antibodies in serum using an enzyme-linked immunosorbent assay (ELISA). Because the SRBC is a T-dependent antigen, these assays are often referred to as T-dependent antibody response (TDAR) assays. Both T and B cells, as well as antigen presenting cells, must be functional to have a successful immunization. Suppression of this response is highly predictive of suppression of other immune function tests and also correlates well with tests that assess resistance to challenge with an infectious agent or tumor cells. The disadvan-

IgM Response to SRBCs

1. Inject SRBC IV

2. Remove spleen 4 days

2. Ordraw blood for ELISA

3. Assess antibody forming cells

Figure 19.3 Assessing chemicals for immunosuppressive effects: The most common approach to accomplish this goal is to inject chemical and vehicle-treated mice or rats with antigen and assess the antibody response. Most often, the antigen injected is sheep red blood cells (SRBC); 4 days later, slides are made with a single cell suspension of spleen cells, sheep red blood cells, and complement immobilized in agar. Slides are incubated and spleen cells making antibody against SRBC lyse the surrounding RBCs generating plaques. Plaques are counted to determine the number of antibody-forming cells. Alternatively, serum can be obtained and an ELISA assay performed to detect SRBC specific antibody.

TABLE 19.4   Tier II More In-Depth Evaluation of Immunosuppressive Chemicals[a]

Immunopathology	Quantitation of B and T cell numbers using flow cytometry
Antibody-mediated immunity	IgG PFC to SRBC
	IgM PFC to T cell-independent antigen (e.g., TNP-LPS)
Cell-mediated immunity	Cytotoxic T lymphocyte (CTL) cytolysis
	Delayed hypersensitivity response (DHR)
Nonspecific immunity	Macrophage: phagocytosis, bactericidal/tumoricidal activity)
	Neutrophil: function (phagocytosis and bactericidal activity)
Host resistance models	Response to challenge with infectious agent or tumor cells

[a]For details on specific assays, see Luster et al. *Fund. Appl. Toxicol.* **10**:2–19, 1988.

tage to this test is that it usually requires a dedicated set of animals because of the antigen challenge. The most common approach has been to treat the animals for 14–28 days with the xenobiotic of interest, inject the antigen at the end of that exposure, and collect spleen or serum 4–5 days later. Unlike the tests for antibody-mediated immunity, tier 1 tests for cell-mediated immunity and natural killer cell activity can be done *ex vivo* and do not require a dedicated set of animals. However, these tests focus on one cell type and are not as predictive of overall immunocompetence as the TDAR.

When immunosuppressive effects are noted in tier 1, an in depth evaluation using more sophisticated tests may be carried out (tier 2, Table 19.4). This might include enumeration of lymphocyte subsets (B cells, total T cells, and CD4+ and CD8+ T cells) using flow cytometry or assessment of the IgM response to a T-independent antigen in an effort to determine what portion of the immune response is the actual target. Unlike tier 1, tests of cell-mediated immunity in tier 2 require administration of an antigen and subsequent test for cytotoxic T cells (e.g., against an immunizing tumor cell) or a delayed type hypersensitivity response (similar to the response to a tuberculin test). In order to understand the mechanisms underlying immune suppression, a host of other tests can be carried out, including expression of an assortment of cytokines.

Tier 2 also include host resistance models, tests in which an animal is exposed to a xenobiotic and then challenged with an infectious agent or tumor cells. This is considered the ultimate test for an adverse effect on the immune system. However, it should be noted that the amount of immune suppression that can be tolerated is greatly dependent on the dose and virulence of the challenging agent, as well as the genetics of the host. Manipulation of these variables can affect greatly results obtained in host resistance tests.

The tiered approach assesses systemic effects on immune responses. For inhaled compounds and dermally applied compounds, it is also important to consider effects on local immune responses in the lung and skin, respectively. For example, the alveolar macrophage provides the first line of defense against inhaled microbes, and a number of air pollutants suppress this response leading to increased risk of respiratory infections that would not necessarily be detected using the classic tiered approach. Ultraviolet radiation and some chemicals suppress local responses in the skin and or cause the release of mediators from the skin that affect cell-mediated but not antibody-mediated responses.

As in animal studies, human clinical data obtained from routine hematology (differential cell counts) and clinical chemistry (serum immunoglobulin levels) may provide general information on the status of the immune system in humans. However, as with the animal studies, these may not be as sensitive nor as informative as assays that target specific components of the immune system and/or assess function. The assessment of certain lymphocyte surface antigens has been successfully used in the clinic to detect and monitor the progression or regression of leukemias, lymphomas, and HIV infections, all diseases associated with severe immunosuppression. However, there is considerable variability in the "normal" human population, such that the clinical significance of slight to moderate quantitative changes in the numbers of immune cell populations is difficult to interpret. There is consensus within the immunotoxicology community that tests that measure the response to an actual antigen challenge are likely to be more reliable predictors of immunotoxicity than flow cytometric assays for cell surface markers because the latter generally only assesses the state of the immune system at rest. For ethical reasons, it is not possible to immunize humans with SRBC. One approach is to assess responses to vaccines in chemically exposed populations. This approach has been used successfully to demonstrate a link between mild, stress-induced suppression of the antibody response to influenza vaccine and enhanced risk of infectious disease. There are also a limited number of studies where this approach has been used to demonstrate immune suppression in populations exposed to toxic chemicals. Most notably, a semiquantitative relationship between polychlorinated biphenyls (PCBs) exposure and suppressed responses to diphtheria and tetanus toxoid was demonstrated in a population of children exposed *in utero* and postnatally. Increased susceptibility to infection has also been demonstrated in children exposed to PCBs.

There is some debate over how to interpret immunotoxicity data with respect to adversity. The most conservative interpretation is that any significant suppression of an immune response is adverse because a linear relationship between immune suppression and susceptibility is assumed. Supporting this notion is the fact that apparently immunocompetent individuals suffer from infections, suggesting that adverse effects can occur even when known immune suppression is zero. Others argue that there is clearly redundancy and reserve capacity in the immune response and that some suppression should be tolerable. It is impossible to establish a quantitative relationship between immune suppression and increased risk of infection because both the genetics of the host and the virulence and dose of the infectious agent will influence this relationship. Immunocompetence in a population can probably be represented as a bell-shaped curve, such that a portion of the population is highly susceptible to infection, a portion is highly resistant, and the remaining population falls somewhere in between (Figure 19.4). Genetics, age, nutritional status, and preexisting disease all contribute to the risk represented by this curve. In addition, the portion of the population at risk is determined by the dose and virulence of any infectious agent that might be encountered. The higher the dose and the virulence, the more people are at risk. Exposure to an immunosuppressive agent shifts the whole bell-shaped curve to the left, thus increasing the population at risk. Unfortunately, it is difficult to determine more quantitatively the relationship between small decrements in immune responsiveness and the degree of change in the population at risk.

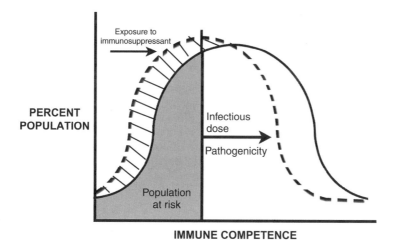

Figure 19.4 Adverse effect of immune suppression: Immune competence is represented by a bell-shaped curve. The shaded area represents the population at risk of infection, which increases or decreases depending on the dose and virulence of infectious agents that are encountered. Exposure to an immunosuppressant shifts the whole curve to the left, such that a larger population is at risk for any given infectious challenge.

19.4 CLASSIFICATION OF IMMUNE-MEDIATED INJURY (HYPERSENSITIVITY)

Under certain circumstances, immune responses can produce tissue damage. These deleterious reactions are collectively known as hypersensitivity or allergy. Hypersensitivity reactions have been divided into four types (originally proposed by Gell and Coombs) based on mechanism (Table 19.5). In all cases, the adverse effects of hypersensitivity develop in two stages: (1) Induction (sensitization) requires a sufficient or cumulative exposure dose of the sensitizing agent to induce immune responses that cause no obvious symptoms. (2) Elicitation occurs in sensitized individuals upon subsequent exposure to the antigen and results in adverse antigen-specific responses that include inflammation.

Type I hypersensitivity (sometimes referred to as atopy) is mediated by antigen-specific cytophilic antibody (usually IgE) that binds to mast cells and basophils. On subsequent exposure, the allergen binds to these cell-bound antibodies and cross-links IgE molecules, causing the release of mediators such as histamine and slow-reacting substance of anaphylaxis (SRS-A). These mediators cause vasodilation and leakage of fluid into the tissues, plus sensory nerve stimulation (leading to itching,

TABLE 19.5 Classification of Hypersensitivity Reactions

Type	Mechanisms		Example
	Induction (Initial Exposure to Antigen)	Elicitation (Re-Exposure to Antigen)	
I (immediate)	Clonal expansion of B cells; Cytophilic antibody (IgE) generated; binds to mast cells	Antigen binds to cell-bound antibody, cross-links receptors, causing release of mediators	Anaphylactic response to bee sting or food allergen
II (cytolytic)	Clonal expansion of B cells; IgM, IgG generated. Antigen binds to cell membrane	Anamnestic[a] Ig binds to cell-bound antigen, and in the presence of complement or activated macrophages, cell lysis occurs	Rh factor incompatability, Hemolytic anemia in reaction to drug treatment
III (Arthus)	Clonal expansion of B cells; IgM, IgG generated	Anamnestic Ig response; antigen antibody complexes form in some tissues leading to inflammation	Glomerular nephritis, Bacterial endocarditis, Farmer's lung
IV (delayed)	Clonal expansion of antigen-specific T cells occurs	T cells activated, release cytokines, activate macrophages, inflammation	Contact dermatitis

[a]Heightened response on re-exposure to antigen.

sneezing, and cough). Type I is also called immediate-type hypersensitivity because reactions occur within minutes after exposure of a previously sensitized individual to the offending antigen. Type I reactions include immediate asthmatic responses to allergen, allergic rhinitis (hay fever), atopic dermatitis (eczema), and acute urticaria (hives). The most severe form is systemic anaphylaxis (e.g., in response to a bee sting), which results in anaphylactic shock and, potentially, death.

Type II hypersensitivity is the result of antibody-mediated cytotoxicity that occurs when a foreign antigen or hapten is incorporated into the cell membrane and antibodies respond to this cell surface antigens. Antibodies bound to antigen on the cell surface activate the complement system and/or macrophages leading to lysis of the target cell. Frequently, blood cells are the targets, as in the case of an incompatible blood transfusion or Rh blood incompatibility between mother and child. The basement membrane of the kidney or lung may also be a target. Autoimmune diseases can result from drug treatments with penicillin, quinidine, quinine, or acetaminophen. Apparently, these drugs interact with the cell membrane such that the immune system detects "foreign" antigens on the cell surface. This type of autoimmune disease may also have unknown etiologies.

Type III reactions are the result of antigen–antibody (IgG) complexes that accumulate in tissues or the circulation, activate macrophages and the complement system, and trigger the influx of granulocytes and lymphocytes (inflammation). This is sometimes referred to as the Arthus reaction and includes postinfection sequelae, such as bacterial endocarditis. Another example, serum sickness, occurs when poorly catabolized foreign antigens are injected in large quantities, which can occur in response to antivenom treatment for snake bites or following treatments that involve monoclonal antibodies. Arthus reactions can also occur when inhaled antigens provoke an IgG rather than an IgE response. An example is Farmer's lung, a pneumonitis caused by molds, which appears to involve both type III and type IV responses. Unlike the preceding three types, Type IV or delayed-type hypersensitivity (DTH), involves T cells and macrophages, not antibodies. Activated T cells release cytokines that cause accumulation and activation of macrophages that in turn cause local damage. This type of reaction is very important in defense against intracellular infections such as tuberculosis, but is also responsible for contact hypersensitivity responses (allergic contact dermatitis) such as the response to poison ivy. Inhalation of beryllium can result in a range of pathologies including acute pneumonitis, tracheobronchitis, and chronic beryllium disease, all of which appear to be due to type IV beryllium-specific immune responses. The expression of type IV responses following challenge is delayed, occurring 24–48 h after exposure. Type IV responses can be further subdivided into three groups based on knowledge of T cell biology obtained since Gel and Coombs devised their classification. CD4 Th1 cells respond to soluble antigens presented by MHC II and activate macrophages to produce the classic tuberculin type reaction. CD4 Th2 cells also respond to soluble antigens presented by MHC II but activate eosinophils leading to inflammation typically seen in allergy and asthma. CD8 cells respond to cell-associated antigens presented by MHC I and are directly cytotoxic.

The different types of immune mediated injury are not mutually exclusive. More than one hypersensitivity mechanism may be involved in the response to a particular antigen. Also, the resulting pathology, particularly that caused by type III and IV reactions, may appear very similar although the mechanisms leading to the effect are different.

19.5   EFFECTS OF CHEMICALS ON ALLERGIC DISEASE

Xenobiotics can affect allergic disease in one of two ways. They can themselves act as antigens and elicit hypersensitivity responses, or they can enhance the development or expression of allergic responses to commonly encountered allergens, such as dust mite. Chemicals that act as allergens include certain proteins that can by themselves induce an immune response and low molecular weight chemicals (known as haptens) that are too small to induce a specific immune response, but may react with a protein to induce an immune response that is then hapten specific. Haptens have been associated with both allergic contact dermatitis (ACD), sometimes called contact hypersensitivity (CHS), and respiratory hypersensitivity. Systemic hypersensitivity, the most extreme manifestation of which is anaphylaxis, can also occur in response to low molecular weight compounds, most notably drugs. Proteins have been associated with respiratory hypersensitivity and systemic hypersensitivity responses such as food allergy. When a chemical is an allergen or a hapten, there are two doses of concern, the sensitizing dose and the elicitation dose. In general, the dose required for sensitization is greater than that required to elicit a response in a sensitized individual. However, the two are not completely independent of one another. When the sensitizing dose is high, the elicitation dose may be lower than when the sensitizing dose is low. Chemicals that enhance the development of allergic sensitization are referred to as adjuvants. Air pollutants have been associated both with enhanced sensitization and exacerbation of allergic respiratory symptoms.

19.5.1   ACD

ACD or CHS is one of the most common occupational health problems and hence is one of the most common problems associated with immunotoxicity. It is a type IV response that occurs as a result of dermal exposure to chemicals that are haptens. Following dermal exposure, the chemical reacts with host cell protein at the surface of the skin and is picked up by epidermal dendritic cells known as Langerhans cells. Cytokines released from the epidermal keratinocytes and from Langerhans cells cause maturation and mobilization of the Langerhans cells, which travel to the draining local lymph node and presents antigen to lymphocytes. Clonal expansion occurs enlarging the number of T lymphocytes specific for that allergen and generating memory cells that in addition to specificity for the allergen have the propensity to home to the skin. On reexposure to the chemical, these specific T cells are activated, proliferate, home rapidly to the site of exposure, and produce erythema and edema typical of a type IV response. This elicitation response is mediated by Th1 cells which produce various cytokines that contribute to the inflammatory response, CD8 cells which are directly cytotoxic to keratinocytes, and regulatory cells that serve to control the response. The reaction to poison ivy is the classic example.

Methods to assess chemicals (including drugs, pesticides, dyes, cosmetics, and household products) for potential to induce CHS are well established, and several protocols using guinea pigs have been in use since the 1930s. These protocols assess the actual disease end point, skin erythema and edema, following sensitization and challenge with the test agent. Two commonly used tests are the guinea pig maximization test and the Buehler occluded patch test. The sensitization procedure for the maximization tests includes intradermal injection of the test chemical with an adjuvant (intended to enhance the sensitization process) as well as topical application. The Buehler test relies on topical sensitization alone. In both cases, after approximately 2 weeks, animals are challenged at a different site on the skin and erythema and edema are assessed 24–48 h later. This assessment is somewhat subjective, and these tests are fairly expensive. A chemical is considered to be a sensitizer if 30% (maximization) or 15% (Buehler) of the animals respond. Recently, a more economical, less subjective, test for CHS has been developed using mice. This test, the local lymph node assay (LLNA), assesses the proliferative response of lymphocytes in the draining lymph node following application of the agent to the ear and is based on our understanding of the immunologic mechanisms underlying CHS; that is, clonal expansion has to occur in the draining lymph node if there is to be allergic sensitization (Figure 19.5). In most cases, the LLNA is accepted as a stand-alone alternative to the guinea pig tests and is generally the assay of choice. Finally, structure activity approaches have recently been developed to identify contact sensitizers. This approach is based on the concept that the biologic mechanisms that determine a chemical's effect are related to its structure and hence, chemicals with similar structures will have similar effects. Computer models have been developed to compare the structure of an unknown chemical to structures in a database for known contact sensitizers. CHS lends itself to this approach because there is a large database of chemicals known to cause it, and there is a reasonable understanding of chemical characteristics that facilitate skin penetration, chemical reactivity with host proteins, and immune reactivity.

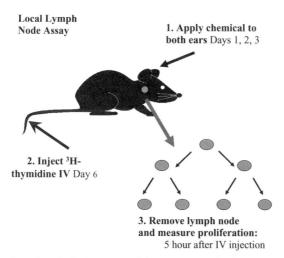

Local Lymph Node Assay

1. **Apply chemical to both ears** Days 1, 2, 3

2. **Inject ^{3}H-thymidine IV** Day 6

3. **Remove lymph node and measure proliferation:**
5 hour after IV injection

Figure 19.5 Assessing chemicals for potential contact sensitivity: In the local lymph node assay, the chemical in question is applied to both ears on three consecutive days. Control mice are treated with vehicle. Radioisotope is injected intravenously on day 6. The draining lymph nodes are removed 5 h later, and the proliferative response is measured by the incorporation of radio isotope. Results are frequently presented as a stimulation index (counts per minute [cpm] for the test chemical/cpm for control). Adapted from Sailstad, D. *Lab. Anim.* **31**:36, 2002.

Because nonspecific inflammatory responses also can occur following chemical exposure to the skin, a distinction must be made between an irritant and a sensitizer. An irritant is an agent that causes local inflammatory effects but induces no immunological memory. Therefore, on subsequent exposures, local inflammation will again result, but there is no enhancement of the magnitude of the response and no change in the dose required to induce the response. In immunologically mediated inflammation (hypersensitivity), there may be no response to a sensitizer during the induction stage, but responses to subsequent exposures are exacerbated. The dose required for elicitation is usually less than that required to achieve sensitization.

19.5.2 Respiratory Allergens

There is evidence that both occupational and environmental exposures to chemicals (both proteins and haptens) can result in the induction or exacerbation of respiratory allergies (Table 19.6). Of particular concern is the induction of allergic asthma. In sensitized asthmatic individuals, antigen challenge generally causes a type I (IgE-mediated) immediate hypersensitivity response with release of mediators responsible for bronchoconstriction. Between 2 and 8 h after the immediate response, asthmatics experience a more severe and prolonged (late phase) reaction that is characterized by mucus hypersecretion, bronchoconstriction, airway hyperresponsiveness to a variety of nonspecific stimuli (e.g., histamine, methacholine), and airway inflammation characterized by eosinophils. This later response is not mediated by IgE.

TABLE 19.6   Example of Chemicals Associated with Respiratory Allergy

Proteins
   Enzymes
   Latex
   Animal dander
   Dust mite
   Molds
   Cockroach
   Microbial pesticides
Low molecular weight (<3000)-haptens
   Toluene diisocyanate
   Diphenylmethane diisocyanate
   Phthalic anhydride
   Trimellitic anhydride
   Platinum salts
   Reactive dyes
Adjuvants
   Ozone
   Nitrogen dioxide
   Diesel exhaust
   Residual oil fly ash

Although proteins are generally immunogens, not all proteins are allergens, and there is a range of potencies for those that are. There is also a strong genetic component associated with susceptibility to develop allergic reactions to proteins. Susceptible individuals are called atopic. There is at present no structural motif that can be used to characterize a protein as an allergen for hazard identification. Examples of occupational protein exposures associated with respiratory allergy and asthma include enzymes, latex, flour (both the grain itself and fungal contaminants), and animal dander. Environmental (mostly indoor) exposures including molds, spores, dust mite, animal dander, and cockroach have also been associated with this type of respiratory disease. Because this is a type I response, cytophilic antibodies (IgE) specific for the allergen are frequently used to identify proteins that may cause this effect. For example, in order to determine the etiology of occupational asthma in human subjects, the skin prick test is often used. Different proteins are injected under the skin to test for the presence of cytophilic antibodies in order to identify which proteins are causing a response in an individual. Serum may also be tested for protein-specific IgE. Because IgE can sometimes be detected in the absence of respiratory responses, a positive IgE test may be followed by an assessment of respiratory responses. Under very controlled situations, patients may be exposed via the respiratory route to suspect allergens (bronchoprovocation test) and respiratory function monitored to pinpoint the offending allergen. Guinea pigs and mice have been used to test proteins for potential allergenicity. Animals are usually sensitized by the respiratory route and monitored for the development of cytophilic antibody (IgG1 in guinea pigs; IgE in mice) as well as increased respiratory rate and other changes in pulmonary function. The guinea pig intratracheal test has been used to establish the relative potency of different detergent enzymes and establish safe occupational exposure levels. As the name implies, guinea pigs are sensitized by intratracheal exposure and induction of cytophilic antibodies are assessed. Dose responses obtained for new enzymes are compared to a reference enzyme for which safe exposure levels have been established. The relative potency of the new enzyme to this reference can be used to establish a safe exposure level for the new enzyme.

Exposure to certain low (<3000) molecular weight compounds (haptens) has also been associated with the development of occupational asthma. Highly reactive compounds such as the diisocyanates or acid anhydrides have the capacity to react with protein and induce an immune response. Toluene diisocyanate (TDI) and trimellitic anhydride are the compounds that have been most extensively studied in this regard. There is a great deal of interest in developing a test to screen chemicals for this type of effect in order to avoid induction of immune responses that could lead to occupational asthma. Although specific IgE antibodies have been detected in some individuals with TDI asthma, it has not been uniformly present, and some of these individuals exhibit the late phase but not the immediate response. Hence, unlike proteins, there is less certainty about the mechanisms underlying respiratory allergic responses to low molecular weight compounds. Structure activity approaches similar to those described for contact sensitizers have been developed, but this approach has limitations because the database of known respiratory sensitizers is small compared to contact sensitizers, and the underlying mechanisms are less well defined. At the other extreme, guinea pigs have been exposed by inhalation for a number of days, rested, and then challenged at a later date by inhalation with subsequent monitoring of respiratory responses. Although this approach has produced a good model of TDI asthma, it is too cumbersome and expensive for routine testing. Because the capacity to interact with protein is a prerequisite to allergenicity, it has been suggested that testing for protein reactivity *in vitro* could provide an initial screening test for chemicals. Also, because it appears that respiratory sensitizers are a subset of chemicals that produce positive results in a contact sensitivity test, it has been suggested that the LLNA test be used as the first tier in screening chemicals for this effect. The problem then becomes separating chemicals that are strictly contact sensitizers from those that have the capacity to cause respiratory sensitization. Efforts have been made to determine whether differences in responses to dermal application of these chemicals could provide a means for making this distinction. One proposal is to assess total serum IgE following dermal exposure, assuming that respiratory sensitizers would produce a bigger IgE signal. Another approach has been to assess cytokine profiles in the draining lymph node following dermal exposure. Different subsets of Th cells have been associated with type I immediate (Th2) and type IV delayed (Th1) responses. These different populations of T cells are distinguished by different cytokine profile, and efforts are underway to use these differing profiles to distinguish respiratory from contact sensitizers. However, there is as yet no well-validated, well-accepted test to assess low molecular weight chemicals for the capacity to induce respiratory allergy. This remains a subject of research.

19.5.3 Adjuvants

An adjuvant is a compound administered in conjunction with an antigen that nonspecifically enhances the immune response to that antigen. Adjuvants are used in vaccines to promote immunogenicity. There is now growing concern that chemicals in our environment (particularly air pollutants) might act as adjuvants for allergic sensitization to common allergens such as dust mite and pollen. Laboratory rodents have been used to show that nitrogen dioxide, residual oil fly ash, and diesel exhaust enhance allergic sensitization and disease. Enhanced sensitization to an allergen has also been demonstrated in rhesus monkeys exposed to ozone and humans exposed to diesel exhaust. The significance of these findings in terms of enhanced burden of respiratory allergies in the human population is unclear. As in other areas of toxicology, simultaneous environmental exposures to agents that are not the agent of immediate concern can certainly influence outcomes. Adjuvancy is a concern that likely extends beyond air pollution and type I responses.

19.5.4 Systemic Hypersensitivity

All allergic responses are systemic in that sensitized immune cells can circulate throughout the body and can respond when challenge occurs at sites other than the original site of sensitization. However, for the allergic diseases described above, the response to challenge is usually localized around the site of challenge. Food allergy is an example of a more systemic response. IgE-mediated food allergies can cause symptoms in the skin, the upper and lower respiratory tract, as well as the gastrointestinal tract. Food allergens have been reported to be one of the leading causes of systemic anaphylaxis seen in emergency departments. Hymenoptera stings and administered drugs are the other common causes of anaphylactic reactions seen in medical facilities. Immune-related problems are the largest single area of adverse events that are not detected by preclinical testing of drugs. Many of these events are dermal reactions associated with systemic administration of drugs although multiorgan reactions are a more worrisome occurrence. Good methods for assessing the potential for drugs to elicit such responses are not available.

Toxicologists have recently been drawn into the area of food allergy by advances in biotechnology and the need to assess the safety of genetically modified foods in terms of potential allergenicity. There is concern that insertion of a novel gene into a food crop (e.g., to increase yield or pest resistance) might inadvertently introduce a new allergen into the food supply. Food allergies are relatively rare, affecting approximately 5% of children and 2–3% of adults, and even in these individuals, most proteins are not allergens. However, when food allergy does occur, the consequences can be severe. Anaphylactic (life-threatening) reactions to peanuts provide the best example. Unfortunately, the mechanisms underlying food allergies and oral tolerance which protects most of people from developing reactions to the foreign proteins they eat are poorly understood. Also, the characteristics that make a protein a food allergen, and the characteristics that make an individual susceptible to food allergies are unclear. These are some of the issues that need to be resolved in order to develop appropriate safety assessment tools.

19.6   OTHER ISSUES: AUTOIMMUNITY AND THE DEVELOPING IMMUNE SYSTEM

Autoimmune diseases result from a breakdown of immunological tolerance leading to immune responses against self-molecules that involve activation of both innate and adaptive immune responses. Autoimmune disorders can affect virtually any site in the body, and present as a spectrum of diseases. Autoimmune diseases affect about 3% of the population and comprise a diverse array of both organ-specific (e.g., type I diabetes, thyroiditis) and systemic (systemic lupus erythematosus) diseases. Epidemiologic studies suggest associations with specific genetic loci, and environmental factors, including exposures to certain drugs, chemicals, and infectious agents. In many cases, women appear to be more vulnerable than men. Xenobiotics have the potential to affect the development, progression, or severity of autoimmune diseases. A variety of mechanisms could contribute to xenobiotic effects on the development and maintenance of immune tolerance or unmasking or modification of self-proteins. There is evidence that exposure to certain drugs, heavy metals, silica, asbestos, and endocrine disruptors are a concern in this regard. Although there are now a variety of animal models designed to mimic different types of autoimmune diseases, much of the information we have to date comes from associations based on human epidemiology. Current research includes both human and animals studies to determine the extent of risk and ways to assess and control it.

Finally, there is growing concern that the developing immune system may be particularly vulnerable to xenobiotic exposures and that perinatal and/or *in utero* exposures may have a lifelong impact on susceptibility to infectious, allergic, or autoimmune diseases. As in other areas of toxicology, tests designed to assess the risk of immunotoxicity for adults may not be sufficient to protect children, and research is currently underway to determine how best to meet this need.

Clearly, exposure to xenobiotics can have a number of effects on the immune system that in turn can affect an array of health outcomes. In some areas of immunotoxicology, significant progress has been made in terms of identifying and understanding the risks associated with xenobiotic exposure. In other areas, more research is needed.

BIBLIOGRAPHY AND SUGGESTED READING

Coico, R. and G. Sunshine. *Immunology: A Short Course*, 6th ed. Hoboken, NJ: Wiley-Blackwell, 2009.

Delves, P., S. Martin, D. Burton, et al. *Roitt's Essentials in Immunology*, 11th ed. London: Wiley-Blackwell, 2006.

Luebke, R., R. House, and I. Kimber. *Immunotoxicology and Immunopharmacology*, 3rd ed. New York: Taylor and Francis, 2007.

Luster, M. I., A. E. Munson, P. T. Thomas, et al. Development of a testing battery to assess chemical-induced immunotoxicity: National Toxicology Program's guidelines for immunotoxicity evaluation in mice. *Fundam. Appl. Toxicol.* **10**:2–19, 1988.

Luster, M. I., C. Portier , D. M. Pail, et al. Risk assessment in immunotoxicology I. Sensitivity and predictability of immune tests. *Fundam. Appl. Toxicol.* **18**:200–210, 1992.

Luster, M. I., C. Portier, D. M. Pail, et al. II. Relationships between immune and host resistance tests. *Fundam. Appl. Toxicol.* **21**:71–82, 1993.

Sailstad, D. Murine local lymph node assay: an alternative test method for skin hypersensitivity testing. *Lab. Anim.* **31**:36, 2002.

SAMPLE QUESTIONS

1. Innate (nonspecific) responses include
   a. activation of toll receptors
   b. clonal expansion and differentiation of plasma cells
   c. inflammation
   d. natural killer cell responses
   e. immunoglobulin production

2. T-helper (Th) cell subsets
   a. are distinguished by the cytokines they produce
   b. are mutually antagonistic
   c. provide "help" in the production of various classes of antibodies
   d. All of the above is true
   e. None of the above is true

3. Which of the following tests are the most commonly used to assess potential immunotoxic outcomes either allergenic or immune suppression
   a. host resistance to infectious challenge
   b. T cell cytotoxicity
   c. local lymph node assay
   d. cytokine profiling
   e. IgM response to sheep red blood cells

4. Mechanisms involved in immune suppression could include
   a. disruption of T cell development in the thymus
   b. upset in Th1/Th2 balance
   c. phagocytosis
   d. cytotoxicity
   e. lymphocyte proliferation

5. Which of the following have been associated with immune suppression in humans?
   a. poison ivy
   b. environmental tobacco smoke
   c. polychlorinated biphenyls (PCBs)
   d. diisocyanates
   e. arsenic

6. The consequences of immune suppression are influenced by
   a. pathogenicity of infectious agent
   b. degree of immune deficit
   c. exposure dose of infectious agent
   d. genetics of the host
   e. all of the above

7. Which of the following responses is mediated by IgE antibodies?
   a. Type I immediate-type hypersensitivity
   b. Type II hypersensitivity (cytolytic)
   c. Type III hypersensitivity (immune complex)
   d. Type IV delayed-type hypersensitivity
   e. None of the above

8. Which of the following responses are associated with asthma?
   a. IgM responses
   b. eosinophilic inflammation
   c. Th1 responses
   d. all of the above
   e. none of the above

9. Which of the following are associated with allergic contact dermatitis (contact hypersensitivity)?
   a. lymphocyte proliferation
   b. IgG antibody responses
   c. erythema and edema
   d. Th2 responses
   e. mobilization of Langerhans (dendritic) cells

APPLIED TOXICOLOGY

Toxicity Testing

ERNEST HODGSON and HELEN CUNNY

20.1 INTRODUCTION

The purpose of this chapter is not to describe all available tests for chemical toxicity but rather to summarize those that are, or can be, currently required by regulatory agencies. Although not without difficulties, either in execution or injustification, the tests described have been validated and used extensively. Tests for endocrine disruption have been under development and validation for several years and will probably be required in the near future (see Chapter 29).

As a consequence of the large number of chemicals to be tested under a number of federal statutes and the extensive backlog of untested chemicals, it will be necessary to develop high throughput, rapid assays that can handle many chemicals simultaneously. From both cost and animals welfare considerations, it will not be possible to use whole animal assays, and assays based on human cell lines will doubtless be used. Testing systems based on quantitative structure–activity relationships, using engineered human cell lines and the techniques of genomics, proteomics, metabolomics, and systems biology are currently being developed (see Chapters 28 and 29). However, at least for some considerable time, these new systems will be used to establish priorities rather than make final regulatory decisions, and the tests described in this chapter will still be used.

Although current testing for toxicity, usually for the purposes of human health risk assessment, might be expected to be one of the more routine aspects of toxicology, it is actually one of the more controversial. Among the many areas of controversy are the use of animals for testing and the welfare of the animals; extrapolation of animal data to humans; extrapolation from high-dose to low-dose effects; and the increasing cost and complexity of testing protocols relative to the benefits expected.

Most testing can be subdivided into *in vivo* tests for acute, subchronic, or chronic effects and *in vitro* tests for genotoxicity or cell transformation, although other tests are used and are described in this chapter.

A Textbook of Modern Toxicology, Fourth Edition. Edited by Ernest Hodgson
Copyright © 2010 John Wiley & Sons, Inc.

TABLE 20.1   Some Agencies and Statutes Involved in Regulation of Toxic Chemicals in the United States

Food and Drug Administration (FDA)
    Food, Drug and Cosmetic Act
Labor Department
    Occupational Safety and Health Act
Consumer Products Safety Commission
Environmental Protection Agency (EPA)
    Federal Insecticide, Fungicide, and Rodenticide Act
    Clean Air Act
    Federal Water Pollution Control Act
    Safe Drinking Water Act
    Toxic Substances Control Act
    Resource Conservation and Recovery Act
State governments
    Various state and local laws
    Enforcement of certain aspects of federal law delegated
       to states

Toxicity assessment is the determination of the potential of any substance to act as a poison, the conditions under which this potential will be realized, and the characterization of its action. *Risk assessment*, however, is a quantitative assessment of the probability of deleterious effects under given exposure conditions. Both are involved in the regulation of toxic chemicals. *Regulation* is the control, by statute, of the manufacture, transportation, sale, or disposal of chemicals deemed to be toxic after testing and according to criteria laid down in the law in question.

Testing in the United States is carried out by many groups: industrial, governmental, academic, and others. Regulation, however, is carried out by a narrow range of governmental agencies, each charged with the formulation of regulations under a particular law or laws and with the administration of those regulations. The principal regulatory agencies for the United States are shown in Table 20.1. Other industrialized countries have counterpart laws and agencies for the regulation of toxic chemicals.

Although the objective of most toxicity testing is the elimination of potential risks to humans, most of the testing is carried out on experimental animals. This is necessary because current knowledge of quantitative structure–activity relationships (QSAR) does not permit accurate extrapolation to new compounds. Human data are difficult to obtain experimentally for ethical reasons, but are necessary for such deleterious effects as irritation, nausea, allergies, odor evaluation, and some higher nervous system functions. Some insight may be obtained in certain cases from occupational exposure data, although this often tends to be irregular in time and not clearly defined as to the composition of the toxicant or the exposure levels. Clearly, any experiments involving humans must be carried out under carefully defined conditions after other testing is complete.

Although extrapolation from experimental animals to humans presents problems due to a variety of reasons, including differences in metabolic pathways, dermal

TABLE 20.2   A Summary of Toxicity Tests and Related End Points

I   Chemical and physical properties
   For the chemical in question, probable contaminants from synthesis as well as
      intermediates and waste products from the synthetic process.

II   Exposure and environmental fate
   a. Degradation studies—hydrolysis, photodegradation, etc.
   b. Degradation in soil, water, under various conditions
   c. Mobility and dissipation in soil water and air
   d. Accumulation in plants, aquatic animals, wild terrestrial animals, food plants and
      animals, etc.

III   *In vivo* tests
   a. Acute
      1. LD_{50} and LC_{50}—oral, dermal, or inhaled
      2. Eye irritation
      3. Dermal irritation
      4. Dermal sensitization
   b. Subchronic
      1. 90-day feeding
      2. 30- to 90-day dermal or inhalation exposure
   c. Chronic/Reproduction
      1. Chronic feeding (including oncogenicity tests)
      2. Teratogenicity
      3. Reproduction (multigeneration)
   d. Special tests
      1. Neurotoxicity
      2. Potentiation
      3. Metabolism
      4. Pharmacodynamics
      5. Behavior

IV   *In vitro* tests
   a. Mutagenicity—prokaryote (Ames test)
   b. Mutagenicity—eukaryote (*Drosophila*, mouse, etc.)
   c. Chromosome aberration (*Drosophila*, sister chromatid exchange, etc.)

V   Effects on wildlife
   Selected species of wild mammals, birds, fish, and invertebrates: acute toxicity,
      accumulation, and reproduction in laboratory simulated field conditions.

penetration, mode of action, and others, experimental animals present numerous
advantages in testing procedures. These advantages include the possibility of clearly
defined genetic constitution and their amenity to controlled exposure, controlled
duration of exposure, and the possibility of detailed examination of all tissues following necropsy.

Although not all tests are required for all potentially toxic chemicals, any of the
tests shown in Table 20.2 may be required. The particular set of tests required
depends on the predicted or actual use of the chemical, the predicted or actual route
of exposure, and the chemical and physical properties of the chemical.

20.2 EXPERIMENTAL ADMINISTRATION OF TOXICANTS

20.2.1 Introduction

Regardless of the chemical tested and whether the test is for acute or chronic toxicity, all *in vivo* testing requires the reproducible administration of a known dose of the chemical under test that is generally related to the expected route of humans exposure. The nature and degree of the toxic effect can be affected by the route of administration (Table 20.3). This may be related to differences at the portals of entry or to effects on pharmacokinetic processes. In the latter case, one route (e.g., intravenous) may give rise to a concentration high enough to saturate some rate-limiting process, whereas another (e.g., subcutaneous [SC]) may distribute the dose over a longer time and avoid such saturation. Another key question is that of appropriate controls. To identify effects of handling and other stresses as well as the effects of the solvents or other carriers, it is usually better to compare treated animals with both solvent-treated and untreated or possibly sham-treated controls.

20.2.2 Routes of Administration

Oral Oral administration is often referred to as administration per os (PO). Compounds can be administered mixed in the diet, dissolved in drinking water, by gastric gavage, by controlled-release capsules, or by gelatin capsules. In the first two cases, either a measured amount can be provided or access can be *ad libitum* (available 24h per day), with the dose estimated from consumption measurements. For certain tests pair-feeding of controls should be considered; that is, controls are permitted only the amount of food consumed by treated animals, and, in any case, it is essential to consider possible nutritional effects caused by reduction of food intake due to distasteful or repellent test materials. In the case of gastric gavage, the test material is administered through a stomach tube or gavage needle; if a

TABLE 20.3 Variation in Toxicity by Route of Exposure

Chemical	Species/Gender	Route	LD50 (mg/kg)
N-methyl-*N*	Mouse/M	Oral	371
-(1-naphthyl)		Dermal	402
fluoroacetamide[a]		Subcutaneous	250
	Rat/M	Oral	115
		Dermal	300
		Subcutaneous	78
Chlordane[b]	Rat/M	Oral	335
		Dermal	840
Endrin[b]	Rat/M	Oral	18
		Dermal	18

[a] Data from Hashimoto, Y., et al. *Toxicol. Appl. Pharmacol.* **12**:536–547, 1968.
[b] Data from Allen, J. R., et al. *Pharmacol. Ther.* **7**:513–547, 1979.

solvent is necessary for preparation of dosing solutions or suspensions, the vehicle is administered also to control animals.

Dermal Dermal administration is required for estimation of toxicity of chemicals that may be absorbed through the skin, as well as for estimation of skin irritation and photosensitization. Compounds are applied, either directly or in a suitable solvent, to the skin of experimental animals after hair has been removed by clipping. Often dry materials are mixed with water to make a thick paste that can be applied in a manner to assure adequate contact with the skin. Frequently, the animals must be restrained to prevent licking and hence oral uptake of the material. Solvent and restraint controls should be considered when stress is involved. Skin irritancy tests may be conducted on either animals or humans, using volunteer test panels for human tests.

Inhalation The respiratory system is an important portal of entry and, for evaluation purposes, animals must be exposed to atmospheres containing potential toxicants. The generation and control of the physical characteristics of such contaminated atmospheres is technically complex and expensive in practice. The alternative—direct instillation into the lung through the trachea—presents problems of reproducibility as well as stress, and for these reasons is generally unsatisfactory.

Inhalation toxicity studies are conducted in inhalation chambers. The complete system contains an apparatus for the generation of aerosol particles, dusts, or gas mixtures of defined composition and particle size, a chamber for the exposure of experimental animals, and a sampling apparatus for the determination of the actual concentration within the chamber. All these devices present technical problems that are difficult to resolve. For rat studies, a particle size of 4μ is usually targeted.

Animals are normally exposed for a fixed number of hours each day and a fixed number of days each week. Exposure may be nose only, in which the nose of the animal is inserted into the chamber through an airtight ring, or whole body, in which animals are placed inside the chamber. In the latter case, variations due to unequal distribution in the test atmosphere are minimized by rotation of the position of the cages in the chamber during subsequent exposures. Whole body exposure results are usually less satisfactory due to test material accumulation on the fur of the animals and subsequent ingestion during grooming. Figure 20.1 shows a typical inhalation system and supporting equipment.

Injection Except in the case of certain pharmaceuticals and drugs of abuse, injection (parenteral administration) does not correspond to any of the expected modes of exposure. It may be useful, however, in studies of mechanism or in QSAR studies in order to bypass absorption and/or permit rapid action. Methods of injection include intravenous (IV), intramuscular (IM), intraperitoneal (IP), and SC. Infusion of test materials over an extended period is also possible. Again, both solvent controls and untreated controls are necessary for proper interpretation of the results.

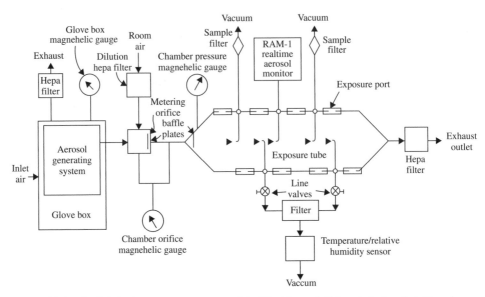

Figure 20.1   An inhalation exposure system. Modified from Adkins et al. *Am. Int. Hyg. Assoc. J.* **41**:494, 1980.

20.3   CHEMICAL AND PHYSICAL PROPERTIES

Although the determination of chemical and physical properties of known or potential toxicants does not constitute a test for toxicity, it is an essential preliminary for such tests.

The information obtained can be used as follows: Structure activity comparisons with other known toxicants may indicate the most probable hazards. These comparisons may also aid in identification in subsequent poisoning episodes. Determination of stability to light, heat, freezing, and oxidizing or reducing agents, may enable preliminary estimates of persistence in the environment as well as indicate the most likely breakdown products that may also require testing for toxicity. Establishing such properties as the lipid solubility or octanol/water partition coefficient may enable preliminary estimates of rate of uptake and persistence in living organisms. Vapor pressure may indicate whether the respiratory system is a probable route of entry. Acquiring knowledge of the chemical and physical properties is needed to develop analytical methods for the measurement of the compound and its degradation products. If the chemical is being produced for commercial use, similar information is needed on intermediates in the synthesis or by products of the process because both are possible contaminants in the final product.

20.4   EXPOSURE AND ENVIRONMENTAL FATE

Data on exposure and environmental fate are needed, not to determine toxicity, but to provide information that may be useful in the prediction of possible exposure in the event that the chemical is toxic. Primarily useful for chemicals released into the environment such as pesticides, these tests include the rate of breakdown under aerobic and anaerobic conditions in soils of various types, the rates of leaching into surface water from soils of various types, or the rate of movement toward groundwater. The effect of physical factors on degradation through photolysis and hydrolysis studies and the identification of the product formed can indicate the rate of loss of the hazardous chemical or the possible formation of hazardous degradation products. Tests for accumulation in plants and animals and movement within the ecosystem are considered in Section 20.7.

20.5 *IN VIVO* TESTS

Traditionally, the basis for the determination of toxicity has been administration of the test compound, *in vivo*, to one or more species of experimental animal, followed by examination for clinical signs of toxicity and/or mortality in acute tests. In addition, pathological examination for tissue abnormalities is also performed, especially in tests of longer duration. The results of these tests are then used, by a variety of extrapolation techniques, to estimate hazard to humans. These tests are summarized in the remainder of this section. While these tests offer many advantages and are widely used, they suffer from a number of disadvantages: they require the use of experimental animals, the numbers of which are often deemed unnecessary by both animal rights and animal welfare advocates; they are extremely expensive to conduct; and they are time-consuming. As a result, they have been supplemented by many specialized *in vitro* tests, some of which are summarized in Section 20.6, and research is ongoing to further develop tests with fewer disadvantages.

20.5.1 Acute Toxicity

Acute toxicity test methods measure the adverse effects that occur within a short time of administration of a single dose of a test substance. This testing is performed principally in rodents and is usually done early in the development of a new chemical or product to provide information on its potential toxicity. This information is used to protect individuals who are working with the new material and to develop safe handling procedures for transport and disposal. The information gained also serves as the basis for hazard classification and labeling of chemicals in commerce. Acute toxicity data can help identify the mode of toxic action of a substance and may provide information on doses associated with target-organ toxicity and lethality that can be used in setting dose levels for repeated-dose studies. This information may also be extrapolated for use in the diagnosis and treatment of toxic reactions in humans. The results from acute toxicity tests can also provide information for comparison of toxicity and dose–response among members of chemical classes and help in the selection of candidate materials for further work.

The results of acute toxicity tests also have a wide variety of regulatory applications. These include determination of the need for childproof packaging, determination of reentry intervals after pesticide application, establishment of the requirement and basis for training workers in chemical use, determination of requirements for protective equipment and clothing, and decision making about general registration of pesticides or their restriction for use by certified applicators. Acute oral toxicity may be used in risk assessments of chemicals for humans and nontarget environmental organisms.

The various national and international regulatory authorities have used different hazard classification systems in the past. In light of the importance of hazard classification, the Organisation for Economic Co-operation and Development (OECD) recently harmonized criteria for hazard classification for global use. For example, the five harmonized categories for acute oral toxicity (in mg/kg body weight) are: 0–5, 5–50, 50–300, 300–2000, and 2000–5000.

Acute Oral Testing   Traditionally, acute oral toxicity testing focused on determining the dose that kills half the animals (i.e., the median lethal dose or LD_{50}), the timing of lethality following acute chemical exposure, as well as observing the onset, nature, severity, and reversibility of toxicity. The LD_{50} concept was developed by Trevan in 1927. Original testing methods were designed to characterize the dose–response curve by using several animals (usually at least 5/gender) at each of several test doses. Data from a minimum of three doses is required. The LD_{50} values are presented as estimated doses (mg/kg) with confidence limits. The simplest method for the determination of the LD_{50} is a graphic one and is based on the assumption that the effect is a quantal one (all or none), that the percentage responding in an experimental group is dose-related, and that the cumulative effect follows a normal distribution. Data from a typical example, its analysis and implications, are discussed in Chapter 10, Acute Toxicity.

As a result of much recent controversy, the LD50 test has been the subject of considerable regulatory attention and recent changes in requirements have been promulgated. These changes are intended to obtain more information but, at the same time, use fewer animals.

Criticism of the LD_{50} Test   The criticisms of the test include:

- Used uncritically, it is an expression of lethality only, not reflecting other acute effects.
- It requires large numbers of experimental animals to obtain statistically acceptable values. Moreover, the results of LD_{50} tests are known to vary with species, strain, gender, age, and so on (Table 20.4); thus, the values are seldom closely similar from one laboratory to another, in spite of the numbers used.
- Because, for regulatory purposes, the most important information needed concerns chronic toxicity, little useful information is derived from the LD_{50} test. The small amount of information that is acquired could be acquired as well from an approximation requiring only a small number of animals.
- Extrapolation to humans is difficult.

TABLE 20.4   Factors Causing Variation in LD_{50} Values

Species	Health	Temperature
Strain	Nutrition	Time of day
Age	Gut contents	Season
Weight	Route of administration	Human error
Gender	Housing	

Support of the LD$_{50}$ Test   Continued use of the test has been advocated, however, on the grounds that it is of use in the following ways:

- Properly conducted, acute toxicity tests yield not only the LD$_{50}$, but also information on other acute effects such as cause of death, time of death, symptomatology, nonlethal acute effects, organs affected, and reversibility of nonlethal effects.
- Information concerning mode of action and metabolic detoxication can be inferred form the slope of the mortality curve.
- The results can form the basis for the design of subsequent subchronic studies.
- The test is useful as a first approximation of hazards to workers.
- The test is rapidly completed.

For the previously listed reasons, there has been a concerted effort in recent years to modify the concept of acute toxicity testing as it is embodied in the regulations of many countries and to substitute more meaningful methods that use fewer experimental animals. The article by Zbinden and Flury-Roversi is an excellent summary of the factors affecting LD$_{50}$ determinations, the advantages and disadvantages of requiring such tests, and the nature and value of the information derived. It concludes that the acute toxicity test (single-dose toxicity) is still of considerable importance for the assessment of risk posed by new chemical substances, and for a better control of natural and synthetic agents in the human environment. It is not permissible, however, to regard a routine determination of the LD$_{50}$ in various animal species as a valid substitute for an acute toxicity study.

Current Status   Recently, attention has been focused on developing alternatives to the classical LD$_{50}$ test to reduce the number of animals used or refine procedures to make exposures less stressful to animals. OECD adopted several alternative methods for determining acute oral toxicity: a limit test for materials with anticipated low toxicity, a fixed-dose procedure, an acute toxic class method, and an up-and-down procedure. The fixed-dose procedure and the acute toxic class method estimate the LD$_{50}$ within a dose range for use in classification and labeling. The up-and-down procedure generates point estimates and confidence intervals of the LD$_{50}$ and therefore, may be useful in a wider set of applications.

The fixed-dose procedure (Guideline OECD 420) aims to identify the appropriate hazard class for new chemicals; it does not provide a point estimate of the LD$_{50}$. This method calls for testing animals sequentially at one of four doses: 5, 50, 300, or 2000 mg/kg body weight. The test begins with a sighting study in which animals are tested, one at a time, at doses selected from the set doses. Once clear signs of toxicity appear, additional animals (females or the more sensitive sex) are dosed at that level for a total of five animals. Subsequent groups of animals may receive doses at higher or lower levels, if necessary, depending on the outcome of the previous group. Decision criteria based on the number of animals surviving or showing evident toxicity provide for classification decisions.

The acute toxic class method (Guideline OECD 423) aims to identify the appropriate hazard and labeling classification and provides a range for lethality rather than a point estimate of the LD$_{50}$. Groups of three animals (females or the more sensitive sex) receive one of four or five doses: 5, 50, 300, 2000 and, if necessary, 5000 mg/kg body weight. Depending on the survival or mortality of the first group of animals, three or more animals may receive the same or a higher or lower dose. The number of animals that survive or die determines the classification decisions.

The up-and-down procedure (Guideline OECD 425) employs sequential dosing, using only a single animal at each step, the dosage depending on whether the previously dosed animal lives or dies. The test provides a point estimate of lethality and confidence intervals, and can be used to evaluate lethality up to 5000 mg/kg. The main test incorporates elements of range finding and uses a flexible stopping point. A sequential limit test uses up to five animals. Default dose spacing is 3.2 times the previous dose. The starting dose should be slightly below the estimated LD_{50}. If no information is available to estimate the LD_{50}, the starting dose is 175 mg/kg. A computer program was developed by the U.S. Environmental Protection Agency (U.S. EPA) to simplify both the experimental phase of the test and the calculation of the LD_{50} and confidence intervals.

For all three guidelines, selection of a starting dose close to the actual LD_{50} should decrease the number of animals necessary, reduce study duration, and decrease the amount of test substance needed. Therefore, it is desirable that all available information on the test substance be made available to the testing laboratory for consideration prior to conducting the study. Such information includes the identity and chemical structure of the substance, its physicochemical properties, the results of any other toxicity test tests on the substance, toxicological data on structurally related substances, the anticipated uses of the substance, or cytotoxicity data on the substance. This information will aid the testing laboratory in selecting the most appropriate test to satisfy regulatory requirements and in choosing the starting dose.

As with the traditional acute oral toxicity methods, the alternative tests involve the administration of a single-bolus dose of a test substance to fasted healthy young adult rodents by oral gavage, observation for morbidity/mortality for up to 14 days after dosing, with recording of body weight (weekly) and clinical signs (daily), and a necropsy at study termination. At the time of dosing, each animal should be between 8 and 12 weeks old and its weight should fall in an interval within ±20% of the mean weight of all previously dosed animals taken on their day of dosing. Observation of the postdosing effects on each animal should be for at least 48 h or until it is clear whether the dosed animals will survive. However, depending on the characteristics of the test material, investigators can vary this time between dosing, so long as the interval is sufficient. Only when the results are clear can a decision be made about whether an additional dose is necessary, and if so, whether to dose the next animal or group of animals at the same, higher, or lower dose. The information from every animal, even those that die after the initial observation period, is used in the final determination of the test outcome.

These newer methods call for testing to be done in a single gender to reduce variability in the test population. This reduction in variability in turn minimizes the number of animals needed. Normally, females are used. Although there is usually little difference in sensitivity between males and females, in those cases where there are observable differences, females are most commonly the more sensitive gender. Normally, animal suppliers have an excess of female rats because many researchers order only male rats to avoid physiological changes associated with estrus cycling in females; therefore, preferential use of female animals for acute testing should not result in excess male animals.

Eye Irritation   Because of the prospect of permanent blindness, ocular toxicity has long been a subject of both interest and concern. Although all regions of the eye are subject to systemic toxicity, usually chronic but sometimes acute, the tests of concern in this section are tests for irritancy of compounds applied topically to the eye. The tests used are all variations of the Draize test, and the preferred experimental animal is the albino rabbit.

The test consists of placing the material to be tested directly into the conjunctival sac of one eye, with the other eye serving as the control. The lids are held together for a few seconds, and the material is left in the eye for at least 24 h. After that time, it may be rinsed out, but in any case, the eye is examined and graded after 1, 2, and 3 days. Grading is subjective and is based on the appearance of the cornea, particularly as regards opacity; the iris, as regards both appearance and reaction to light; the conjunctiva, as regards redness and effects on blood vessels; and the eyelids, as regards swelling. Fluorescein dye may be used to assist visual examination because the dye is more readily absorbed by damaged tissues, which then fluoresce when the eye is illuminated. Each end point in the evaluation is scored on a numerical scale and chemicals are compared on this basis. In addition to the "no-rinse" test, some protocols also investigate the effect of rinsing the eye 1 min after exposure to determine if this reduces the potential for irritation. In addition, eyes may be graded for up to 21 days after administration of an irritating test material to evaluate recovery.

The eye irritation test is probably the most criticized by advocates of animal rights and animal welfare, primarily on the grounds that it is inhumane. It has also been criticized on narrower scientific grounds in that both concentration and volumes used are unrealistically high, and that the results, because of high variability and the greater sensitivity of the rabbit eye, may not be applicable to humans. It is clear, however, that because of great significance of visual impairment, tests for ocular toxicity will continue.

Attempts to solve the dilemma have taken two forms: to find substitute *in vitro* tests and to modify the Draize test so that it becomes not only more humane but also more predictive for humans. Substitute tests consist of attempts to use cultured cells or eyes from slaughtered food animals, but neither method is yet acceptable as a routine test. Modifications consist primarily of using fewer animals. Usually, one animal is tested first and, if the material is severely irritating, no further eye testing is conducted. EPA has reduced the required number of animals from six to three. In addition, eye irritation should never be carried out on materials with a pH of less than 2 or more than 10 as these materials can be assumed to be potential eye irritants.

Dermal Irritation and Sensitization   These are tests for dermal irritation caused by topical application of chemicals and fall into four general categories: primary irritation, cutaneous sensitization, phototoxicity, and photosensitization. Because many foreign chemicals come into direct contact with the skin, including cosmetics, detergents, bleaches, and many others, these tests are considered essential to the proper regulation of such products. Less commonly, dermal effects may be caused by systemic toxicants.

In the typical primary irritation test, the backs of albino rabbits are clipped free of hair and an area of about $5\,cm^2$ on each rabbit is used in the test. This area is then treated with either $0.5\,mL$ or $0.5\,g$ of the compound to be tested and then covered with a gauze pad. The entire trunk of the rabbit is wrapped to prevent ingestion. After 4–24 h, the tape and gauze are removed, the treated areas are evaluated for erythematous lesions (redness of the skin produced by congestion of the capillaries) and edematous lesions (accumulation of excess fluid in SC tissue), each of which is expressed on a numerical scale. After an additional 24–48 h, the treated areas are again evaluated.

Skin sensitization tests are designed to test the ability of chemicals to affect the immune system in such a way that a subsequent contact causes a more severe reaction than the first contact. The latter may be elicited at a much lower concentration and in areas beyond the area of initial contact. The antigen involved is presumed to be formed by the binding of the chemical to body proteins, the ligand–protein complex then being recognized as a foreign protein to which antibodies can be formed. Subsequent exposure may then give rise to an allergic reaction. Skin sensitization tests generally follow protocols that are modifications of the Buchler (dermal inductions) method or the Magnesson and Kligman (intradermal inductions) method. The test animal commonly used in skin sensitization tests is the guinea pig; animals are treated with the test compound in a suitable vehicle, with the vehicle alone, or with a positive control such as 2, 4-dinitrochlorobenzene (a relatively strong sensitizer) or cinnamaldehyde (a relatively weak sensitizer) in the same vehicle. During the induction phase, the animals are treated for each of 3 days evenly spaced during a 2-week period. This is followed by a 2-week rest period followed by the challenge phase of the test. This consists of a 24-h topical treatment carried out as described for primary skin irritation tests. The lesions are scored on the basis of severity and the number of animals responding (incidence). If there is a greater skin reaction in the animals given induction doses compared to those given the test material for the first time, the compound is considered to be a dermal sensitizer.

Other test methods include those in which the induction phase is conducted by intradermal injection together with Freund's adjuvant (a chemical mixture that enhances the antigenic response) and the challenge by dermal application, or tests in which both induction and challenge doses are topical but the former is accompanied by intradermal injections of Freund's adjuvant. It is important that compounds that cause primary skin irritation be tested for skin sensitization at concentrations low enough that the two effects are not confused.

Phototoxicity tests are designed to evaluate the combined dermal effects of light (primarily ultra violet [UV] light) and the chemical in question. Tests have been developed for both phototoxicity and photoallergy. In both cases, the light energy is believed to cause a transient excitation of the toxicant molecule, which, on returning to the lower energy state, generates a reactive, free-radical intermediate. In phototoxicity, these organic radicals act directly on the cells to cause lesions, whereas in photoallergy, they bind to body proteins. These modified proteins then stimulate the immune system to produce antibodies, because the modifications cause them to be recognized as foreign or "nonself" proteins. These tests are basically modifications of the tests for primary irritation and sensitization except that, following application of the test chemical, the treated area is irradiated with UV light. The differences between the animals treated and irradiated and those treated and not irradiated is a measure of the phototoxic effect.

Safety Pharmacology Studies Safety pharmacology studies investigate the potential undesirable pharmacodynamic effects of a test article on physiological functions in relationship to exposure. These tests are typically conducted as part of the development of new drugs. The objectives of safety pharmacology studies are threefold. First, to identify undesirable effects of a test article which may have relevance to its use in humans. Second, to evaluate a test article for possible effects observed in toxicology or clinical studies. And third, to investigate the mechanism underlying any undesirable effects of the test article. Safety pharmacology consists of a core battery of studies with follow-up studies as indicated by preliminary findings. The core studies are designed to target vital organ systems, particularly the central nervous system, cardiovascular system, and pulmonary system. These studies are typically conducted using small numbers of rats and dogs. In the study for pulmonary function, end points measured are respiratory rate, minute volume, and tidal volume. In the cardiovascular telemetry study, end points include heart rate, blood pressure, and electrocardiogram evaluation. In telemetry studies, a radio transmitter is implanted in all animals to permit continuous monitoring for 24 h pretest and 24 h after dosing. A cardiopulmonary study can also be conducted in which respiratory rate, minute volume, tidal volume, blood pressure, heart rate, electrocardiogram, and body temperature are monitored in restrained animals for typically 2 h after dosing.

20.5.2 Subchronic Tests

Subchronic tests examine toxicity caused by repeated dosing over an extended period, but not one that that constitutes a significant portion of the normal life span of the species tested. A 28- or 90-day oral study in the rat or dog would be typical of this type of study, as would a 21- to 28-day dermal application study or a 28- to 90-day inhalation study. Such tests provide information on essentially all types of chronic toxicity other than carcinogenicity and are usually believed essential for establishing the dose regimens for prolonged chronic studies. They are frequently used as the basis for the determination of the no observable effect level (NOEL). This value is often defined as the highest dose level at which no deleterious or abnormal effect can be measured, and is used in risk assessments. Subchronic tests are also useful in providing information on target organs and on the potential of the test chemical to accumulate in the organism.

Ninety-Day Tests Chemicals are usually tested by administration in the diet, less commonly in the drinking water, and only when absolutely necessary, by gavage, because the last process involves much handling and subsequent stress. Numerous experimental variables must be controlled and biologic variables must be evaluated. In addition, the number of end points that can be measured is large and, as a consequence, record keeping and data analysis must be carefully planned. If all is done with care, much may be learned from such tests.

Experimental (Nonbiologic) Variables   Several environmental variables may affect toxicity evaluations, some directly and others by their effects on animal health. Major deviations from the optimum temperature and humidity for the species in question can cause stress reactions. Stress can also be caused by housing more than one species of experimental animal in the same room. Many toxic or metabolic effects show diurnal variations that are related to photoperiod. Cage design and the nature of the bedding have also been shown to affect the toxic response. Thus, the optimum housing conditions are clean rooms, each containing a single species, with the temperature, humidity, and photoperiod being constant and optimized for the species in question. Cages should be the optimum design for the species, bedding should be inert (not cause enzyme induction or other metabolic effect), and cages should not be overcrowded, with individual caging whenever possible.

Dose selection, preparation, and administration are all important variables. Subchronic studies are usually conducted using three (less often, four) dose levels. The highest should produce obvious toxicity but not high mortality and the lowest no toxicity (NOEL), whereas the intermediate dose should give effects clearly intermediate between these two extremes. Although the dose can be extrapolated from acute tests, such extrapolation is difficult, particularly in the case of compounds that accumulate in the body; and frequently, a 14-day range finding study is made. Although the route of administration should ideally mimic the expected route of exposure in humans, in practice, the chemical is usually administered *ad libitum* in the diet, because this is, on average, most appropriate. Diets containing known amounts of the test material are presented to the animals. Measurement of food consumption is recommended to provide an estimate of the test material consumed. In cases in which a highly accurate measurement of dose is an important factor in the experimental design, the animals may be treated by gavage or by capsules containing the test material.

To avoid effects from nonspecific variations on the diet, enough feed from the same batch should be obtained for the entire study. Part is set aside for the controls, and the remainder is mixed with the test chemical at the various dose levels. Care should be taken to store all food in such a way that not only does the test chemical remain stable, but the nutritional value is also maintained. The identity and concentration of the test chemical should be checked periodically by chemical analysis. Treated diets may be prepared at set intervals, such as weekly, depending upon the stability of the test material in the diet.

Subchronic studies are usually conducted with 10–20 males and 10–20 females of a rodent species at each dose level and 4–8 of each gender of a larger species, such as the dog, at each dose level. Animals should be drawn from a larger group and assigned to control or treatment groups by a random process, but the larger group should not vary so much that the mean weights and ages of the subgroups vary significantly at the beginning of the experiment.

Biologic Variables   Subchronic studies should be conducted on two species, ideally a rodent, and a nonrodent. Ideally, the species chosen should be those with the greatest pharmacokinetic and metabolic similarity to humans; for the compound in question, this information is seldom available. In practice, the most common rodent used is the rat, and the most common nonrodent used is the dog. It has long been held that inbred rodent strains should be used to reduce variability. This and the search for strains that were sensitive to chemical carcinogenesis but did not have an unacceptably high spontaneous tumor rate led to widespread use of the F344 rat and B6C3F1 mouse. Other researchers believe that an outbred strain such as the Sprague Dawley rat is more robust and prefer to use them.

Although ideally the age should be matched to the expected exposure period in terms of the stage of human development, this is not often done. Young adult or adolescent animals that are still growing are preferred in almost all cases, and both sexes are routinely used.

Good animal care is critical at all times because toxicity has been shown to vary with diet, disease, and environmental factors. Animals should be quarantined for some time before being admitted to the test area, their diet should be optimum for the test species, and the facility should be kept clean at all times. Regular inspection by a veterinarian is essential, and any animal showing unusual symptoms not related to the treatment (e.g., in controls or in low-dose but not high-dose animals) should be removed from the test and autopsied.

Results   Although the information required from subchronic tests varies somewhat from one regulatory agency to another, the requirements are basically similar (Table 20.5).

For explanatory purposes, the data obtained from these tests can be described as two types: that which can be obtained from living animals during the course of the test and that which is obtained from animals sacrificed either during or at the end of the test period. Many of the tests performed on living animals can be carried out first before the test period begins to provide a baseline for comparison to subsequent measurement. A satellite group of treated animals can be added to the test for evaluation of "recovery." For these animals, the treated food would be removed at the end of the test period, and they would be returned to the control diet for 21–28 days while the various end points are followed. This is done to establish whether any effects noted are reversible. Autopsies should be performed on all animals found dead or moribund during the course of the test. The following is a list of end points that may be measured during a 90-day oral toxicity study.

A. In-Life Tests. Interim tests are carried out at intervals before the study, to establish baselines at intervals during the study, and at the end of the study.
   1. Appearance—Mortality and morbidity as well as the condition of the skin, fur, mucous membranes, and orifices should be checked at least daily. Presence of palpable masses or external lesions should be noted.
   2. Eyes—Ophthalmologic examination of both cornea and retina should be carried out at the beginning and at the end of the study.
   3. *Food consumption*
   4. *Body weight*
   5. *Behavioral abnormalities*
   6. *Respiration rate*
   7. ECG—Particularly with the larger animals
   8. EEG—particularly with the larger animals.

TABLE 20.5 Summary of Subchronic Test Guidelines by Regulatory Agency

Character of Tests	EPA Pesticide Assessment Guidelines	FDA "Red Book"	FDA IND/NDA Pharmacology Review Guidelines	OECD	EPA Health Effects Guidelines	NTP
Purpose	Pesticide registration support	Food and color additives; safety assessment	IND/NDR pharmacology review guidelines	Assessment and evaluation of toxic characteristics	Select chronic dose levels	Predict dose range for chronic study
	No observed effect level	No observed adverse effects, no effect level	Characterize pharmacology, toxicology, pharmacokinetics, and metabolism of drugs for precautionary clinical decisions	Select chronic dose levels. Use information and permissible human exposure	Establish safety criteria for human exposure, no observed effect level	
Species	Rat, dog, mouse	Rat, dog	Rat, mouse, other rodents, dog, monkey, other non-rodents	Rat, dog	Rat, dog	Fischer 344 rats, B6CF1 mice
Doses	3 dose levels	3 dose levels	3 dose levels	3 dose levels	3 dose levels	5 dose levels
End points	Clinical signs	Clinical signs				
	Ophthalmology	Ophthalmology	Ophthalmology	Ophthalmology	Ophthalmology	Ophthalmology
	Hematology	Hematology	Hematology	Hematology	Hematology	Hematology
	Clinical chemistry	Clinical chemistry	Clinical chemistry	Clinical chemistry	Clinical chemistry	Clinical chemistry
						Weight loss
	Histopathology	Histopathology	Histopathology	Histopathology	Histology	Histopathology
	Target organs	Target organs	Target organs	Target organs	Target organs	Target organs
			Behavioral and pharmacological effects			

Source: National Toxicology Program, Washington, DC: Department of Health and Human Services. Report of the NTP Ad hoc Panel on Chemical Carcinogenesis Testing and Evaluation.

EPA, Environmental Protection Agency; FDA, Food and Drug Administration; IND/NDA, investigative new drug/new drug assessment; OECD, Organization for Economic Cooperation and Development; NTP, National Toxicology Program.

9. Hematology—Assessment should be made prior to chemical administration (pretest) and at least prior to termination. Hemoglobin, hematocrit, RBC, WBC, differential counts, platelets, reticulocytes, and clotting parameters should be assessed.

10. Blood Chemistry—Should be done pretest and at least prior to termination. Electrolytes and electrolyte balance; acid–base balance; glucose; urea nitrogen; serum lipids; serum proteins (albumin-globulin ratio); enzymes indicative of organ damage such as transaminases and phosphatases; also, plasma and RBC cholinesterase levels should be measured. Toxicant and metabolite levels should be assessed as needed.

11. Urinalysis—Should be done pretest and at least prior to termination. Microscopic appearance (sediment, cells, stones, etc.), pH, specific gravity, chemical analysis for reducing sugars, proteins, ketones, bilirubin etc., as well as toxicant and metabolite levels should all be assessed.

12. Fecal Analysis—Occult blood, fluid content, and toxicant and metabolite levels should be assessed.

B. Termination Tests. Because the number of tissues that may be sampled is large (Table 20.6) and the number of microscopic methods is also large, it is necessary to consider all previous results before carrying out the pathological examination. For example, clinical tests or blood chemistry analyses may implicate a particular target organ that can then be examined in greater detail. All control and high dose animals are examined in detail. If lesions are found, the next lowest dose group is examined for these lesions, and this method continues until a no effect group is reached.

Because pathology is largely a descriptive science with a complex terminology that varies form one practitioner to another, it is critical that the terminology be defined at the beginning of the study and that the same pathologist examine the slides form both treated and control animals. Pathologists are not in agreement on the necessity or the wisdom of coding slides so that the assessor is not aware of the treatment given the animal from which a particular slide

TABLE 20.6 Tissues and Organs to be Examined Histologically in Chronic and Subchronic Toxicity Tests

Adrenals	Larynx	Salivary gland
Bone and bone marrow	Liver	Sciatic nerve
Brain	Lungs and bronchi	Seminal vesicles
Cartilage	Lymph nodes	Skin
Cecum	Mammary glands	Spinal cord
Colon	Mandibular lymph node	Spleen
Duodenum	Mesenteric lymph node	Stomach
Esophagus	Nasal cavity	Testes
Eyes	Ovaries	Thigh muscle
Gallbladder	Parathyroids	Thymus
Ileum	Pituitary	Urinary bladder
Jejunum	Prostate	Uterus
Kidneys	Rectum	

is derived. Such coding, however, eliminates unintentional bias, a hazard in a procedure that depends on subjective evaluation. Other items of utmost importance are quality control, slide identification, and data recording. Many tissues may be examined; consequently, an even larger number of tissue blocks must be prepared. Because each of these may yield many slides to be stained, comparable quality of staining and the accurate correlation of a particular slide with its parent block, tissue, and animal is critical.

1. *Necropsy*—This must be conducted with care to avoid postmortem damage to the specimens. Tissues are removed, weighed, and examined closely for gross lesions, masses, etc. Tissues are then fixed in buffered formalin for subsequent histologic examination.

2. *Histology*—The tissues listed in Table 20.6 plus any lesions, masses, or abnormal tissues are embedded, sectioned, and strained for light microscopy. Parafin embedding and staining with hemotoxylin and eosin are the preferred routine methods, but special stains may be used for particular tissues or for a more specific examination of certain lesions. Electron microscopy may also be used for more specific examination of lesions or cellular changes after their initial localization by more routine methods.

Repeated Dose Dermal Tests   Twenty-one- to twenty-eight-day dermal tests are particularly important when the expected route of human exposure is by contact with the skin, as is the case with many industrial chemicals or pesticides. Compounds to be tested are usually applied daily to clipped areas on the back of the animal, either undiluted or in a suitable vehicle. In the latter case, if a vehicle is used, it is also applied to the controls. Selection of a suitable solvent is difficult because many affect the skin, causing either drying or irritation, whereas others may markedly affect the rate of penetration of the test chemical. Corn oil, ethanol, or carboxymethyl cellulose are preferred to dimethyl sulfoxide (DMSO) or acetone. It should also be considered that some of the test chemical may be ingested as a result of grooming by the animal, although this can be controlled to some extent by use of restraining collars and/or wrapping.

The criteria for environment, dose selection, species selection, and so on, are not greatly different to the criteria used for 90-day feeding tests, although the list of end points to be examined is often shorter (e.g., fewer organs may be examined). It is necessary, however, to pay close attention to the skin at the point of application because local effects may be as important as systemic ones.

Twenty-Eight to Ninety-Day Inhalation Tests   Inhalation studies are indicated whenever the route of exposure is expected to be through the lungs. Animals are commonly exposed for 6–8 h each day, 5 days each week, in chambers of the type previously discussed. Even in those cases in which the animals are maintained in the inhalation chambers during nonexposure hours, food is always removed during exposure. In spite of this, exposure tends to be in part dermal and, due to grooming of the fur, in part oral. Environmental and biologic parameters are the same as for other subchronic tests, as are the routine end points to be measured before, during, and after the test period. Particular attention must be paid, however, to effects on the tissues of the nasal cavity and the lungs, because these are the areas of maximum exposure.

If the test material is particulate, consideration must be given to the particle size and its inhalation potential. Particles of 4μ in size are considered to be inhalable; larger particles will be cleared from the respiratory tract by ciliary action and subsequently swallowed (oral exposure) or expelled by sneezing or expectoration.

20.5.3   Chronic Tests

Chronic tests are those conducted over a significant part of the life span of the test animal. The duration of a chronic study is generally 1 year or more. Typically, rat and dog are the preferred species; for carcinogenicity studies, rats and mice are used.

Chronic Toxicity and Carcinogenicity   Descriptions of tests for both chronic toxicity and carcinogenicity are included here because the design is similar—so similar in fact that they can be combined into one test. Chronic toxicity tests are designed to discover any of numerous toxic effects and to define safety margins to be used in the regulation of chemicals. As with subchronic tests, two species are usually used, one of which is either a rat or a mouse strain, in which case the tests are run for 2 years or 1.5–2.0 years, respectively. Data is gathered after 1 year to determine chronic effects without potential confounding effects of aging. Data are gathered after 1.5 years (mouse) or 2 years (rat) to determine carcinogenic potential. The non-rodent species used may be the dog, a nonhuman primate, or, rarely, a small carnivore such as the ferret. Chronic toxicity tests may involve administration in the food, in the drinking water, by capsule, or by inhalation, the first being the most common. Gavage is rarely used. The dose used is the maximum tolerated dose (MTD) and usually two lower doses, perhaps 0.25 MTD and 0.125 MTD with the lowest dose being a predicted no effect level.

MTD   The MTD has been defined for testing purposes by the U.S. EPA as:

> the highest dose that causes no more than a 10% weight decrement, as compared to the appropriate control groups; and does not produce mortality, clinical signs of toxicity, or pathologic lesions (other than those that may be related to a neoplastic response) that would be predicted to shorten the animals' natural life span.

This dose is determined by extrapolation from subchronic studies.

The requirements for animal facilities, housing, and environmental conditions are as described for subchronic studies. Special attention must be paid to diet formulation because it is impractical to formulate all of the diets for a 2-year study form a single batch. In general, semisynthetic diets of specified components should be formulated regularly and analyzed before use for test material content.

The end points used in these studies are those described for the subchronic study: appearance, ophthalmology, food consumption, body weight, clinical signs, behavioral signs, hematology, blood chemistry, urinalysis, organ weights, and pathology. Some animals may be killed at fixed intervals during the test (e.g., 6, 12, or 18 months) for histologic examination. Particular attention is paid to any organs or tests that showed compound-related changes in the subchronic tests.

Carcinogenicity tests have many requirements in common (physical facilities, diets, etc.) with both chronic and subchronic toxicity tests as previously described. Because of the numbers and time required, these tests are usually carried out using rats and/or mice, but in some cases, a non-rodent species may also be used. The chemical under test may be administered in the food, in the drinking water, by dermal application, by gavage or by inhalation, the first two methods being the most common. Because the oncongenic potency of chemicals varies through extreme limits, the purity of the test chemical is of great concern. A 1% contaminant need only be 100 times as potent as the test chemical to have an equivalent effect, and differences of this magnitude and greater are not unheard of.

Dosing is carried out over the major part of the life span for rodents, beginning at or shortly after weaning. The highest dose used is the MTD. The principal end point is tumor incidence as determined by histologic examination. The statistical problem of distinguishing between spontaneous tumor occurrence in the controls and chemical-related tumor incidence in the treated animals is great; for that reason, large numbers of animals are used. A typical test involves 50 or more rats or mice of each gender in each treatment group. Some animals are necropsied at intermediate stages of the test (e.g., at 12 months), as are all animals found dead or moribund. All surviving animals are necropsied at the end of the test. Tissues to be examined are listed in Table 20.6, with particular attention being paid to abnormal masses and lesions.

Reproductive Toxicity and Teratogenicity   The aim of developmental and reproductive testing is to examine the potential for a compound to interfere with the ability of an organism to reproduce. This includes testing to assess reproductive risk to mature adults as well as the developing individual at various stages of life, from conception to sexual maturity. Traditionally, animal studies have been conducted in three "segments": (I) in adults, treatment during a premating period and optionally continuation for the female through implantation or lactation; (II) in pregnant animals treatment during the major period of organogenesis; and (III) treatment of pregnant/lactating animals from the completion of organogenesis through lactation (peri- and postnatal study). Although guidelines addressing treatment regimens have been rather similar throughout the world, required end points measured in adults and developing organisms have varied. International harmonization of guidelines has shown a need for flexibility in testing for reproductive and developmental toxicity, and toxicologists are now often challenged to design unique studies to examine potential effects on all the parameters considered in the classical segment I, II, and III studies. In adults, these include development of mature egg and sperm, fertilization, implantation, delivery of offspring (parturition), and lactation. In the developing organism, these include early embryonic development, major organ formation, fetal development and growth, postnatal growth including behavioral assessments, and attainment of full reproductive function. These evaluations are usually best carried out in several separate studies.

Some Definitions in Reproductive Biology   At this point, some discussion of reproductive biology is helpful in the understanding of study designs to evaluate reproductive and developmental toxicity. Tests to assess general reproductive performance and fertility are generally conducted using rats. In the rat, multiple eggs are ovulated from mature follicles in the ovary. The follicle that an egg leaves behind develops into glandular tissue known as a corpus luteum. The corpora lutea secrete progesterone, a hormone needed to maintain pregnancy (unlike humans in which progesterone is secreted by the placenta). Corpora lutea are visible as blister-like protuberances on the ovary. A count of the corpora lutea in the ovary allows one to determine the maximum number of potential offspring for that pregnancy. Fertilized eggs develop into zygotes that may attach to the wall of the uterus (implantation). The discrete areas of implantation may be observed and counted upon examination of the uterus at C-section. Calculation of pre- and postimplantation loss are important end points in a reproductive toxicity study. Preimplantation loss is the death of a fertilized ova prior to implantation in the uterine wall. Postimplantation loss (i.e., resorption and/or fetal death) is the death of the conceptus after implantation in the uterine wall and prior to parturition. Postimplantation loss can be broken down into early and late resorptions and fetal death. A late resorption has discernable features such as limbs, eyes, and nose, whereas an early resorption has none of these features.

Single and Multiple Generation Tests   Fertility and general reproductive performance can be evaluated in single and multiple generation tests. These tests are usually conducted using rats. Fertility is defined as the ability to produce a pregnancy while the ability to produce live offspring is known as fecundity. An abbreviated protocol for a single generation test is shown in Figure 20.2.

In typical tests, 25 males per dose group are treated for 70 days prior to mating and 25 females per dose group are treated for 14 days premating. The number of animals is chosen to yield at least 20 pregnant females per dose group including controls. The treatment durations are selected to coincide with critical times during which spermatogenesis and ovulation occur. It takes approximately 70 days in the rat for spermatogonial cells to become mature sperm capable of fertilization. In the female rat, the estrus cycle length is 4–5 days and a 14-day dosing period is considered sufficient time to detect potential effects on hormonal or other systems which may effect ovulation. In some study designs, both males and females are treated for 70 days premating. Treatment of the females is continued through pregnancy (21 days) and until the pups are weaned. Pups are usually 21 days of age. The test compound is administered at three dose levels either in the feed, in drinking water, or by gavage. The high dose is chosen to cause some, but not excessive, maternal toxicity (e.g., an approximate 10% decrease in body weight gain, or effects on target organs). Low doses are generally expected to be no-effect levels.

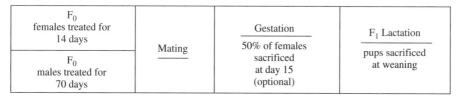

Figure 20.2   Abbreviated protocol for a one generation reproductive toxicity test.

After the premating period, the rats are placed in cohabitation, with one male and one female caged together. Mating is confirmed by the appearance of spermatozoa in a daily vaginal smear. Day 1 of gestation is the day insemination is confirmed. The females bear and nurse their pups. After birth, the pups are counted, weighed, and examined for external abnormalities. The litters are frequently culled to a constant number (usually 8–10) after 4 days. At weaning, the pups are killed and autopsied for gross and internal abnormalities. In a multigeneration study, approximately 25 of each gender/group are saved to produce the next generation. Brother-sister pairings are avoided. Treatment is continuous throughout the test, which can be carried out for two, sometimes, three generations. An abbreviated protocol for a multiple generation test can be seen in Figure 20.3. Note that the parental generation is known as the F_0 generation and the offspring are known as the F_1's and F_2's. In some studies, parents produce two litters, for example the F_1A and F_1B litters.

Because both males and females are treated in this type of study design, it is not possible to distinguish between maternal and paternal effects in the reproductive

| F_0 Females treated for 70 days | Mating #1 | Gestation | F_1A Lactation — pups sacrificed at weaning |
| F_0 Males treated for 70 days | Mating #2 | Gestation | F_1B Lactation — pups sacrificed at weaning—enough left for next generation |

| F_1B Females continued on test | Mating #1 | Gestation | F_2A Lactation — pups sacrificed at weaning |
| F_1B Males continued on test | Mating #2 | Gestation | F_2B Lactation — pups sacrificed at weaning—enough left for next generation |

| F_2B Females continued on test | Mating #1 | Gestation | F_3A Lactation — pups sacrificed at weaning |
| F_2B Males continued on test | Mating #2 | Gestation | F_3B Lactation — pups sacrificed at weaning—complete histology |

Figure 20.3   Abbreviated protocol for a multigeneration reproductive toxicity test.

performance. To permit this separation, it is necessary to treat additional animal to the stage of mating and then breed them to untreated members of the opposite sex. Similarly, if effects are seen postnatally, it may not be possible to distinguish between effects mediated *in utero* or mediated by lactation. This distinction can be made by "cross-fostering" the offspring of treated females to untreated females and vice versa.

The end points observed in these types of tests, depending on study design, are as follows:

1. Fertility index, the number of pregnancies relative to the number of matings
2. The number of live births, relative to the number of total births
3. Preimplantation death, or number of corpora lutea in the ovaries relative to the number of implantation sites
4. Postimplantation death, or the number of resorption sites in the uterus relative to the number of implantation sites
5. Duration of gestation
6. Effects on male or female reproductive systems
7. Litter size and condition, gross morphology of pups at birth, gender and ano-genital distance
8. Survival of pups
9. Weight gain and performance of pups
10. Time of occurance of developmental landmarks, for example, eye opening, tooth eruption, vaginal opening in females, preputial separation in males
11. Morphological abnormalities in weanlings

Results from single and multiple generation tests provide important information for assessment of test materials that may perturb a variety of systems including the endocrine system. A number of variations of the single and multiple generation tests exist. For example, a number of weanlings may be left to develop and be tested later for behavioral and/or physiological defects (e.g., developmental neurotoxicity testing).

Teratology is the study of abnormal fetal development. For an agent to be labeled a teratogen, it must significantly increase the occurence of adverse structural or functional abnormalities in offspring after its administration to the female during pregnancy or directly to the developing organism. In teratology testing, exposure to the test chemical may be from implantation to parturition, although it has also been restricted to the period of major organogenesis, the most sensitive period for inducing structural malformations. Observations may be extended throughout life, but usually, they are made immediately prior to birth after a C-section. The end points observed are mainly morphologic (structural changes and malformations), although embryo-fetal mortality is also used as an end point. Figure 20.4 shows an outline of a typical teratology study.

Teratology studies are carried out in two species, a rodent species (usually the rat) and in another species such as the rabbit (rarely in the dog or primate). Enough females should be used so that, given normal fertility for the strain, there are 20 pregnant females in each dosage group. Traditionally, the timing of compound

Teratology

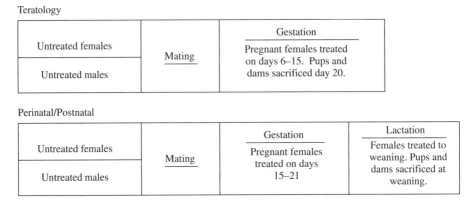

Untreated females	Mating	Gestation Pregnant females treated on days 6–15. Pups and dams sacrificed day 20.
Untreated males		

Perinatal/Postnatal

Untreated females	Mating	Gestation Pregnant females treated on days 15–21	Lactation Females treated to weaning. Pups and dams sacrificed at weaning.
Untreated males			

Figure 20.4 Abbreviated protocol for a teratology test and for a perinatal/postnatal toxicity test.

administration has been such that the dam is exposed during the period of major organogenesis, that is, days 6 through 15 in the rat or mouse and days 6 through 18 for the rabbit. Newer study designs call for dosing until C-section. Day 1 is the day spermatozoa appear in the vagina in the case of rats, or the day of mating in the rabbit.

The test chemical is typically administered directly into the stomach by gavage which is a requirement of the EPA and some other regulatory agencies. This method of dosing allows a precise calculation of the amount of test material received by the animal. Studies typically have three dose levels and a control group that receives the vehicle used for test material delivery. The high dose level is chosen to be one at which some maternal toxicity is known to occur, but never one that would cause more than 10% mortality. The low dose should be one at which no maternal toxicity is apparent, and the intermediate dose(s) should be chosen as a predicted low effect level.

The test is terminated by performing a C-section on the day before normal delivery is expected. The uterus is examined for implantation and resorption sites and for live and dead fetuses and the ovaries are examined for corpora lutea. In rodent studies, half of the fetuses are examined for soft tissue malformations, and the remaining are examined for skeletal malformations. In nonrodents, all fetuses are examined for both soft tissue and skeletal malformations. The various end points that may be examined include maternal toxicity, embryo-fetal toxicity, external malformations, and soft tissue and skeletal malformations.

Careful evaluation of maternal toxicity is necessary in assessing the validity of the high-dose level and the possibility that maternal toxicity is involved in subsequent events. The parameters evaluated include body weight, food consumption, clinical signs, and necropsy data such as organ weights. Because exposure starts after implantation, conception and implantation rates should be the same in controls and all treatment levels. If not, the test is suspect, with a possible error in the timing of the dose or use of animals from a source unsuitable for this type of testing.

Embryo-fetal toxicity is determined from the number of dead fetuses and resorption sites relative to the number of implantation sites. In addition to the possibility of lethal malformations, such toxicity can be due to maternal toxicity, stress, or direct toxicity to the embryo or fetus that is not related to developmental malformations. Fetal weight and fetal size may also be a measure of toxicity but should not be confused with the variations seen as a result of differences in the number of pups per litter. Smaller litters tend to have larger pups while larger litters have smaller pups.

Anomalies may be regarded as either variations that may not adversely affect the fetus and not have a fetal outcome, or as malformations that are considered to have adverse effects on the fetus. For some findings, there is disagreement as to which class it belongs, such as the number of ribs in the rabbit which inherently has a large amount of variability. Common external anomalies are listed in Table 20.7

TABLE 20.7   External Malformations Commonly Seen in Teratogenicity Tests

Brain, cranium, spinal cord
  Encephalocele—protrusion of brain through an opening of the skull.
  Cerebrum is well formed and covered by transparent connective tissue.
  Exencephaly—lack of skull with disorganized outward growth of the brain.
  Microcephaly—small head on normal sized body.
  Hydrocephaly—marked enlargement of the ventricles of the cerebrum.
  Craniorachischisis—exposed brain and spinal cord.
  Spina bifida—Nonfusion of spinal processes. Usually ectoderm covering is missing and
    spinal cord is evident.
Nose
  Enlarged naris—enlarged nasal cavities
  Single naris—a single naris, usually median
Eye
  Microphthalmia—small eye
  Anophthalmia—lack of eye
  Open eye—no apparent eyelid, eye is open
Ear
  Anotia—absence of the external ear
  Microtia—small ear
Jaw
  Micrognathia—small lower jaw
  Agnathia—absence of lower jaw
  Aglossia—absence of tongue
  Astomia—lack of mouth opening
  Bifid tongue—forked tongue
  Cleft lip—either unilateral or bilateral cleft of upper lip
Palate
  Cleft palate—a cleft or separation of the median portion of the palate
Limbs
  Clubfoot—foot that has grown in a twisted manner, resulting in an abnormal shape or
    position. It is possible to have a malposition of the whole limb.
  Micromelia—abnormal shortness of the limb.
  Hemimelia—absence of any of the long bones, resulting in a shortened limb.
  Phomelia—absence of all of the long bones of a limb, the limb is attached directly to the
    body.

and are determined by examination of fetuses at C-section. Visceral anomalities are determined by examination of fetuses after fixation using either the dissection method of Staples or by the hand-sectioning method of Wilson. Common visceral findings are listed in Table 20.8. Fetal skeletons are examined after first fixing the fetus and then staining the bone with Alizarin Red. Numerous skeletal variations occur in controls and may not have an adverse effect on the fetus (Table 20.9).

TABLE 20.8 Some Common Visceral Anomalies Seen in Teratogenicity Tests

Intestines
  Umbilical hernia—protrusion of the intestines into the umbilical cord
  Ectopic intestines—extrusion of the intestines outside the body wall
Heart
  Dextrocardia—rotation of the heart axis to the right
  Enlarged heart—either the atrium or the ventricle may be enlarged
Lung
  Enlarged lung—all lobes are usually enlarged
  Small lung—all lobes are usually small. Lung may appear immature
Uterus/testes
  Undescended testes—testes are located anterior to the bladder instead of lateral; may be bilateral or unilateral
  Agenesis of testes—one or both testes may be missing
  Agenesis of uterus—one or both horns of the uterus may be missing
Kidney
  Hydronephrosis—fluid-filled kidney, often grossly enlarged; may be accompanied by a hydroureter (enlarged, fluid-filled ureter)
  Fused—kidneys fused, appearing as one misshapen kidney with two ureters
  Agenesis—one or both kidneys missing
  Misshapen—small, enlarged (usually internally), or odd-shaped kidneys

TABLE 20.9 Skeletal Abnormalities Commonly Seen in Teratogenicity Tests

Digits
  Polydactyly—presence of extra digits, in mouse six or more, instead of five
  Syndactyly—fusion of two or more digits
  Oligodactyly—absence of one or more digits
  Brachydactyly—smallness of one or more digits
Ribs
  Wavy—ribs may be any aberrant shape
  Extra—may have extra ribs on either side
  Fused—may be fused anywhere along the length of the rib
  Branched—single base and branched
Tail
  Short—short tail, usually lack of vertebrae
  Missing—absence of tail
  Corkscrew—corkscrew-shaped tail

Their frequency of occurrence may, however, be dose related and should be evaluated.

Almost all chemically induced malformations have been observed in control animals, and most malformations are known to be produced by more than one cause. Thus, it is obvious that great care is necessary in the interpretation of teratology studies. For an agent to be classified as a development toxicant or teratogen, it must produce adverse effects on the conceptus at exposure levels that do not induce toxicity in the mother. Signs of maternal toxicity include reduction in weight gain, changes in eating patterns, hypo or hyperactivity, neurotoxic signs, and organ weight changes. Adverse effects on development under these conditions may be secondary to stress on the maternal system. Findings in the fetus, at dose levels that produce maternal toxicity cannot be easily separated from the maternal toxicity. Compounds can be deliberately administered at maternally toxic dose levels to determine the threshold for adverse effects on the offspring. In such cases, conclusions can be qualified to indicate that adverse effects on the offspring were found at maternally toxic dose levels and may not be indicative of selective or unique developmental toxicity.

Effect of Chemicals in Late Pregnancy and Lactation (Perinatal and Postnatal Effects)   These tests are usually carried out on rats, and 20 pregnant females per dosage group are treated during the final third of gestation and through lactation to weaning (day 15 of pregnancy through day 21 postpartum) (Figure 20.4). The duration of gestation, parturition problems, and the number and size of pups in the naturally delivered litter are observed, as is the growth performance or the offspring. Variations of this test are the inclusion of groups treated only to parturition and only postpartum in order to separate prenatal and postnatal effects. Cross-fostering of pups to untreated dams may also be used to the same end. Behavioral testing of the pups has been suggested, and this and other physiological testing are to be recommended.

20.5.4 Special Tests

This general heading is used to include brief assessments of tests that are not always required but that may be required in particular cases or have been suggested as useful adjuncts to current testing protocols.

Neurotoxicity   The nervous system is complex, both structurally and functionally, and toxicants can affect one or more units of this system in selective fashion. It is necessary, therefore, to devise tests, or sequences of tests that measure not only changes in overall function but that also indicate which basic unit is affected and how the toxicant interacts with its target. This is complicated by the fact that the nervous system has a considerable functional reserve, and specific observable damage may not affect overall function until it becomes even more extensive. Types of damage to the nervous system are classified in various ways but include neuronal toxicity, axonopathy, toxic interruption of impulse transmission, myelinopathy, and synaptic alterations in transmitter release or receptor function. Signs of neuropathy are frequently revealed by the acute, subchronic, chronic, and other tests that are required by regulatory agencies. Neurotoxicity is of great significance in toxicology, however, and tests have been devised to supplement those routinely required. These include acute and subchronic neurotoxicity studies as well as developmental neurotoxicity studies.

Behavioral and Pharmacological Tests Behavioral and pharmacological tests involve the observation of clinical signs and behavior. These include signs of changes in awareness, mood, motor activity, central nervous system excitation, posture, motor incoordination, muscle tone, reflexes, and autonomic functions. If these tests so indicate, more specialized tests can be carried out that evaluate spontaneous motor activity, conditioned avoidance responses, operant conditioning, as well as tests for motor incoordination such as the inclined plane or rotarod tests.

Tests for specific classes of chemicals include the measurement of transmitter stimulated adenyl cyclase and Na/K-ATPase for chemicals that affect receptor function or cholinesterase inhibition for organophosphates or carbamates. Electrophysiological techniques may detect chemicals such as DDT or pyrethroids, which affect impulse transmission.

Acute and Subchronic Neurotoxicity Tests Acute and Subchronic Neurotoxicity Tests may be designed to assess a wide range of effects including CNS stimulation or depression, reflex perturbation, peripheral nerve damage, cognitive effects on learning and memory, motor activity effects, and neuropathology. These tests are conducted using rats or sometimes mice.

In the acute neurotoxicity study, approximately 10–15 animals per sex per dose group are administered a single gavage (bolus) dose of the test material. There are usually three dose groups and a control. Behavioral assessments are made on the day of dosing, and at 1 and 2 weeks post dose. The assessments include tests on motor activity, a functional observation battery (FOB), and neuropathology (at termination). The FOB screens for sensorimotor, neuromuscular, autonomic, and general physiological effects of a test compound. Table 20.10 depicts component tests of the FOB. These functional tests have the advantage over biochemical

TABLE 20.10 Example of Behavioral Procedures Included in a Functional Observation Battery

Home-Cage and Open-Field	Manipulative	Physiological
Arousal	Ease of removal	Body temperature
Gait	Ease of handling	Body weight
Posture	Touch response	
Vocalizations	Righting response	
Piloerection	Hindlimb foot splay	
Lacrimation	Forelimb grip strength	
Salivation	Hindlimb grip strength	
Urination/defecation	Finger-snap response	
Grooming behavior	Catalepsy	
Rearing	Palpebral closure	
Abnormal movements	Pupil function	
Tremors, convulsions		

measures that they permit repeated evaluation of individual animals over time to determine the onset, progression, duration, and reversibility of neurotoxic effects. Motor activity is also measured over time and can be evaluated by a variety of devices. One such device that has been frequently used is the figure 8 maze which consists of a series of interconnected alleys converging on a central open area and covered with transparent acrylic plastic. Motor activity is detected by photobeams, and an activity count is registered each time a photobeam is interrupted by the animal. Motor activity sessions are generally 60 min in length and each session is divided into 5- to 10-min reporting intervals (epochs). "Habituation" is an end point evaluated in the motor activity test and this is defined as a decrement in activity during the test session. Activity is expected to decrease toward the end of the test as the animal's exploratory activity normally lessens as the time in the maze increases. Neuropathological examinations are the same as those described below for the subchronic neurotoxicity test.

Before the acute neurotoxicity study is conducted, it is necessary to conduct a preliminary test to determine the time of peak effect after dosing of the test material. Preliminary tests may evaluate a selected group of end points in the FOB or other sensitive end points if known for a particular test material. Results of this preliminary test will determine the time when observations are performed on the day of dosing in the acute neurotoxicity study.

In the subchronic neurotoxicity study, end points measured are similar to those measured in the acute neurotoxicity study. However, the duration of dosing is 90 days and exposure to the test material is usually via the diet. As for the acute neurotoxicity study, these studies consist of three test groups and a control group. The FOB and motor activity tests are conducted at selected intervals such as weeks 5, 9, and 13, as well as pretest. At test termination, at least six animals per group are perfused via the heart with fixative to ensure optimal fixation of nervous tissues for histopathology examination. Nervous tissues examined include brain, spinal cord (various segments), and selected nerves such as the optic, sciatic, tibial, and sural nerves.

For all behavioral tests, it is important that the person making the actual observations is unaware of the treatment group for each animal ("blind" to dose group assignment). In addition, laboratories that conduct neurotoxicity studies for regulatory agencies must demonstrate that their methods are validated. Therefore, these laboratories must conduct positive control studies using known neurotoxicants and provide this information to regulators as necessary. Also, since it is not feasible for one person to perform the observations on all animals on all test occasions, laboratories must maintain evidence of interobserver reliability (agreement) for individuals who are involved with performing the FOBs.

Developmental Neurotoxicity Testing   Developmental neurotoxicity testing is a separate component of developmental toxicology that focuses on potential behavioral or morphological modifications resulting from exposures to toxicants during early development. These studies track the outcome of such exposures through the postnatal period and into early adulthood. In a developmental neurotoxicity study, at least 20 pregnant female rats for each of three treatment groups plus a control are administered test material from gestation day (GD) 6 through weaning on lactation day 21. FOBs are conducted on the maternal animals at selected intervals such as GD 6, GD 17, lactation day 11, and lactation day 21. Evaluations include observations in the home cage, during handling and outside the home cage in an open field. Body weights and food consumption are also monitored in the maternal animals. After birth, the offspring are counted, weighed, and gender is determined. On postnatal day (PND) 4, litters are culled to eight pups per litter. Following culling, at least 10 pups of each gender/group are assigned to one of the following tests: learning and memory, motor activity, or acoustic startle. Additional pups are assigned for neuropathology and brain weight evaluations on PND 11 and PND 70 (10/sex/group). FOBs are performed on the offspring at selected intervals such as PND 11, 17, 21, 35, and 60/70. Indicators of physical development such as preputial separation (male sexual maturation) and vaginal patency (female sexual maturation) are evaluated for all offspring as well as body weight and food consumption. Learning and memory can be evaluated with a variety of tests. Frequently, a water maze is used where the rat learns to swim through a series of alleys to find a platform it can use to climb out of the water. The time it takes for the animal to swim through the maze to the platform (trial latency) and the number of mistakes made are some of the end points evaluated in this test. The trials are conducted over a series of days and the assay provides an index of the development of both working memory (with-in day performance) as well as reference memory (between-day performance). The startle test measures the animal's response to a burst of loud noise and also how quickly it becomes habituated to 10 pulses of startle-eliciting tones in five blocks. For the startle test, special chambers lined with sound-attenuating and vibration-absorbing material are used. These chambers can measure the force exerted on a platform on which the animal stands during the test procedure. The startle test is conducted at two time points such as PND 22 and 60. When the offspring are approximately 70 days old, the test is terminated. Selected animals are perfused with fixative for neuropathology assessments. In addition to detailed microscopic evaluation of at least five different sections of the brain, simple morphometric analysis of the cerebrum, hippocampus, and cerebellum are conducted.

Delayed Neuropathy (organophosphate-induced delayed neuropathy [OPIDN])
The delayed neurotoxic potential of certain organophosphates such as tri-*o*-cresyl phosphate (TOCP) is usually evaluated by observation of clinical signs (paralysis of leg muscles in hens) or pathology (degeneration of the motor nerves in hens), but a biochemical test involving the ratio between the ability to inhibit cholinesterase and the ability to inhibit an enzyme that has been referred to as the neurotoxic esterase (NTE) has been suggested. The preferred test organism is the mature hen, because the clinical signs are similar to those in humans and such symptoms cannot be readily elicited in the common laboratory rodents.

Potentiation   Potentiation and synergism represent interactions between toxicants that are potential sources of hazard because neither humans nor other species are usually exposed to one chemical at a time. The enormous number of possible combinations of chemicals makes routine screening for all such effects impossible.

One of the classic cases is the potentiation of the insecticide malathion by another insecticide, EPN, the LD_{50} of the mixture being dramatically lower than that of either compound alone. This potentiation can also be seen between malathion and certain contaminants that are formed during synthesis, such as isomalathion. For this reason, quality control during manufacture is essential. This example of potentiation involves inhibition, by EPN or isomalathion, of the carboxylesterase responsible for the detoxication of malathion in mammals.

It is practical to test for potentiation only when there has been some preliminary indication that it might occur or when either or both compounds belong to chemical classes previously known to cause potentiation. Such a test can be conducted by comparing the LD_{50}, or any other appropriate toxic end point, of a mixture of equitoxic doses of the chemicals in question with the same end point measured with the two chemicals administered alone.

In the case of synergism, in which one of the compounds is relatively nontoxic when given alone, the toxicity of the toxic compound can be measured when administered alone or after a relatively large dose of the nontoxic compound.

Toxicokinetics and Metabolism   Routine toxicity testing without regard to the mechanisms involved is likely to be wasteful of time and of human, animal, and financial resources. A knowledge of toxicokinetics and metabolism can give valuable insights and provide for testing that is both more efficient and more informative. Such knowledge provides the necessary background to make the most appropriate selection of test animal species and of dose levels, and the most appropriate method for extrapolating from animal studies to the assessment of human hazard. Moreover, they may provide information on possible reactive intermediates as well as information on induction or inhibition of the enzymes of xenobiotic metabolism, the latter being critical to an assessment of possible interaction.

The nature of metabolic reactions and their variations between species is detailed in Chapters 7 and 8, with some aspects of toxicokinetics in Chapter 5. The methods used for the measurement of toxicants and their metabolites are detailed in Chapter 24. The present section is concerned with the general principles, use, and need for metabolic and toxicokinetics studies in toxicity testing.

Toxicokinetics studies are designed to measure the amount and rate of the absorption, distribution, metabolism, and excretion of a toxicant. These data are used to construct predictive mathematical models so that the distribution and excretion of other doses can be simulated. Such studies are usually carried out using radiolabeled compounds to facilitate measurement and total recovery of the administered dose. This can be done entirely *in vivo* by measuring levels in blood, expired air, feces, and urine; these procedures can be done relatively noninvasively and continuously in the same animal. Tissue levels can be measured by sequential sacrifice and analysis of organ levels. It is important to measure not only the compound administered but also its metabolites, because simple radioactivity counting does not differentiate among them.

The metabolic study, considered separately, consists of treatment of the animal with the radiolabeled compound followed by chemical analysis of all metabolites formed *in vivo* and excreted via the lungs, kidneys, or bile. Although reactive intermediates are unlikely to be isolated, the chemical structure of the end products may provide vital clues to the nature of the intermediates involved in their formation. The use of tissue homogenates, subcellular fractions, and purified enzymes may serve to clarify events occurring during metabolic sequences leading to the end products.

Information of importance in test animal selection is the similarity in toxicodynamics and metabolism to that of humans. Although all of the necessary information may not be available for humans, it can often be inferred with reference to metabolism and excretion of related compounds, but it is clearly ill advised to use an animal that differs from most others in the toxicokinetics or metabolism of the compound in question or that differs from humans in the nature of the end products. Dose selection is influenced by knowledge of whether a particular dose saturates a physiological process such as excretion or whether it is likely to accumulate in a particular tissue because these factors are likely to become increasingly important the longer a chronic study continues.

Behavior   Although the primary emphasis in toxicity testing has long been the estimation of morphologic changes, much recent interest has focused on more fundamental evaluations. One such aspect has been the evaluation of chemical effects on behavior.

The categories of methods used in behavioral toxicology fall into two principal classes, stimulus-oriented behavior, and internally generated behavior. The former includes two types of conditioned behavior: operant conditioning, in which animals are trained to perform a task in order to obtain a reward or to avoid a punishment, and classical conditioning, in which an animal learns to associate a conditioning stimulus with a reflex action. Stimulus-oriented behavior also involves unconditioned responses in which the animal's response to a particular stimulus is recorded.

Internally generated behavior includes observation of animal behavior in response to various experimental situations, and includes exploratory behavior, circadian activity, social behavior, and so on. The performance of animals treated with a particular chemical is compared with that of untreated controls as a measure of the effect of the chemical.

Many of the variables associated with other types of testing must also be controlled in behavioral tests: sex, age, species, environment, diet, and animal husbandry. Behavior may vary with all of these. Norton describes a series of four tests that may form an appropriate series inasmuch as they represent four different types of behavior; the series should therefore reflect different types of nervous system activity. They are as follows:

1. Passive avoidance. This test involves the use of a shuttle box, in which animals can move between a light side and a dark side. After an acclimatization period, in which the animal can move freely between the two sides, it receives a mild electric shock while in the dark (preferred) side. During subsequent trials, the time spent in the "safe side" is recorded.

2. Auditory startle. This test involves the response (movement) to a sound stimulus either without, or preceded by, a light-flash stimulus.

3. Residential maze. Movements of animals in a residential maze are automatically recorded during both light and dark photoperiods.

4. Walking patterns. Gait is measured in walking animals, including such characteristics as the length and width of stride and the angles formed by the placement of the feet.

Problems associated with behavioral toxicology include the functional reserve and adaptability of the nervous system. Frequently, behavior is maintained in spite of clearly observable injury. Other problems are the statistical ones associated with multiple tests, multiple measurements, and the inherently large variability in behavior.

The use of human subjects occupationally exposed to chemicals is often attempted, but such tests are complicated by the subjective nature of the end points (e.g. dizziness).

Covalent Binding   Toxicity has been associated with covalent binding in a number of ways. Organ-specific toxicants administered *in vivo* bind covalently to macromolecules, usually at a higher level in the target tissues that in nontarget tissues. Examples include acetaminophen in the liver, carbon tetrachloride in the liver, *p*-aminophenol in the kidney, and ipomeanol in the lung. Similarly, many carcinogens are known to give rise to DNA adducts. In general, covalent binding occurs as a result of metabolism of the toxicant to highly reactive intermediates, usually, but not always, by cytochrome P450. Because these intermediates are highly reactive electrophiles, they bind to many nucleophilic sites on DNA, RNA, or protein molecules, not just the site of toxic action. Thus, measurement of covalent binding may be a measure of toxic potential rather than a specific measurement, related directly to a mechanism of action. The occurrence of covalent binding at the same time as toxicity is so common an occurrence, however, that a measurement of covalent binding of a chemical may be regarded as an excellent although perhaps not infallible indication of potential for toxicity.

The measurement of DNA adducts is an indirect indication of genotoxic (carcinogenic) potential, and DNA adducts in the urine are an indication, obtained by a noninvasive technique, of recent exposure. Protein adducts give an integrated measure of exposure because they accumulate over the life span of the protein and, at the same time, indicate possible organ toxicity.

Tissue protein adducts are usually demonstrated in experimental animals following injection of radiolabeled chemicals and, after a period to time, the organs are removed, homogenized and, by rigorous extraction, all the noncovalently bound material is removed. Extraction methods include lipid solvents, acids and bases, concentrated urea solutions, and solubilization and precipitation of the proteins. They tend to underestimate the extent of covalent binding because even covalent bonds may be broken by the rigorous procedures used. Newer methods involving dialysis against detergents and separation of adducted proteins will probably prove more appropriate.

Blood proteins, such as hemoglobin, may be used in tests of human exposure because blood is readily and safely accessible. For example, the exposure of mice to ethylene oxide or dimethylnitrosamine was estimated by measuring alkylated residues in hemoglobin. The method was subsequently extended to people exposed occupationally to ethylene oxide by measuring N-3-(2-hydroxyethyl) histidine residues in hemoglobin. Similarly, methyl cysteine residues in hemoglobin can be used as a measure of methylation.

DNA–RNA adducts can also be measured in various ways, including rigorous extraction, separation, and precipitation following administration of labeled compounds *in vivo*, or use of antibodies raised to chemically modified DNA or RNA.

Although many compounds of different chemical classes have been shown to bind covalently when activated by microsomal preparations *in vitro* (e.g., aflatoxin, ipomeanol, stilbene, vinyl chloride), these observations have not been developed into routine testing procedures. Such procedures could be useful in predicting toxic potential.

Immunotoxicity   Immunotoxicology comprises two distinct types of toxic effects: the involvement of the immune system in mediating the toxic effect of a chemical and the toxic effects of chemicals on the immune system. The former is shown, for example, in tests for cutaneous sensitization, whereas the latter is shown in impairment of the ability to resist infection.

Tests for immunotoxicity are not required by all regulatory agencies, but it is an area of great interest, both in the fundamental mechanisms of immune function and in the design of tests to measure impairment of immune function. Both of these aspects are discussed in detail in Chapter 19.

20.6   *IN VITRO* AND OTHER SHORT-TERM TESTS

20.6.1   Introduction

The toxicity tests that follow are tests conducted largely *in vitro* with isolated cell systems. Some are short-term tests carried out *in vivo* or are combinations of *in vivo* and *in vitro* systems. The latter are included because of similarities in approach, mechanism, or intent. In general, these tests measure effects on the genome or cell transformation; their importance lies in the relationship between such effects and the mechanism of chemical carcinogenesis. Mutagenicity of cells in the germ line is itself an expression of toxicity, however, and the mutant genes can be inherited and expressed in the next or subsequent generations.

The theory that the initiating step of chemical carcinogenesis is a somatic mutation is well recognized, and considerable evidence shows that mutagenic potential is correlated with carcinogenic potential. Thus, the intent of much of this type of testing is to provide early warning of carcinogenic potential without the delay involved in conducting lifetime chronic feeding studies in experimental animals. In spite of the numerous tests that have been devised, regulatory agencies have not yet seen fit to substitute any of them, or any combination of them, for chronic feeding studies. Instead, they have been added as additional testing requirements. One function of such tests should be to identify those compounds with the greatest potential for toxicity and enable the amount of chronic testing to be reduced to more manageable proportions.

20.6.2   Prokaryote Mutagenicity

Ames Test   The Ames test, developed by Bruce Ames and his coworkers of the University of California, Berkeley, depends on the ability of mutagenic chemicals to bring about reverse mutations in *Salmonella typhimurium* strains that have defects in the histidine biosynthesis pathway. These strains will not grow in the absence of histidine but can be caused to mutate back to the wild type, which can synthesize histidine and hence can grow in its absence. The postmitochondrial supernatant (*S*-9 fraction), obtained from homogenates of livers of rats previously treated with polychlorinated biphenyls (PCBs) in order to induce certain cytochrome P450 isoforms, is also included in order to provide the activating enzymes involved in the production of the potent electrophiles often involved in the toxicity of chemicals to animals.

Bacterial tester strains have been developed that can test for either base-pair (e.g., strain TA-1531) or frameshift (e.g., strains TA-1537, TA-1538) mutations. Other, more sensitive strains such as TA-98 and TA-100 are also used, although they may be less specific with regard to the type of mutation caused.

In brief, the test is carried out (Figure 20.5) by mixing a suspension of bacterial cells with molten top agar. This also contains cofactors, *S*-9 fraction, and the material to be tested. The mixture is poured onto Petri plates containing hardened minimal agar. The number of bacteria that revert and acquire the wild-type ability to grow in the absence of histidine can be estimated by counting the colonies that develop on incubation. To provide a valid test, a number of concentrations are tested, and positive controls with known mutagens are included along with negative controls that lack only the test compound. The entire test is replicated often enough to satisfy appropriate statistical tests for significance. Parallel tests without the *S*-9 fraction may help distinguish between chemicals with intrinsic mutagenic potential and those that require metabolic activation.

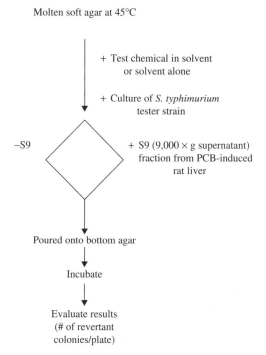

Molten soft agar at 45°C

+ Test chemical in solvent
or solvent alone

+ Culture of *S. typhimurium*
tester strain

−S9

+ S9 (9,000 × g supernatant)
fraction from PCB-induced
rat liver

Poured onto bottom agar

Incubate

Evaluate results
(# of revertant
colonies/plate)

Figure 20.5 Protocol for the Ames test for mutagenicity.

The question of correlation between mutagenicity and carcinogenicity is crucial in any consideration of the utility of this or similar tests. In general, this appears to be high, although a small proportion of both false positives and false negatives occurs. For example, certain base analogs and inorganics such as manganese are not carcinogens but are mutagens in the Ames test, whereas diethylstilbestrol (DES) is a carcinogen but not a bacterial mutagen (see Chapter 11 for additional detail).

Related Tests   Related tests include tests based on reverse mutations, as in the Ames test, as well as tests based on forward mutations. Examples include:

1. Reverse mutations in *Escherichia coli*. This test is similar to the Ames test and depends on reversion of tryptophane mutants, which cannot synthesize this amino acid, to the wild type, which can. The *S*-9 fraction from the liver of induced rats can also be used as an activating system in this test. Other *E. coli* reverse mutation tests utilize nicotinic acid and arginine mutants.

2. Forward mutations in *S. typhimurium*. One such assay, dependent on the appearance of a mutation conferring resistance to 8-azaguanine in a histidine revertant strain, has been developed and is said to be as sensitive as the reverse-mutation tests.

3. Forward mutations in *E. coli*. These mutations depend on mutation of galactose non-fermenting *E. coli* to galactose fermenting *E. coli* or the change from 5-methyltryptophane to 5-methyltryptophane resistance.

4. DNA repair. Polymerase-deficient, and thus DNA repair-deficient, *E. coli* has provided the basis for a test that depends on the fact that the growth of a deficient strain is inhibited more by a DNA-damaging agent than is that of a repair-competent strain. The recombinant assay using *Bacillus subtilis* is conducted in much the same way, because recombinant deficient strains are more sensitive to DNA-damaging agents.

20.6.3   Eukaryote Mutagenicity

Mammalian Cell Mutation   The development of cell culture techniques that permit both survival and replication have led to many advances in cell biology, including the use of certain of these cell lines for detection of mutagens. Although such cells, if derived from mammals, would seem ideal for testing for toxicity toward mammals, there are several problems. Primary cells, which generally resemble those of the tissue of origin, are difficult to culture and have poor cloning ability. Because of these difficulties, certain established cell lines are usually used. These cells, such as Chinese hamster ovary cells and mouse lymphoma cells, clone readily and do not become senescent with passage through many cell generations. Unfortunately, they have little metabolic activity toward xenobiotics and thus do not readily activate toxicants. Moreover, they usually show chromosome changes, such as aneuploidy (i.e., more or fewer than the usual diploid number of chromosomes).

The characteristics usually involved in these assays are resistance to 8-azaguanine or 6-thioguanine (the hypoxanthine guanine phosphoribosyl transferase or HGPRT locus), resistance to bromodeoxyuridine or trifluorothymidine (the thymidine kinase or TK locus) or resistance to ouabain (the OU or Na/K-ATPase locus). HGPRT is responsible for incorporation of purines from the medium into the nucleic acid synthesis pathway. Its loss prevents uptake of normal purines and also of toxic purines such as 8-azaguanine, which would kill the cell. Thus, mutation at this locus confers resistance to these toxic purine analogs. Similarly, TK permits pyrimidine transport, and its loss prevents uptake of toxic pyrimidine analogs and confers resistance to them. In the absence of HGPRT or TK, the cells can grow by *de novo* synthesis of purines and pyrimidines. Ouabain kills cells by combining with the Na/K-ATPase. Mutation at the OU locus alters the ouabain-binding site in a way that prevents inhibition and thus confers resistance.

A typical test system is the analysis of the TK locus in mouse lymphoma cells for mutations that confer resistance to bromodeoxyuracil. The tests are conducted with and without the *S*-9 fraction from induced rat liver because the lymphoma cells have little activating ability. Both positive and negative controls are included, and the parameter measured is the number of cells formed that are capable of forming colonies in the presence of bromodeoxyuridine.

***Drosophila* Sex-Linked Recessive Lethal Test**   The advantages of *Drosophila* tests are that they involve an intact eukaryotic organism with all of its interrelated organ systems and activation mechanisms but, at the same time, are fast, relatively easy to perform, and do not involve mammals as test animals. The most obvious disadvantages are that the hormonal and immune systems of insects are significantly different from those of mammals and that the nature, specificity, and inducibility of the cytochrome P450s are not as well understood in insects as they are in mammals.

In a typical test, males that are 2 days postpuparium and that were raised from eggs laid within a short time period (usually 24 h) are treated with the test compound in water to which sucrose has been added to increase palatability. Males from a strain carrying a gene for yellow body on the X chromosome are used. Preliminary tests determine that the number of offspring of the survivors of the treatment doses (usually $0.25 \, LD_{50}$ and $0.5 \, LD_{50}$) are adequate for future crosses. Appropriate controls, including a solvent control (with emulsifier if one was necessary to prepare the test solution), and a positive control, such as ethyl methanesulfonate, are routinely included with each test. Individual crosses of each surviving treated male with a series of three females are made on a 0- to 2-, 3- to 5-, and 6- to 8-day schedule. The progeny of each female is reared separately, and the males and females of the F_1 generation are mated in brother-sister matings. If there are no males with yellow bodies in a particular set of progeny, it should be assumed that a lethal mutation was present on the treated X chromosomes. A comparison of the F_2 progeny derived from females inseminated by males at different times after treatment allows a distinction to be made between effects on spermatozoa, spermatids, and spermatocytes.

In the Basc (Muller-5) test shown in Figure 20.6, the strain used for the females in the F_1 cross is a multiple-marked strain that carries a dominant gene for bar eyes and recessive genes for apricot eyes and a reduction of bristles on the thorax (scute gene). (Basc is an acronym for bar, apricot, and scute.).

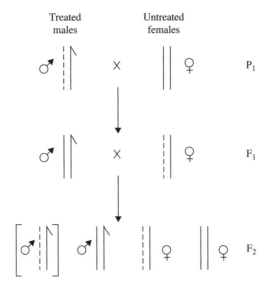

Treated males   Untreated females

Figure 20.6 The Basc (Muller-5) mating scheme. Dashed lines represent the treated X chromosome of males. Brackets indicate males with yellow bodies, which would be absent if a lethal mutation occurred on the X chromosome of the treated male.

Related Tests   Many tests related to the two types of eukaryote-mutation tests are discussed in Mammalian Cell Mutation section and in *Drosophila* Sex-Linked Recessive Lethal Test section, and many of them are simply variations of the tests described. Two distinct classes are worthy of mention: the first uses yeasts as the test organisms, and the second is the spot test for mutations in mice.

One group of tests using yeasts includes tests for gene mutations and strains that can be used to detect forward mutations, in genes that code for enzymes in the purine biosynthetic pathway: other strains can be used to detect reversions. Yeasts can also be used to test for recombinant events such as reciprocal mitotic recombination (mitotic crossing over) and nonreciprocal mitotic recombination. *Saccharomyces cerevisiae* is the preferred organism in almost all these tests. Although they possess cytochrome P450s capable of metabolizing xenobiotics, their specificity and sensitivity are limited as compared with those of mammals, and an *S*-9 fraction is often included, as in the Ames test, to enhance activation.

The gene mutation test systems in mice include the specific locus test, in which wild-type treated males are crossed with females carrying recessive mutations for visible phenotypic effects. The F_1 progeny have the same phenotype as the wild-type parent unless a mutation, corresponding to a recessive mutant marker, has occurred. Such tests are accurate, and the spontaneous (background) mutation rate is very low, making them sound tests which are predictive for other mammals. Unfortunately, the large number of animals required has prevented extensive use. Similar tests involving the activity and electrophoretic mobility of various enzymes in the blood or other tissues in the F_1 progeny from treated males and untreated females have been developed. In the previously mentioned tests, as with many others, sequential mating of males with different females can provide information about the stage of sperm development at which the mutational event occurred.

20.6.4 DNA Damage and Repair

Many of the end points for tests described in this chapter, including gene mutation, chromosome damage, and oncogenicity, develop as a consequence of damage to or chemical modification of DNA. Most of these tests, however, also involve metabolic events that occur both prior to and subsequent to the modification of DNA. Some tests, however, use events at the DNA level as end points. One of these, the unscheduled synthesis of DNA in mammalian cells, is described in some detail; the others are summarized briefly.

Unscheduled DNA Synthesis in Mammalian Cells   The principle of this test is that it measures the repair that follows DNA damage and is thus a reflection of the damage itself. It depends on the autoradiographic measurement of the incorporation of tritiated thymidine into the nuclei of cells previously treated with the test chemical.

The preferred cells are usually primary hepatocytes in cultures derived from adult male rats, the cells of which are dispersed and allowed to attach themselves to glass coverslips. From this point on, the test is carried out on the attached cells. Both positive controls with agents known to stimulate unscheduled DNA synthesis, such as the carcinogen aflatoxin B1 or 2-acetylaminofluorene, and negative controls, which are processed through all procedures except exposure to the test compound, are performed routinely with every test. Cells are exposed by replacing the medium for a short time with one containing the test chemical. The dose levels are determined by a preliminary cell viability test (trypan blue exclusion test) and consist of several concentrations that span the range from no apparent loss of viability to almost complete loss of viability. Following exposure, the medium is removed and the cells are washed by several changes of fresh medium and finally placed in a medium containing tritiated thymidine. The cells are fixed and dried, and the coverslip with the cells attached is coated with photographic emulsion. After a suitable exposure period (usually several weeks), the emulsion is developed and the cells are stained with hemotoxylin and eosin. The number of grains in the nuclear region is corrected by subtracting nonnuclear grains, and the net grain count in the nuclear area is compared between treated and untreated cells.

This test has several advantages in that primary liver cells have considerable activation capacity and the test measures an event at the DNA level. It does not, however, distinguish between error-free repair and error-prone repair, the latter being itself a mutagenic process. Thus, it cannot distinguish between events that might lead to toxic sequelae and those that do not. A modification of this test measures *in vivo* unscheduled DNA synthesis. In this modification, animals are first treated *in vivo*, and primary hepatocytes are then prepared and treated as already described.

Related Tests   Tests for the measurement of binding of the test material to DNA have already been discussed under covalent binding. Another method of assessing DNA damage is the estimation of DNA breakage following exposure to the test chemical; the DNA-strand length is estimated by using alkaline elution or sucrose density gradient centrifugation. This has been done with a number of cell lines and with freshly prepared hepatocytes, in the latter case following treatment either *in vivo* or *in vitro*. It may be regarded as promising but not yet fully validated. The polymerase-deficient *E. coli* tests as well as recombinant tests using yeasts are also related to DNA repair.

20.6.5 Chromosome Aberrations

Tests for chromosome aberrations involve the estimation of effects on extended regions of whole chromosomes rather than on single or small numbers of genes. Primarily, they concern chromosome breaks and the exchange of material between chromosomes.

Sister Chromatid Exchange Sister chromatid exchange (SCE) occurs between the sister chromatids that together make up a chromosome. It occurs at the same locus in each chromatid and is thus a symmetrical exchange of chromosome material. In this regard, it is not strictly an aberration because the products do not differ in morphology from normal chromosome. SCE, however, is susceptible to chemical induction and appears to be correlated with the genotoxic potential of chemicals as well as with their oncogenic potential. The exchange is visualized by permitting the treated cells to pass through two DNA replication cycles in the presence of 5-bromo-2'-deoxyuridine, which is incorporated in the replicated DNA. The cells are then stained with a fluorescent dye and irradiated with UV light, which permits differentiation between chromatids that contain bromodeoxyuridine and those that do not (Figure 20.7).

The test can be carried out on cultured cells or on cells from animals treated *in vivo*. In the former case, the test chemical is usually evaluated in the presence and absence of the *S*-9 activation system from rat liver. Typically, cells from a Chinese hamster ovary cell line are incubated in a liquid medium and exposed to several concentration of the test chemical, either with or without the *S*-9 fraction, for about 2 h. Positive controls, such as ethyl methanesulfonate (a direct-acting compound) or dimethylnitrosamine (one that requires activation), as well as negative controls are also included. Test concentrations are based on cell toxicity levels determined by prior experiment and are selected in such a way that even at the highest dose excess

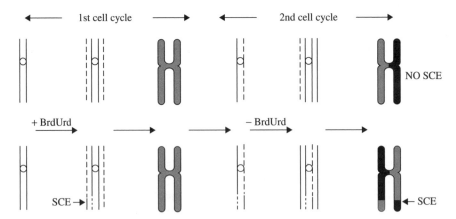

Figure 20.7   Visualization of sister chromatid exchange.

growth does not occur. At the end of the treatment period, the cells are washed, bromodeoxyuridine is added, and the cells are incubated for ≥24 h or more. The cells are then fixed, stained with a fluorescent dye, and irradiated with UV light. Second division cells are then scored under the microscope for SCEs (Figure 20.7).

The test can also be carried out on cells treated *in vivo*, and analyses have been made of SCEs in lymphocytes from cancer patients treated with chemotherapeutic drugs, smokers, and workers exposed occupationally; in several cases, increased incidence of SCEs has been noted. This is a sensitive test for compounds that alkylate DNA, with few false positives. It may be useful for detecting promoters such as phorbol esters.

Micronucleus Test   The micronucleus test is an *in vivo* test usually carried out in mice. The animals are treated *in vivo*, and the erythrocyte stem cells from the bone marrow are stained and examined for micronuclei. Micronuclei represent chromosome fragments or chromosomes left behind at anaphase. It is basically a test for compounds that cause chromosome breaks (clastogenic agents) and compounds that interfere with normal mitotic cell division, including compounds that affect spindle fiber function.

Male and female mice from an outbred strain are handled by the best animal husbandry techniques, as described for acute, subchronic, and chronic tests, and are treated either with the solvent, 0.5 LD_{50}, or 0.1 LD_{50} of the test chemical. Animals are killed at several time intervals up to 2 days; the bone marrow is extracted, placed on microscope slides, dried, and stained. The presence of micronuclei is scored visually under the microscope.

Dominant Lethal Test in Rodents   The dominant lethal test, which is performed using rats, mice, or hamsters, is an *in vivo* test to determine the germ-cell risk from a suspected mutagen. The test consists of treating males with the test compound for several days, followed by mating to different females each week for enough weeks to cover the period required for a complete spermatogenic cycle. Animals are maintained under optimal conditions of animal husbandry and are dosed, usually by a gavage, with several doses of less than 0.1 LD_{50}. The females are killed after 2 weeks of gestation and dissected; corpora lutea and living and dead implantations are counted. The end points used to determine the occurrence of dominant lethal mutations in the treated males are the fertility index (ratio of pregnant females to mated females), preimplantation losses (the number of implantations relative to the number of corpora lutes), the number of females with dead implantations relative to the total number of pregnant females, and the number of dead implantations relative to the total number of implantations. Mutations in sperm that are dominant and lethal do not result in viable offspring.

Related Tests   Many cells exposed to test chemicals can be scored for chromosome aberrations by staining procedures followed by visual examination with the aid of the microscope. These include Chinese hamster ovary cells in culture treated in a protocol very similar to that used in the test for SCEs, bone marrow cells from animals treated *in vivo*, or lymphocytes from animals treated *in vivo*. The types of aberrations evaluated include chromatid gaps, breaks, and deletions; chromosome gaps, breaks, and deletions; chromosome fragments; translocations; and ploidy.

Heritable translocations can be detected by direct examination of cells from male or female offspring in various stages of development or by crossing the treated animals to untreated animals and evaluating fertility, with males with reduced fertility being examined for translocations, and so on. Progeny from this or other tests, such as those for dominant lethals, can be permitted to survive and then examined for translocations and other abnormalities.

20.6.6   Mammalian Cell Transformation

Most cell transformation assays utilize fibroblast cultures derived from embryonic tissue. The original studies showed that cells from C3H mouse fibroblast cultures developed morphologic changes and changes in growth patterns when treated with carcinogens. Later, similar studies were made with Syrian hamster embryo cells. The direct relationship of these changes to carcinogenesis was demonstrated by transplantation of the cells into a host animal and the subsequent development of tumors. The recent development of practical assay procedures involves two cell lines from mouse embryos, Balb/3T3 and C3h/10T1/2, in which transformation is easily recognized and scored. In a typical assay situation, cells, such as Balb/3T3 mouse fibroblasts, will multiply in culture until a monolayer is formed. At this point, they cease dividing unless transformed. Chemicals that are transforming agents will, however, cause growth to occur in thicker layers above the monolayer. These clumps of transformed cells are known as foci. In spite of many recommended controls, the assay is only semiquantitative. The doses are selected from the results of a preliminary experiment and range from a high dose that reduces colony formation (but not by >50%) to a low dose that has no measurable effect on colony formation. After exposure to the test chemical for 1–3 days, the cells are washed and incubation is continued for up to 4 weeks. At that time, the monolayers are fixed, stained, and scored for transformed foci.

Transformation assays have several distinct advantages. Because transplanted foci give rise to tumors in congenic hosts (those from the same inbred strain from which the cells were derived) whereas untransformed cells do not, cell transformation is believed to be illustrative of the overall expression of carcinogenesis in mammalian tissues. The two cell types used most (Balb/3T3 and C3H/10T1/2) respond to promoters in the manner predicted by the multistage model for carcinogenesis *in vivo* and may eventually be useful in the development of assays for promotion. Unfortunately, a large number of false-negative results are obtained because these cell lines do not show much activation capacity; it has not proved practical to combine them with the *S*-9 activation system. Furthermore, the cells are aneuploidy and may be preneoplastic in the untreated state. Syrian hamster cells, which do have considerable activation capacity, have proved difficult to use in test procedures and are difficult to score.

20.6.7   General Considerations and Testing Sequences

Considering all of the tests for acute and chronic toxicity, long and short term, *in vivo*, and *in vitro*, it is clearly impractical to apply a complete series of tests to all commercial chemicals and all their derivatives in food, water, and the environment. The challenge of toxicity testing is to identify the most effective set or sequence of tests necessary to describe the apparent and potential toxicity of a particular chemical or mixture of chemicals. The enormous emphasis on *in vitro* or short-term tests that has occurred since the mid-1970s had its roots in the need to find substitutes for lifetime feeding studies in experimental animals or, at the very least, to suggest a sequence of tests that would enable priorities to be set for which chemicals should be subjected to chronic tests. Such tests might also be used to eliminate the need for chronic testing for chemicals that either clearly possessed the potential for toxicity or clearly do not. Although there has been much success in test development, the challenge outlined here has not been met, primarily because of the failure of scientists and regulatory agencies, worldwide, to agree on test sequences or on the circumstances in which short-term tests may substitute for chronic tests. Thus, not only are short-term tests often required; these are in addition to long-term tests. As an example, the U.S. EPA requirements for the Federal Insecticide, Fungicide, and Rodenticide Act (FIFRA) include, in addition to a full battery of acute, subchronic, and chronic tests, tests to address the following three categories: gene mutations, structural chromosome aberrations, and other genotoxic tests as appropriate (such as DNA damage and repair and chromosome aberrations). It is important, however, that test sequences have been suggested and considered by regulatory agencies and, in addition, it must be considered that short-term tests do not provide all of the information needed from the longer-term tests.

20.7   ECOLOGICAL EFFECTS

Tests for ecological effects include those designed to address the potential of chemicals to affect ecosystems and population dynamics in the environment. Such tests are designed to estimate effects on field populations of vertebrates, invertebrates, and plants. The use of these tests in environmental risk assessment is discussed in detail in Chapter 27.

20.7.1   Laboratory Tests

There are two types of laboratory tests: toxicity determinations on wildlife and aquatic organisms and the use of model ecosystems to measure bioaccumulation and transport of toxicants and their degradation products.

Among the tests included in the first category are the avian oral LD_{50}, the avian dietary lethal concentration 50 (LC_{50}), wild mammal toxicity, and avian reproduction. The avian tests are usually carried out on bobwhite quail or mallard ducks, whereas the wild mammals may be species such as the pine mouse, *Paramyscus*. The tests are similar to those described under acute and chronic testing procedures but suffer from some drawbacks; the standards of animal husbandry used with rats and mice are probably unattainable with birds or wild mammals even through bobwhite quail and mallards are easily reared in captivity. The genetics of the birds and mammals used are much more variable than are those of the traditional laboratory rodent strains.

Similar tests can be carried out with aquatic organisms (e.g., the LC_{50} for freshwater fish such as rainbow trout and bluegills), the LC_{50} for estuarine and marine organisms, the LC_{50} for invertebrates such as *Daphnia,* and the effect of chemicals on the early stages of fish and various invertebrates.

Model systems, first developed by ecologists to study basic ecological processes, have been adapted to toxicological testing. In toxicology, these models were first used to determine the movement and concentration of pesticides. Typically, the model has a water phase containing vertebrates and invertebrates, and a terrestrial phase containing at least one plant species and one herbivore species. First, the ^{14}C-labeled pesticide or other environmental contaminant is applied to the leaves of the terrestrial plant sorghum (*Sorgum halpense*) then salt marsh caterpillars (*Estigmene acrea*) are placed on the plants. The larvae eat the plants and contaminate the water with feces and their dead bodies. The aquatic food chain is simulated with plankton (diatoms, rotifers, etc.), water fleas (*Daphnia*), mosquito larvae (*Culex pipiens*), and fish (*Gambusia affinis*). From an analysis of the plants, animals, and substrates for the ^{14}C-labeled compound and its degradation products, the biologic magnification or rate of degradation can be calculated.

More complex models involving several compartments, simulated rain, simulated soil drainage, simulated tidal flow, and so on, have been constructed and their properties investigated, but none have been brought to the stage of use in routine testing. Similarly, aquatic models using static, recirculating, and continuous flow have also been used, as have entirely terrestrial models: again, none have been developed for routine testing.

20.7.2 Simulated Field Tests

Simulated field tests may be quite simple, consisting of feeding treated prey to predators and studying the toxic effects on the predator, enabling some predictions concerning effects to nontarget organisms. In general, however, the term is used for greenhouse, small plot, small artificial pond, or small natural pond tests. These serve to test biologic accumulation and degradation under conditions somewhat more natural than in model ecosystems and the test chemicals are exposed to environmental as well as biologic degradation. Population effects may be noted, but these methods are more useful for soil invertebrates, plants, and aquatic organisms because other organisms are not easily contained in small plots.

20.7.3 Field Tests

In field-test situations, test chemicals are applied to large areas under natural conditions. The areas are at least several acres and may be either natural or part of some agroecosystem. Because the area is large and in the open, radiolabeled compounds cannot be used, and it is not possible to obtain a balance between material applied and material recovered.

The effects are followed over a long period of time and two types of control may be used: first, a comparison with a similar area that is untreated; and second, a comparison with the same area before treatment. In the first, case it is difficult, if not impossible, to duplicate exactly a large natural area, and in the second, changes can occur that are unrelated to the test material.

In either case, studies of populations are the most important focus of this type of testing, although the disappearance of the test material, its accumulation in various life forms, and the appearance, accumulation, and disappearance of its degradation products are also important. The population of soil organisms, terrestrial organisms, and aquatic organisms as well as plants all must be surveyed and characterized, both qualitatively and quantitatively. Following application of the test material, the populations can be followed through two or more annual cycles to determine both acute and long-term population effects.

20.8 RISK ANALYSIS

The preceding tests for various kinds of toxicity can be used to measure adverse effects of many different chemical compounds in different species, organ, tissues, cells, or even populations, and under many different conditions. This information can be used to predict possible toxicity of related compounds from QSAR or of the same chemical under different conditions (e.g., mutagenicity as a predictor of carcinogenicity). It is considerably more difficult to use this information to predict possible risk to other species, such as humans, because little experimental data on this species is available. Some methods are available to predict risk to humans and to provide the risk factor in the risk–benefit assessment that provides the basis for regulatory action, however. Human health risk assessment is discussed in detail in Chapter 24. The benefit factor is largely economic in nature, and the final regulatory action is not, in the narrow sense, a scientific one. It also involves political and legal aspects and, *in toto*, represents society's evaluation of the amount of risk that can be tolerated in any particular case.

20.9 THE FUTURE OF TOXICITY TESTING

Because of the public awareness of the potentially harmful effects of chemicals, it is clear that toxicity testing will continue to be an important activity and that it will be required by regulatory agencies before the use of a particular chemical is permitted either in commercial processes or for use by the public. Because of the proliferation of testing procedures, the number of experimental species and other test systems available, as well as the high dose rates usually used, it is clear that eventually, some expression of some type of toxicity will be obtained for most exogenous chemicals. Thus, the identification of toxic effects with the intent of banning any chemical causing such effects is no longer a productive mode of attack. The aim of toxicity testing should be to identify those compounds that present an unacceptable potential for risk to humans or to the environment and thus ought to be banned, but, at the same time, provide an accurate assessment of the risk to humans and the environment of less toxic compounds so that their use may be regulated.

Subjecting all chemicals to all possible tests is logistically impossible, and the future of toxicity testing must lie in the development of techniques that will narrow the testing process so that highly toxic and relatively nontoxic compounds can be identified early and either banned or permitted unrestricted use without undue waste of time, funds, and human resources. These vital commodities could then be concentrated on compounds whose fate and effects are less predictable.

Such progress will come from further development and validation of the newer testing procedures and the development of techniques to select, for any given chemical, the most suitable testing methods. Perhaps of most importance is the development of integrated test sequences that permit decisions to be made at each step, thereby either abbreviating the sequence or making the next step more effective and efficient. As more data are developed and analyzed, structure–activity models should become more predictive. Some current models for predicting the potential for carcinogenesis are accurate in about 90% of cases.

BIBLIOGRAPHY AND SUGGESTED READING

Adkins, B., G. H. Lugenbuhl, and D. E. Gardner. Acute exposure of laboratory mice to manganese oxide. *Am. Ind. Hyg. Assoc. J.* **41**:494–500, 1980.

Allen, J. R., W. A. Hargreaves, M. T. S. Hsia, et al. Comparative toxicology of chlorinated compounds on mammalian species. *Pharmacol. Ther.* **7**:513–547, 1979.

Balls, M., R. J. Riddell, and A. N. Worden. *Animals and Alternatives in Toxicity Testing.* London: Academic Press, 1983.

Brown, V. K. H. Acute toxicity testing—A critique. In *Testing for Toxicity*, ed. J. W. Gorrod. London: Taylor & Francis, 1981.

Brusick, D. Genetic toxicology. In *Principles and Methods of Toxicology*, 2nd ed., ed. A. W. Hayes. New York: Raven Press, 1989.

Burger, G. T. and L. C. Miller. Animal care and facilities. In *Principles and Methods of Toxicology*, 2nd ed., ed. A. W. Hayes. New York: Raven Press, 1989.

Chan, P. K. and A. W. Hayes. Principles and methods for acute and eye irritancy. In *Principles and Methods of Toxicology*, 2nd ed., ed. A. W. Hayes. New York: Raven Press, 1989.

Clark, B. and D. A. Smith. Pharmacokinetics and toxicity testing. *CRC Crit. Rev. Toxicol.* **12**:343, 1984.

Cobb, L. M. Pulmonary toxicity. In *Testing for Toxicity*, ed. J. W. Gorrod. London: Taylor & Francis, 1981.

Couch, J. A. and W. J. Hargis Jr. Aquatic animals in toxicity testing. *J. Am. Coll. Toxicol.* **3**:331, 1984.

de Serres, F. J. and J. Ashby, eds. *Evaluation of Short Term Tests for Carcinogens.* New York: Elsevier, 1981.

Dean, J. H., M. I. Luster, M. J. Murray, et al. Approaches and methodology for examining the immunological effects of xenobiotics. *Immunotoxicology* **7**:205, 1983.

Dean, J. H., J. B. Cornacoff, G. J. Rosenthal, et al. Immune system: Evaluation of injury. In *Principles and Methods of Toxicology*, 2nd ed., ed. A. W. Hayes. New York: Raven Press, 1989.

Dewar, A. J. Neurotoxicity testing—With particular reference to biochemical methods. In *Testing for Toxicity*, ed. J. W. Gorrod. London: Taylor & Francis, 1981.

Ecobichon, D. J. *The Basis of Toxicity Testing.* Boca Raton, FL: CRC Press, 1992.

Enslein, K. and P. N. Craig. Carcinogenesis: A predictive structure-activity model. *J. Toxicol. Environ. Health* **10**:521, 1982.

Gorrod, J. W., ed. *Testing for Toxicity*. London: Taylor & Francis, 1981.

Gorrod, J. W. Covalent binding as an indication of drug toxicity. In *Testing for Toxicity*, ed. J. W. Gorrod. London: Taylor & Francis, 1981.

Hashimoto, Y., T. Makita, H. Miyata, T. Noguchi, and G. Ohta. Acute and subchronic toxicity of a new fluorine pesticide, N-methyl-N-(1-naphthyl) fluoroacetamide. *Toxicol. Appl. Pharmacol.* **12**:536–547, 1968.

Hayes, A. W., ed. *Principles and Methods of Toxicology*, 2nd ed. New York: Raven Press, 1989.

Hogan, M. D. and D. G. Hoel. *Extrapolation to man. In Principles and Methods of Toxicology*, 2nd ed., ed. A. W. Hayes. New York: Raven Press, 1989.

Kennedy, G. L. Jr. Inhalation toxicology. In *Principles and Methods of Toxicology*, 2nd ed., ed. A. W. Hayes. New York: Raven Press, 1989.

Manson, J. M. and Y. J. Kang. Test methods for assessing female reproductive and developmental toxicology. In *Principles and Methods of Toxicology*, 2nd ed., ed. A.W. Hayes. New York: Raven Press, 1989.

Mosberg, A. T. and A. W. Hayes. Subchronic toxicity testing. In *Principles and Methods of Toxicology*, 2nd ed., ed. A. W. Hayes. New York: Raven Press, 1989.

Moser, V. C., G. C. Becking, V. Cuomo, et al. The IPCS collaborative study on neurobehavioral screening methods: IV. Control data. *Neurotoxicology* **18**:947–967, 1997.

National Toxicology Program. Report of the NTP Ad hoc Panel on Chemical Carcinogenesis Testing and Evaluation. National Toxicology Program. Washington, DC: Department of Health and Human Services, 1998.

Norton, S. Methods for behavioral toxicology. In *Principles and Methods of Toxicology*, 2nd ed., ed. A. W. Hayes. New York: Raven Press, 1989.

Parish, W. E. Immunological tests to predict toxicological hazards to man. In *Testing for Toxicity*, ed. J. W. Gorrod. London: Taylor & Francis, 1981.

Patrick, E. and H. I. Maiback. Dermatotoxicology. In *Principles and Methods of Toxicology*, 2nd ed., ed. A. W. Hayes. New York: Raven Press, 1989.

Renwick, A. G. Pharmacokinetics in toxicology. In *Principles and Methods of Toxicology*, 2nd ed., ed. A. W. Hayes. New York: Raven Press, 1989.

Roberts, J. F., W. W. Piegorsch, and R. L. Schueler. Methods in testing for carcinogenicity. In *Principles and Methods of Toxicology*, 2nd ed., ed. A. W. Hayes. New York: Raven Press, 1989.

Roe, F. J. C. Testing in vivo for general chronic toxicity and carcinogenicity. In *Testing for Toxicity*, ed. J. W. Gorrod. London: Taylor & Francis, 1981.

Stevens, K. R. and M. A. Gallo. Practical considerations in the conduct of chronic toxicity studies. In *Principles and Methods of Toxicology*, 2nd ed., ed. A. W. Hayes. New York: Raven Press, 1989.

Styles, J. A. Other short-term tests in carcinogenesis studies. In *Testing for Toxicity*, ed. J. W. Gorrod. London: Taylor & Francis, 1981.

Venitt, S. Microbial tests in carcinogenesis studies. In *Testing for Toxicity*, ed. J. W. Gorrod. London: Taylor & Francis, 1981.

Weiss, B. and D. Cory-Slechta. Assessment of behavioral toxicity. In *Principles and Methods of Toxicology*, 3rd ed., ed. A. Wallace Hayes. New York: Raven Press, 1994.

Zenick, H. and E. D. Clegg. Assessment of male reproductive toxicology: A risk assessment approach. In *Principles and Methods of Toxicology*, 2nd ed., ed. A. W. Hayes. New York: Raven Press, 1989.

SAMPLE QUESTIONS

1. Determination of the chemical and physical properties of a test chemical are not tests for toxicity. Why is knowledge of these properties an essential preliminary for such tests?

2. The determination of the LD_{50} as a measure of acute toxicity has been much criticized. Outline the nature of these criticisms.

3. Relative to the life span of the experimental animal, what time periods for dosing would be characteristic of (a) an acute toxicity test, (b) a subchronic toxicity test, and (c) a chronic toxicity test?

4. In addition to acute, subchronic, and chronic toxicity tests, other tests may be required by regulatory agencies on a case-by-case basis or are known to be useful adjuncts to current mandatory testing protocols. Name as many of these tests as you can.

5. Briefly describe the underlying principal of the Ames test.

6. A number of tests have been devised for the measurement of the frequency and nature of chromosome aberrations. What are they?

Forensic and Clinical Toxicology

SHARON A. MEYER and BONITA L. BLAKE

21.1 INTRODUCTION

Forensic toxicology and clinical toxicology are specialized applications of the basic principles of toxicology covered earlier in this book. Forensic toxicology utilizes toxicological principles within the context of legal issues and is a component of forensic science. Toxicology results can provide evidence for determination of cause of death and time of lethal exposure; source of contamination of foods, water bodies, and adulterated pharmaceuticals; role of licit and illicit drug use; and impairment in vehicular collisions, workplace accidents, and domestic disputes, among other litigation-related matters. Clinical toxicology addresses treatment and prevention of chemical poisonings of both humans and domestic and companion animals, and includes aspects of occupational and emergency medicine, poison control, and public health. Recently, a heightened role for clinical toxicology has resulted from its contribution to emergency preparedness and homeland defense. Both clinical and forensic toxicology rely heavily upon analytical chemistry, usually of analytes in complex biological matrices, and utilize many fundamentals of toxicokinetics. Exposures to specific chemicals are surmised from recognition of clinical symptoms in a process operationally the reverse of the hazard identification. Both are represented by professional societies that support mechanisms for licensure and facilities accreditation and are taught in dedicated postbaccalaureate curriculum. Each have emphasis areas characteristic of the specialization as detailed in the material that follows.

21.2 FORENSIC TOXICOLOGY

21.2.1 Overview

Empirical knowledge of acute toxicity of chemicals has existed since antiquity. Chemical poisoning was a favorite plot element of the old masters—Shakespeare's poisonings in *Hamlet* and *Romeo and Juliet* and Chaucer's recounting of poisoned wine used in *The Parson's Tale* are a few examples—and various pesticides (e.g.,

A Textbook of Modern Toxicology, *Fourth Edition*. Edited by Ernest Hodgson
Copyright © 2010 John Wiley & Sons, Inc.

arsenical rat's bane, henbane with alkaloids hyoscyamine, and scopolamine) were readily available from medieval apothecaries. However, formal recognition of poison as a murder weapon began with presentation of postmortem residue analysis as legal evidence by Mathieu Orfila of Sorbonne University, Paris, and included application, in 1840, of the Marsh arsenic test on tissues of the deceased spouse of Madame LaFarge. As in today's application of forensic toxicology, this evidence was accepted only after reasonable doubt, introduced by Orfila's own observation of detectable baseline levels of tissue arsenic, was negated by his demonstration of Mr. LaFarge's unusually high body burden. With acceptance of an arsenic detection method and passage of the British Arsenic Law (1851) restricting purchase of arsenical rat poison, aspiring poisoners adopted neurogenic phytotoxins as a new murder weapon. This led to another milestone in the late nineteenth century with development of the Stas–Otto extraction for plant alkaloids, whose premise of pH-dependent solubility continues to be used today (Figure 21.1). The launch of forensic toxicology in the United States coincides with the 1918 appointment of A.O. Gettler in the Medical Examiner's office of the city of New York. Dr. Gettler's extensive contributions ranged from diagnosis of osteonecrosis from the popular

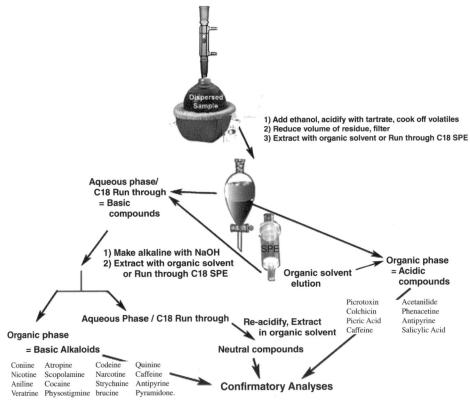

Figure 21.1   Partitioning of chemical classes in hydrophobic organic and hydrophilic aqueous phases as affected by pH as employed in the Stas–Otto procedure. Fractionation with a separatory funnel and solid phase extraction (SPE) cartridge are shown. Adapted from Autenrieth, 1921.

radium-based patent medicine "Radithor," development of methodology to quantify blood ethanol, and quality control recommendations on serotyping of degraded blood. Of broader impact was Dr. Gettler's extensive involvement in training of forensic toxicologists.

Twentieth-century advances in technology moved the field from see-taste-smell qualitative end points to instrumental measurements. The latter third of the twentieth century was marked by improved capability of and accessibility to analytical instrumentation enabled by incorporation of transistors, microchip electronics, and automation. On-site data collection enabled by equipment miniaturization eliminated some problems associated with field collection, then transport and lab analysis. Technological and electronic advances have also made it feasible to employ sophisticated instrumentation in the crime labs that previously were available only for research. Specificity of residue analysis has improved, and predictions from quantitative analysis became harder to refute. Recent expansion of forensic analysis in human performance certifications and drug use compliance has resulted in a shift in workload from predominantly postmortem chemical analysis to biomarker analysis of living tissues. Today's modern crime lab includes state-of-the-art instrumentation for analyses of a range of chemicals most relevant to forensics and is run by highly trained professionals nearly as efficient as portrayed in the popular medium.

21.2.2 Evidentiary Requirements

As with other accredited and licensed applications, performance standards require that proscribed procedures be followed. Crime labs must operate under Good Laboratory Practices (GLP). Isolation, identification, and annotation are critical for sample collection and are embodied with other information that becomes part of the sample's accession number. Chain-of-custody records documenting date of possession of signatories are mandatory for admissibility as evidence in litigation. Confidentiality requirements are strictly enforced. Most forensic laboratories operate as medico-legal components of state or local investigative agencies. Primary oversight for compliance with best practices results from facilities accreditation in the United States by the American Board of Forensic Toxicologists. If interstate activity is involved, the Department of Justice through the FBI may act in cooperation with state law enforcement agencies. Other federal agencies may have a role depending upon context, such as by FDA in cases of food and drug adulteration. Specific certification is required for conduct of mandatory urine testing of applicants for federal job positions. A recent National Academy of Science report has recommended expansion and standardization of forensic laboratory oversight (NRC, 2009).

Toxicological information relevant to litigation can be used formally as evidence in criminal and civil cases or informally in consultation with Counsel. Toxicology reports presented in court have recently been judged to be "testimonial evidence" by the U.S. Supreme Court and thus, those who generate the data in these reports may be cross-examined by defense exercising their sixth amendment rights (*Melendez-Diaz v. Massachusetts*, 2009). Forensic toxicologists that testify as expert witnesses are subject to provisions of Article VII of the Federal Rules of Evidence, especially Rule 702. This rule guides the acceptance of a witness as an "expert" based upon "knowledge, skill, experience, training, or education" in a specialty as determined by the trial judge hearing the case, a condition decided in *Daubert v. Merrell Dow Pharmaceuticals, Inc. (1993)*. Consenting expert witnesses may be called by prosecution, defense, or be appointed by the court. Once admitted as an expert, the rule specifies that that individual's testimony be based upon facts interpreted from application of sound toxicological principles and methods. The expert toxicologist may testify to his opinion on facts presented at or before the hearing and, if the latter, may be requested to disclose those facts upon request of the court or upon cross-examination.

21.2.3 Sample Type and Chemical Classes Analyzed in Forensic Toxicology

Tissue types taken for forensic analyses depend upon the nature of needed information. Results intended for human performance assessment and determination of illicit drug use are from readily accessible, vital tissues—almost invariably blood and urine plus occasionally tears, expired air, perspiration, breast milk, feces, vaginal mucus, semen, and saliva. In addition, a wider range of tissues including internal organs can be taken during autopsy for postmortem examination. Brain, liver, and kidney are routinely collected solid tissues. Postmortem redistribution of blood requires comparison of analyte values from heart and a peripheral site, although availability of unclotted blood may be problematic at some sites. High tissue concentrations of chemical may suggest route of administration, while concentrations of parent compound or toxic metabolite sufficient to cause damage to critical organs can indicate cause of death. Stomach contents are examined in suspected oral overdoses. Hair and nails can be especially informative since their continuous growth during life and localized deposition of certain acute toxicants near time of exposure can be used to define a time interval from exposure to death. Banding in hair and nails has been used to establish time of poisoning by arsenic and heavy metals and to dispute their presence in exhumed corpses as a consequence of soil contamination. Vitreous humor is an essential sample that can be analyzed to confirm values in a clinical panel for blood, which are more susceptible to postmortem changes. For example, ethanol can be produced in decaying tissue and enter blood, thus compromising prediction of alcohol abuse from blood ethanol content. The vitreous chamber is somewhat protected from putrefaction and is thus not as susceptible to microbial ethanol contamination. Disposition of certain chemicals can elucidate circumstances of death, such as hemoglobin-bound CO below threshold (50%) in a body recovered from a burned building that would indicate death occurred for reasons other than smoke inhalation.

The presence of blood can be detected from the fluorescent product produced by hemoglobin peroxidase activity with luminal. Promising new laser-based biospectroscopic methods are in development that will enable detection of other biological fluids in the field. Blood is analyzed soon after collection and is transported to the laboratory for cell number and type and blood gases; thereafter, frozen or spotted, dried samples are prepared for archiving. Urine is chilled if not immediately analyzed to prevent loss of volatiles. Survey analyses are qualitative screens against a standardized background control with simplified procedures that sometimes can be done in the field. If a positive sample is detected, more rigorous quantitative analysis may be done with more complex and costly instrumentation in a wet laboratory. More frequently employed analytical panels detect ethanol, various abused and therapeutic drugs (e.g., amphetamines, barbiturates, benzodiazepines, cannabinoids, opiates), common homicidal and suicidal agents (e.g., arsenic, cyanide, CO, pesticides, and thallium, the poison *du jour*), and performance enhancers (e.g., anabolic steroids, erythropoietin). Occasional environmental contaminants are monitored for specific types of litigation, such as establishing liability for illegal dumping of hazardous materials or off-site public exposure to releases implicated in causing illness or violation of policies limiting worker exposures. A well-publicized example includes hexavalent chromium contamination of drinking water portrayed in the movie *Erin Brockovich* (2000) and, more recently, exposure of National Guardsmen at a water pumping station in southern Iraq where sodium dichromate had been used prior to U.S. involvement to prevent pipe corrosion (Mancuso et al., 2008). A class of compounds of more recent interest is CWAs (chemical weapons agents, e.g., ricin, organophosphate nerve gases, phosgene, sulfur mustard).

A unique issue of forensic toxicology arises when information is needed after the decedent's body is embalmed (Levine, 2003). Components of embalming fluids, aldehyde fixatives, and disinfectants (phenol, methanol, ethanol), plus case-specific additives such as disinfectant benzaldehyde, fungicide parachlorobenzene, boric acid buffer, and eosin dyes, complicate toxicant analysis. However, in the case of the 1991 death of Kay Sybers, complete postmortem degradation of succinylcholine was evidently slowed by immediate embalming arranged by her husband who erroneously believed the preservatives would mask the poison he had earlier injected. Subsequent events can occasionally justify analysis on an exhumed corpse, such as confirmation of a *modus operandi* of a serial poisoner. Aside from previously mentioned complications associated with putrefaction and soil contamination, degradation of organic constituents by anaerobic microbial metabolism may limit their detection. Minerals will have more permanence and may be concentrated in longer-lasting remains such as bone, hair, and nails. Examples of homicidal use of selenium, arsenic, and heavy metals have been determined with exhumed material.

Residue analyses on nonbiological samples are also the subject of forensic toxicology. Chemical analysis of nitrogen is the basis of detection of gunshot residue. Product adulteration by poisonous additives is a constant problem in commerce, as evidenced by the recent supplementation of pet foods with the protein mimic, melamine. At accidental industrial releases, both state environmental officials and company representatives collect samples for future determination of liability. Analysis of accelerants by thermal desorption/gas chromatography (GC) is common in suspected cases of arson. The presence of specific sets of chemicals in a sample associated with the crime scene can give a fingerprint useful for identification of material associated with a potential suspect, for example, a particular source of chemicals used in an illicit drug synthesis lab, vegetation carried from the suspect's residence, explosives characteristic of an identified bomb maker. For these applications, sample collection must be as controlled as possible to avoid contamination, and worker safety of those obtaining samples must be of high priority.

21.3 CLINICAL TOXICOLOGY

21.3.1 Overview

Clinical toxicology is the application of toxicological principles for the purposes of diagnosing, treating, and preventing medical issues that arise from exposure to pharmaceutical agents, alcohol, and illicit drugs, biological toxins (e.g., snake and spider bites), or chemicals in the household, workplace, or environment. The clinical toxicologist is a specialist in the interaction of drugs and chemicals with the body. He or she must be familiar with the adverse effects of a large variety of therapeutic and toxic agents, and must have a thorough knowledge of the appropriate interventions for acute and chronic toxicant exposures.

21.3.2 Clinical Toxicology and Health Care

The term "clinical toxicology" is often used synonymously with "medical toxicology." After World War II, a profusion of new drugs and chemicals in the home and workplace prompted a marked increase in the rate of toxic incidents that were observed by physicians and hospitals. Recognition of the severity of the problem by the medical community was led by the American Academy of Pediatrics, who organized a nationwide committee on accident prevention to educate and assist physicians in the treatment and prevention of childhood poisoning. Largely out of these efforts, the first poison control center was established in Chicago in 1953. The classification of clinical toxicology as a subspecialty of medicine is thus relatively recent, and the role of the clinical toxicologist in health care is still evolving. Today, clinical toxicologists are doctors, pharmacists, and other members of the health-care community with special training in toxicology (see below). In hospitals and tertiary care centers (also known as specialty or referral centers), clinical toxicologists evaluate patient history and symptoms, order and interpret appropriate laboratory analyses, and consult with other members of the medical team to diagnose and plan treatment strategies for the patient. They also coordinate the practice of therapeutic drug monitoring (TDM), measuring and evaluating blood levels of medications in patients. TDM is particularly important for drugs with a low therapeutic index (TI) (see below), allowing drug dosages to be continuously adjusted so that adverse side effects can be minimized while therapeutic effectiveness is maintained.

Poison control centers are staffed by clinical toxicologists who provide 24-h, free access to information about poison exposure management and prevention to the public, as well as diagnostic and treatment recommendations to health-care providers. The American Association of Poison Control Centers (AAPCC) is an organization that certifies poison control centers and their personnel. The AAPCC also provides public and professional education about toxic agents and maintains the only poison information and surveillance database in the United States. The surveillance provided by the National Poison Data System (NPDS) allows for the monitoring of public health hazards, unusual exposure patterns, and outbreaks of public health emergencies, thus facilitating the early detection and elimination of hazardous chemical products and exposure incidents. The information obtained by the NPDS is used by public health and emergency preparedness specialists, and regulatory agencies such as the Food and Drug Administration, Environmental

Protection Agency, Consumer Product Safety Commission, and the Drug Enforcement Agency. It also serves as a valuable resource of toxicity exposure information for interested nongovernmental parties such as company product safety departments and academic medical centers.

Increasingly, clinical toxicologists are finding roles outside the hospital or poison control center in which to apply their training. Contract testing laboratories are often led by a clinical or forensic toxicologist skilled in analytical techniques. Many clinical toxicologists participate in occupational, environmental, and medical or medicolegal consulting, while others work in governmental and regulatory agencies such as the Food and Drug Administration, Centers for Disease Control, Public Health Service, Department of Homeland Security, and local health departments. Large companies—particularly in the chemical and pharmaceutical industries— employ clinical toxicologists as medical officers, health and safety officials, or in product development. Finally, due to the constant emergence of new and potentially hazardous drugs and chemicals, the prevalence of drug abuse, and the threat of terrorist attacks, there is a growing need for clinical toxicologists in academic research and teaching as our society prepares for its future.

21.3.3 Training and Certification

Most clinical toxicologists are trained as medical doctors, doctors of pharmacy, or nurses, although roles for pharmacologists and laboratory medicine specialists, or poison information specialists are increasingly being developed. Physicians who specialize in clinical toxicology must complete medical school and residency, and then complete a 2-year fellowship in affiliation with an academic medical center and/or poison control center. The expanding role of hospital pharmacists on healthcare teams requires advanced knowledge of adverse drug reactions, and most pharmacy training programs offer a clinical toxicology elective. Pharmacy students interested in specializing in clinical toxicology may choose a rotation in a poison control center as their senior practice experience. Some PharmD graduates complete residencies that focus on clinical toxicology or emergency medicine, although no formal accredited residency programs in those specialties are available. Registered nurses who wish to specialize in toxicology may complete training in an advanced practice specialty to become nurse practitioners. Generally, this requires a master's or a doctoral degree beyond the Bachelor of Science degree in nursing, although accelerated programs are available. Some programs require at least 1–2 years of clinical experience as a registered nurse for admission.

Educational requirements for specific positions in clinical toxicology vary from position to position and from state to state. For example, most poison control centers employ poison information specialists to field calls from health professionals and the public. The requirements for this position usually include certification by the American Association of Poison Control Centers as a certified specialist in poison information (CSPI). To be certified, the association requires licensure as a registered pharmacist within the state of certification, with a minimum of a bachelor's degree in pharmacy, or a registered nurse within the state, with a minimum of a bachelor's degree in nursing. On the other hand, to be a manager or director of a poison control center, usually, one must be a licensed physician, pharmacist, or nurse (although sometimes, advanced degrees in a biomedical discipline are sufficient). Managers and directors must be board certified; physicians by the American Board of Medical Toxicology or a related discipline, and other professionals by the American Board of Applied Toxicology. Each of these boards maintains further requirements to be eligible to take their certification exams.

21.3.4 Clinical Management of Toxicant Exposure

Toxicant exposures may present in a variety of ways, ranging from an acute medical emergency to chronic conditions that have grown debilitating over time. In any emergency, basic life support is the most critical priority. Once the ABC's of airway, breathing, and circulation are addressed, supportive care (e.g., intravenous fluids, glucose, and seizure control if necessary), diagnostic evaluation and therapeutic interventions can be initiated. It is important to note at the outset that in any case where there is doubt about the diagnosis or treatment of a suspected poisoning, a call to the regional poison control center is well-advised. Cases of rare poisoning events and threats to public health should always be reported to a poison control center.

Evaluation A thorough history can provide some of the most valuable diagnostic information that can be obtained about a poisoning incident. Often, the patient is aware of the agent that he or she has been exposed to and can give details important to therapeutic planning such as the time since exposure and the dose received. In patients who are unresponsive or uncooperative, the diagnosis of poisoning must rely on the results of the physical exam and laboratory tests. General signs may be present that provide clues to the source of toxicity. For example, characteristic odors such as almonds (cyanide), garlic (organophosphates, metals such as arsenic, thallium, or tellurium), fruitiness (isopropanol, alcoholic ketoacidosis) may be detectable. An unusual appearance of the skin, such as bruising, blisters, or jaundice may also be informative. Vital signs such as temperature, pulse, respiratory rate, and blood pressure are always collected and provide diagnostic clues as well as indicate the progression or resolution of toxic effects. Neurologic signs such as confusion, agitation, sleepiness, seizures, or coma may be present depending on the agent involved. Often, patients display several signs at once that point to a specific class of poisons. This constellation of typical symptoms that are characteristic of a class of poisons is called a toxidrome. For example, the toxidrome for organophosphate toxicity can be recalled by a mnemonic known as DUMBELS, which stands for diarrhea, urination, miosis (pinpoint pupils), bronchorrhea (excessive secretions in the bronchi), bradycardia (slowed heart rate), emesis (vomiting), lacrimation (tearing), and salivation. Rapid recognition of toxidromes can help determine whether the issue is a poison and the class of poison that has been encountered.

Standard laboratory tests may reveal alterations in electrolytes, blood gases, or blood glucose concentrations, acid–base disturbances, or other changes that when considered alone would not be diagnostic of poisoning. However, these findings often contribute to toxidromes and thus aid in identifying the class of toxicant that might be present. Other clinical blood chemistry analyses such as liver and kidney function tests may be altered by specific toxicants; monitoring these throughout the therapeutic period is useful for weighing therapeutic options and developing prognoses. Electrocardiogram abnormalities are common with a wide variety of drugs and toxins and occasionally bear characteristic signatures that help diagnose exposure to specific toxicants.

Most diagnoses of toxicity are made based on patient history and the clinical presentation (i.e., the signs, symptoms, and results of general lab tests). Since the mainstay of treatment in chemical poisoning is supportive care, these basic findings are often sufficient to begin therapeutic intervention. On the other hand, if the physician suspects a toxicant or class of toxicant for which a specific antidote exists, he or she may desire additional confirmative information. Furthermore, for patients in acute distress, stat (urgent) diagnostic testing is necessary. Because of this need, several qualitative and semiquantitative tests have been developed to rapidly identify selected toxicants. These tests are used with the understanding that there is a balance between the need for quick decision making and the need for accuracy. The National Academy of Clinical Biochemistry has prepared a set of guidelines for two tiers of laboratory tests that should be available for patients in emergency departments. The first tier consists of "stat" (i.e., with an ideal reporting turnaround time of 1 h or less) tests for selected toxicants that are commonly encountered and for which a semiquantitative test is available (see Table 21.1). Usually, these tests are based on immunoassay technology (described in Section 21.4). The second tier of tests consists of comprehensive or broad-spectrum analyses for toxicants that may not have been identified in the first tier, or are unavailable on a stat basis, or

TABLE 21.1 Recommended "Stat" Toxicology Assays[a]

Serum (Quantitative or Semiquantitative)	Urine (Qualitative)
Acetaminophen	Cocaine
Lithium	Opiates
Salicylates	Barbiturates
Theophylline	Amphetamines
Valproic acid	Propoxyphene (Darvon)
Carbamazepine	PCP
Digoxin	Tricyclic antidepressants
Phenobarbital	
Iron	
Ethanol	
Methanol	
Ethylene glycol	
Methemoglobin, carboxyhemoglobin[b]	

[a] Adapted from Wu et al. National Academy of Clinical Biochemistry Laboratory Medicine Practice guidelines: Recommendations for the use of laboratory tests to support poisoned patients who present to the emergency department. *Clin. Chem.* **49** (3): 357–379, 2003.

[b] Not specific for any of the multiple toxicants that produce methemoglobinemia or carboxyhemoglobinemia. Detected by co-oximetry, which measures oxygenated hemoglobin as a percentage of total hemoglobin.

for which quantitative results are needed for long-term patient management. These tests entail more quantitative analytical tools such as GC/mass spectrometry (MS) and high-performance liquid chromatography (HPLC). Smaller hospitals that do not have adequate resources to perform these tests usually send samples to a local or regional reference laboratory or toxicology laboratory for analysis. For purposes of monitoring therapeutic effectiveness, it is sometimes more practical to rely on other biomarkers of toxic effect such as acetylcholinesterase activity or coagulation, rather than on the concentration of the toxicant itself.

Treatment   The first goal in the treatment of a poisoned patient is to minimize further exposure. This involves removal of any material that has not been absorbed by means that are appropriate to the route of exposure, and by enhancing elimination of material that has been absorbed. As might be expected, the use of these techniques is guided by pharmaco- and toxicokinetic principles (for further discussion of these principles, see Chapters 5–9), and the choice of which method to use depends on the kinetic limitations presented by the suspected toxicant. In addition, elimination therapies are most effective when administered within the first few hours after intoxication.

Poisons that are ingested may be removed by inducing vomiting or gastric lavage with a stomach tube and saline. However, these methods are inappropriate if significant time has elapsed since ingestion (since most toxicants will have been absorbed within the first hour or so), if the patient is suspected to have ingested corrosive agents (as this promotes further damage to the mucosal lining of the esophagus), or when central nervous system function is compromised (since the patient could aspirate the stomach contents). Furthermore, vomiting may preclude the use of oral therapeutics. Rather than induce vomiting, most hospitals prefer to use chemical adsorption with agents such as activated charcoal or whole bowel irrigation. These techniques not only prevent absorption by removing the agent from the gut, but also enhance elimination by inhibiting the enterohepatic recirculation of toxicants that are normally taken up by the liver. Activated charcoal is administered as a slurry either orally or by using a nasogastric tube in single or multiple doses. A wide variety of poisons are effectively adsorbed by charcoal; notable exceptions include alcohols and heavy metals, to which it binds poorly. These agents may be removed by whole bowel irrigation. It has been proposed that whole bowel irrigation promotes a beneficial dialysis-like effect, in which low molecular weight compounds move by a diffusion gradient from the circulation back into the gut. Cathartics such as sorbitol and magnesium sulfate may be used to decrease the absorption of substances by accelerating gastrointestinal motility. Although cathartics are sometimes added to activated charcoal slurry, few clinical data exist to demonstrate their effectiveness. The elimination of weak acids and agents that are substantially excreted by the kidneys can be aided by alkaline diuresis, involving the addition of sodium bicarbonate to intravenous fluids. This method is particularly effective for reducing circulating levels of salicylates (e.g., aspirin) and chlorphenoxy herbicides. Enhancing clearance by hemodialysis (in effect, ultrafiltration) is useful for highly water-soluble, low molecular weight compounds, and those with a low apparent volume of distribution (Vd) that tend to remain in circulation rather than distribute to the tissues. Ethanol, ethylene glycol, salicylates, and lithium are often removed by dialysis. Hemoperfusion, in which large volumes of patient blood is passed over adsorbent activated charcoal, is appropriate for more lipid-soluble drugs such as barbiturates, acetaminophen, and insecticides.

The second goal of therapeutic management of toxicant exposure is to reduce toxic effects by preventing the interaction of the toxicant with its target sites, or altering its biotransformation to toxic metabolites. Some antidotes act by binding specific toxicants and sequestering them in body compartments where they cannot reach their target sites. In the case of chelating agents, this also enhances the elimination of heavy metals. Chelating agents are negatively charged molecules that bind to the metal forming a stable complex, which is excreted in the urine. Dimercaprol, also known as BAL (British anti-lewisite), was developed for use in warfare as an antidote to the organoarsenical agent lewisite. It is still used today for the treatment of acutely symptomatic patients who have ingested arsenic, lead, and other metals. Other dithiol-containing agents used for chelation include succimer (dimercapto-succinic acid) and deferoxamine. Calcium edentate (EDTA) is also used therapeutically to chelate metals (particularly lead), as well as calcium in cases of hypercalcemia resulting from digitalis toxicity. The synthetic pigment Prussian blue is approved by the FDA for medical use as a chelator of thallium and radioactive cesium. Immunotherapeutics (i.e., antibodies and Fab fragments of antibodies) also act by sequestering toxicants away from their site of action. Fab fragment antidotes are available for treating overdoses with certain drugs such as digoxin, and for many natural, peptide-based toxins such as snake and spider venoms.

The mechanisms of action of antidotes are as varied as those of toxicants themselves. Pharmacological antagonists, competitive inhibitors or activators of drug metabolizing enzymes, decoys, redox agents, and substances that otherwise reduce the impact of toxicants may be used as antidotes. Despite the wide variety of antidotal mechanisms, however, the actual number of effective antidotes available for clinical use is limited. Selected examples of common toxicants and their antidotes are presented in Table 21.2.

TABLE 21.2   Common Toxicant Antidotes and Their Mechanisms of Action

Toxicant	Antidote	Mechanism of Action
Acetaminophen	*N*-acetylcysteine	Enhances glutathione synthesis, may directly bind reactive metabolites
Crotalid snake venom	Fab antibody fragments	Neutralize toxin by high affinity binding and sequestration from its target
Ethylene glycol	Fomepizole (4-methylprazole)	Inhibits alcohol dehydrogenase, the primary enzyme in the biotransformation of ethylene glycol to toxic metabolites
Nitrates/Nitrites	Methylene blue	Accelerates the conversion of methemoglobin (Fe^{3+}) to hemoglobin (Fe^{2+})
Opioids	Naloxone	Opioid receptor antagonist
Organophosphates	Atropine	Muscarinic acetylcholine receptor antagonist
	Oximes	Reactivate acetylcholinesterase
Warfarin	Vitamin K	Restores synthesis of coagulation factors

Additional Considerations Health-care professionals that attend to poisoned patients must be prepared to assess toxic situations and decide on treatment options rapidly. As described above, several factors must be weighed, including the type of poison, the dose, route and time since exposure, the clinical status of the patient (including age, general health, concurrent use of prescribed/abused drugs), and the availability of medical resources and facilities. Decisions such as those regarding when and how often to order quantitative analyses, the appropriate use of elimination enhancers, and the optimal duration of antidote therapy are guided by awareness of these factors, along with a thorough working knowledge of pharmacokinetic and toxicokinetic principles. These principles are described in more detail in Part III of this book. For clinicians, predictive kinetic calculations should include consideration of factors such as the metabolic activation of compounds to toxic metabolites, the continued redistribution of the toxicant between the blood-stream and the target tissues, and the effects of the toxicant on its own clearance (such as by high-dose saturation of plasma protein binding or metabolic and excretion mechanisms). An important precaution is that poisons often adversely affect organ function, causing disproportionate changes in their blood concentrations. For example, barbiturates lower blood pressure significantly, reducing hepatic and renal perfusion and prolonging their toxicity. Finally, while the goal of therapy is to change the kinetics (absorption, distribution, metabolism, and excretion) or dynamics (interactions with target sites) of the overdosed drug or poison, it is important to note that therapeutic interventions act by their own pharmacokinetic principles. Thus, short-acting antidotes such as naloxone, an opioid receptor antagonist, must be dosed multiple times over the course of treatment, since their half-life is much shorter than that of most of the opiates with which they are intended to compete.

For some drugs, relatively small changes in systemic concentration can produce very marked changes in therapeutic and toxic response. These drugs are said to have a low TI, quantitatively represented as the ratio lethal dose 50/effective dose 50 (LD_{50}/ED_{50}). Drugs that are lethal to 50% of a test animal population at a relatively low dose (the LD50), compared to their minimum effective dose for 50% of the population (the ED50) would have a low TI and thus be likely to cause toxicity in patients within a range very close to therapeutic levels. Lithium, warfarin, phenytoin, and digoxin are examples of drug with a low TI.

Genetic factors are often unknown players in the disposition of toxicants in acute cases. Rarely is the genetic profile of an individual known at the time of treatment for poisoning. Nevertheless, interindividual variation in the response to drugs, poisons, and their antidotes may affect the clinical course and response to treatment in acute exposures. The role of genomics in toxicology is discussed in Chapter 28. With respect to clinical toxicology, it is important to keep in mind that not only may genetic variability affect toxic outcomes due to differences in metabolic and clearance enzymes, but it may also determine adverse responses in other ways. For example, polymorphisms in opioid receptor genes mediate the adverse effects (e.g., nausea) that some patients experience at therapeutic doses of morphine. Similarly, genetic variations at the receptor and even downstream of receptors at the second messenger level are thought to influence the development tolerance and dependence to opiates.

21.4   ANALYTICAL METHODS IN FORENSIC AND CLINICAL TOXICOLOGY

Qualitative screens are often solid-state antibody- or enzyme-based techniques with colorimetric end points, similar to dip-stick products used for home testing for pregnancy and blood glucose. Screens compare analytes in test sample matrices, with very little preliminary workup, to cut-off values and often have rapid responses necessary to provide information "stat" for clinical toxicology (Table 21.1). Specificity is dependent upon cross-reactivity of the antibody and is generally good for chemical class, such as amphetamines, but may be compromised for detection of a specific member of a class, such as methamphetamine. Basic principles and conduct of such immunoassays is outlined in Chapter 2. Urine is often used in screens since urinary concentrations of parent compound or metabolites are usually not well correlated with clinical symptoms. Standardized immunoassays are typically available for several analgesics and antidepressant fluoxetine, causes of the first and eighth most common human poisoning in the United States (Bronstein et al., 2008), in addition to benzodiazepines, barbiturates, and others. Urinary creatinine is measured for all samples because values below 20 mg/dL indicate potential sample dilution, an adulteration often encountered with drug testing. In clinical toxicology, urinary and blood creatinine are used to determine renal clearance, and abnormally low values indicate impaired kidney function. Another common screen used in the field for forensics relies upon photoelectric detection of ethanol oxidation in air exhaled into a breathalyzer.

Since rapid turnaround is not as critical in forensic as in clinical toxicology, while quality control is more demanding, positive samples in screens undergo follow-up confirmatory analyses. If amounts are limited, the choice of which analyses are performed is guided by prior knowledge based upon additional case history. In general, these procedures involve a preliminary preparative stage to remove analyte from its biological matrix, such as filtration to remove solids, tissue homogenization, pH adjustment, and protein precipitation, then some type of fractionation followed by instrumental measurement of an output related to concentration. Fractionation can be either low-resolution segregation into chemical classes (Figure 21.1) or high-resolution chromatographic separation (Figure 21.2). Concentration-dependent output will depend upon what type of instrument is hyphenated with the fractionation technology and is normalized to certified standards. Blood is the most frequent tissue used for such analyses and considerable guidance exists relating blood values and adverse effects, such as lethality and behavioral effects. For example, comprehensive nomograms exist for acetaminophen relating time after ingestion, plasma concentration, and severity of hepatotoxicity.

Bulk fractionation into polar versus nonpolar solvents as effected by pH is a historical procedure, but has been largely replaced by solid phase extraction. Both rely upon greater distribution of the charged conjugate base or protonated acid of Lewis acids and bases, respectively, in more polar solvents or on solid adsorbants (Figure 21.1). Elution into apolar medium is then achieved with a change in pH relative to a compound's pKa. The process is repeated until tissue components are segregated into fractions containing originally acidic, neutral, and basic compounds. Since many drugs, licit and illicit, have pKas in a range accommodated by this

Figure 21.2 General example of analyte separation from a mixture based upon differential affinities for stationary and mobile phase (Φ) constituents. Separated analyte stream is then fed into hyphenated instrument for confirmation, such as the gas chromatograph/mass spectrometer of this example.

procedure, it can be used as a preparatory method for identification of drugs in the resultant fractions by chromatographic techniques.

Blood analysis for clinical toxicology typically starts with a general assessment of complete blood counts (CBCs) and panels of biomarkers, including intermediary metabolites, acute phase proteins, and tissue selective enzymes. Chemical assays and more targeted diagnostic tests may be performed early in treatment if specific information is available from case histories. A CBC analyzer is a dedicated flow cytometer that measures and classifies blood cells with respect to size and intracellular granularity and can distinguish erythrocytes, lymphocytes, granulocytes classes, and platelets. Hemoglobin is measured simultaneously, enabling classification of various types of anemia. Various biomarkers are measured with automated, continuous flow bioanalyzers/spectrophotometers; examples are alanine aminotransferase (ALT; increased with liver damage), albumin (decreased with inflammation), glucose (elevated with diabetes), urea (blood urea nitrogen [BUN]; elevated with poor kidney function), and electrolytes. The latter is used to detect the presence of an imbalance between serum anions and cations, the so-called anion gap calculated as $([Na^+] - ([Cl^-] + [HCO_3^-]))$, and whose elevation is diagnostic for compounds that cause loss of bicarbonate, that is, metabolic acidosis. Methanol toxicity is associated with a high anion gap (>16 mEq/L) because of its metabolism to formic acid.

If the identity of poison and time or amount of administration is known, then quantitative determination of serum values will aid emergency room medical personnel to make a prognosis of outcome and chart an appropriate course of treatment. For example, nomograms for salicylate (aspirin) and acetaminophen are commonly used. Serum analyses are also used to monitor effective levels of therapeutics given as antidotes, especially those with a low TI such as theophylline. Occasionally, other tissues are analyzed on a case-specific basis, such as neonatal meconium to determine maternal drug abuse. Clinical laboratories may also be utilized to biomonitor hazardous materials as they relate to occupational medicine, for example, as encountered by first responders with time after an accidental chemical release. Similar facilities with equivalent capabilities exist for forensic and clinical toxicological applications to veterinary medicine (Galey and Talcott, 2005).

Principles of separation for many of the laboratory instruments used in any analytical laboratory, including that for forensic and clinical toxicology, are detailed in Chapter 24 of this book. Additionally, details of quality control necessary to be considered "sound toxicological methodology" are outlined. In general, analytes adsorb with varying affinity based upon lipophilicity, volatility, size, and/or pH onto an immobile phase and are eluted into a mobile phase as conditions progressively change. Common solid phase materials are modified silicas, resins, cellulose, and other polymers of varying porosity. Compounds differentially eluted into mobile phase of either gas (GC) or liquid (liquid chromatography, LC) are separated by distance from origin on a solid support (thin layer chromatography, TLC) or time to appear before a detector for mobile phase freely flowing through a column format. The resulting chromatogram separates and localizes compounds depending upon their specific, relevant properties, which are often known from databases and can be used for identification. Both parent compound and metabolites are simultaneously detected with chromatographic procedures, and their relative abundance will depend upon type of sample. For example, psychoactive agent of marijuana, Δ^9-tetrahydrocannabinol is lipophilic (calculated K_{ow} ~10^7) and readily deposits in lipid-rich brain, while oxidized metabolite 11-*nor*-9-carboxylic acid-Δ^9-tetrahydrocannabinol is excreted in urine. For testosterone, normal amounts of androgenic C-17 S isomer are 1–4 times that of the R isomeric epitestosterone in human male urine. Ratios higher than 4:1 indicate administration of exogenous testosterone for performance enhancement. The presence of exogenous urinary testosterone is confirmed by lowered carbon isotope ratios. Testicular synthesis of testosterone from precursors with typical environmental ratios of ^{12}C:^{13}C of ~100:1 causes enrichment of product in the more reactive lighter isotope, while chemically synthesized testosterone has the same ^{12}C:^{13}C as the feedstock and is excreted into urine with the carbon isotope ratio unchanged.

The workhorse of quantitative forensic analysis is capillary GC in which deposition of stationary phase as a thin liquid or polymer film on the walls of a glass capillary greatly provides a highly efficient format. Volatile components, such as short-chain alcohols and sulfide decomposition products, can be off-gassed from intact samples into headspace that is directly analyzed by GC. Vapors and gases trapped on sampling sorbents can be injected directly from a thermal desorption attachment coupled to a GC sample port. Analysis by LC, with mobile phase propelled by high pressure (HPLC), is also frequently used. Detection methods vary from those based upon formation of colored or fluorescent derivatives, absorption spectroscopy in the ultraviolet and visible ranges, redox-coupled coulometry, and electron expulsion or capture. GC and HPLC analyses of specific chemicals in blood are also critical components of the clinical toxicology laboratory. MS hyphenated with either GC or HPLC provides a powerful, versatile tool for simultaneous detection plus compound identification. Recent improvements in durability and size reduction of instruments performing these chromatographic techniques and analyte detection, largely funded by the Departments of Defense and Homeland Security to counter threats from CWAs, have been developed to provide essential real-time monitoring. Extensive mass spectral libraries for compounds typically encountered in forensic and clinical toxicology are readily available.

Metals can be analyzed directly in blood and urine. Solid tissue is typically ashed in a furnace or digested with strong acid and residue is redissolved for analysis. Contamination with extraneous metals is minimized by using acid-washed glassware and metal-free reagents. If remains are recovered from a metal-rich environment, such as soil, surrounding environment must be concurrently analyzed using the same workup procedures. For quantification by atomic absorption spectroscopy (AAS), dissolved metal ions are then atomized by high temperature, either generated via an oxyacetylene flame, graphite furnace, or electrothermally, and energy absorbed to excite metal valence electrons by a transecting beam of visible light is measured. Incident light of a very narrow wavelength characteristic of the specific metal analyte is produced by a hollow cathode lamp, and photon absorption is measured by a downstream spectrophotometer. A detection method that can simultaneously measure dozens of metals and metalloids, including As, Cd, Cr, Hg, Pb, Se, Tl, and U, is inductively coupled plasma–mass spectrometry (ICP-MS). This procedure also atomizes metal ions but uses an argon plasma to achieve temperatures so high that the valence electron ionizes from the metal atom. The resulting cations are then introduced into and resolved by a mass spectrometer. Sensitivity of ICP-MS approaches pg/mL (picograms per milliliter) for urine and blood, ~10-fold greater than that of AAS. Another versatile technique recently introduced for forensic analysis of metals is X-ray fluorescence (XRF) in which inner shell electrons excited to higher orbitals emit photons upon decay back to ground state that are characteristic of the specific metals in a sample. This also is a multi-element technique and can achieve resolution of ~50 μ as determined by the size of the incident X-ray beam. A unique feature of XRF analysis is the ability to determine metal speciation, that is, identification of what ligands are coordinated with a given metal.

BIBLIOGRAPHY AND SUGGESTED READING

Autenrieth, W. Non-volatile poisons. In *Laboratory Manual for the Detection of Poisons and Powerful Drugs*, ed. and trans. W. H. Warren, pp. 61–147. Philadelphia: P. Blakiston's Son, 1921.

Brent, J., K. L. Wallace, K. K. Burkhart, et al., eds. *Critical Care Toxicology*. Philadelphia: Elsevier Mosby, 2005.

Bronstein, A. C., D. A. Spyker, L. R. Cantilena Jr., et al. 2007 Annual report of the American Association of Poison Control Centers' National Poison Data System (NPDS): 25th annual report. *Clin. Toxicol* **46**:927–1057, 2008. http://www.aapcc.org/DNN/Portals/0/NPDS%20reports/2008%20AAPCC%20Annual%20Report.pdf

Daubert v. Merrell Dow Pharmaceuticals, 509 U.S. 579, 1993.

Galey, F. D. and P. A. Talcott. Effective use of a diagnostic laboratory. In *Small Animal Toxicology*, eds. M. E. Peterson and P. A. Talcott, pp. 154–164. Philadelphia: Saunders, 2005.

Levine, B. Postmortem forensic toxicology. In *Principles of Forensic Toxicology*, ed. B. Levine, pp. 3–13. Washington, DC: AACC Press, 2003.

Mancuso, J. D., M. Ostafin, and M. Lovell, Postdeployment evaluaton of health risk communication after exposure to a toxic industrial chemical. *Mil. Med.* **173**:369–374, 2008.

Melendez-Diaz v. Massachusetts. 557 U.S._____. 2009. http://www.supremecourtus.gov/opinions/08pdf/07-591.pdf

National Research Council (NRC). *Strengthening Forensic Science in the United States: A Path Forward*. National Academy of Sciences, 2009. http://www.nap.edu/catalog.php?record_id=12589

Wu, A. H. B., McKay, C., Broussard, L. A., et al. National Academy of Clinical Biochemistry Laboratory Medicine Practice guidelines: Recommendations for the use of laboratory tests to support poisoned patients who present to the emergency department. *Clin. Chem.* **49** (3): 357–379, 2003.

SAMPLE QUESTIONS

1. The first judicial application of forensic toxicology was
   a. the state-sponsored execution of Socrates
   b. a result of one of the provisions of the Magna Carta
   c. conducted by Orfila to convict Marie Lafarge of homicide using arsenic poisoning
   d. to provide support for Occupational Safety and Health Administration (OSHA) standards for benzidine dye workers

2. A comatose teenager is brought to a hospital emergency room and a screening immunoassay detects a urinary barbiturate well above cutoff value. The patient's mother reports that neither the teen nor her siblings have a history of epilepsy and that her daughter was awake and alert only an hour ago. Secobarbital has a half-life of ~30 h and a volume of distribution (Vd) of ~1.5 L/kg, while long-acting phenobarbital's half-life is ~100 h and Vd ~0.7 L/kg. Both are commonly abused street drugs.

   The patient's respiration is stabilized with a ventilator and blood pressure is maintained with IV fluids. There are no specific antidotes for barbiturates. What should be the next clinical treatment for this patient?

3. Compare how toxicokinetic information is applied to clinical toxicology versus forensic toxicology.

4. Cadaverine (1,5-diaminopentane) is a volatile decomposition product contributing to the malodor of decaying corpses. It is a basic aliphatic compound with pKa ~10. If a forensic toxicologist sought to determine whether a body contained residues of strychnine using fractions generated with the Stas–Otto procedure, would cadaverine contaminate the strychnine-containing fraction? Briefly explain your answer.

Prevention of Toxicity

ERNEST HODGSON

22.1 INTRODUCTION

Regardless of the results of hazard assessment (Chapter 20) toxicity is always a consequence of exposure, and without exposure there cannot be a toxic effect. However, if both hazard and exposure are verified, and the risk appears to be significant, there are a range of possible actions available to reduce that risk. These actions range from outright banning of both production and use of the chemical in question through legislative means, through measures to reduce exposure, to measures to restrict effect.

Exposure can be restricted by prevention of manufacture, control of use patterns, control of application techniques, by environmental manipulation and/or by education. Effects can be restricted by prophylactic and therapeutic methods and by education. Many of these approaches are controlled in whole, or in part, by legislation while many are simply the use of common sense in the cause of good domestic and industrial hygiene. In many circumstances, particularly in the home and workplace, wisdom dictates courses of action not necessarily prescribed by law. All of these aspects, taken together, comprise the subject matter of this chapter.

Laws and regulations provide the framework for organized efforts to prevent toxicity, and sanctions are necessary to prevent those without social conscience from deliberately exposing their fellows to risks from toxic hazards. However, without a population educated to toxic hazards and their prevention, the laws cannot be properly administered. The key to toxicity prevention lies in information and education with legislation, regulation, and penalties as final safeguards. In all probability, the better educated and informed the general population is, the less likely are laws to be necessary.

22.2 LEGISLATION AND REGULATION

In the best sense, legislation provides an enabling act describing the areas to be covered under the particular law and the general manner in which they are to

A Textbook of Modern Toxicology, Fourth Edition. Edited by Ernest Hodgson
Copyright © 2010 John Wiley & Sons, Inc.

be regulated, while designating an executive agency to write and enforce specific regulations within the intent of the legislative body. For example, the Toxic Substances Control Act (TSCA) was passed by Congress to regulate the introduction of chemicals into commerce, to determine their hazards to the human population and the environment, and to regulate or ban those deemed hazardous. The task of writing and enforcing specific regulations was assigned to the United States Environmental Protection Agency (USEPA).

Legislative attempts to write specific regulations into laws usually fail. The resultant laws lack flexibility and, because they are generally not written by toxicologists, are frequently ambiguous and seldom address the problems in a scientifically rigorous manner.

It should be borne in mind that legislation is a synthesis of science, politics, and public and private pressure. It represents a society's best estimate, at that moment, of the risks it is prepared to take and those it wishes to avoid, as well as the price it is prepared to pay. Such decisions properly include more than science. The task of the toxicologist is to see that the science that is included is accurate and is interpreted logically.

This section is based primarily on regulations in the United States, not because these are necessarily the best but because, in toto, they are the most comprehensive. In many respects, they are a complex mixture of overlapping laws and jurisdictions, providing unnecessary work for the legal profession. At the same time, few, if any, toxic hazards in the home, workplace, or environment are not addressed.

22.2.1 Federal Government

The following is a summary of the most important federal statues concerned in whole or in part with the regulation of toxic substances. Further details can be obtained from the appropriate websites (e.g., www.epa.gov), including titles of all Acts administered, and actions taken under those acts.

Clean Air Act   The Clean Air Act is administered by the EPA. Although the principal enforcement provisions are the responsibility of local governments, overall administrative responsibility rests with the EPA. The Clean Air Act is the law that defines EPA's responsibilities for protecting and improving the nation's air quality and the stratospheric ozone layer. Since the last major change in the law, the Clean Air Act Amendments of 1990, legislation has been passed to provide for several minor changes. The Clean Air Act, like other laws enacted by Congress, was incorporated into the United States Code (Title 42, Chapter 85).

This act requires criteria documents for air pollutants and sets both national air quality standards and standards for sources that create air pollutants, such as motor vehicles and power plants. An important action previously taken under this law was setting the standards for the now completed elimination of lead in gasoline.

Clean Water Act   The Clean Water Act, which amends the Federal Water Pollution Control Act, is also administered b the EPA and provides for the protection of surface water quality, in large part through funding of municipal sewage treatment plants. However, with respect to toxicity prevention, it is more important that the act regulates emissions from municipal and industrial sources. It has as its goal the elimination of discharges of pollutants and the protection of rivers so that they are "swimmable and fishable" and applies to "waters of the United States" subsequently defined to include all waters that reach navigable waters, wetland, and intermittent streams. It is not directly concerned with the quality of drinking since this is regulated under the Safe Drinking Water Act, also by the USEPA. While previous actions dealt primarily with point sources discharges, such as municipal sewage plants and industrial plant discharges, current actions also deal with so-called wet-weather sources including runoff from streets and farms. This act allows the federal government to recover cleanup and other costs as damages from the polluting agency, company, or individual.

Safe Drinking Water Act (1974, 1986, 1996)   Specifically applied to water supplied for human consumption, this act requires the EPA to set maximum levels for contaminants in water delivered to users of public water systems but excludes private wells serving less than twenty-five people. Two criteria are established for a particular contaminant: the *maximum containment level goal (MCGL)* and the *maximum contaminant level (MCL)*. The former, the MCLG, is the level at which no known or anticipated adverse effects on the health of persons occur and that allows an adequate margin of safety. The latter, the MCL, is the maximum permissible level of a contaminant in water that is delivered to any user of a public water system. MCLs are expected to be as close to the MCLG as is feasible.

Originally focused on water treatment since the 1996 amendments, the law has also focused on source water protection and funding for water system improvements thus extending the reach of the law from source to tap.

Comprehensive Environmental Response, Compensation, and Liability Act (CERCLA)   Commonly known as Superfund, CERCLA was enacted in 1980 and was amended in 1986 to deal with the many waste sites that exist across the nation. It covers remedial action, including the establishment of a National Priorities List to identify those sites that should have a high priority for remediation. This act authorizes the cleanup of hazardous waste sites, including those containing pesticides, that threaten human health or the environment. If they can be identified, USEPA is authorized to recover cleanup costs from those parties responsible for the contamination. CERCLA provides a fund to pay for the cleanup of contaminated sites when no other parties are able to conduct the cleanup. The *Superfund Amendments and Reauthorization Act (SARA) (1986)* is an amendment to CERCLA that enables USEPA to identify and cleanup inactive hazardous waste sites and to recover reimbursement of cleanup costs. One section of CERCLA authorizes the EPA to act whenever there is a release or substantial threat of release of a hazardous substance or "any pollutant or contaminant that may present an imminent or substantial danger to the public health or welfare" into the environment.

Consumer Products Safety Act (CPSA) and Consumer Products Safety Commission Improvements Act (CPSCIA)   CPSA is administered by the Consumer Products Safety Commission (CPSC). The CPSA is designed to protect the public against risk of injury from consumer products and to set safety standards for such products. It provides the authority to ban products where no feasible standard can be set and to recall products that present a substantial risk. Unfortunately, the Act is weakened by exclusion of many categories of consumer product from CPSC's authority. Such products, include food, drugs, cosmetics, medical devices, tobacco products, firearms, motor vehicles, pesticides, aircraft, and boats. These products are covered by other laws and jurisdictions, the USEPA and the Food and Drug Administration (FDA) being the most common, making a unified approach difficult, if not impossible.

Controlled Substances Act (CSA) The CSA not only strengthens law enforcement in the field of drug abuse but also provides for research into the prevention and treatment of drug abuse. CSA is administrated by the Drug Enforcement Authority (DEA) which, with the FDA, decides which substances are added to or deleted from the five schedules (classifications) of controlled substances. DEA deals primarily with law enforcement; thus, its involvement with toxicological aspects is limited.

Federal Food, Drug, and Cosmetic Act (FD&C Act) First passed in 1938 following an incident in which over 100 people died as a result of a then legal medication in which ethylene glycol was used as a solvent, the FD&C Act is administered by the FDA. It establishes limits for food additives and cosmetic components, sets criteria for drug safety for both human and animal use, and requires the manufacturer to prove both safety and efficacy. The FDA is authorized to define the required toxicity testing for each product. This act contains the Delaney clause, which states that food additives that cause cancer in humans or animals at any level shall not be considered safe and are, therefore, prohibited from such use. This clause has recently been modified to permit the agency to use more flexible risk/benefit-based guidelines. Under the Food Quality Protection Act (FQPA) of 1966 (see below) the Delaney clause is no longer applied to pesticide residues in food. This law also empowers the FDA to establish and modify the generally recognized as safe (GRAS) list and to establish good laboratory practice (GLP) rules.

Occupational Safety and Health Act This act, enacted in 1970, created the Occupational Safety and Health Administration (OSHA) and is administered by OSHA. The act concerns health and safety in the workplace. OSHA sets standards for worker exposure to specific chemicals, for air concentration values, and for monitoring procedures. Enforcement by OSHA is also provided for under the act. Construction and environmental controls also come under this act. This act provides for research, information, education, and training in occupational safety and health.

By establishing the National Institute for Occupational Safety and Health (NIOSH), the act provided for appropriate studies to be conducted so that regulatory decisions could be based on the best available information.

National Environmental Policy Act The National Environmental Policy Act (NEPA) of 1970 is an umbrella act covering all U.S. government agencies, requiring them to prepare environmental impact statements for all federal actions affecting the quality of the human environment. Environmental impact statements must include not only an assessment of the effect of the proposed action on the environment, but also alternatives to the proposed action, the relationship between local short-term use and enhancements of long-term productivity, and a statement of irreversible commitment of resources. This act also created the Council on Environmental Quality, which acts in an advisory capacity to the president on matters affecting or promoting environmental quality.

Resource Conservation and Recovery Act (RCRA)

Administered by the EPA, the RCRA is the most important act governing the disposal of hazardous wastes; it promulgates standards for identification of hazardous wastes, their transportation, treatment storage, and disposal. Included in the latter are siting and construction criteria for landfills and other disposal facilities as well as the regulation of owners and operators of such facilities. The three principal areas covered are hazardous wastes, nonhazardous solid wastes, and underground storage tanks. Farmers and commercial pesticide applicators are subject to penalties if they fail to store or dispose of pesticides and pesticide containers properly. The EPA is responsible for enforcement.

The Federal Hazardous and Solid Waste Amendments (1984) to RCRA focus on waste minimization and the phasing out of land disposal of hazardous wastes. These amendments also strengthened enforcement authority for EPA and provided for more stringent waste management standards.

Toxic Substances Control Act (TSCA)

Administered by the EPA, the TSCA is mammoth, covering almost all chemicals manufactured in the United States for industrial and other purposes, excluding certain compounds covered under other laws such as Federal Insecticide, Fungicide and Rodenticide Act (FIFRA) and the FD&C Act. The EPA may control or stop production of compounds deemed hazardous. Producers must give notice or intent to manufacture new chemicals or decrease significantly the production of existing chemicals. They may be required to conduct toxicity and other tests. Due to the enormous number of existing chemicals that must be evaluated, this law may never be completely applied or at least, not fully applied until new and more efficient toxicity tests are implemented. Once extensively applied, however, it will be the most important statute involving toxicology.

Statutes Affecting the Manufacture and Use of Agricultural Chemicals

Because of the intense interest and concern over the use of agricultural chemicals, especially pesticides, and their possible effects on human health, they are, perhaps, the most overregulated group of commercial xenobiotics in use today. A number of laws deal almost exclusively with this use class while several others also deal with them, to a greater or lesser extent. The first law directed specifically toward pesticides in the United States was The *Insecticide Act of 1910*. This act was passed to ensure that the percentages of ingredients were as stated and that the product was efficacious. Surprisingly, it was 37 years before a law was written to replace the 1910 Act. This replacement was the *FIFRA*. First passed in 1947 and amended many times since, this act is now administered by the EPA. FIFRA regulates all pesticides and other agricultural chemicals, such as plant growth regulators, used in the United States. Establishing the requirement "that the burden of proof of a product's acceptability rested with the manufacturer," it includes the authority to establish registration requirements, with appropriate chemical and toxicological tests prescribed by the agency. This act also permits the agency to specify labels, to restrict application to certified applicators, and to deny, rescind, or modify registration. Under this act, the EPA also establishes tolerances for residues on raw agricultural products. FIFRA was amended in 1988 requiring a reevaluation of all pesticides manufactured prior to 1984. The purposes of the 1988 amendment were to remove hazardous pesticides, and to require additional testing, primarily toxicity tests, which were not available when these early compounds were registered. Section 19 of the 1988 FIFRA amendments greatly expanded the Agency's authority to regulate pesticide storage, transport and disposal of pesticides, containers and rinsates of containers. The *FQPA* of 1996 is an amendment to FIFRA and provides a new standard for evaluating pesticides applied to food crops, in that there be "reasonable certainty of no harm" from residues found on food. USEPA is required to perform an aggregate risk assessment that combines dietary risk from a specific pesticide with those from residues in drinking water and from residential exposure. As a result of this law, USEPA is required to reevaluate all existing food tolerance residue levels based upon a number of criteria. One of these is to determine the cumulative (combined) risk of exposure to classes of pesticides having the same mechanism of toxicity, with special emphasis on infants and children. In some instances, this requires adding an additional safety factor (default value of 10) to the risk assessment for certain compounds to ensure the safety of children. This additional factor is in addition to the safety factor of 100 covering differences due to species and individual variation. Thus, if typical residue levels on a food crop is 1.0 ppm, then a tolerance of 0.01 ppm could be established, and if the additional factor of 10 were added, the tolerance could be set at 0.001 ppm. Currently, organophosphorus insecticides and several other chemical classes are undergoing this reassessment process.

While the intent of FQPA is praiseworthy, it is an excellent example of the effect of confusing legislation and regulation since specific regulations that are difficult to put on a toxicological basis are written into the law. The difficulties include deriving safety factors from toxicological data rather than using default values and the problems involved in determining (combined) risk of exposure to classes of pesticides having the same mechanism of toxicity.

The Act established the Tolerance Reassessment Advisory Committee (TRAC), composed of individuals with a variety of backgrounds and interests, to consult and make recommendations to both the USEPA and the United States Department of Agriculture (USDA). When this committee went out of existence in 1999, USEPA and USDA established a new advisory committee, the Committee to Advise on Reassessment and Transition (CARAT) to provide strategic advice on issues raised by this Act.

An Endocrine Disruptor Screening and Testing Advisory Committee (EDSTAC) was established to develop a comprehensive screening and testing program for pesticides and other compounds to determine potential estrogenic effects on both humans and on wildlife. EDSTAC's Final Report was presented to EPA in September 1998. EPA outlined the Endocrine Disruptor Screening Program (EDSP), which incorporated many of EDSTAC's recommendations.

FQPA is one of the most significant and far-reaching amendments ever made to FIFRA and, no doubt, will continue to generate controversy as it is put into effect.

The *Worker Protection Standard for Agricultural Pesticides* (1994) was written to protect workers from pesticide exposures. Responsibility lies with the employer and involves two types of employees: agricultural workers (e.g., harvesters) and pesticide handlers (e.g., mixers). It requires that these people be provided safety training, access to labels, and that medical treatment be made available prior to and 30 days after the re-entry interval (REI) has expired. The types of protection offered include notification prior to applying pesticides, exclusion during applications and during an REI, and monitoring the worker's personal protective equipment (PPE). In addition, the employer is required to provide a decontamination site equipped with water and a clean change of clothes.

The Act was written to cover pesticide use on farms, forests, nurseries, and greenhouses. It does not include applications to pastures, golf courses, parks, livestock, right-of-way or home gardens, nor does it cover treatments for mosquito abatement and rodent control.

Many other legislative acts impact in whole or in part on pesticide use. They include the *Endangered Species Act* of 1973, an act written to protect endangered wildlife, and regulates pesticide use around wildlife sanctuaries. Pesticides might injure or kill endangered species if allowed to drift onto habitat, or runoff into streams, lakes, or wetlands might be found to significantly degrade endangered wildlife habitat. Also included are the Clean Water Act, the Safe Drinking Water Act, RCRA, CERCLA, and SARA, all discussed above.

Other Statutes with Relevance to the Prevention of Toxicity   It should be noted that some of these statutes have been superseded by others, either in whole or in part.

- Comprehensive Employment and Training Act
- Dangerous Cargo Act
- Federal Coal Mine Safety and Health Amendment Act
- Federal Caustic Poison Act
- Federal Railroad Safety Authorization Act
- Hazardous Materials Transport Act
- Lead-Based Paint Poison Prevention Act
- Marine Protection Research and Sanctuaries Act
- Poison Prevention Packaging Act
- Ports and Waterways Safety Act

22.2.2   State Governments

Within the United States, states are free to adopt legislation with toxicological significance although their jurisdiction does not extend beyond their geographic boundaries. In other cases, the states may enforce federal statutes under certain circumstances. For example, if state regulations concerning hazardous waste disposal is neither less comprehensive nor less rigorous than the federal statute, enforcements is delegated to the states. Similarly, certain aspects of FIFRA are enforced by individual states. In some cases (California is notable in this respect), states have passed laws considerably more comprehensive and more rigorous than the corresponding federal stature.

22.2.3   Legislation and Regulation in Other Countries

It would serve little purpose to enumerate all the laws affecting toxicology, toxicity testing, and the prevention of toxicity that have been promulgated in all countries that have such laws. Legislation in this area has been adopted in most of the countries of Western Europe and in Japan. However, the recently implemented (June 2007) European Union Registration, Evaluation, Authorization and Registration of Chemicals (REACH) law is an example of the approach of another geographic region.

Although the laws in use in the United States are a complex mixture of overlapping statutes and enforcement agencies, they are probably the most comprehensive set of such laws in existence. Most other industrialized countries have legislation in the same areas, although the emphasis varies widely from one country to another. Many underdeveloped countries, due to the lack of both trained manpower and financial resources, are unable to write and enforce their own code of regulations and instead many adopt the regulatory decisions of either the United States or some other industrialized nation. For example, they will permit the use, in their own territory, of pesticides registered under FIFRA by the USEPA and will prohibit the use of pesticides not so registered.

22.3   PREVENTION IN DIFFERENT ENVIRONMENTS

Humans spend their time in many different but sometimes overlapping environments. Homes vary with climate, family income, and personal choice. The workplace varies from pristine mountains to industrial jungles, and the outdoor environment from which recreation, food, and water are derived varies through the same extremes. Each of these environments has its own specific complex of hazards, and thus requires its own set of rules and recommendations if these hazards are to be avoided.

22.3.1   Home

Approximately 50% of all accidental poisoning fatalities in the United States involve preschool children. Thus, prevention of toxicity is particularly important in homes with young children.

Prescription drugs should always be kept in the original container (in the United States and in some other countries, these are now required to have safety closures). They should be taken only by the person for whom they were prescribed, and excess drugs should be discarded safely when the illness is resolved. When children are present, prescription drugs should be kept in a locked cabinet, because few cabinets are inaccessible to a determined child. Although nonprescription drugs are usually less hazardous, they are frequently flavored in an attractive way. Thus, it is prudent to follow the same rules as for prescription drugs.

Household chemicals such as lye, polishes, and kerosene should be kept in locked storage if possible; if not, they should be kept in as secure a place as possible, out of the reach of children. Such chemicals should never be stored in anything but the original containers. Certainly they should never be stored in beverage bottles, kitchen containers, and so on. Unnecessary materials should be disposed of safely in appropriate disposal sites.

Certain household operations such as interior painting should be done only with adequate ventilation. Insecticide treatment should be done precisely in accordance with instructions on the label.

The cyclic increases in fuel costs often cause changes in lifestyle, and some of these changes carry potential toxic hazards. They include increased burning of wood and coal and the construction of heavily insulated and sealed houses with a concomitant reduction in ventilation. In the latter circumstances, improperly burning furnaces can generate high levels of CO and aromatic hydrocarbons, while even those burning properly may still generate oxides of nitrogen (NO_x) at levels high enough to cause respiratory tract irritation in sensitive individuals. These effects can be avoided by ensuring that all heating equipment (e.g., furnaces, wood stoves, heaters) is properly ventilated, maintained, and checked regularly. In addition, some ventilation of the building itself should always be provided. Less ventilation may be desirable when the temperature is either excessively high or excessively low, and more when the temperature is in the midrange, but under no circumstances should the homeowner strive for a completely sealed house.

22.3.2 Workplace

Exposure levels of hazardous chemicals in the air of work environments are mandated by OSHA as exposure limit values. The studies necessary to establish these limits are carried out by NIOSH. However, the more complete list of the better-known threshold limit values (TLVs) is established by the American Conference of Governmental Industrial Hygienists. Although TLVs are not binding in law, they are an excellent guide to the employer. In fact, they are often adopted by OSHA as permissible exposure limits (PELs). The concentrations thus expressed are the weighted average concentrations normally considered safe for an exposure of 8h/day, 5days/week. Absolute upper limits (excursion values) may also be included. Some exposure limits are shown in Table 22.1.

TABLE 22.1   Some Selected Threshold Limit Values (1991)

Chemical	TLV-TWA ppm	TLV-STEL ppm	TLV-C ppm
Acetaldehyde	100	150	—
Boron trifluoride	—	—	1
O-dichlorobenzene	—	—	50
p-dichlorobenzene	75	110	—
N-ethylmorpholine	5	20	—
Fluorine	1	2	—
Phosgene	0.1	—	—
Trichloroethylene	50	200	—

TLV-TWA, threshold limit value–time-weighted average concentration for a normal 8-h workday and 40-h workweek to which nearly all workers may be repeatedly exposed without adverse effect; TLV-STEL, threshold limit value–short-term exposure limit concentration. This time-weighted 15-min average exposure should not be exceeded at any time during a workday even if the TLV-TWA is within limits. Intended as supplement to TLV-TWA; TLV-C, threshold limit value-ceiling, concentration that should not be exceeded at any time.

Concentrations at or lower than those normal or working exposures are usually maintained by environmental engineering controls. Operations that generate large amounts of dusts or vapors are conducted in enclosed spaces that are vented separately or are under hoods. Other spaces are ventilated adequately, and temperature and humidity controls are installed where necessary.

Other precautions must be taken to prevent accidental or occasional increases in concentrations. Materials should be transported in "safe" containers, spilled material removed rapidly, and floor and wall materials selected to prevent contamination and allow easy cleaning.

Additional methods for the prevention of toxicity in the workplace include the use of personal safety equipment—protective clothing, gloves, and goggles are the most important. In particularly hazardous operations, closed-circuit air masks, gas masks, and so on, may also be necessary.

Preemployment instruction and preemployment physical examinations are of critical importance in many work situations involving hazardous chemicals. The former should make clear the hazards involved the need to avoid exposure under normal working conditions, and the mechanisms by which exposure is limited. Furthermore, employees should understand how and when to contain spills and how and when to evacuate the area around the spill. Locations and use of emergency equipment, showers, eye washes, and so on, should also be given, and the most important procedures should be posted in the work area.

22.3.3 Pollution of Air, Water, and Land

The toxicological significance of pollution of the environment may be work related, as in the case of agricultural workers, or related to the outside environment encountered in daily life. In the case of agricultural workers, numerous precautions are necessary for the prevention of toxicity. For example:

- Pesticides and other agricultural chemicals should be kept only in the original container, carrying the labels prescribed by EPA under FIFRA.
- Empty containers and excess chemicals should be disposed of properly in safe hazardous waste disposal sites, incinerated when possible or, in some cases, decontaminated.
- Workers should not reenter treated areas until the safe reentry period has elapsed.
- Certain workers such as applicators, those preparing tank mixes, should wear appropriate protection clothing, gloves, face masks, and so on. The development of closed systems for mixing pesticides should help protect mixers and loaders of pesticides from exposure.
- Spraying operations should be carried out in such a way as to minimize drift, contamination of water, and so on.

Pesticides have caused a number of fatalities in the past. The current practice in some countries (including the United States) of restricting the most hazardous chemicals for use only by certified operators should greatly minimize pesticide poisoning in these locations.

Individuals can do little to protect themselves from poisoning by chemicals that pollute the air and water except to insist that discharge of toxicants into the environment be minimized. The exposure levels are low compared with those in acute toxicity cases, and the effects may be indirect, as in the increase in preexisting respiratory irritation during smog. Thus, these effects can be determined only at the epidemiologic level. Because many persons are not affected or may not be affected for years, it is often argued that environmental contamination is not very important. However, a small percentage increase may represent a large number of people when the whole population is considered. Furthermore, chronic toxicity is not often reversible. Since in most industrialized countries laws already exist to control emission problems, if such problems exist in these countries, they are usually problems of enforcement.

One of the most critical areas for the prevention of toxicity caused by environmental contamination is that of disposal of hazardous wastes. It is now apparent that past practices in many industrialized countries have created large numbers of waste sites in which the waste is often unidentified, improperly stored, and leaching into the environment. The task of rectifying these past errors is an enormous one only now being addressed.

The ideal situation for current and future practices is to reduce chemical waste to an irreducible minimum and then to place the remainder in secure storage. Waste reduction can be accomplished in many ways.

- Refine plant processes so that less waste is produced.
- Recycle waste into useful products.
- Concentrate wastes.
- Incineration. The technology is available to incinerate essentially all waste to inorganic slag. Unfortunately, the technology is sophisticated and expensive. Inadequate incineration is itself a hazard because of the risk of generating dioxins and other toxicants and releasing them into the environment. Less complex and more easily maintained incinerators will be essential if this technology is to play a prominent role in waste reduction.

Safe storage for the remaining waste may be in dump sites or in aboveground storage. In either case, such storage ideally should be properly sited, constructed, maintained, and monitored.

Because of the nature of commerce, probably none of these measures will be successful unless the laws, penalties, and incentives are manipulated in such a way as to make safe disposal more attractive economically than unsafe disposal.

22.4   EDUCATION

Because chemicals, many of them hazardous, are an inevitable part of life in industrialized countries, education is probably the most important method for the prevention of toxicity. Unfortunately, it is also one of the most neglected. In a typical public debate concerning a possible chemical hazard, the principle protagonists tend to fall into two extreme groups: the "everything is OK" protagonists and the "ban it completely" protagonists. The media seldom seem to educate the public, usually serving only to add fuel to the flames.

The educational role of the toxicologist should be the voice of reason, presenting a balanced view of risks and benefits, and outlining alternatives whenever possible. The simple lesson that science deals not in certainty, but rather degrees of certitude, must be learned by all involved.

In terms of ongoing educational programs, there should be opportunities at all levels: elementary schools, high schools, university, adult education, and media education. Several approaches can be used to educate the general public in ideal situations:

- *Elementary schools*—Teach the rudiments of first aid and environmental concerns (e.g., proper disposal).
- *High school*—Teach concepts of toxicology (dose response) and environmental toxicology (bioaccumulation). These concepts can be introduced into general science courses.
- *University*—In addition to toxicology degrees, general courses for nontoxicology and/or nonscience majors should stress a balanced approach, with both responsible use and toxicity prevention as desirable end points. General Toxicology should be a required course in all chemically related academic programs such as chemistry, chemical engineering, and so on.
- *Media*—Encourage a balanced approach to toxicity problems. Toxicologists should be available to media representatives and, where appropriate, should be involved directly.

BIBLIOGRAPHY AND SUGGESTED READING

American Conference of Governmental Industrial Hygienists. *TLVs—Threshold Limit Values for Chemical Substances and Physical Agents in the Work Environment with Intended Changes.* Cincinnati, OH (published annually).

American Conference of Governmental Industrial Hygienists. *Documentation of the Threshold Limit Values for Substances in Workroom Air.* Cincinnati, OH (published annually).

Doull, J. Recommended limits for exposure to chemicals. In *Casarett & Doull's Toxicology: The Basic Science of Poisons*, 6th ed., ed. C. D. Klaassen, pp. 1115–1176. New York: McGraw-Hill, 2001.

Ellenhorn, M. J. and D. G. Barceloux. *Medical Toxicology: Diagnosis and Treatment of Human Poisoning.* New York: Elsevier, 1998.

Fan, A. M. and L. W. Chang, eds. *Toxicology and Risk Assessment: Principles, Methods and Applications.* New York: Marcel Dekker, 1996.

Merrill, R. A. Regulatory toxicology. In *Casarett & Doull's Toxicology, the Basic Science of Poisons*, 6th ed., ed. C. D. Klaassen, pp. 1141–1153. New York: McGraw-Hill, 2001.

True, B.-L. and R. H. Dreisbach. *Handbook of Poisoning: Prevention, Diagnosis, and Treatment*, 13th ed. London: Parthenon, 2002.

SAMPLE QUESTIONS

1. Name the federal statute most important for chemical safety in the workplace. Which agency is responsible for its administration?

2. Name three federal acts important for prevention of chemical toxicity that are administered by the United States Environmental Protection Agency.

3. One means of preventive chemical toxicity is waste reduction. In general, what are the means by which this can be improved?

4. Education is important in toxicity prevention. What approaches are likely to be effective at various levels of the education process?

Human Health Risk Assessment

RONALD E. BAYNES

23.1 INTRODUCTION

We often perform toxicological research to better under the mechanism and associated health risk following exposure to hazardous agents. Risk assessment is a systematic scientific characterization of potential *adverse health effects* following exposure to these hazardous agents. Risk assessment activities are designed to *identify, describe*, and *measure qualities and quantities* from these toxicological studies, which are often conducted with homogenous animal models at doses and exposure duration often not encountered in a more heterogeneous human population. Herein lies the challenge of risk assessment. Due to the use of default assumptions because of some level of uncertainty in our extrapolations across species, doses, routes, and interindividual variability, the risk assessment process is often perceived as lacking scientific rigor. This chapter will describe traditional practices as well as new and novel approaches that utilize more of the available scientific data to identify and reduce uncertainty in the process. The advent of powerful computers and sophisticated software programs has allowed the development of quantitative models that better describe the dose–response relationship, refine biologically relevant dose estimates in the risk assessment process, and encourage departure from traditional default approaches (Conolly et al., 1999). Although the focus of this chapter is on current and novel risk assessment methods that are scientifically based, it is critical at this point that the reader be aware of the differences between risk assessment and risk management which are summarized in Table 23.1.

Results from the risk assessment are used to inform *risk management*. The risk manager then uses this information in conjunction with factors such as the social importance of the risk, the social acceptability of the risk, the economic impacts of risk reduction, engineering, and legislative mandates, when deciding on and implementing risk management approaches.

The risk assessment may be perceived as the source of a risk management decision, when in fact, social concerns, international issues, trade, public perception, or

---

A Textbook of Modern Toxicology, Fourth Edition. Edited by Ernest Hodgson
Copyright © 2010 John Wiley & Sons, Inc.

TABLE 23.1 Comparison of Risk Assessment and Risk Management Activities

Risk Assessment	Risk Management
Nature of effects	Social importance of risk
Potency of agent	Acceptable risk
Exposure	Reduce/not reduce risk
Population at risk	Stringency of reduction
Average risk	Economics
High-end risk	Priority of concern
Sensitive groups	Legislative mandates
Uncertainties of science	Legal issues
Uncertainties of analysis	Risk perception
Identify	*Evaluate*
Describe	*Decide*
Measure	*Implement*

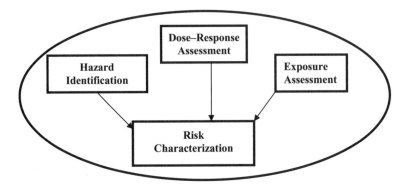

Figure 23.1 Risk assessment paradigm as per NAS and USEPA.

other nonrisk considerations may be taken into consideration. Finally, there is one activity known as *risk communication*, which involves making the risk assessment and risk management information comprehensible to lawyers, politicians, judges, business and labor, environmentalists, and community groups.

23.2 RISK ASSESSMENT METHODS

According to the National Research Council of the National Academy of Science, risk assessment consists of four broad but *interrelated* components: hazard identification; dose–response assessment; exposure assessment; and risk characterization, as depicted in Figure 23.1. The reader should, however, be aware that these risk assessment activities can provide research needs that improve the accuracy of estimating the "risk" or probability of an adverse outcome.

23.2.1 Hazard Identification

In this first component of risk assessment, the question of causality in a qualitative sense is addressed; that is, the degree to which evidence suggests that an agent elicits a given effect in an exposed population. Among many factors, the quality of the studies and the severity of the health effects should be evaluated at this stage. The following are evaluated: (1) validity of the toxicity data; (2) a weight-of-evidence summary of the relationship between the substance and toxic effects; and (3) estimates of the generalizability of data to exposed populations. Where there are limited *in vivo* toxicity data, *structural activity relationships* (SARs) and *short-term assays* may be indicative of a chemical hazard. Key molecular structures such as *n*-nitroso or aromatic amine groups and azo dye structures can be used for prioritizing chemical agents for further testing. SARs are useful in assessing relative toxicity of chemically related compounds, but there are several limitations. For example, toxicity equivalent factors (TEFs) based on induction of aryl hydrocarbon (Ah) receptor by dioxins demonstrated that SARs may not always be predictive. Inexpensive in vitro short-term tests such as bacterial mutation assays can help *identify* carcinogens, and there are other short-term test that can help identify chemicals that potentially can be associated with neurotoxicity, developmental effects, or immunotoxicity. Many of these *in vitro* studies can provide some insight into mechanism(s) of action, but there may be some *false positives* and *false negatives*. Animal studies are usually route specific and relevant to human exposure, and animal testing usually involves two species, both sexes, 50 animals/dose group, and near-lifetime exposures. Doses are usually 90, 50, and 10–25% of the maximum tolerated dose (MTD). In carcinogenicity studies, the aim is to observe significant increases in number of tumors, induction of rare tumors, and earlier induction of observed tumors. However, rodent bioassays may not be predictive of human carcinogenicity because of mechanistic differences. For example, renal tumors in male rats is associated with $\alpha_{2\mu}$-globulin-chemical binding and accumulation leading to neoplasia; however, $\alpha_{2\mu}$-globulin is not found in humans, mice, or monkeys. There are differences in susceptibility to aflatoxin-induced tumors between rats and mice which can be explained by genetic differences in expression of cytochrome P450 and glutathione s- transferase (GST) isoenzymes. Whereas humans may be as sensitive as rats to AFB_1-induced liver tumors, mice may not be predictive of AFB_1-induced tumors in humans. Epidemiological data from human epidemiological studies are the most convincing of an association between chemical exposure and disease and therefore can be very useful for hazard identification. Exposures are not often well-defined and retrospective, and confounding factors such as genetic variations in a population and human lifestyle differences (e.g., smoking) present a further challenge. The three major types of epidemiological studies available are: (1) *cross-sectional studies* involve sampling without regard to exposure or disease status, and these studies identify risk factors (exposure) and disease, but are not useful for establishing cause–effect relationships; (2) *cohort studies* involves sampling on the basis of exposure status, and they target individuals exposed and unexposed to chemical agent and monitored for development of disease, and these are *prospective studies*; (3) *Case-control studies* involve sampling on the basis of disease status. These are retrospective studies, where diseased individuals are matched with disease-free individuals.

23.2.2 Exposure Assessment

This process is an integral part of the risk assessment process; however, this will be briefly introduced in this chapter. The reader is encouraged to consult Chapter 27 in this text as well as numerous other text that describe this process in more depth. In brief, exposure assessment attempts to identify potential or completed exposure pathways resulting in contact between the agent and at-risk populations. It also includes demographic analysis of at-risk populations describing properties and characteristics of the population that potentiate or mitigate concern and description of the magnitude, duration, and frequency of exposure. The reader should be aware that exposure may be aggregate (single event added across all media) and/or cumulative (multiple compounds that share a similar mechanism of toxicity). Various techniques such as biomonitoring, model development, and computations can be used to arrive at an estimate of chemical dose taken up by humans, that is, chemical exposure. For example, the lifetime average daily dose (LADD) is a calculation for individuals exposed at levels near the middle of the exposure distribution.

$$LADD = \frac{\begin{array}{c}(\text{Conc. in Media}) \times (\text{Contact Rate}) \\ \times (\text{Contact Fraction}) \times (\text{Exposure Duration})\end{array}}{(\text{Body weight}) \times (\text{Lifetime})}$$

Biological monitoring of blood and air samples represents new ways of reducing uncertainty in these extrapolations. For occupational exposures, there are occupational exposure limits (OELs) that are guidelines or recommendations aimed at protecting the worker over their entire working lifetime (40 years) for 8 h/day, 5 days/week work schedule. Most OELs are presented as a time-weighted average concentration for an 8-h day for a 40-h work week. There are threshold limit values (TLVs) which refer to airborne concentrations and conditions under which workers may be exposed daily but do not develop adverse health effects. The short-term exposure limit (STEL) are recommended when exposures are of short duration to high concentrations known to cause acute toxicity.

23.2.3 Dose Response and Risk Characterization

Dose response is a quantitative risk assessment process, and primarily involves characterizing the relationship between chemical potency and incidence of adverse health effect. Approaches to characterizing dose–response relationships include effect levels such as lethal dose 50 (LD_{50}), lethal concentration 50 (LC_{50}), effective dose 50 (ED_{50}), No observed adverse effect levels (NOAELs), margins of safety, and therapeutic index. The dose–response relationship provides an estimation of the relationship between the dose of a chemical agent and incidence of effects in a population. Intuitively, a steep dose–response curve may be indicative of a homogenous population response, while less steep or almost flat slope may be indicative of greater distribution in response. In extrapolating from relatively high levels of exposure in experimental exposures (usually animals) to significantly lower levels that are characteristic of the ambient environment for humans, it is important to note the shape of the dose–response function below the experimentally observable range and therefore the range of inference. The shape of the slope may be linear

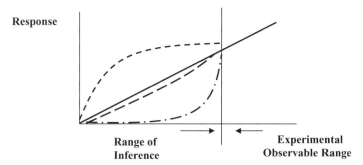

Figure 23.2 Dose–response curve, with emphasis on the shape of the dose–response function below the experimentally observable range and therefore the range of inference where people are realistically exposed.

or curvilinear, and it should be noted that the focus of risk assessment is generally on these lower regions of the dose response curve (Figure 23.2).

There is a class of curvilinear dose–response relationships in toxicological and epidemiological studies that may be described as *U-shaped* or *J-shaped curves*. Other terms such as biphasic and more recently, *hormesis* has been used to refer to paradoxical effects of low-level toxicants. In brief, these dose–response curves reflect an apparent improvement or reversal in the effect of an otherwise toxic agent. These U-shaped effects can be explained in terms of homeostatic adjustments or overcorrections in the operation of feedback mechanisms. Examples of studies with data fitting a U-shaped curve include the hormetic effect of organic lead on body growth in rats (Cragg and Rees, 1984) and peripheral nerve conduction velocity in children at low doses (Ewert et al., 1986). Similar relationships have been observed with alcohol and nicotine in humans. It has been proposed that because thresholds are inherent in U-shaped dose–response curves, the linear no-threshold extrapolation method is not an appropriate approach for regulating hormetic agents. The current risk assessment paradigm used by United States Environmental Protection Agency (USEPA) and other federal agencies does not conflict with the concept of hormesis, but it has been proposed that the risk assessor's analyses make an active consideration of the data and the application of that data in the low-dose portion of the dose–response curve for hormetic agents.

23.3 NONCANCER RISK ASSESSMENT

The noncancer risk assessment process assumes a *threshold*. For many noncarcinogenic effects, protective mechanisms are believed to exist that must be overcome before an adverse effect is manifested. At the cellular level for some toxicant, a range of exposures exists from *zero* to some finite value that can be tolerated by the organism with essentially no chance of expression of adverse effects. The aim here in risk assessment is to identify the upper bound of this tolerance range (i.e., the maximum subthreshold level). This approach involves obtaining the no observable adverse effect level (NOAEL). The NOAEL is the highest dose level that *does not produce a significant* elevated increase in an adverse response. Significance

TABLE 23.2   Comparison of Less Serious Effects and Serious Effects

Less Serious	Serious
Reversible cellular changes	Death
Necrosis, metaplasia, or atrophy	Cancer
	Clinically significant organ impairment
Delayed ossification	Visceral or skelatal abnormalities
Alteration in offspring weight	Cleft palate, fused ribs
Altered T cell activity	Necrosis in immunologic components
Auditory disorders	Visual disorders
50% reduction in offspring	Abnormal sperm

refers to biological and statistical criteria and is dependent on dose levels tested, number of animals, background incidence in the nonexposed control groups. Sometimes, there is insufficient data to arrive at a NOAEL, and a LOAEL (lowest observed adverse effect level) is derived. The NOAEL is the key datum obtained from the study of the dose–response relationship. The NOAEL is used to calculate reference doses (RfD) for chronic oral exposures and reference concentrations (RfC) for chronic inhalation exposures as per EPA. Other agencies such as the Agency for Toxic Substances and Disease Registry (ATSDR) and World Health Organization (WHO) use the NOAEL to calculate *minimum risk levels* (MRLs) and *acceptable daily intakes* (ADI). The USEPA describes the RfD as an estimate, with uncertainty spanning perhaps an order of magnitude, of a daily exposure to the human population, including sensitive subgroups that is likely to be without appreciable deleterious effects during a lifetime. In deriving reference doses, ADIs, or MRLs, the NOAEL is divided by uncertainty factors (UF) as per EPA (EPA, 1989) and ATSDR (ATSDR, 1993) and by modifying factors (MF) as per EPA.

$$RfD = NOAEL/(UF \times MF) \ldots\ldots\ldots\ldots US\ EPA$$

$$MRL = NOAEL/UF \ldots\ldots\ldots\ldots ATSDR$$

The calculated RfD or RfC is based on the selected critical study and selected critical end point. The risk assessor may obtain numerous studies where the toxicant may have more than one toxic end point and thus there may be many NOAELs to choose from the literature. In some instances, poor data quality may be used to exclude those end point from consideration. Also at issue is determining what is considered an adverse effect, and this has been summarized with a few examples in Table 23.2. In effect, the MRL or RfD is based on the less serious effects and no serious effects. The following are examples of effects not used in obtaining a NOAEL: decrease in body weight less than 10%; enzyme induction with no pathologic changes; changes in organ weight with no pathologic changes; increased mortality over controls that is not significant ($p > 0.05$); and hyperplasia or hypertrophy with or with out changes in organ weights.

23.3.1   Default Uncertainty and Modifying Factors

Many of the extrapolations from animal experimental data in the risk assessment process require the utilization of various uncertainty factors, because we are not certain how to extrapolate across species, with species for the most sensitive population, and across duration. To account for variation in the general population and intent to protect sensitive subpopulations, an uncertainty factor of 10 is used by EPA and ATSDR. The value of 10 is derived from a threefold factor for differences in toxicokinetics and for threefold factor for toxicodynamics. To extrapolate from animals to humans and account for interspecies variability between humans and other mammals, an uncertainty factor of 10 is used by EPA and ATSDR, and as with intraspecies extrapolations, this 10-fold factor is assumed to be associated within toxicodynamics and toxicokinetics. An uncertainty factor of 10 is used when a NOAEL derived from a subchronic study instead of a chronic study is used as the basis for a calculation of a chronic RfD (EPA only). Note that ATSDR does not perform this extrapolation as they do derive chronic and subchronic MRLs. An uncertainty factor of 10 is used when deriving an RfD or MRL from a LOAEL and a NOAEL is not available. It should be noted that there are no reference doses for dermal exposure; however, when there are insufficient dermal absorption data, the EPA uses a default factor of 10% to estimate bioavailability for dermal absorption. A modifying factor ranging from 1–10 is included by EPA only to reflect a qualitative professional assessment of additional uncertainties in the critical study and in the entire database for the chemical not explicitly addressed by preceding uncertainty factors.

Refinements of the RfC have utilized mechanistic data to modify the interspecies uncertainty factor of 10 (Jarabek, 1995). The reader should appreciate that with the inhalation route of exposure dosimetric adjustments are necessary as this can impact extrapolations of toxicity data of inhaled agents for human health risk assessment. The EPA has included dosimetry modeling in RfC calculations, and the resulting dosimetric adjustment factor (DAF) used in determining the RfC is dependent on physiochemical properties of the inhaled toxicant as well as type of dosimetry model ranging from rudimentary to optimal model structures. In essence, the use of the DAF can reduce the default uncertainty factor for interspecies extrapolation from 10 to 3.16.

The 1996 Food Quality Protection Act (FQPA) now requires that an additional safety factor of 10 be used in the risk assessment process of pesticides to assure the safety of infants and children unless the EPA can show that an adequate margin of safety is assured without it (Scheuplein, 2000). The rationale behind this additional safety factor is that infants and children have different dietary consumption patterns than adults and infants and children are more susceptible to toxicants than adults. We do know from pharmacokinetic studies with various human pharmaceuticals that drug elimination is slower in infants up to 6 months of age than in adults, and therefore the potential exists for greater tissue concentrations and vulnerability for neonatal and postnatal effects. Based on these observations, the USEPA supports a default safety factor greater or less than 10 which may be used on the basis of reliable data. However, there are few scientific data from humans or animals to compare sensitivities on the differential responses of children and adults, but there are some examples such as lead where children are the more sensitive population. It some cases, qualitative differences in age-related susceptibility are small beyond 6 months of age, and quantitative differences in toxicity between children and adults can sometimes be less than a factor of 2 or 3.

Much of the research efforts in risk assessment are therefore aimed at reducing the need to use these default uncertainty factors, although the risk assessor is limited by data quality of the chemical of interest. With sufficient data and the advent of sophisticated and validated physiologically based pharmacokinetic models and biologically based dose response models (Conolly and Butterworth, 1995), these default values can be replaced with science-based factors. In some instances, there may be sufficient data to be able to obtain distributions rather than point estimates.

23.3.2 Derivation of Developmental Toxicant RfD

Developmental toxicity includes any detrimental effect produced by exposures during embryonic development, and the effect may be temporary or overt physical malformation. Adverse effects include death, structural abnormalities, altered growth, and functional deficiencies. Maternal toxicity is also considered. The evidence is assessed and assigned a weight-of-evidence designation as follows: Category A, Category B, Category C, and Category D. The scheme takes into account the ratio of minimum maternotoxic dose to minimum teratogenic dose, the incidence of malformations and thus the shape of the dose response curve or dose relatedness of each malformation, and types of malformations at low doses. A range of uncertainty factors are also utilized according to designated category as follows: Category A = 1–400; Category B = 1–300; Category C = 1–250; and Category D = 1–100. Developmental RfDs are based a short duration of exposure and therefore cannot be applied to lifetime exposure.

23.3.3 Determination of RfD and RfC of Naphthalene using the NOAEL Approach

The inhalation RfC for naphthalene was $0.003\,mg/m^3$, and this RfC was derived from a chronic (2-year) National Toxicology Program (NTP) inhalation study in mice using exposures of *0, 10*, or *30ppm* (NTP, 1992). Groups of mice were exposed for 5 days per week and 6h per day. This study identified a *LOAEL of 10ppm*. A dose-related incidence of chronic inflammation of the epithelium of the nasal passages and lungs was observed. This LOAEL concentration was normalized by adjusting for the 6-h-per-day and 5-day-per-week exposure pattern. A LOAEL of $9.3\,mg/m^3$ was derived by converting 10ppm first to mg/m^3 and then duration-adjusted levels for 6h/day and 5days/week for 103 weeks. A UF of 3000 was used, where 10 was for the interspecies (mice to humans) extrapolations; 10 for intraspecies variation in humans; 10 for using a LOAEL instead of a NOAEL; and 3 for database deficiencies.

The oral RfD for naphthalene was 0.02 mg/kg/day, and a study by Battelle (1980) was used to calculate the RfD. Decreased body weight was the most sensitive end point in groups of Fischer 344 rats given 0, 25, 50, 100, 200, or 400mg/kg for 5days/week for 13 weeks. These doses were also duration-adjusted to 0, 17.9, 35.7, 71.4, 142.9, and 285.7mg/kg/day, respectively. The NOAEL for a >10% decrease in body weight in this study was 71mg/kg/day. The UF of 3000 was based on 10 for rats to humans extrapolation; 10 for human variation; 10 to extrapolate from subchronic to chronic and 3 for database deficiencies, including lack of chronic oral exposure studies.

23.3.4 Benchmark Dose Approach

There are several problems associated with using the NOAEL approach to estimate RfDs and RfCs. The first obvious constraint is that the NOAEL must, by definition, be one of the experimental doses tested. Once this dose is identified, the rest of the dose–response curve is ignored. In some experimental designs where there is no identifiable NOAEL but LOAEL, the dose response curve is again ignored, and the NOAEL is derived by application of uncertainty factors as described earlier. This NOAEL approach does not account for the variability in the estimate of the dose response, and furthermore, experiments that test fewer animals result in larger NOAELs and thus larger RfDs and RfCs.

An alternative approach known as the benchmark dose (BMD) approach has been developed and implemented by risk assessors as an alternative to the NOAEL approach to estimate RfDs and RfCs. This approach is not as constrained by experimental design as the NOAEL approach, and incorporates information on the sample size and shape of the dose-response curve. In fact, this approach can be used for both threshold and nonthreshold adverse effects as well as continuous and quantal data sets. This requires use of Benchmark Dose Software where the does–response is modeled and the lower confidence bound for a dose at a specified response level (benchmark response) is calculated. The benchmark response is usually specified as a 1–10% response; that is, it corresponds to a dose associated with a low level of risk, such as 1–10%.

Figure 23.3 shows how an effective dose that corresponds to a specific change of effect/response (e.g., 10%) over background and a 95% lower confidence bound on the dose is calculated. The latter is often referred to as the BMDL or LBMD, as opposed to the BMD, which does not have this confidence limited, associated with it.

Because the benchmark represents a statistical lower limit, larger experiments will tend on average to give larger benchmarks, thus rewarding good experimentation. This is not the case with NOAELs, as there is an inverse relationship between NOAEL and size of experiments. For example, poorer experiments possessing less sensitivity for detecting statistically significant increases in risk inappropriately result in higher NOAELs and RfDs which may have an unknown unacceptable level of risk. In essence, the NOAEL is very sensitive to sample size and there can also

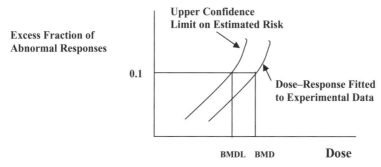

Figure 23.3 Benchmark dose determination from dose response relationship with the BMDL corresponding to the lower end of a one-sided 95% confidence interval for the BMD.

be high variability between experiments. With the benchmark dose approach, all the doses and slope of the curve influences the calculations, variability of the data is considered, and the BMD is less variable between experiments. In the BMD approach, quantitative toxicological data such as continuous data (e.g., organ weights, serum levels) and quantal or incidence data (e.g., pathology findings, genetic anomalies) are fitted to numerous dose–response models described in the literature. The resulting benchmark dose that, for example, corresponds to a tumor risk of 10% generally can be estimated with adequate precision and not particularly dependent on the dose–response model used to fit the data. Note that dose intervals are not required for BMD estimation. This will be greatly appreciated in the cancer risk assessment section of this chapter.

23.3.5 Determination of BMD and BMDL for ETU

The BMD method has been used quite extensively in assessing quantal data, and very often, this has involved analysis of data from developmental and reproductive toxicity studies. In this study example (Crump, 1984), rats were exposed to ethylenethiourea (ETU) at 0, 5, 10, 20, 40, and 80 mg/kg doses, and the number affected with fetal anomalies per number of rats were 0/167, 0/132, 1/138, 14/81, 142/178, and 24/24, respectively. The benchmark dose computation can involve utilization of any given dose–response probability model, but in this example, the quantal Weibull model was used, and the specified effect was set at 0.01 (1%) with confidence level of 0.95. The BMD was determined to be 8.9 mg/kg, and the BMDL was 6.9 mg/kg. This value is close to the NOAEL which is 5 mg/kg, but it does demonstrate that the NOAEL approximates a lower confidence limit on the BMD corresponding to an excess risk of about 1% for proportions of fetal anomalies. In fact, an empirical analysis of some 486 developmental toxicity studies have demonstrated that the NOAEL can result in an excess risk of 5% for proportions of dead or malformed fetuses per litter. The reader should at this stage recognize that the BMD approach can also be used in cancer risk assessment as we are often times working with quantal data which is ideally suited for BMD modeling.

23.3.6 Quantifying Risk for Noncarcinogenic Effects: Hazard Quotient

The measure used to describe the potential for noncarcinogenic toxicity to occur is not expressed as the probability. Probabilistic approach is used in cancer RA. For noncancer RA, the potential for noncarcinogenic effects is evaluated by comparing an exposure level (E) over a specified time period with a reference dose (RfD). This ratio is called a hazard quotient:

$$\text{Hazard Quotient} = E/RfD$$

In general, the greater the value of E/RfD exceeds unity, the greater the level of concern. Note that this is a ratio and not to be interpreted as a statistical probability.

23.3.7 Chemical Mixtures

Human populations are more likely to be exposed simultaneously or sequentially to a mixture of chemicals rather than one single chemical. Standard default approaches to mixture risk assessment consider doses and responses of the mixture components to be additive. However, it should also be recognized that components in the mixture can also result in synergistic, antagonistic, or no toxicological effect following exposure to a chemical mixture. Therefore, mixture toxicity cannot always be predicted even if we know the mechanism of all toxic components in a defined mixture. Furthermore, tissue dosimetry can be complicated by interactions at the route of entry (e.g, gastrointestinal tract [GIT], skin surface) and clearance mechanisms in the body. In essence, there are considerable uncertainties involved in trying to extrapolate effects following exposure to chemical mixtures. Several physiologically based pharmacokinetic (PBPK) models have been used to quantitate these effects and also provide some information useful for risk assessment of chemical mixtures (Krishnan et al., 1994; Haddad et al., 2001).

The 1996 FQPA has also mandated that the EPA should also consider implementing cumulative risk assessments for pesticides. Cumulative risk assessments usually involve integration of the hazard and cumulative exposure analysis, and it primarily involves cumulative nonoccupational exposure by multiple routes or pathways to two or more pesticides or chemicals sharing a common mechanism of toxicity.

Calculation procedures differ for carcinogenic and noncarcinogenic effects, but both sets of procedures *assume dose additivity* in the absence of information on mixtures:

$$\text{Cancer risk equation for mixtures: Risk}_T = \Sigma \text{Risk}_I$$

$$\text{Noncancer Hazard Index} = E_1/\text{RfD}_1 + E_2/\text{RfD}_2 + \ldots + E_i/\text{RfD}_i$$

This hazard index (HI) approach as well as others (e.g., Relative Potency Factors) is applied for mixture components that induce the same toxic effect by identical mechanism of action. In cases where there are different mechanisms, separate HI values can be calculated for each end point of concern. As the above equation indicates, the HI is easy to calculate as there is simply scaling of individual component exposure concentrations by a measure of relative potency such as the RfD or RfC, and adding scaled concentrations to get an indicator of risk from exposure to the mixture of concern. However, as noted above, this additivity approach does not take into account tissue dosimetry and pharmacokinetic interactions. Recent published risk assessments have utilized mixture PBPK models to account for multiple pharmacokinetic interactions among mixture constituents. These interaction-based PBPK models can quantify change in tissue dose metrics of chemicals during exposure to mixtures and thus improve the mechanistic basis of mixture risk assessment. Finally, the reader should be aware that this HI is different from the a term known as the Margin of Safety (MOS) which is the ratio of the critical or chronic NOAEL for a specific toxicological end point to an estimate of human exposure. MOS values greater than 100 are generally considered protective if the NOAEL is derived from animal data.

23.4 CANCER RISK ASSESSMENT

For cancer risk assessment, an assumption is held that a threshold for an adverse effect does not exist with most individual chemicals. It is assumed that a small number of molecular events can evoke changes in a single cell that can lead to uncontrolled cellular proliferation and eventually to a clinical state of disease. This mechanism is referred to as "nonthreshold" because there is believed to be essentially no level of exposure to such a chemical that does not pose a finite probability, however small, of generating a carcinogenic response. That is, no dose is thought to be risk free. Therefore ,in evaluating cancer risks, an effect threshold cannot be estimated. For carcinogenic effects, the U.S. EPA uses a two-part evaluation: (1) the substance is first assigned a weight-of-evidence classification and then (2) a slope factor is calculated.

1. *Assigning a weight-of-evidence:* The aim here is to determine the likelihood that the agent is a human carcinogen. The *evidence* is characterized separately for human studies and animal studies as *sufficient, limited, inadequate, no data,* or *evidence of no effect.* Based on this characterization and on the extent to which the chemical has been shown to be a carcinogen in animals or humans or both, the chemical is given a provisional *weight-of-evidence* classification.

The U.S. EPA classification system (EPA, 1986) shown in Table 23.3 below has been revised in the EPA (1996) proposed guidance and more recent draft guidance (EPA, 1999).

This system was also adapted from the approach taken by the International Agency for Research on Cancer (IARC). This alphanumeric classification system has been replaced with a narrative and the following descriptor categories: "*known/ likely,*" "*cannot be determined,*" or "*not likely.*" These EPA (1996) guidelines indicate that not only are tumor findings an important consideration, but also structure– activity relationships, modes of action of carcinogenic agents at cellular or subcellular level as well as toxicokinetic and metabolic processes. These revised guidelines also indicate that the weighing of evidence should address the conditions under which the agent may be expressed. For example, an agent may "*likely*" be carcinogenic via inhalation exposure, but "*not likely*" via oral exposure. The narrative will summarize much of this information as well as the mode of action information.

TABLE 23.3 Weight of Evidence Designation Based on EPA(1986) Guidelines

Group	Description
A	Human carcinogen
B1 or B2	Probable human carcinogen
C	Possible human carcinogen
D	Not classifiable as to human carcinogenicity
E	Evidence of noncarcinogenicity for humans

Note: B1 indicates that limited human data are available; B2 indicates sufficient evidence in animals and inadequate or no evidence in humans.

2. *Quantifying Risk for Carcinogenic Effects:* In the second part of the evaluation, the EPA (1986) guidelines required that quantitative risk be based on
the evaluation that the chemical is a known or probable human carcinogen, a
toxicity value that defined quantitatively the relationship between dose and
response (*slope factor*) is calculated. Slope factors have been calculated for
chemicals in classes A, B1, and B2. Sometimes, a value is derived for those in
Class C on a case-by-case basis. The slope factor is a plausible upper-bound
estimate of the probability of a response per unit intake of chemical over
a lifetime. Slope factors have been accompanied by the weight-of-evidence
classification to indicate the strength of evidence that the chemical is a human
carcinogen.

Development of a slope factor entails applying a model to the available data set
and using the model to extrapolate from high doses to lower exposure levels expected
for human contact. There are a number of low-dose extrapolation models which can
be divided into distribution models (e.g., log-probit, Weibull) and mechanistic
models (e.g., one-hit, multihit, and *linearized multistage*). EPA 1986 guidelines for
carcinogen risk assessment are currently being revised, and it is very likely that the
new guidelines will encourage the use of bioloically based models for cancer risk
assessment. The previous guidelines (EPA, 1986) recommended that the linearized
multistage model, which is a mechanistic model, be employed as the default model
in most cases. Most of the other models are less conservative. The proposed biologically based models attempt to incorporate as much mechanistic information as
possible to arrive at an estimate of slope factors. In essence, after the data are fit to
the selected model, the upper 95th percent confidence limit of the slope of the resulting dose response curve is calculated. *This represents the probability of a response
per unit intake over a lifetime* or that there is a 5% chance that the probability of a
response could be greater than the estimated value on the basis of experimental data
and model used. In some cases, slope factor based on human dose–response data
are based on "best" estimate instead of upper 95th percent confidence limit. The
toxicity values for carcinogenic effects can be expressed in several ways:
The slope factor is expressed as q_1*

$$\text{Slope factor} = \text{risk per unit dose} = \text{risk per mg/kg/day}$$

The slope factor can therefore be used to calculate the upper bound estimate on
risk (R)

$$\text{Risk} = q_1 * \left[\text{risk} \times (\text{mg/kg/day})^1 \right] \times \text{exposure}\,(\text{mg/kg/day})$$

Here, risk is a unit-less probability (e.g., 2×10^{-5}) of an individual developing
cancer and exposure is really chronic daily intake averaged over 70 years: mg/kg/
day, and this can be determined if we can determine the slope factor and human
exposure at the waste site or occupational site. The EPA usually sets a goal of
limiting lifetime cancer risks in the range of 10^{-6}–10^4 for chemical exposures, while
the FDA typically aims for risks below 10^{-6} for general population exposure. It
is therefore quite likely for very high exposures for the accepted EPA range of
risk to be exceeded. The EPA range is considered protective of the general and

sensitive human population. It should be noted that these orders of magnitude are substantially greater than those used in estimating RfD and RfCs in noncancer risk assessment.

Because relatively low intakes (compared to those experienced by test animals) are most likely from environmental exposure at Superfund sites, it generally can be assumed that the dose–response relationship will be linear on the low-dose portion of the multistage model dose–response curve. The above equation can apply to these linear low-dose situations. This linear equation is valid only at low risk levels (i.e., below estimated risk of 0.01). For risk above 0.01 then, the one-hit equation should be used:

$$\text{Risk} = 1 - \exp(-\text{exposure} \times \text{slope factor})$$

As indicated above, biologically based extrapolation models are the preferred approach for quantifying risk to carcinogens, although it is possible that all the necessary data would not be available for many chemicals. The EPA (1986) guidelines have been modified to include the response data on effects of the agent on carcinogenic processes in addition to data on tumor incidence. Precursor effects and tumor incidence data may be combined to extend the dose–response curve below the tumor data, that is, below the range of observation. Thus, a biologically based or case-specific dose response model is developed when there is sufficient data or a standard default procedure when there is insufficient data to adequately curve-fit the data. In brief, the dose–response assessment is considered in two parts or steps: range of observation and range of extrapolation, and the overriding preferred approach is to use the biologically based or case-specific model for both of these ranges. In the first step of this process, the lower 95% confidence limit on a dose associated with an estimated 10% increase in tumor or nontumor response (LED_{10}) is identified. When human real-world exposures are outside the range of the observed or experimental data, this serves as the point of departure or marking the beginning for the extrapolating to these low environmental exposure levels. Please note that these procedures are very similar to the benchmark procedure for quantitating risk to noncarcinogenic chemicals. In the second step, the biologically based or case-specific models is preferred to be used in extrapolations to lower dose levels provided there is sufficient data. If the latter is not the case, then default approaches consistent with agent chemical mode of action are implemented with the assumption of linearity or nonlinearity of the dose–response relationship. The linear default approach is a departure from the 1986 guidelines which used the linearized multistage (LMS) procedure but is based on mode of action of action or alternatively there is insufficient data to support a nonlinear mode of action. In brief, it thus involves drawing a straight line from the point of departure (LED_{10}) to the origin (i.e., zero). When there is no evidence of linearity or there is a nonlinear mode of action, the default approach is the margin of exposure (MOE) analysis. The MOE approach computes the ratio between the LED_{10} and the environmental exposure, and the analysis begins from the point of departure that is adjusted for toxicokinetic differences between species to give a human equivalent dose.

Finally, it should be noted that prior to the FQPA in 1996, the Delaney clause prohibited the establishment of tolerances or maximum allowable levels for food additives if it has been shown to induce cancer in man or animals. This is an important change in regulations because pesticide residues were considered as food additives. Because of the FQPA, pesticide residues are no longer regarded as food additives, and there is no prohibition against setting tolerances for carcinogens.

23.5 PBPK MODELING

PBPK modeling has been used in risk assessment to making more scientifically based extrapolations, and at the same time help to explore and reduce inherent uncertainties. Historically, pharmacokinetics has relied on empirical models, and in many instances, this process offers little insight into mechanisms of absorption, distribution, and clearance of hazardous agents and does not facilitate translation from animal experiments to human exposures. For example, dose scaling using by body weight or size may often overestimate or underestimate toxicant levels at the target tissue. PBPK models can help predict tissue concentrations in different species under various conditions based on *independent* anatomical, physiological, and biochemical parameters. In these analyses, physiological parameters such as organ volumes, tissue–blood partition coefficients, and blood flow to specific tissue compartments described by the model, are calculated or obtained from the literature and integrated into the model. Monte Carlo analysis, a form of uncertainty analysis, can now be preformed, and this allows for the propagation of uncertainty through a model which results in estimation of the variance of model output. This can be achieved by randomly sampling model parameters from defined distributions; some parameters such as cardiac output, metabolic, and log P parameters, may have a log normal distribution, while other parameters may be normal or uniform. In essence, the Monte Carlo analysis, when coupled with PBPK, characterizes the distribution of potential risk in a population by using a *range* of potential values for each input parameter (not single values) as well as an estimate of how these values are distributed (Clewell and Andersen, 1996). Using these approaches, uncertainty is identifiable and quantifiable, and this can reduce inappropriate levels of concern in reporting the risk of chemical exposure. These mathematical modeling approaches also help identify areas of potential scientific research that could improve the human health assessment.

In recent years, there have been significant efforts at harmonization noncancer and cancer risk assessment (Barton et al., 1998; Clewell et al., 2002), and in this respect, PKPD modeling can be a very useful tool in the risk assessment process. For example, recall that noncancer risk assessment addresses variability in a population by dividing the NOAEL by 10, whereas the cancer risk assessment does not address this quantitatively. PBPK modeling coupled with Monte Carlo analysis is one approach as described in the previous paragraph that will help address this level of uncertainty in the risk assessment. In conclusion, it should be noted that the use of PBPK modeling has been utilized with very few toxicants, but it is hoped that risk assessment policy would encourage the use of this tool as well as other appropriate models to integrate mechanistic information and the pharmacokietics (dosimetry), and pharmacodynamics (dose response) of toxicants. These improved quantitative risk assessments would ultimately provide scientifically sound information that will influence the risk management decision process.

BIBLIOGRAPHY AND SUGGESTED READING

ATSDR. *Guidance for the Preparation of Toxicological Profiles*. Atlanta, GA: ATSDR, US Public Health Service, 1993.

Battelle. Subchronic Toxicity Study: Naphthalene (C52904), Fischer 344 Rats. Report to U.S. Department of Health and Human Services, National Toxicology Program, Research Triangle Park, NC. Columbus, OH: Battelle Columbus Laboratories, 1980.

Barton, H. A., M. E. Anderson, and H. J. Clewell. Harmonization: Developing consistent guidelines for applying mode of action and dosimetry information for applying mode of action and dosimetry information to cancer and noncancer risk assessment. *Hum. Ecol. Risk Assess.* **4**:75–115, 1998.

Clewell, H. J. and M. E. Andersen. Use of physiologically based pharmacokinetic modeling to investigate individual versus population risk. *Toxicol.* **111**(1–3):315–329, 1996.

Clewell, H. J., M. E. Andersen, and H. A. Barton. A consistent approach for the application of pharmacokinetic modeling in cancer and noncancer risk assessment. *Envion. Health Perspect.* **110**:85–93, 2002.

Conolly, R. B., B. D. Beck, and J. I. Goodman. Stimulating research to improve the scientific basis of risk assessment. *Toxicol. Sci.* **49**:1–4, 1999.

Conolly, R. B. and B. E. Butterworth. Biologically based dose–response model for hepatic toxicity: A mechanistically based replacement for traditional estimates of noncancer risk. *Toxicol. Lett.* **82–83**:901–906, 1995.

Cragg, B. and S. Rees. Increased body: brain weight ratio in developing rats after low exposure to organic lead, *Exp Neurol.* **86**:113–121, 1984.

Crump, K. S. A new method for determining allowable daily intakes. *Fundam. Appl. Toxicol.* **4**:854–871, 1984.

EPA. Guidelines for carcinogen risk assessment. *Fed. Reg.* **51**:33992–34003, 1986.

EPA. *Risk Assessment Guidance for Superfund. Volume 1. Human Health Evaluation Manual (Part A)*. Washington, DC: US EPA, 1989.

EPA. Proposed guidelines for carcinogen risk assessment. *Fed. Reg.* **61**:17960–18011, 1996.

EPA. Guidelines for carcinogen risk assessment. Risk Assessment Forum, U.S. Environmental Protection Agency, NCEA-F-0644, July 1999, Review Draft, 1999.

Ewert, T., U. Beginn, F. Winneke, B. Hofferberth, and J. Jorg, Sensory neurography, visual and somatosensory evoked potentials (VEP and SEP) in lead-exposed children, *Nervenarzt.* **57**:465–471, 1986.

Haddad, S., M. Beliveau, R. Tardif, et al. A PBPK modeling-based approach to account for interactions in the health risk assessment of chemical mixtures. *Toxicol. Sci.* **63**:125–131, 2001.

Jarabek, A. M. The application of dosimetry models to identify key processes and parameters for default dose–response assessment approaches. *Toxicol. Lett.* **79**(1–3):171–184, 1995.

Krishnan, K., H. J. Clewell, and M. E. Andersen. Physiologically based pharmacokinetic analyses of simple mixtures. *Environ. Health Perspect.* **102**(Suppl. 9):151–155, 1994.

NTP. National Toxicology Program. Technical report series No. 410. Toxicology and carcinogenesis studies of naphthalene (CAS No. 91-20-3) in B6C3F1 mice (inhalation studies). Research Triangle Park, NC: U.S. Department of Health and Human Services, Public Health Services, National Institutes of Health, 1992. NIH Publication No. 92-3141.

Page, N. P., D. V. Singh, W. Farland. Implementation of EPA revised cancer assessment guidelines: Incorporation of mechanistic and pharmacokinetic data. *Fundam. Appl. Toxicol.* **37**:16–36, 1997.

Scheuplein, R. J. Pesticides and infant risk: Is there a need for an additional safety margin? *Regul. Toxicol. Pharmacol.* **31**(3):267–279, 2000.

Sensory neurography, visual and somatosensory evoked potentials (VEP and SEP) in lead-exposed children. *Nervenarzt.* **57**: 465–471, 1986.

SAMPLE QUESTIONS

1. a. Cancer risk assessment does not assume a threshold dose.
  b. PBPK models can reduce the uncertainty factor in an RfC calculation.
  c. The bench mark dose approach is an improvement on the NOAEL approach.

2. a. Obtaining the slope factor is critical to cancer risk assessment.
  b. The uncertainty factor only takes into account animal to human extrapolation.
  c. Hazard identification takes into account bioassays and epidemiological data.

3. Define the following terms: NOEL; NOAEL; LOAEL; RfD.

4. The following dose–response from a chronic mouse study is being considered for use in a noncancer risk assessment. Use the data set to answer the questions below.

Data set.	Dose (mg/kg)	0	1	2	4	8	16
	Response	0	0	0	X	X	X*

X* is a significant increase in hepatic necrosis.

  a. What is the NOAEL?
  b. What is the LOAEL?
  c. What is the RfD?

ENVIRONMENTAL TOXICOLOGY

Toxicant Analysis: Analytical Methods and Quality Assurance

CHRIS HOFELT

24.1 INTRODUCTION

Some estimates suggest that there are some 200,000 chemicals synthesized annually worldwide. At the most fundamental level, the risk of one of these compounds causing harm to a living organism is a function of two things: the toxicity of that compound to the species of interest and the organism's exposure to the compound. The former has largely been the subject of this text so far. This chapter will begin to examine the latter.

Today's analytical chemist has access to analytical tools for the detection and quantification of a wide variety of potentially toxic compounds, at concentrations that are increasingly miniscule. Analytical laboratories routinely report concentrations of chemicals in the subparts-per-billion range. To try and envision this infinitesimally small number, consider the following analogy. A part per million is akin to a single white ping-pong ball in a railroad hopper car full of one million colored ping-pong balls. A part per billion is akin to a single white ping-pong ball in 1000 railroad hopper cars (i.e., a train stretching for ~9.5 mi) full of colored ping-pong balls. Although these advances have tremendous benefits, this can be a double-edged sword. Many chemicals can now be detected at concentrations far below any level of toxicological significance, whereas there is a public perception that if a chemical can be detected (e.g., in human serum), then it must be causing deleterious effects.

Although new techniques and instruments continue to enter the commercial market, the basic analytical process has not changed:

- Define the research goal(s).
- Identify appropriate techniques and methods.
- Develop a sampling scheme to obtain representative samples.
- Isolate the compound(s) of interest.
- Remove potential interfering components.
- Quantify and evaluate the data in relation to the original research goals.

A Textbook of Modern Toxicology, Fourth Edition. Edited by Ernest Hodgson
Copyright © 2010 John Wiley & Sons, Inc.

Based on the data generated, many options are available. For example, was the sampling scheme complete? Would further refinement of the analytical procedure be required? Should other sample types be analyzed? Thus, it is obvious that within these general categories, particular methods vary considerably depending on the chemical characteristics of the toxicant. This chapter is concerned with the sampling, isolation, separation, and measurement of toxicants, including the various quality assurance (QA) and quality control (QC) measures employed to assure the accuracy and precision of the data.

24.2 ENVIRONMENTAL SAMPLE COLLECTION METHODS

Even with the most sophisticated analytical equipment available, the resulting data are only as representative as the samples from which the results are derived. This is particularly true for environmental samples. A great deal of the error in measurements, such as the ones discussed in this chapter, is introduced in the sample collection preservation and storage. Therefore, a great deal of effort should go into the planning of this phase of the study, and care must be taken to ensure that the resulting data meet the objectives of the study. Often, special attention to sampling procedures is necessary. Ultimately, the initial hypothesis and goals of the study will determine what the most appropriate sampling schemes and analytical methods will be.

24.2.1 Sampling Schemes

The goal in developing a sampling scheme is to come up with a sample that truly reflects the composition of the matrix to be analyzed within the context of the study aims; in other words, a *representative sample*. This can be challenging as contamination is rarely, if ever, uniformly distributed. This is particularly true of solid matrices. Finally, not only can contamination vary spatially but temporally as well. Sampling schemes generally fall into three categories: *random*, *systematic*, and *judgmental* (Figure 24.1).

Random Sampling From a statistical standpoint, a random sampling design is really the only choice. In a random design, sampling locations within the area of interest (AOI) are determined using a method that prevents any bias in the selection. For example, one could use a random number generator to generate latitude/longitude coordinates within the confines of the area. This type of design is useful if one is trying to determine baseline conditions at a reference site. In this case, a robust random sampling design, with a sufficient number of samples, will have a high degree of statistical power, giving the investigator confidence to draw conclusions about the overall site conditions.

There are, however, a number of challenges that may limit the utility of this type of design. First, to achieve a high degree of statistical power, a large number of samples are often required. Given that collection and analysis of environmental samples is often quite expensive, the number of Samples required can make it cost prohibitive. Second, in a very heterogeneous AOI, random sampling locations can easily fall in locations that are not able to be sampled or would not make sense to sample (e.g., in the middle of a street or building). Finally, the nature of this type of

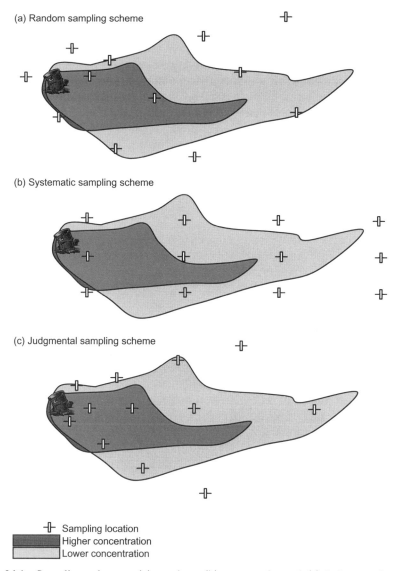

Figure 24.1   Sampling schemes: (a) random, (b) systematic, and (c) judgmental sampling designs.

design can lead to "holes" in the sampling scheme with other areas having large numbers of samples. This last problem can be alleviated with the use of a systematic sampling scheme.

Systematic Sampling   In this type of sampling scheme, the AOI is blocked off with a grid and a sample is taken from within each of the grid squares. This type of scheme is useful if one wants to be assured of consistent sample coverage of the AOI. The disadvantage here is that one loses most (if not all) of that ability to perform statistical analyses of the data.

Judgmental Sampling   Often, with environmental contamination, the source of the contamination is known. What needs to be determined is the nature and extent of the contamination (e.g., what are the contaminants and how far have they traveled from the source). Therefore, this is one of the most common sampling schemes used in environmental toxicological and environmental exposure assessment settings. The sampling scheme is designed such that it maximizes the number of samples taken in and around the source of contamination. Sampling locations can then move out from the source along transects, with increasing distance between samples as one moves further from the source. The disadvantage with a judgmental sampling design is that there may be secondary sources of contamination that are not known and could thus be missed.

24.2.2   Environmental Matrices

The most common environmental matrices include water, air, solid (soil, sediment, or sludge), biota, and vegetation. With the exception of air samples, each of these sample types consists of a complex matrix that can include many and varied constituents in addition to the analytes of interest. Each of these sample types requires unique strategies to collect a representative sample.

Water   One of the most common types of environmental matrices is water. Many factors must be considered to obtain representative samples of water. The most important factor is the contaminant of potential concern (COPC) and the point at which it entered the aquatic environment. Pollutants can be contributed by agricultural, industrial, municipal, or other sources, such as spills. Other factors include the velocity of stream or river flow (or velocity and direction in the case of groundwater), temperature, thermal and salinity stratification, and sediment content.

The simplest method of collecting water is the "grab" technique, whereby a container is lowered into the water, rinsed, filled, and capped. Specialized samplers can be used to obtain water at greater depths. In the case of groundwater, samples are obtained using a groundwater well or a piezometer, the latter being easier and cheaper to install. The collection of groundwater samples is far more involved and expensive than collecting a surface water sample; therefore, much more effort is put into determining locations for sampling. Obtaining a representative sample is affected not only by where but also when to collect a sample. Sources of contamination that are intermittent (e.g., pesticide runoff) will vary greatly by the time of year that a sample is taken. For this reason, the use of passive sampling devices (PSDs) is increasingly common. PSDs come in a variety of configurations, but the general design principles are the same: a clean adsorptive phase is placed in contact with the media of interest (water) for a period of time (generally, about 1 month). Contaminants in the water that have an affinity for the adsorbent will partition into the PSD at a constant rate known as the sampling rate. Using the known sampling rate, one can calculate the average water concentration over the time period that the PSD was deployed. In this way, the PSD acts as an "integrative sampler." The PSD has the added advantage that it acts to concentrate ultra-trace levels of chemicals present in the water, making them detectable with conventional laboratory methods. Finally, PSDs deployed in water will only sequester contaminants that are freely dissolved; therefore, they can provide an estimation of the bioavailability of contaminants.

Air Most pollutants entering the atmosphere come from fuel combustion, industrial processes, and solid waste disposal. Additional miscellaneous sources, such as forest fires, dusts, volcanoes, natural gaseous emissions, agricultural burning, and pesticide drift, contribute to the level of atmospheric pollution. To affect terrestrial animals and plants, particulate pollutants must be in a size range that allows them to enter the body and remain there; that is, they must be in an aerosol (defined as an airborne suspension of liquid droplets) or on solid particles small enough to possess a low settling velocity (see Chapter 18). Suspensions can be classified as liquids including fogs (small particles) and mists (large particles) produced from atomization, condensation, or entrapment of liquids by gases; and solids including dusts, fumes, and smoke produced by crushing, metal vaporization, and combustion of organic materials, respectively.

Air samplers have been miniaturized and adsorbents have been developed to collect either particulate matter in the size range most detrimental to humans or to "trap" organic toxicants from air. An air sampler generally consists of an inlet to direct air through a filter (to entrap particles that might be of interest e.g., dust); through the adsorbent (which collects organic vapors); a flowmeter/valve to calibrate airflow and a pump to pull air through the system. Personnel samplers are run by battery power and can be attached to an individual's clothing, thus allowing continual monitoring while performing assigned tasks in the work environment. This allows the estimation of individual exposure.

Many air samplers use various types of filters to collect solid particulate matter, such as asbestos, which is collected on glass fiber filters with pores 20 μm or less in diameter. Membrane filters with pores 0.01–10.0 μm in diameter are used to collect dusts and silica. Liquid-containing collectors, called impingers, are used to trap mineral dusts and pesticides. Mineral dusts are collected in large impingers that have flow rates of 10–50 L of air per minute, and insecticides can be collected in smaller "midget" impingers that handle flows of 2.0–4.5 L of air per minute. The solvent within the impinger will vary depending on the COPCs. Because of the ease of handling and the rapid desorption of compounds, polyurethane foam (PUF) has become a popular trapping medium for pesticides and is rapidly replacing the use of midget impingers. Finally, PSDs have also been used to sample airborne contaminants.

Soil When environmental pollutants are deposited on land areas, their subsequent behavior is complicated by a series of simultaneous interactions with organic and inorganic components, existing liquid–gas phases, microscopic organisms, and other soil constituents. Depending on the chemical composition and physical structure, pollutants might remain in one location for varying periods of time, be absorbed into plant tissue, or move through the soil profile from random molecular motion. Movement is also affected by mass flow as a result of external forces such as the pollutant being dissolved in or suspended in water or adsorbed onto both inorganic and organic soil components. Thus, sampling for pollutants in soils is complex and statistical approaches must be taken to ensure representative samples.

To obtain such samples, the chemical and physical characteristics of the site(s) must be considered, as well as possible reactions between the compound(s) of interest and soil components and the degree of variability (i.e., variation in soil profiles) within the sampling site. With these data, the site(s) can then be divided into homogeneous areas and the required number of samples can be collected. The required number of samples depends on the functions of variance and the degree of accuracy. Once the correct procedure has been determined, sampling can proceed.

Many types of soil samplers are available, but coring devices are preferable because this collection method allows determination of a pollutant's vertical distribution. These devices can be either stainless steel tubes, varying in both diameter from 2.5 to 7.6 cm and length from 60 to 100 cm (hand operated). Large, mechanically operated boring tubes, 200 cm in length, are also used. It is possible to sample to uniform depths with these devices, and one can subdivide the cores into specific depths (e.g., 0–7 cm and 7–15 cm) to determine movement. Another type of coring device is a wheel to which are attached tubes so that large numbers of small subsamples can be collected, thus allowing a more uniform sampling over a given area. Soils from specific depths can be collected using a large diameter cylinder (~25 cm) that incorporates a blade to slice a core of soil after placing the sampler at the desired depth.

Biota   The collection of biota for the purpose of environmental monitoring is primarily done in the aquatic environment, although terrestrial and avian species may be sampled as well. Sampling techniques fall into three general categories. First is the collection of nonlethal samples, for example, hair or feathers for mercury analysis. The second category is the collection of feral organisms such as fish or mussels. An example of this is the National Oceanic and Atmospheric Administration's (NOAA) National Status and Trends Mussel Watch program. In this program, mussel and sediment samples have been collected for analysis of over 100 organic and inorganic constituents, from 300 coastal sites around the United States. This program has been operating since 1986. This type of sampling has the advantage of giving a direct measure of the actual body burden of contaminants for organisms in the AOI. The final category is the use of bioassays, where caged "clean" organisms (fish or mussels) are placed in the sampling area for a period of time to allow them to accumulate site-related contaminants. This type of assay has the advantage of reducing the potential spatial variability errors introduced from trapping feral organisms such as fish. The disadvantage here is that caged organisms occasionally suffer from low survivorship and/or succumb to predation.

24.3   ANALYTICAL TECHNIQUES

Once the environmental samples have been collected, they have to be prepared for analysis. First, the analytes of interest have to be extracted from the bulk medium (water, soil, etc.). During this process, many analytically interfering compounds will also be extracted from the bulk medium. Therefore, some type of cleaning and concentration step has to be performed. Finally, the clean and concentrated sample extract is analyzed using an appropriate technique/instrument, and the mass of the chemicals present in the original sample can be estimated.

24.3.1 Extraction Techniques

In most cases, the analysis of the COPC depends on its physical removal from the sample medium. In order to ensure that the sample used is homogeneous, it is chopped, ground, or blended to a uniform consistency and then subsampled. This subsample is extracted, which involves bringing a suitable solvent into intimate contact with the sample, generally in a ratio of 5–25 vol of solvent to 1 vol of sample. One or more of five different procedures can be used, depending on the chemical and physical characteristics of the toxicant and the sample matrix. Other extraction methods such as boiling, grinding, or distilling the sample with appropriate solvents are used less frequently.

Blending The use of an electric or air-driven blender is currently the most common method of extraction of biologic materials. The weighed sample is placed in a container; solvent is added; and the tissue is homogenized by motor-driven blades. Blending for 5–15 min followed by a repeat blending will extract most environmental toxicants. A homogenate in an organic solvent can be filtered through anhydrous sodium sulfate to remove water that might cause problems in the quantification phase of the analysis. The use of sonication is a popular method for extracting tissue samples, particularly when the binding of toxicants to subcellular fractions is of interest. Sonicator probes rupture cells rapidly, thus allowing the solvent to come into intimate contact with all cell components. Large wattage (e.g., 450 W) sonicators are used to extract compounds from environmental samples, and several United States Environmental Protection Agency (EPA) methods list sonification as a valid method of extraction.

Shaking Pollutants are generally extracted from water samples, and in some cases, soil samples, by shaking with an appropriate solvent or solvent combination. Mechanical shakers are used to handle several water or soil samples at once. These devices allow the analyst to conduct long-term extractions (e.g., 24 h) if required. Two or more shakings normally are required for complete removal (i.e., >98%) of the toxicant from the sample matrix.

Solid-Phase Extraction (SPE) In SPE, water samples are filtered through a cartridge or filter disk made of material such as C-18. Analytes of interest are retained on the SPE filter and can be collected by eluting with different solvents or solvent mixtures. This general process will be discussed further in the next section.

Continuous Extraction The procedure, called Soxhlet extraction, is performed on solid samples (e.g., soil) and involves the use of an organic solvent or a combination of solvents. The sample is weighed into a cup (thimble) of specialized porous material such as cellulose or fiberglass and is placed in the apparatus. This consists of a boiling flask in which the solvent is placed, an extractor, which holds the thimble, and a water-jacketed condenser. When heated to the boiling point, the solvent vaporizes, is condensed, and fills the extractor, thus bathing the sample and extracting the toxicant. A siphoning action drains the solvent back into the boiling flask, and the cycle begins again. Depending on the nature of the toxicant and the sample matrix, the extraction can be completed in as little as 2 h but may take as long as 3–4 days. Automated instruments have been introduced that perform the same operation in a shorter period of time (e.g., 30 min) and use much less solvent (e.g., 15–30 mL compared with 250 mL). In addition, specialized glassware can be used to perform continuous liquid–liquid extractions for water samples. The general process is the same as for a Soxhlet extractor.

Supercritical Fluid Extraction   Conditions can be generated that allow materials to behave differently from their native state. For example, boiling points are defined as that temperature at which a liquid changes to a gas. If the liquid is contained and pressure is exerted, the boiling point changes. For a particular liquid, a combination of pressure and temperature will be reached, called the critical point, at which the material is neither a liquid nor a gas. Above this point exists a region, called the supercritical region, at which increases in both pressure and temperature will have no effect on the material (i.e., it will neither condense nor boil). This so-called supercritical fluid will exhibit properties of both a liquid and a gas. The supercritical fluid penetrates materials as if it was a gas but has the solvent properties of a liquid.

Of all the materials available for use as a supercritical fluid, CO_2 has become the material of choice because of its chemical properties. Instruments have been developed to utilize the principles described to effect extractions of compounds from a variety of sample matrices including asphalt, plant material, and soils. The supercritical fluid is pumped through the sample, through a filter or column to a trap where the fluid vaporizes and solvent is added to transfer the analyses to a vial for analysis. More recent instruments combine the supercritical fluid extraction system with a variety of columns and detectors to acquire data from complex samples.

24.3.2   Sample Cleanup and Enrichment

A little over 100 years ago, a Russian botanist by the name of M. S. Tswett published a paper describing a new method he had devised for separating out plant pigments by percolating a plant extract through a column of $CaCO_3$ with petroleum ether. He called the process *chromatography*, or "color writing." The fundamental principles of chromatography are the same, regardless of whether it is preparative chromatography or analytical chromatography. There is a stationary phase and a mobile phase, and the separation of sample components is achieved through the differential interaction between the two phases. Compounds that have a higher affinity for the mobile phase will move through the system very quickly, whereas compounds that have a high affinity for the stationary phase will move very slowly through the system (Figure 24.2). This process is the basis for the vast majority of environmental analytical chemistry techniques in use today.

Using one of the extraction techniques described above will remove the analytes of interest from the bulk medium but will also extract other matrix constituents (e.g., wax and lipid inorganic components). These interfering compounds must be removed prior to analysis, and there are various chromatographic methods available to separate the desired components from the matrix interferences.

- High affinity for stationary phase

- Affinity for both stationary phase and mobile phase

- High affinity for moblie phase

Figure 24.2 Column chromatography. Compounds with high affinity for the mobile phase will be the first to elute from the column.

Thin-Layer Chromatography (TLC) Many toxicants and their metabolites can be separated from interfering substances with TLC. In this form of chromatography, the adsorbent is spread as a thin layer (250–2000 μm) on glass or on resistant plastic backings. When the extract is placed near the bottom of the plate and the plate is placed in a tank containing a solvent system, the solvent migrates up the plate, and the toxicant and other constituents move with the solvent; differential rates of movement result in separation. The compounds can be scraped from the plate and eluted from the adsorbent with suitable solvents. This technique is not used much for environmental chemistry applications.

Column: Adsorption, Hydrophobic, Ion Exchange A large number of adsorbents are available to the analyst. The adsorbent can be activated charcoal, aluminum oxide, Florisil, silica, silicic acid, or mixed adsorbents. The characteristics of the toxicant determine the choice of adsorbent. When choosing an adsorbent, select conditions that either bind the co-extractives to it, allowing the compound of interest to elute, and vice versa. The efficiency of separation depends on the flow rate of solvent through the column (cartridge) and the capacity of the adsorbent to handle the extract placed on it. This amount depends on the type and quantity of adsorbent, the capacity factor (k') and concentration of sample components, and the type and strength of the solvents used to elute the compound of interest. Many environmental samples contain a sufficient amount of interfering materials so that the analyst must prepare a column using a glass chromatography tube into which the adsorbent is added. In the most common sequence, the column is packed in an organic solvent of low polarity; the sample is added in the same solvent; and the column is then developed with a sequence of solvents or solvent mixtures of increasing polarity. Such a sequence might include (in order of increasing polarity) hexane, benzene, chloroform, acetone, and methanol. Once removed, the eluate containing the toxicant is reduced to a small volume for quantification (Figure 24.2).

However, cartridge technologies are improving to allow similar concentrations of sample to be added that result in a less expensive and more rapid analysis. A number of miniaturized columns have been introduced since the early 1980s. Most contain 0.5–2.0 g of the adsorbent in a plastic tube with fitted ends. The columns can be attached to standard Luer Lock syringes. Other companies have designed vacuum manifolds that hold the collecting device. The column is placed on the apparatus, a vacuum is applied, and the solvent is drawn through the column. Some advantages of these systems include pre-weighed amounts of adsorbent for uniformity, easy disposal of the co-extractives remaining in the cartridge, no breakage, and decreased cost of the analysis because less solvent and adsorbent are used. Other forms of column chromatography can be used. They include ion exchange chromatography and affinity chromatography. Ion exchange chromatography depends on the attraction between charged molecules and opposite charges on the ion exchanger, usually a resin. Compounds so bound are eluted by changes in pH and, because the net charge depends on the relationship between the pH of the solution and the isoelectric point of the compounds, compounds of different isoelectric points can be eluted sequentially. Both ionic and anionic exchangers are available. Affinity chromatography is a potent tool for biologically active macromolecules but is seldom used for purifying small molecules, such as most toxicants. It depends on the affinity of an enzyme for a substrate (or substrate analogue) that has been incorporated into a column matrix or the affinity of a receptor for a ligand.

Size Exclusion Chromatography   Also referred to as gel permeation chromatography (GPC), this technique is primarily used during the analysis of biological samples. When tissue samples are extracted with a nonpolar solvent, a good deal of lipids is also extracted. In addition, some configurations of PSDs contain a lipid matrix and would be prepared via GPC as well. GPC columns are packed with a cross-linked polymer material that is very porous. Cross-linked dextrans such as Sephadex or agarose (Sepharose) are commonly used materials. Small molecules can get into the pores and are thus retained for longer periods of time on the column, whereas large molecules (e.g., lipids) cannot and therefore pass through the column very quickly. The GPC material is available in varying pore sizes depending on the application for which it will be used. When using this type of cleanup, the lipid fraction is retained, dried, and weighed. Thus, analytical results from the sample can be reported on a lipid weight basis where appropriate.

24.3.3 Analysis

Once the samples have been extracted, cleaned up and concentrated, the next step is to analyze the extract using a variety of techniques, depending on the analytes of interest.

Gas–Liquid Chromatography   Gas–liquid chromatography, generally referred to as gas chromatography (GC), is used most commonly for the separation and quantification of organic toxicants. This system consists of an injector port, oven, detector, amplifier (electrometer), and supporting electronics (Figure 24.3). Gas chromatographs use a capillary column to effect separation of complex mixtures of organic molecules. The stationary phase is coated onto the inside of the capillary column. The mobile phase in this system is an inert gas (called the carrier gas), usually helium or nitrogen, that passes through the column. The term "gas–liquid chromatography" derives from the fact that the polymer coating that acts as the stationary phase is technically a liquid.

When a sample is injected, the injector port is at a temperature sufficient to vaporize the sample components. Based on the solubility and volatility of these components with respect to the stationary phase, the components separate and are swept through the column by the carrier gas to a detector, which responds to the concentration of each component. The column is contained within an oven that can be programmed by the analyst. Similar to the way the solvent systems can be changed in column chromatography, the temperature program can be altered to

Figure 24.3   Diagram of a typical GC system. Reprinted with permission from *Capillary Gas Chromatography,* D. W. Grant, New York: Wiley, 1996.

maximize the analyte separation while minimizing the run time per sample. The electronic signal produced as the component passes through the detector is amplified by the electrometer, and the resulting signal is sent to a computer or other electronic data-collecting devices for quantification. The time at which a specific compound exits the column for a given set of conditions within the instrument is called the retention time. Standard mixtures are run under the given conditions to determine the retention time for each analyte of interest. This is then used to compare with the retention time of peaks in the unknown samples.

Increased sensitivity and component resolution have resulted from advances in solid-state electronics and column and detector technologies. In the field of column technology, the capillary column has revolutionized toxicant detection in complex samples. This column is generally made of fused silica 5–60 m in length with a very narrow inner diameter (0.23–0.75 mm) to which a thin layer (e.g., 1.0 μm) of polymer is bonded. The polymer acts as the stationary phase. The carrier gas flows through the column at rates of 1–2 mL/min. Two types of capillary columns are used: the support-coated, open tubular (SCOT) column and the wall-coated, open tubular (WCOT) column. The SCOT column has a very fine layer of diatomaceous earth coated with liquid phase, which is deposited on the inside wall. The WCOT column is pretreated and then coated with a thin film of liquid phase. Of the two columns, the SCOT is claimed to be more universally applicable because of large sample capacity, simplicity in connecting it to the chromatograph, and lower cost. However, for difficult separations or highly complex mixtures, the WCOT is more efficient and is used to a much greater extent.

Detectors   Five detectors are used widely in toxicant detection: flame ionization detector (FID), nitrogen–phosphorous detector (NPD), flame photometric detector (FPD), electron capture detector (ECD), and mass spectrometer (MS) detector.

The FID operates on the principle of ion formation from compounds being burned in a hydrogen flame as they elute from a column. The concentrations of ions formed are several orders of magnitude greater than those formed in the uncontaminated flame. The ions cause a current to flow between two electrodes held at a constant potential, thus sending a signal to the electrometer. The NPD detects the presence of nitrogen- and phosphorous-containing compounds and functions similarly to an FID. In the NPD, a heated rubidium silicate bead emits ions when nitrogen and phosphorus-containing compounds pass over it, and this signal is passed to the electrometer.

The FPD is a specific detector in that it detects either phosphorous- or sulfur-containing compounds. When atoms of a given element are burned in a hydrogen-rich flame, the excitation energy supplied to these atoms produces a unique emission spectrum. The intensity of the wavelengths of light emitted by these atoms is directly proportional to the number of atoms excited. Larger concentrations cause a greater number of atoms to reach the excitation energy level, thus increasing the intensity of the emission spectrum. The change in intensity is detected by a photomultiplier, amplified by the electrometer, and recorded.

The ECD is used to detect halogen-containing compounds, although it will produce a response to any electronegative compound. When a negative DC voltage is applied to a radioactive source (e.g., ^{63}Ni, ^{3}H), low-energy β particles are emitted, producing secondary electrons by ionizing the carrier gas as it passes through the detector. The secondary electron stream flows from the source (cathode) to a collector (anode), where the amount of current generated (called a standing current) is amplified and recorded. As electronegative compounds pass from the column into the detector, electrons are removed or "captured," and the standing current is reduced. The reduction is related to both the concentration and the electronegativity of the compound passing through, and this produces a response that is recorded. The sensitivity of ECD is greater than that of any other detectors currently available and can be used for ultra-trace detection of compounds such as Polychlorinated biphenyls (PCBs) and other chlorinated hydrocarbons.

Although the detectors mentioned above have some degree of specificity (e.g., they detect halogenated compounds or nitrogen-containing compounds), they are nonspecific detectors in that they cannot identify what the compound is, only that it shares a retention time with a specific analyte from the standard. Occasionally, this is remedied through the use of dual-column confirmation methods. In this technique, the same sample is run on two different columns, thus producing two different retention times. In this way, the unknown compound can be confirmed. However, the only truly specific detector is the MS. This detector, discussed in more detail below, bombards the molecule and breaks it apart. The resulting fragments are separated within the detector based on their mass-to-charge ratio (m/z). Thus, regardless of retention time, the mass spectrum of a given compound is unique, much like a fingerprint.

High-Performance Liquid Chromatography (HPLC)   HPLC has become very popular in the field of analytical chemistry for the following reasons: it can be run at ambient temperatures; it is nondestructive to the compounds of interest, which can be collected intact; in many instances, derivatization is not necessary for response; and columns can be loaded with large quantities of the material for detection of low levels. However, the most important advance was the development of the MS detector that could be coupled with HPLC.

The instrument consists of a solvent reservoir, a gradient-forming device, a high-pressure pumping device, an injector, a column, and a detector (Figure 24.4). The principle of operation is very similar to that of GC except that the mobile phase is a liquid instead of a gas. The composition of the mobile phase and its flow rate affect separations (recall that in the GC, the oven temperature was controlled to affect separation). The columns being developed for HPLC are too numerous to discuss in detail. Most use finely divided packing (3–10 µm in diameter); some have bonded phases, and others are packed with alumina or silica. The columns normally are 15–25 cm in length, with small diameters. (~4.6 mm number diameter). A high-pressure pump is required to force the solvent through this type of column. The major detectors presently used for HPLC are UV or fluorescent spectrophotometers as well as mass spectrometers.

Capillary Electrophoresis (CE)   A relatively new analytical technique, CE, is receiving considerable attention in the field of toxicology, and methods have been developed to analyze a diversity of compounds, including DNA adducts, drugs, small aromatic compounds, and pesticides. Commercial instruments are available that are composed of an autosampler, a high-voltage power supply, two buffer reservoirs, the capillary (approximately 70 cm × 75 µm in diameter) and a detector (Figure 24.5a). The versatility of the process lies in the ability to separate compounds of interest by a number of modes, including affinity, charge/mass ratios,

Figure 24.4   High-performance liquid chromatography (HPLC) system. Reprinted with permission from *Chromatographic Methods*, A. Braithwaite and J. F. Smith, New York: Springer, 1996.

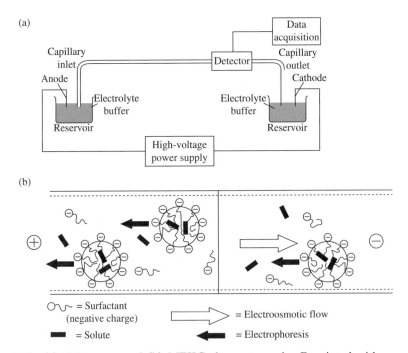

Figure 24.5   (a) CE system and (b) MEKC chromatography. Reprinted with permission from *Instant Notes: Analytical Chemistry*, D. Kealy and P. J. Haines, New York: Garland Science/Taylor & Francis, 2002.

chirality of the compounds, hydrophobicity, and size. The theory of operation is simple. Because the capillary is composed of silica, silanol groups are exposed in the internal surface, which can become ionized as the pH of the eluting buffer is increased. The ionization attracts cations to the silica surface, and when current is applied, these cations migrate toward the cathode, which causes a fluid migration through the capillary. This flow can be adjusted by changing the dielectric strength of the buffer, altering the pH, adjusting the voltage, or changing the viscosity.

Under these conditions, both anions and cations are separated in a single separation, with cations eluting first. Neutral molecules (e.g., pesticides) can be separated by adding a detergent (e.g., sodium dodecyl sulfate) to the buffer, forming micelles into which neutral molecules will partition based on their hydrophobicity. Because the micelles are attracted to the anode, they move toward the cathode at a slower rate than does the remainder of fluid in the capillary, thus allowing separation. This process is called micellar electrokinetic capillary chromatography (MEKC) (Figure 24.5b). Many of these analyses can be carried out in 5–10 min with sensitivities in the low parts-per-billion range. A UV detector is usually used, but greatly sensitivities can be obtained using laser-induced fluorescence detectors.

Spectroscopy   In certain experiments involving radiation, observed results cannot be explained on the basis of the wave theory of radiation. It must be assumed that radiation comes in discrete units, called quanta. Each quantum of energy has a definite frequency, v, and the quantum energy can be calculated by the equation $E = hv$, where h is Planck's constant (6.6×10^{-27} erg-s). Matter absorbs radiation one quantum at a time, and the energy of radiation absorbed becomes greater as either the frequency of radiation increases or the wavelength decreases. Therefore, radiation of shorter wavelength causes more drastic changes in a molecule than does that of longer wavelength. Spectroscopy is concerned with the changes in atoms and molecules when electromagnetic radiation is absorbed or emitted. Instruments have been designed to detect these changes, and these instruments are important to the field of toxicant analysis. Discussions of atomic absorption (AA) spectroscopy, atomic emission spectroscopy, inductively coupled plasma (ICP), and mass spectroscopy (MS) follow.

AA Spectroscopy   One of the more sensitive instruments used to detect metal-containing toxicants is the AA spectrophotometer. Samples are vaporized either by aspiration into an acetylene flame (flame AA) or by carbon rod atomization in a graphite cup or tube (graphite furnace AA). The atomic vapor formed contains free atoms of an element in their ground state, and when illuminated by a light source that radiates light of a frequency characteristic of that element, the atom absorbs a photon of wavelength corresponding to its AA spectrum, thus exciting it. The amount of absorption is a function of concentration. The graphite furnace instruments are much more sensitive than conventional flame AA. For example, arsenic can be detected at levels of 0.1 ng/mL and selenium at 0.2 mg/mL, which represent sensitivity three orders of magnitude greater than that of conventional flame AA. The disadvantage of AA methods is that only one metal is analyzed at a time.

ICP Spectrometry   An even more sensitive instrument has been developed to detect and quantitate, simultaneously, all inorganic species contained with a sample matrix. With ICP, a stream of argon passed through an induction coil, producing temperatures of up to 10,000 K. When the sample is heated, an emission spectrum is given off and the spectral lines are observed. One such system is the inductively coupled plasma–optical emission spectrometer (ICP-OES). The ICP-OES takes an aliquot of sample that has been acid digested and mixes it with a gas (e.g., argon) forming a plasma (i.e., an ionized gas) that is channeled into a nebulizer. Energy is applied to excite the atoms that are converted by the optics of the instrument into individual wavelengths. The spectra are captured by a charge-coupled device (CCD) that converts the light to measurable electrons at specific wavelengths. Wavelength coverage ranges from 175 to 785 nm. In addition, the ICP can be coupled with a mass spectrometer (ICP-MS) to collect information on the analyte being sought within the sample matrix. These instruments utilize high throughput of samples and are used in both research and industrial settings.

Mass Spectroscopy (MS)   The mass spectrometer is an outstanding instrument for the identification of compounds. In toxicant analysis, MS is widely used as a highly sensitive detection method for GC and is increasingly used with HPLC, CE, and ICP because these instruments can be interfaced to the mass spectrometer. Chromatographic techniques (e.g., GC, CE, HPLC) are used to separate individual components as previously described. A portion of the column effluent passes into the mass spectrometer, where it is bombarded by an electron beam. Electrons or negative groups are removed by this process, and the ions produced are accelerated. After acceleration, they pass through a magnetic field, where the ion species are separated by the different curvatures of their paths under gravity. The resulting pattern is characteristic of the molecule under study. Two detectors are used primarily in pollutant analysis: the quadrapole (Figure 24.6) and the ion trap. Both produce

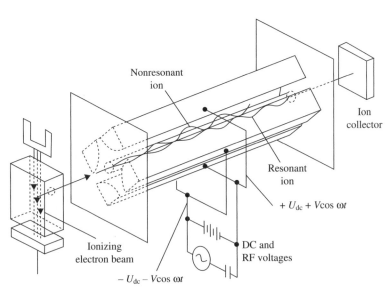

Figure 24.6 Quadrupole mass spectrometer. Reprinted with permission from *Instant Notes: Analytical Chemistry*, D. Kealy and P. J. Haines, New York: Garland Science/Taylor & Francis, 2002.

reliable and reproducible data, and if routine maintenance is performed, both are reliable. Computer libraries of mass spectral data continue to expand, and data are generated rapidly with current software.

Bimolecular Interaction Analysis–Mass Spectrometry (BIA-MS)   A new field that is utilizing mass spectrometry as a tool in biological and toxicological research to investigate protein interactions is that of proteomics. This rapidly expanding science explores proteins within the cellular environment, their various forms, interacting partners (e.g., cofactors), and those processes that affect their regulation and processing. The BIA-MS can determine such things as the kinetics of protein interactions, selectively retrieve and concentrate specific proteins from biological media, quantify target proteins, identify protein–ligand interactions, and recognize protein variants (e.g., point mutations). BIA-MS uses two technologies, surface plasmon resonance (SPR) sensing and matrix-assisted laser desorption/ionization time-of-flight mass spectrometry (MALDI-TOF MS). Cells are fragmented and come in contact with a gold-plated glass slide called a chip. The chip has highly defined sites containing a number of immobilized ligands to which the proteins of interest bind and are quantified by SPR that monitors the interaction and quantifies the amount of protein localized at precise locations on the surface of the chip. The chip is then subjected to MALDI-TOF MS, which yields the masses of retained analytes and other bound biomolecules.

24.4   QUANTIFICATION, QA, AND QC

Toxicants are generally found at low concentrations (e.g., parts per million or parts per billion) regardless of the sample matrix being evaluated. These concentrations are based on the measurement of a response from some instrument to the compound(s) of interest from an extract of the sample matrix. Thus, it is necessary to have a system capable of measuring the compound of interest, and in order to ensure the reliability of the data, the analytical process (instrument and analytical method) must be monitored closely.

This measurement process involves much more than injecting some amount of the extracted sample and comparing its response to that of a standard of known concentration. Analytical standards must be prepared, weighed, and diluted carefully to ensure that the concentrations reported reflect those found in the sample analyzed. In addition, the analytical instrument used must be calibrated properly to ensure accuracy. Essentially, this involves two processes: (1) calibration of the detector against the compound of interest in order to eliminate or to minimize any deviation (bias) of response in one direction or another from that expected from previous experience or expected results and (2) calibration of the total analytical system using statistical approaches to minimize bias in determining the presence or absence of the analyte being sought.

24.4.1   Quantification Approaches and Techniques

Analytical Instrument Calibration   In setting up instrument parameters, consider what is involved in determining residue levels of an analyte. The data produced are only as good as the extract derived from the original sample. If the analyte is distributed uniformly over the area sampled, the concentrations found will be equal, regardless of where the sample is taken. Along these same lines, the analytical procedure will result in uniform residue values if all procedures and instrument parameters remain the same. Based on experience, we know that this distribution of residue over an area will vary as will the analytical procedures and instrument parameters. If we increase the number of samples collected and analyzed, the differences observed will tend to get smaller, resulting in a mean or average value that locates the center of the distribution. Ideally, this distribution is called a normal distribution or Gaussian distribution, and looks like a bell (the classic "bell-shaped curve") when the parameters being measured are plotted on a graph (i.e., frequency vs. concentration). Second, the difference in individual measurement, called the standard deviation (σ), defines the variation found in individual measurements. Equations and tables have been developed to determine the significance of suspected deviations and are used to confirm the presence of a suspected problem. If an infinite number of samples from the area are collected and analyzed, the variation in 95% of the samples will cover the true population percentage.

Quantification   Quantification of unknown concentrations of COPCs is the ultimate goal of the residue chemist. Residue analysis involves the removal of the compound of interest from some sample matrix. As discussed previously, the sample is collected in the field, and that matrix is then homogenized, subsampled, extracted, cleaned up, concentrated, and finally run on an appropriate instrument. Analytical instruments are calibrated such that they give a consistent reproducible response for a given mass of analyte. The most fundamental decision made is whether the analyte is present or absent, particularly when its concentration is at or close to its detection limit. Because the measurements are derivations of a known relationship between the analyte concentration and the magnitude of the signal made by the instrument, there is an additional signal (noise) generated from the presence of co-extractives, column bleed, and the like. The analyst uses this "contaminated" signal to decide whether the analyte is present or absent and selects one of these choices. The decision process is subject to two types of errors: the analyte is present when actually it is not, and the analyte is absent when actually it is present. The terminology for these decision processes are commonly called "false positives" and "false negatives," respectively. As the mass of analyte increases, the analyst moves beyond simple presence/absence determinations to quantification of concentrations. Using a set of standards to establish a three- or five-point calibration curve, the mass of analyte in the unknown can be extrapolated from the detector response (Figure 24.7).

24.4.2 QA and QC

Over the last 20 years, the reliability of data produced by analytical laboratories has increased dramatically. Strict requirements have ensured that the data were produced under defined standards of quality with a stated level of confidence. The routine day-to-day activities (e.g., matrix fortifications) to control, assess, and ensure the quality of generated data are the QCs associated with analytical processes.

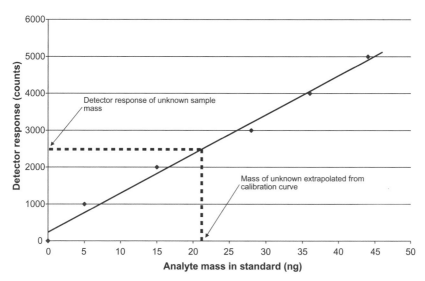

Figure 24.7 A five-point calibration curve. Unknown analyte concentrations are determined by extrapolating from the detector response to the mass.

The management of the system that ensures that these processes are in place and functional is the QA portion of the laboratory program to produce reliable data.

QA is an essential part of analytical protocols. Each laboratory is required to detect and to correct problems in analytical processes and to reduce errors to agreed-upon limits. To produce data that have acceptable quality, all laboratory members must follow established guidelines and protocols. Some of the essential elements that must be included in a QA program are as follows:

1. Laboratory practices (e.g., glass washing protocols) must be developed, reviewed, and updated with the staff's participation on a scheduled basis and followed strictly by all laboratory members.

2. Standard operating procedures (SOPs) (e.g., SOPs monitoring freezer temperatures daily) must be standardized, documented, and supplied to each member of the laboratory staff and updated on a set schedule.

3. Monitoring programs (e.g., surface water monitoring of supplies furnishing public drinking water) must be carefully designed.

4. Maintenance of equipment and instruments must be documented in a laboratory information management system (LIMS) or appropriate maintenance books kept with the equipment.

5. Expiration dates of analytical standards, chemicals, and solvents must be observed and replacements made prior to their expiration date.

6. Good laboratory practices (GLPs) must be implemented as needed.

7. Audits must be performed on a scheduled basis to verify that all aspects of the QA program are operating sufficiently.

QC concerns procedures that maintain a measurement system in a state of statistical control. This does not mean that statistics control the analytical procedures but that statistical evidence is used to ensure that the procedure is working under the conditions set by protocol. The accuracy of an analytical method depends on statistical control being conducted prior to determining any other parameter. How well the basic method will work with the sample matrix being evaluated will depend on the way the QC samples are examined. A comprehensive QC analytical procedure would include the following:

1. Replicated environmental samples to test the precision of the sampling or analytical procedures
2. Replicated analyses conducted on the same sample multiple times in order to determine analytical precision
3. Trip blanks to determine if contaminants are introduced to the processes of collecting, shipping, or storing of samples
4. Matrix-fortified laboratory blanks consisting of solvent and reagent blanks to determine levels of bias due to matrix effects or analytical method problems
5. Sample blanks (a sample matrix that does not contain the toxicant, although this is sometimes difficult to obtain) to ensure no extraneous or interfering peaks; the peaks indicate where a problem might exist in the method used
6. Fortified field blanks to determine the effects that the matrix might have on analyte recovery

Although many of these above procedures may seem mundane, when data are required for decision making, especially in the area of regulatory actions or risk to human health, it is imperative that the data are as reliable and accurate as possible.

24.5 SUMMARY

Whether the exposure analysis is being conducted for a human health or ecological risk assessment, the essential elements are the same: defining the research goal(s), identification of appropriate techniques and methods, development of a sampling scheme to obtain representative samples, isolating the compound(s) of interest, removing potentially interfering components, and quantifying and evaluating the data in relation to the original research goals. These essential elements are critical to ensuring that the data generated answer what is requested and enable some decision affecting environmental or human welfare. Essential decision criteria must be included in the protocol that describes the analytical process in detail, including the objective(s) of the study, the QA/QC requirements, the sample plan, methods of analysis, calculations, documentation, and data reporting. Meaningful data can only be generated with the proper method of analysis.

BIBLIOGRAPHY AND SUGGESTED READING

Bayen, S., T. L. ter Laak, J. Buffle, et al. Dynamic exposure of organisms and passive samplers to hydrophobic chemicals. *Environ. Sci. Technol.* **43**:2206–2215, 2009.

Braithwaite, A. and J. F. Smith. *Chromatographic Methods*. New York: Springer, 1996.

Grant, D. W. *Capillary Gas Chromatography*. New York: Wiley, 1996.

Kealey, D. and P. J. Haines. *Instant Notes: Analytical Chemistry*. Oxford: Garland Science/ Taylor & Francis, 2002.

Koester, C. J. and A. Moulik. Trends in environmental analysis. *Anal. Chem.* **77**:3737–3754, 2005.

Kraly, J., A. Fazal, R. M. Schoenherr, et al. Bioanalytical applications of capillary electrophoresis. *Anal. Chem.* **78**:4097–4110, 2006.

Raynie, D. E. Modern extraction techniques. *Anal. Chem.* **78**:3997–4004, 2006.

Richardson, S. D. Environmental mass spectrometry: Emerging contaminants and current issues *Anal. Chem.* **80**:4373–4402, 2008.

Richardson, S. D. and T. A. Ternes. Water analysis: Emerging contaminants and current issues. *Anal. Chem.* **77**:3807–3838, 2005.

SAMPLE QUESTIONS

1. Describe the basic principles behind chromatographic separation.

2. Name a sampling scheme and describe when one might use it.

3. What are the steps involved in the process of evaluating contamination in the environment?

4. What are three of the essential elements that should be included in a laboratory QA program?

Basics of Environmental Toxicology

GERALD A. LEBLANC and DAVID B. BUCHWALTER

25.1 INTRODUCTION

Industrial and agricultural endeavors are intimately associated with the extensive use of a wide array of chemicals. Historically, chemical wastes generated through industrial processes were disposed of through flagrant release into the environment. Gases quickly dispersed into the atmosphere; liquids were diluted into receiving waters and were efficiently transported away from the site of generation. Similarly, pesticides and other agricultural chemicals revolutionized farm and forest productivity. Potential adverse effects of the application of such chemicals to the environment were viewed as insignificant relative to the benefits bestowed by such practices. Then in 1962, a science writer for the U.S. Fish & Wildlife Service, Rachel Carson, published a book that began by describing a world devoid of birds and from which the title *Silent Spring* was inspired. In her book, Ms. Carson graphically described incidents of massive fish and bird kills resulting from insecticide use in areas ranging from private residences to national forests. Further, she inferred that such pollutant effects on wildlife may be heralding similar incipient effects on human health.

The awakening of the general public to the hazards of chemicals in the environment spurred several landmark activities related to environmental protection including Earth Day, organization of the United States Environmental Protection Agency, and the enactment of several pieces of legislation aimed at regulating and limiting the release of chemicals into the environment. Appropriate regulation of the release of chemicals into the environment without applying unnecessarily stringent limitation on industry and agriculture requires a comprehensive understanding of the toxicological properties and consequences of release of the chemicals into the environment. It was from this need that modern environmental toxicology evolved.

Environmental toxicology is defined as the study of the fate and effects of chemicals in the environment. Though this definition would encompass toxic chemicals naturally found in the environment (i.e., animal venom and microbial and plant toxins), environmental toxicology is typically associated with the study of environmental chemicals of anthropogenic origin. Environmental toxicology can be divided

A Textbook of Modern Toxicology, Fourth Edition. Edited by Ernest Hodgson
Copyright © 2010 John Wiley & Sons, Inc.

into two subcategories: environmental health toxicology and ecotoxicology. Environmental health toxicology is the study of the adverse effects of environmental chemicals on human health, while ecotoxicology focuses upon the effects of environmental contaminants upon ecosystems and constituents thereof (i.e., fish and wildlife). Assessing the toxic effects of chemicals on humans involves the use of standard animal models (i.e., mouse and rat) as well as epidemiological evaluations of exposed human populations (i.e., farmers and factory workers). In contrast, ecotoxicology involves the study of the adverse effects of toxicants on a myriad of organisms that makes up ecosystems ranging from microorganisms to top predators. Further, comprehensive insight into the effects of chemicals in the environment requires assessments ancillary to toxicology such as the fate of the chemical in the environment (Chapter 26) and toxicant interactions with abiotic (nonliving) components of ecosystems. Comprehensive assessments of the adverse effects of environmental chemicals thus utilize expertise from many scientific disciplines. The ultimate goal of these assessments is elucidating the adverse effects of chemicals that are present in the environment (retrospective hazard assessment) and predicting any adverse effects of chemicals before they are discharged into the environment (prospective hazard assessment). The ecological hazard assessment process is discussed in Chapter 27.

Historically, chemicals that have posed major environmental hazards tend to share three insidious characteristics: environmental persistence, the propensity to accumulate in living things, and high toxicity.

25.2 ENVIRONMENTAL PERSISTENCE

Many abiotic and biotic processes that function in concert to eliminate (i.e., degrade) toxic chemicals exist in nature. Accordingly, many chemicals released into the environment pose minimal hazard simply because of their limited life span in the environment. Chemicals that have historically posed environmental hazards (e.g., dichlorodiphenyltrichloroethane (DDT), polychlorinated biphenyls [PCBs], 2,3,7,8-tetrachlorodibenzo-p-dioxin (TCDD)) resist degradative processes and accordingly persist in the environment for extremely long periods of time (Table 25.1). Trace metals represent an extreme case of persistence because as elements, metals cannot be broken down in the environment. Continued disposal of persistent chemicals into the environment can result in their accumulation to environmental levels sufficient to pose toxicity. Such chemicals can continue to pose hazard long

TABLE 25.1 Environmental Half-Life of Some Chemical Contaminants

Contaminant	Half-Life	Media
DDT	10 years	Soil
TCDD	9 years	Soil
Atrazine	25 months	Water
Benzoperylene (polycyclic aromatic hydrocarbon (PAH))	14 months	Soil
Phenanthrene (PAH)	138 days	Soil
Carbofuran	45 days	Water

after their disposal into the environment has ceased. For example, significant contamination of Lake Ontario by the pesticide mirex occurred from the 1950s through the 1970s. Mass balance studies performed 20 years later revealed that 80% of the mirex deposited into the lake persisted. One decade following the contamination of Lake Apopka, Florida with pesticides including DDT and diclofol, populations of alligators continued to experience severe reproductive impairment. Estuarine sediments in the United Kingdom are contaminated with metals that date back to tin mining by the Romans. Both biotic and abiotic processes contribute to the degradation of chemicals.

25.2.1 Abiotic Degradation

A plethora of environmental forces compromise the structural integrity of chemicals in the environment. Many prominent abiotic degradative processes occur due to the influences of light (photolysis) and water (hydrolysis).

Photolysis   Light, primarily in the ultraviolet range, has the potential to break chemical bonds and thus can contribute significantly to the degradation of some chemicals. Photolysis is most likely to occur in the atmosphere, on foliar surfaces, or in shallow surface waters where light intensity is greatest. Photolysis is dependent upon both the intensity of the light and the capacity of the pollutant molecules to absorb the light. Unsaturated aromatic compounds such as polycyclic aromatic hydrocarbons tend to be highly susceptible to photolysis due to their high capacity to absorb light energy. Light energy can also facilitate the oxygenation of environmental contaminants via hydrolytic or oxidative processes. The photooxidation of the organophosphorus pesticide parathion is depicted in Figure 25.1.

Hydrolysis   Water, often in combination with light energy or heat, can break chemical bonds. Hydrolytic reactions commonly result in the insertion of an oxygen atom into the molecule with the commensurate loss of some component of the molecule. Ester bonds, such as those found in organophosphate pesticides (e.g., parathion, Figure 25.1), are highly susceptible to hydrolysis, which dramatically lowers the environmental half-lives of these chemicals. Hydrolytic rates of chemicals are influenced by the temperature and pH of the aqueous media. Rates of hydrolysis increase with increasing temperature and with extremes in pH.

25.2.2 Biotic Degradation

While many environmental contaminants are susceptible to abiotic degradative processes, such processes often occur at extremely slow rates. Environmental degradation of chemical contaminants can occur at greatly accelerated rates through the action of microorganisms. Microorganisms (bacteria, archaea, and fungi) most frequently degrade organic and inorganic compounds by using them as electron donors, electron acceptors, or as sources of nutrients such as nitrogen or sulfur. These biotic degradative processes are enzyme mediated and typically occur at rates that far exceed abiotic degradation. Biotic degradative processes can lead to complete mineralization of chemicals to water, carbon dioxide, and basic inorganic constituents. Biotic degradation includes those processes associated with abiotic

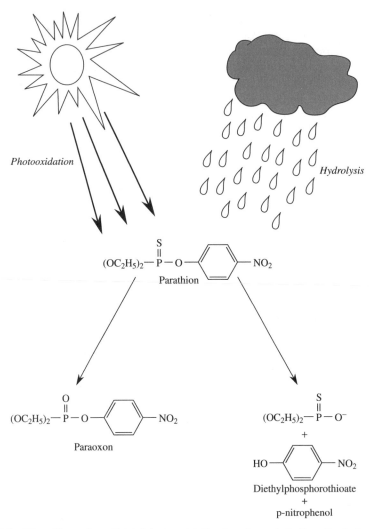

Figure 25.1 The effect of sunlight (photooxidation) and precipitation (hydrolysis) on the degradation of parathion.

degradation (e.g., hydrolysis and oxidation) and processes such as the removal of chlorine atoms (dehalogenation), the scission of ringed structures (ring cleavage), and the removal of carbon chains (dealkylation). The process by which microorganisms are used to facilitate the removal of environmental contaminants is called bioremediation.

25.2.3 Nondegradative Elimination Processes

Many processes that contribute to the regional elimination of a contaminant by altering its distribution are operative in the environment. Contaminants with sufficiently high vapor pressure can evaporate from contaminated terrestrial or aquatic compartments and can be transferred through the atmosphere to new locations. Such processes of global distillation are considered largely responsible for the worldwide distribution of relatively volatile organochlorine pesticides such as lindane and hexachlorobenzene. Entrainment by wind and upper atmospheric currents of contaminant particles or dust onto which the contaminants are sorbed also contributes to contaminant redistribution. Sorption of contaminants to suspended solids in an aquatic environment with commensurate sedimentation can result with the removal of the contaminants from the water column and their redistribution into bottom sediments. Sediment sorption of contaminants greatly reduces bioavailability since the propensity of a lipophilic chemical to partition from sediments to organisms is significantly less than its propensity to partition from water to organisms. Often the amount of organic carbon in sediments is tied to the bioavailability of contaminants. Due to its high affinity to many metals, sulfur can affect metal bioavailability in sediments as well. More highly water-soluble contaminants can be removed and redistributed through runoff and soil percolation. For example, the herbicide atrazine is one of the most abundantly used pesticides in the United States. It is used to control broadleaf and weed grasses in both agriculture and landscaping. Atrazine is ubiquitous in surface waters due to its extensive use. A study of Midwestern states revealed that atrazine was detectable in 92% of the reservoirs assayed. In addition, atrazine has the propensity to migrate into groundwater because of its relatively high water solubility and low predilection to sorb to soil particles. Indeed, field studies have shown that surface application of atrazine typically results in the contamination of the aquifer below the application site. A more detailed account of the fate of chemicals in the environment is presented in Chapter 26.

25.3 BIOACCUMULATION

Environmental persistence alone does not render a chemical problematic in the environment. If the chemical cannot enter the body of organisms, then it would pose no threat of toxicity (see Chapter 5). Once absorbed, the chemical must accumulate in the body to sufficient levels to elicit toxicity. Bioaccumulation is defined as the process by which organisms accumulate chemicals both directly from the abiotic environment (i.e., water, air, soil) and from dietary sources (trophic transfer). Many organic environmental chemicals are largely taken up by organisms by passive diffusion. Metals, on the other hand, cannot simply diffuse across plasma membranes. Instead, metals are usually transported by ion pumps or channels that otherwise would transport essential ions such as calcium or sodium. Primary sites of uptake include membranes of the lungs, gills, and gastrointestinal tract. While integument (skin) and associated structures (scales, feathers, fur, etc.) provide a protective barrier against many environmental insults, significant dermal uptake of some chemicals can occur. Because organic chemicals must traverse the lipid bilayer of membranes to enter the body, the bioaccumulation potential of chemicals is positively correlated with lipid solubility (lipophilicity).

The aquatic environment is the major site at which lipophilic chemicals traverse the barrier between the abiotic environment and the biota. This is because (a) lakes, rivers, and oceans serve as sinks for these chemicals; and (b) aquatic organisms pass tremendous quantities of water across their respiratory membranes (i.e., gills), allowing for the efficient extraction of the chemicals from the water. Aquatic organisms can bioaccumulate lipophilic chemicals and attain body concentrations that are several orders of magnitude greater than the concentration of the chemical found in the environment (Table 25.2). The degree to which aquatic organisms accumulate xenobiotics from the environment is largely dependent upon the lipid content of the organism, since body lipids serve as the primary site of retention of the chemicals (Figure 25.2).

TABLE 25.2   Bioaccumulation of Some Environmental Contaminants by Fish

Chemical	Bioaccumulation Factor[a]
DDT	127,000
TCDD	39,000
Endrin	6800
Pentachlorobenzene	5000
Lepthophos	750
Trichlorobenzene	183

Data derived from LeBlanc, G. A. *Environ. Sci. Technol.* **28**:154–160, 1994.
[a]Bioaccumulation factor is defined as the ratio of the chemical concentration in the fish and in the water at steady-state equilibrium.

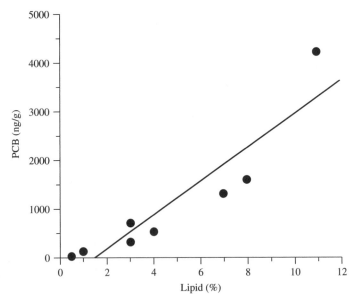

Figure 25.2 Relationship between lipid content of various organisms sampled from Lake Ontario and whole-body PCB concentration. Data derived from Oliver, B. G. and A. J. Niimi. *Environ. Sci. Technol.* **22**:388–397, 1988.

Most aquatic animals directly exchange ions with the surrounding water to maintain salt–water balance. Since many metals have affinities for ion transport systems (usually located on gills or other body surfaces), direct uptake of metals from solution can be an important route of exposure. Freshwater animals are considerably "saltier" than surrounding water, creating a situation where the tendency is to accumulate body water and to lose salts diffusively. (Diffusive salt loss occurs through paracellular channels, not across plasma membranes.) Therefore, freshwater animals are constantly faced with the task of excreting excess body water and of sequestering salts from the surrounding water. Very few freshwater animals drink the surrounding water. Saltwater animals face the opposite challenge. They must conserve their body water and excrete excess salts. To accomplish this, many saltwater organisms drink the surrounding water and produce a small volume of concentrated salty urine. Still, there is direct uptake of ions from the surrounding water to help compensate for salts lost in urine production.

Both organic chemicals and metals can also be transferred along food chains from prey organism to predator (trophic transfer). For highly lipophilic chemicals, this transfer can result in increasing concentrations of the chemical with each progressive link in the food chain (biomagnification). As depicted in Figure 25.3, a chemical that bioaccumulates by a factor of 2, regardless of whether the source of the contaminant is the water or food, would have the potential to magnify at each trophic level, leading to high levels in the birds of prey relative to that found in the abiotic environment. For many compounds, bioaccumulation is typically much greater from water than from food, and it is unlikely that an organism would accumulate a chemical to the same degree from both sources. For some elements such as selenium, bioaccumulation is thought to be primarily derived from the diet. The food chain transfer of DDT was responsible for the decline in many bird-eating raptor populations that contributed to the decision to ban the use of this pesticide in the United States. Methyl mercury and organic selenium species are examples of organometallic compounds that tend to biomagnify in food webs.

For many trace metals, there appears to be a major difference between fish and invertebrates in bioaccumulation patterns. Invertebrates tend to have very high assimilation efficiencies for metals from their diets, making the dietary route of exposure very important to understand in this group. Fish, on the other hand, tend to have poor assimilation efficiencies for many metals. This difference has practical implications in the setting of water quality criteria, because the susceptibility of invertebrates to metals is likely underestimated from toxicity tests that use only dissolved exposures.

Bioaccumulation of lipohilic compounds can lead to a delayed onset of toxicity since the toxicant may be initially sequestered in lipid deposits but is mobilized to target sites of toxicity when these lipid stores are utilized. For example, lipid stores are often mobilized in preparation for reproduction. The loss of the lipid can result in the release of lipophilic toxicants rendering them available for toxic action. Such effects can result in the mortality of adult organisms as they approach reproductive maturity. Lipophilic chemicals also can be transferred to the offspring in lipids associated with the yolk of oviparous organisms or with the milk of mammals, resulting in toxicity to the offspring that was not evident in the parental organisms.

Bioaccumulation of environmental chemicals

Figure 25.3 Bioaccumulation of a chemical along a generic food chain. In this simplistic paradigm, the amount of the chemical in the water is assigned an arbitrary concentration of 1, and it is assumed that the chemical will bioaccumulate either from the water to the fish or from one trophic level to another by a factor of 2. Circled numbers represent the concentration of chemical in the respective compartment. Numbers associated with arrows represent the concentration of chemical transferred from one compartment to another.

25.3.1 Factors That Influence Bioaccumulation

The propensity for an environmental contaminant to bioaccumulate is influenced by several factors. The first consideration is environmental persistence. The degree to which a chemical bioaccumulates is dictated by the concentration present in the environment. Contaminants that are readily eliminated from the environment will generally not be available to bioaccumulate. An exception would be instances where the contaminant is continuously introduced into the environment (e.g., receiving water of an effluent discharge).

As discussed above, lipophilicity is a major determinant of the bioaccumulation potential of a chemical. However, lipophilic chemicals also have greater propensity to sorb to sediments thus rendering them less available to bioaccumulate. For

TABLE 25.3 Measured and Predicted Bioaccumulation Factors in Fish of Chemicals That Differ in Susceptibility to Biotransformation

Chemical	Susceptibility to Biotransformation	Bioaccumulation Factor	
		Predicted	Measured
Chlordane	Low	47,900	38,000
PCB	Low	36,300	42,600
Mirex	Low	21,900	18,200
Pentachloro-phenol	High	4900	780
Tris(2,3-dibromo-propyl)phosphate	High	4570	3

Predicted bioaccumulation factors were based upon their relative lipophilicity as described by D. Mackay. *Environ. Sci. Technol.* **16**:274–278, 1982.

example, sorption of benzo[a]pyrene to humic acids reduced its propensity to bioaccumulate in sunfish by a factor of 3. Fish from oligotrophic lakes, having low suspended solid levels, have been shown to accumulate more DDT than fish from eutrophic lakes that have high suspended solid contents.

Once absorbed by the organism, the fate of the contaminant will influence its bioaccumulation. Chemicals that are readily biotransformed (Chapter 6) are rendered more water soluble and less lipid soluble. The biotransformed chemical is thus less likely to be sequestered in lipid compartments and is more likely to be eliminated from the body. As depicted in Table 25.3, chemicals that are susceptible to biotransformation bioaccumulate much less than would be predicted based upon lipophilicity. Conjugation of xenobiotics to glutathione and glucuronic acid (Chapter 6) can target the xenobiotic for biliary elimination through active transport processes, thus greatly increasing the rate of elimination (Chapter 9). Differences in chemical elimination rates contribute to species differences in bioaccumulation.

25.4 TOXICITY

25.4.1 Acute Toxicity

Acute toxicity is defined as toxicity elicited as a result of short-term exposure to a toxicant. Incidences of acute toxicity in the environment are commonly associated with accident (e.g., derailment of a train resulting in leakage of a chemical into a river) or with imprudent use of the chemical (e.g., aerial drift of a pesticide to nontarget areas). Discharge limits placed upon industrial and municipal wastes, when adhered to, have been generally successful in protecting against acute toxicity to organisms in waste-receiving areas. As discussed in Chapter 10, the acute toxicity of a chemical is commonly quantified as the median lethal concentration (LC_{50}) or median lethal dose (LD_{50}). These measures do not provide any insight into the environmentally acceptable levels of contaminants (a concentration that kills 50% of the exposed organisms is hardly acceptable). However, LC_{50} and LD_{50} values do provide statistically sound, reproducible measures of the relative acute toxicity of chemicals. LC_{50} and LD_{50} ranges for aquatic and terrestrial wildlife, respectively, and their interpretation are presented in Table 25.4.

TABLE 25.4   Ranking Scheme for Assessing the Acute Toxicity of Chemicals to Fish and Wildlife

Fish LC_{50} (mg/L)	Avian/Mammalian (LD_{50}, mg/kg)	Toxicity Rank	Example Contaminant
>100	>5000	Relatively nontoxic	Barium
10–100	500–5000	Moderately toxic	Cadmium
1–10	50–500	Very toxic	1,4-Dichlorobenzene
<1	<50	Extremely toxic	Aldrin

Acute toxicity of environmental chemicals is determined experimentally with select species that serve as representatives of particular levels of trophic organization within an ecosystem (e.g., mammal, bird, fish, invertebrate, vascular plant, algae). For example, the United States Environmental Protection Agency requires acute toxicity tests with representatives of at least eight different species of freshwater and marine organisms (16 tests) that include fish, invertebrates, and plants when establishing water quality criteria for a chemical. Attempts are often made to rank classes of organisms with respect to toxicant sensitivity; however, no organism is consistently more or less susceptible to the acute toxicity of chemicals. Further, the use of standard species in toxicity assessment presumes that these species are "representative" of the sensitivity of other members of that level of ecological organization. Such presumptions are often incorrect.

25.4.2   Mechanisms of Acute Toxicity

Environmental chemicals can elicit acute toxicity by many mechanisms. Provided below are example mechanisms that are particularly relevant to the types of chemicals that are more commonly responsible for acute toxicity in the environment at the present time.

Cholinesterase Inhibition   The inhibition of cholinesterase activity is characteristic of acute toxicity associated with organophosphate and carbamate pesticides (see Chapter 10 for more detail on cholinesterase inhibition). An inhibition of 40–80% of brain cholinesterase activity is typically reported in lethally poisoned fish. Acute toxicity resulting from cholinesterase inhibition is relatively common among incidents of acute poisoning of fish and birds due to the high-volume usage of organophosphates and carbamates in applications such as lawn care, agriculture, and golf course maintenance. Cholinesterase inhibition in fish may occur following heavy rains in aquatic habitats adjacent to areas treated with the pesticides and subject to runoff from these areas. Acute toxicity to birds commonly occurs in birds that feed in areas following application of the pesticides.

Narcosis   A common means by which industrial chemicals elicit acute toxicity, particularly to aquatic organisms, is through narcosis. Narcosis occurs when a chemical accumulates in cellular membranes interfering with the normal function of the membranes. Typical responses to the narcosis are decreased activity, reduced reaction to external stimuli, and increased pigmentation (in fish). The effects are reversible, and non-moribund organisms typically return to normal activity once the chemical is removed from the organism's environment. Prolonged narcosis can result in death. Approximately 60% of industrial chemicals that enter the aquatic environment elicit acute toxicity through narcosis. Chemicals that elicit toxicity via narcosis typically do not elicit toxicity at specific target sites but rather accumulate in the lipid phase or the lipid–aqueous interface of membranes at sufficient levels to disrupt membrane function. Chemicals that induce narcosis include alcohols, ketones, benzenes, ethers, and aldehydes.

Osmoregulatory Disturbance   Some metals have high affinity for ion transport systems. For example, cadmium and zinc have high affinity for calcium transport systems, whereas silver has a high affinity for sodium transporters. If the dissolved concentrations of these metals are high enough, it can lead to ionoregulatory disruption and acute toxicity.

Physical Effects   Perhaps the most graphic among recent incidents of environmental acute toxicity is the physical effects of petroleum following oil spills. Slicks of oil on the surface of contaminated waters result in the coating of animals, such as birds and marine mammals, which frequent the air–water interface. Such a spill of unprecedented magnitude and consequence in the United States occurred on March 24, 1989 when the hull of the *Exxon Valdez* was ruptured on Bligh Reef in Prince William Sound, Alaska. Nearly 11 million gallons of crude oil spilled onto the nearshore waters killing more wildlife than any prior oil spill in history. Thousands of seabirds and mammals succumbed to the acute effects of the oil.

Hypothermia is considered a major cause of death of oiled marine birds and mammals. These organisms insulate themselves from the frigid waters by maintaining a layer of air among the spaces within their coat of fur or feathers. The oil penetrates the fur/feather barrier and purges the insulating air. As a result, the animals rapidly succumb to hypothermia. In addition to hypothermia, these animals can also experience oil toxicosis. Inhalation of oil, as well as ingestion through feeding and preening, can result in the accumulation of hydrocarbons to toxic levels. Toxicity to sea otters has been correlated to degree of oiling and is characterized by pulmonary emphysema (bubbles of air within the connective tissue of the lungs), gastric hemorrhages, and liver damage.

25.4.3 Chronic Toxicity

Chronic toxicity is defined as toxicity elicited as a result of long-term exposure to a toxicant. Sublethal end points are generally associated with chronic toxicity. These include reproductive, immune, endocrine, and developmental dysfunction. However, chronic exposure also can result in direct mortality not observed during acute exposure. For example, chronic exposure of highly lipophilic chemicals can result in the eventual bioaccumulation of the chemical to concentrations that are lethal to the organisms, or as discussed previously, mobilization of lipophilic toxicants from lipid compartments during reproduction may result in lethality. It is important to recognize that, while theoretically, all chemicals elicit acute toxicity at sufficiently

TABLE 25.5   Acute and Chronic Toxicity of Pesticides Measured from Laboratory Exposures of Fish Species

Pesticide	LC_{50} (μg/L)	Acute Toxicity	Chronic Value (μg/L)	ACR	Chronic Toxicity
Endosulfan	166	Extremely toxic	4.3	39.0	Yes
Chlordecone	10	Extremely toxic	0.3	33.0	Yes
Malathion	3000	Very toxic	340.0	8.8	No
Carbaryl	15,000	Moderately toxic	378.0	40.0	Yes

high doses, all chemicals are not chronically toxic. Chronic toxicity is measured by end points such as the highest level of the chemical that does not elicit toxicity during continuous, prolonged exposure (no observed effect level [NOEL]), the lowest level of the chemical that elicits toxicity during continuous, prolonged exposure (lowest observed effect level [LOEL]), or the chronic value (CV), which is the geometric mean of the NOEL and the LOEL. Chronic toxicity of a chemical is often judged by the acute:chronic ratio (ACR), which is calculated by dividing the acute LC_{50} value by the CV. Chemicals that have an ACR of less than 10 typically have low-to-no chronic toxicity associated with them (Table 25.5).

The mechanisms underlying the chronic toxicity of metals are poorly understood, but three major possibilities exist. First, many metals directly generate reactive oxygen species or scavenge thiols that help mediate oxidative damage in the cell. Second, because of the affinity of some metals to sulfur, and the importance of sulfur in many proteins, exposure to some metals may affect protein tertiary structure and function. Third, the displacement of essential metals in metalloenzymes with exogenous toxic metals can render them ineffective and can cause toxicity. This third mechanism is ripe for further study, as metals play a role in approximately one-third of all known enzymes.

The following must always be considered when assessing the chronic toxicity of a chemical:

A. Simple numerical interpretations of chronic toxicity based upon ACRs serve only as gross indicators of the potential chronic toxicity of the chemical. Laboratory exposures designed to establish CVs most often focus upon a few general end points such as survival, growth, and reproductive capacity. Examination of more subtle end points of chronic toxicity may reveal significantly different CVs.

B. Laboratory exposures are conducted with a few test species that are amenable to laboratory manipulation. The establishment of chronic and ACR values with these species should not be considered absolute. Toxicants may elicit chronic toxicity in some species and not in others.

C. Interactions among abiotic and biotic components of the environment may contribute to the chronic toxicity of chemicals, while such interactions may not occur in laboratory assessments of direct chemical toxicity. These considerations are exemplified in the following incidence of chronic toxicity of chemicals in the environment.

TABLE 25.6 Toxicity of Tributyltin to Aquatic Organisms

Species	Acute Toxicity (LC_{50}, μg/L)	Chronic Toxicity (LOEL, μg/L)	Imposex (μg/L)
Daphnid	1.7	—	—
Polychaete worm	—	0.10	—
Copepod	1.0	0.023	—
Oyster	1.3	0.25	—
Dogwhelk	—	—	≤0.0010

25.4.4 Species-Specific Chronic Toxicity

Tributyltin-Induced Imposex in Neogastropods Scientists noted, in the early 1970s, that dogwhelks inhabiting the coast of England exhibited a hermaphroditic-like condition whereby females possessed a penis in addition to normal female genitalia. While hermaphroditism is a reproductive strategy utilized by some molluscan species, dogwhelks are dioecious. This pseudohermaphroditic condition, called imposex, has since been documented worldwide in over 190 species of neogastropods. Imposex has been implicated in the reduced fecundity of neogastropod populations, population declines, and local extinction of affected populations.

The observation that imposex occurred primarily in marinas suggested causality with some contaminant originating from such facilities. Field experiments demonstrated that neogastropods transferred from pristine sites to marinas often developed imposex. Laboratory studies eventually implicated tributyltin, a biocide used in marine paints, as the cause of imposex. Tributyltin is toxic to most marine species evaluated in the laboratory at low parts-per-billion concentrations (Table 25.6). However, exposure of neogastropods to low parts-per-trillion concentrations can cause imposex (Table 25.6). Thus, neogastropods are uniquely sensitive to the toxicity of tributyltin with effects produced that were not evident in standard laboratory toxicity characterizations.

25.4.5 Abiotic and Biotic Interactions

Chlorofluorocarbons-Ozone-ultraviolet radiation-B Radiation–Amphibian Interactions The atmospheric release of chlorofluorocarbons has been implicated in the depletion of the earth's stratospheric ozone layer, which serves as a filter against harmful ultraviolet radiation. Temporal increases in UV-B radiation have been documented and pose increasing risks of a variety of maladies to both plant and animal life.

Commensurate with the increase in UV-B radiation levels at the earth's surface has been the decline in many amphibian populations. Multiple causes may be responsible for these declines including loss of habitat, pollutants, and increased incidence of disease; however, some studies suggest that increases in UV-B radiation may be a major contributor to the decline in some populations. Field surveys in the Cascade Mountains, Oregon, revealed a high incidence of mortality among embryos of the Cascades frog and the Western toad. Incubation of eggs, collected from the environment, in the laboratory along with the pond water in which the eggs were collected resulted in low mortality, suggesting that contaminants or

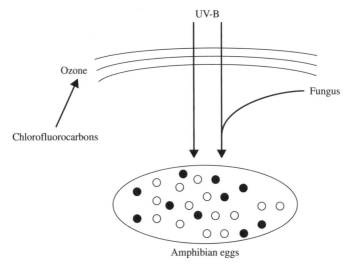

Figure 25.4   Abiotic and biotic interactions leading to the indirect toxicity of chlorofluoro-carbons to amphibians. The atmospheric release of chlorofluorocarbons causes the depletion of the stratospheric ozone layer (abiotic–abiotic interaction). Depleted ozone allows for increased penetration of UV-B radiation (abiotic–abiotic interaction). UV-B radiation alone and in combination with fungus (abiotic–biotic interaction) cause increased mortality of amphibian embryos.

disease organisms in the water were not directly responsible for the mortality. Furthermore, placement of UV-B filters over the embryos, incubated under ambient environmental conditions, significantly increased viability of the embryos.

Several amphibian species were examined for photolyase activity. This enzyme is responsible for the repair of DNA damage caused by UV-B radiation. A >80-fold difference in photolyase activity was observed among the species examined. Photolyase activity was appreciably lower in species known to be experiencing population decline as compared to species showing stable population levels. Recent studies have also suggested that ambient UV-B radiation levels can enhance the susceptibility of amphibian embryos to mortality originating from fungal infection.

These observations suggest that chlorofluorocarbons may be contributing to the decline in amphibian populations. However, this toxicological effect is the result of abiotic interactions (i.e., chlorofluorocarbons depleting atmospheric ozone levels, which increase UV-B radiation penetration resulting in embryo mortality) (Figure 25.4). In addition, abiotic (UV-B) and biotic (fungus) interactions may also be contributing to the toxicity. Such effects would not be predicted from direct laboratory assessments of the toxicity of chlorofluorocarbons to amphibians and highlight the necessity to consider possible indirect toxicity associated with environmental contaminants.

Masculinization of Fish due to Microbial Interactions with Kraft Pulp Mill Effluent   Field surveys of mosquito fish populations in the state of Florida revealed populations containing females that exhibited male traits such as male-type mating

Figure 25.5 Indirect toxicity of kraft pulpmill effluent to mosquito fish. Phytosterols in the mill effluent are converted to C19 steroidal androgens through the action of microorganisms in the environment. These androgens masculinize both anatomy and behavior of female mosquito fish. An arrow identifies the modified anal fin on the masculinized female.

behavior and the modification of the anal fin to resemble the sperm-transmitting gonopodium of males. Masculinized females were found to occur downstream of kraft pulp mill effluents, suggesting that components of the effluent were responsible for the masculinizing effect. Direct toxicity assays performed with the effluent did not produce such effects. However, the inclusion of microorganisms along with the effluent resulted in masculinization. Further studies revealed that phytosterols present in the kraft pulp mill effluent can be converted to androgenic C19 steroids by microorganisms, and these steroids are capable of masculinizing female fish (Figure 25.5). Thus, abiotic (phytosterols)–biotic (microorganisms) interactions in the environment must occur before this occult toxicity associated with the kraft pulp mill effluent is unveiled.

Environmental Contaminants and Disease among Marine Mammals

Worldwide, massive mortality has occurred over the past 20 years among populations of harbor seals, bottlenose dolphins, and other marine mammals. In many instances, this mortality has been attributed to disease. For example, nearly 18,000 harbor seals died in the North, Irish, and Baltic seas in the late 1980s due to phocine distemper virus. Incidences of the disease outbreak were highest in areas containing high levels of pollutants, and seals that succumbed to the disease were found to have high tissue levels of PCBs. PCBs and other organochlorine chemicals such as DDT, hexachlorobenzene, and dieldrin have been shown to immunosuppress laboratory animals, and accumulation of these chemicals by the seals may have increased their susceptibility to the virus. This hypothesis was tested by feeding fish, caught either from a relatively pristine area or from a polluted coastal area, to seals for 93 weeks then by assessing the integrity of the immune system in the seals. Seals fed with the contaminated fish did indeed have impaired immune responses, lending credence to the hypothesis that organochlorine contaminants in the marine environment are rendering some species immunodeficient. Mortality occurs not as a direct result of chemical toxicity but due to increased susceptibility to pathogens.

25.5 CONCLUSION

Environmental toxicologists have learned a great deal about the effects of chemicals in the environment and the characteristics of chemicals that are responsible for the hazards they pose. Much of the information gained has been due to retrospective analyses of the environmental consequences of environmental contamination. Such analyses have resulted in curtailing the release of demonstrated hazardous chemicals into the environment and have provided benchmark information upon which the regulation of chemicals proposed for release into the environment can be based. The recognition that environmentally hazardous chemicals commonly share characteristics of persistence, potential to bioaccumulate, and high toxicity has resulted in the development and use of chemicals that lack one or more of these characteristics yet fulfill societal needs previously served by hazardous chemicals. For example, recognition that persistence and propensity to bioaccumulate was largely responsible for the environmental hazards posed by many organochlorine pesticides led to the development and use of alternative classes of pesticides such as organophosphates, carbamates, and pyrethroids. While these chemicals all possess the toxicity necessary to function as pesticides, their lack of persistence and reduced propensity to bioaccumulate makes them more suitable for use in the environment.

Such advances in our understanding of the fate and effects of chemicals in the environment do not imply that the role of environmental toxicologists in the twenty-first century will diminish. A dearth of information persists in areas vital to the continued protection of natural resources against chemical insult. These include understanding (1) the unique susceptibilities of key species to the toxicity of different classes of chemicals, (2) the interactions of chemical contaminants with abiotic components of the environment that lead to increased toxicity, (3) the toxicological consequences of exposure to complex chemical mixtures, and (4) the consequences of toxicant effects on individuals with respect to ecosystem viability. Additionally, continued research is needed to develop molecular and cellular biomarkers of toxicant exposure and effect that could be used to predict dire consequences to the ecosystem before such effects are manifested at higher levels of biological organization. The role of the environmental toxicologist undoubtedly will increase in prospective activities aimed at reducing the risk associated with chemical contaminants in the environments before problems arise and hopefully will decrease with respect to assessing damage caused by such environmental contaminants.

BIBLIOGRAPHY AND SUGGESTED READING

Banerjee, S. and G. A. Baughman. Bioconcentration factors and lipid solubility. *Environ. Sci. Technol.* **25**:536–539, 1991.

Blaustein, A. R., P. D. Hoffman, D. G. Hokit. UV repair and resistance to solar UV-B in amphibian eggs: A link to population declines? *Proc. Natl. Acad. Sci. USA.* **91**: 1791–1795, 1994.

Colborn, T. and C. Clement, eds. *Chemically-Induced Alterations in Sexual and Functional Development: The Wildlife/Human Connection.* Princeton, NJ: Princeton Scientific, 1992, **403** pp.

Hoffman, D. J., J. Cairns, B. A. Rattner, et al., eds. *Handbook of Ecotoxicology.* New York: Taylor & Francis, 2002.

Landis, W. G. and M.-H. Yu. *Introduction to Environmental Toxicology.* New York: Lewis Publishers, 2003.

LeBlanc, G. A. Trophic-level differences in the bioconcentration of Chemicals. *Environ. Sci. Technol.* **29**:154–160, 1994.

Mackay, D. Correlation of bioconcentration factors. *Environ. Sci. Technol.* **16**:274–278, 1982.

Oliver, B. G. and A. J. Niimi. Trophodynamic analysis of polychlorinated biphenyl congeners and other chlorinated hydrocarbons in the Lake Ontario ecosystem. *Environ. Sci. Technol.* **22**:388–397, 1988.

van Wezel, A. P. and A. Opperhuizen. Narcosis due to environmental pollutants in aquatic organisms: Residue-based toxicity, mechanisms, and membrane burdens. *Crit. Rev. Toxicol.* **25**:255–279, 1995.

Walker, C. H., S. P. Hopkin, R. M. Sibly. *Principles of Ecotoxicology.* New York: Taylor & Francis, 2006.

SAMPLE QUESTIONS

1. What is the major chemical characteristic that is responsible for the bioaccumulation of organic chemicals?

2. What impact does biotransformation have on bioaccumulation? Why?

3. What three characteristics tend to be shared by chemicals that historically have posed toxicological problems in the environment?

4. Is tributyltin a problematic chemical due to its acute toxicity or chronic toxicity? What adverse effect of tributyltin (TBT) occurs at nanogram per liter exposure concentrations?

Transport and Fate of Toxicants in the Environment

DAMIAN SHEA

26.1 INTRODUCTION

More than 100,000 chemicals are released into the global environment every year through their normal production, use, and disposal. To understand and predict the potential risk that this environmental contamination poses to humans and wildlife, we must couple our knowledge on the toxicity of a chemical to our knowledge on how chemicals enter into and behave in the environment. The simple box model shown in Figure 26.1 illustrates the relationship between a toxicant source, its fate in the environment, its effective exposure or dose, and resulting biological effects. A *prospective* or *predictive* assessment of a chemical hazard would begin by characterizing the source of contamination, modeling the chemical's fate to predict exposure, and using exposure/dose response functions to predict effects (moving from left to right in Figure 26.1). A common application would be to assess the potential effects of a new waste discharge. A *retrospective* assessment would proceed in the opposite direction starting with some observed effect and reconstructing events to find a probable cause. Assuming that we have reliable dose/exposure response functions, the key to successful use of this simple relationship is to develop a qualitative description and quantitative model of the sources and fate of toxicants in the environment.

Toxicants are released into the environment in many ways, and they can travel along many pathways during their lifetime. A toxicant present in the environment at a given point in time and space can experience three possible outcomes: it can be *stationary* and add to the toxicant inventory and exposure at that location; it can be *transported* to another location; or it can be *transformed* into another chemical species. Environmental contamination and exposure resulting from the use of a chemical is modified by the transport and transformation of the chemical in the environment. Dilution and degradation can attenuate the source emission, while processes that focus and accumulate the chemical can magnify the source emission. The actual fate of a chemical depends on the chemical's use pattern and

---

A Textbook of Modern Toxicology, Fourth Edition. Edited by Ernest Hodgson
Copyright © 2010 John Wiley & Sons, Inc.

Figure 26.1 Environmental fate models are used to help determine how the environment modifies exposure resulting from various sources of toxicants.

physical–chemical properties, combined with the characteristics of the environment to which it is released.

Conceptually and mathematically, the transport and fate of a toxicant in the environment is very similar to that in a living organism. Toxicants can enter an organism or environmental system by many routes (e.g., dermal, oral, and inhalation vs. smoke stack, discharge pipe, or surface runoff). Toxicants are redistributed from their point of entry by fluid dynamics (blood flow vs. water or air movement) and intermedia transport processes such as partitioning (blood–lipid partitioning vs. water–soil partitioning) and complexation (protein binding vs. binding to natural organic matter). Toxicants are transformed in both humans and the environment to other chemicals by reactions such as hydrolysis, oxidation, and reduction. Many enzymatic processes that detoxify and activate chemicals in humans are very similar to microbial biotransformation pathways in the environment.

In fact, physiologically based pharmacokinetic models are similar to environmental fate models. In both cases, we divide a complicated system into simpler compartments, estimate the rate of transfer between the compartments, and estimate the rate of transformation within each compartment. The obvious difference is that environmental systems are inherently much more complex because they have more routes of entry, more compartments, more variables (each with a greater range of values), and a lack of control over these variables for systematic study. The discussion that follows is a general overview of the transport and transformation of toxicants in the environment in the context of developing qualitative and quantitative models of these processes.

26.2   SOURCES OF TOXICANTS TO THE ENVIRONMENT

Environmental sources of toxicants can be categorized as either *point sources* or *nonpoint sources* (Figure 26.2). Point sources are discrete discharges of chemicals that are usually identifiable and measurable, such as industrial or municipal effluent outfalls, chemical or petroleum spills and dumps, smokestacks, and other stationary atmospheric discharges. Nonpoint sources are more diffuse inputs over large areas with no identifiable single point of entry such as agrochemical (pesticide and fertilizer) runoff, mobile source emissions (automobiles), atmospheric deposition, desorption or leaching from very large areas (contaminated sediments or mine tailings), and groundwater inflow. Nonpoint sources often include multiple smaller point sources, such as septic tanks or automobiles, which are impractical to consider on an individual basis. Thus, the identification and characterization of a source is relative to the environmental compartment or system being considered. For example, there may be dozens of important toxicant sources to a river; each must be considered when assessing the hazards of toxicants to aquatic life in the river or to humans who might drink the water or consume the fish and shellfish. However, these toxicant sources can be well mixed in the river resulting in a rather homogeneous and large point source to a downstream lake or estuary (Figure 26.2).

Figure 26.2 Toxicants enter the environment through many point and nonpoint sources.

The rate (units of gram per hour) at which a toxicant is emitted by a source (*mass emission rate*) can be estimated from the product of the toxicant concentration in the medium (gram per cubic meter) and the flow rate of the medium (cubic meter per hour). This would appear to be relatively simple for point sources, particularly ones that are routinely monitored to meet environmental regulations. However, the measurement of trace concentrations of chemicals in complex effluent matrices is not a trivial task (see Chapter 24). Often the analytical methods prescribed by environmental agencies for monitoring are not sensitive or selective enough to measure important toxicants or their reactive metabolites. Estimating the mass emission rates for nonpoint sources is usually very difficult. For example, the atmospheric deposition of toxicants to a body of water can be highly dependent on both space and time, and high annual loads can result from continuous deposition of trace concentrations that are difficult to measure. The loading of pesticides from an agricultural field to an adjacent body of water also varies with time and space as shown in Figure 26.3 for the herbicide atrazine. Rainfall following the application of atrazine results in drainage ditch loadings more than 100-fold higher than just 2 weeks following the rain. A much smaller but longer-lasting increase in atrazine loading occurs at the edge of the field following the rain. Again, we see the need to define the spatial scale of concern when identifying and characterizing a source. If one is concerned with the fate of atrazine within a field, the source is defined by the

Figure 26.3   The loading of atrazine from an agricultural field to an adjacent body of water is highly dependent on rainfall and on the presence of drainage ditches that collect the chemical and focus its movement in the environment.

application rate. If one is concerned with the fate and exposure of atrazine in an adjacent body of water, the source may be defined as the drainage ditch and/or as runoff from the edge of field. In the latter case, one either needs to take appropriate measurements in the field or to model the transport of atrazine from the field.

26.3 TRANSPORT PROCESSES

Following the release of a toxicant into an environmental compartment, transport processes will determine its spatial and temporal distribution in the environment. The transport medium (or fluid) is usually either air or water, while the toxicant may be in dissolved, gaseous, condensed, or particulate phases. We can categorize physical transport as either *advection* or *diffusion*.

26.3.1 Advection

Advection is the passive movement of a chemical in bulk transport media either within the same medium (intraphase or homogeneous transport) or between different media (interphase or heterogeneous transport). Examples of homogeneous advection include transport of a chemical in air on a windy day or a chemical dissolved in water moving in a flowing stream, in surface runoff (nonpoint source), or in a discharge effluent (point source). Examples of heterogeneous advection include the deposition of a toxicant sorbed to a suspended particle that settles to bottom sediments, atmospheric deposition to soil or water, and even ingestion of contaminated particles or food by an organism (i.e., bioaccumulation). Advection takes place independently from the presence of a chemical; the chemical is simply going along for the ride. Advection is not influenced by diffusion and can transport a chemical either in the same or opposite direction as diffusion. Thus, advection is often called *nondiffusive transport*.

Homogeneous Advection The homogeneous advective transport rate (N, g/h) is simply described in mathematical terms by the product of the chemical concentration in the advecting medium (C, g/m^3) and the flow rate of the medium (G, m^3/h):

$$N = GC.$$

For example, if the flow of water out of a lake is 1000 m^3/h and the concentration of the toxicant is 1 μg/m^3, then the toxicant is being advected from the lake at a rate of 1000 μg/h (or 1 mg/h). The emission rates for many toxicant sources can be calculated in the same way.

As with source emissions, advection of air and water can vary substantially with time and space within a given environmental compartment. Advection in a stream reach might be several orders of magnitude higher during a large rain event compared with a prolonged dry period, while at one point in time, advection within a stagnant pool might be several orders of magnitude lower than a connected stream. Thus, as with source characterization, we must match our estimates of advective transport to the spatial and temporal scales of interest. Again, a good example is the movement of atrazine from an agricultural field (Figure 26.3). Peak flow advective rates that follow the rain might be appropriate for assessing acute toxicity during peak flow periods but not for estimating exposure at other times of the year. Conversely, an annual mean advective rate would underestimate exposure during peak flow but would be more appropriate for assessing chronic toxicity.

In surface waters, advective currents often dominate the transport of toxicants and they can be estimated from hydrodynamic models or current measurements. In many cases, advective flow can be approximated by the volume of water exchanged per unit of time by assuming conservation of mass and by measuring flow into or out of the system. This works only for well-mixed systems that have no or only small volumes of stagnant water. In water bodies that experience density stratification (i.e., thermocline), separate advective models or residence times can be used for each water layer. In air, advection also dominates the transport of chemicals, with air currents being driven by pressure gradients. The direction and magnitude of air velocities are recorded continuously in many areas, and daily, seasonal, or annual means can be used to estimate advective air flow.

Advective air and water currents are much smaller in soil systems but still influence the movement of chemicals that reside in soil. Advection of water in the saturated zone is usually solved numerically from hydrodynamic models. Advection of air and water in the unsaturated zone is complicated by the heterogeneity of these soil systems. Models are usually developed for specific soil property classes, and measurements of these soil properties are made at a specific site to determine which soil model layers to link together.

Heterogeneous Advection Heterogeneous advective transport involves a secondary phase within the bulk advective phase, such as when a particle in air or water acts as a carrier of a chemical. In many cases, we can treat heterogeneous advection the same as homogeneous advection if we know the flow rate of the secondary phase and the concentration of chemical in the secondary phase. Using the lake example above, if the volume fraction of suspended particles in the lake water is 10^{-5}, the flow rate of suspended particles is $0.01\,m^3/h$, and the concentration of the toxicant in the solid particles is $100\,mg/m^3$, then the advective flow of the toxicant on suspended particles would be $1\,mg/h$ or the same as the homogeneous advection via water. Although the flow rate of particles is much lower than that of water, the concentration of the toxicant is much higher in the suspended particles than dissolved in the water. This is typical of a hydrophobic toxicant such as dichlorodiphenyltrichloroethane (DDT) or benzo[*a*]pyrene. In soil and sedimentary systems, colloidal particles (often macromolecular detritus) can play a very important role in heterogeneous advective transport because they have greater mobility than larger particles, and they often have greater capacity to sorb many toxicants because of their higher organic carbon content and higher surface area/mass ratio. In highly contaminated sites, organic co-solvents can be present in the water (usually groundwater) and can act as a high-capacity and high-efficiency carrier of toxicants through heterogeneous advection in the water.

Unfortunately, the dynamics of heterogeneous transport are rarely simple, particularly over shorter scales of time and space. In addition to advection of particles with flowing water, aqueous-phase heterogeneous transport also includes particle settling, resuspension, burial in bottom sediments, and mixing of bottom sediments. Particle settling can be an important mechanism for transporting hydrophobic toxicants from the water to the bottom sediments. Modeling this process can be as simple as using an overall mass transfer coefficient or can include rigorous modeling of particles with different size, density, and organic carbon content. Estimates of particle settling are usually obtained through the use of laboratory settling chambers, *in situ* sediment traps, or by calculation using Stoke's law. Resuspension of bottom sediments occurs when sufficient energy is transferred to the sediment bed from advecting water, internal waves, boats, dredging, fishing, and the movement of sediment-dwelling organisms (i.e., bioturbation). Resuspension rates are difficult to measure and often are highly variable in both time and space. Much as the annual runoff of pesticides from an agricultural field may be dominated by a few rain events, annual resuspension rates can be dominated by a major storm, and in smaller areas by a single boat or a school of bottom fish. Resuspension rates can be estimated from sediment traps deployed just above the sediment surface or from the difference between particle settling and permanent burial or sedimentation. Sedimentation is the net result of particle settling and resuspension and can be measured using radionuclide dating methods (e.g., ^{210}Pb). Sediment dating itself becomes difficult when there is significant mixing of the surface sediments (e.g., through bioturbation). Thus, the heterogeneous transport of toxicants on aqueous particles can be rather complicated, though many aquatic systems have been modeled reasonably well.

Heterogeneous advective transport in air occurs primarily through the absorption of chemicals into falling water droplets (wet deposition) or the sorption of chemicals into solid particles that fall to the earth's surface (dry deposition). Under certain conditions, both processes can be treated as a simple first-order advective transport using a flow rate and concentration in the advecting medium. For example, wet deposition is usually characterized by a washout coefficient, which is proportional to rainfall intensity.

26.3.2 Diffusion

Diffusion is the transport of a chemical by random motion due to a state of disequilibrium. For example, diffusion causes the movement of a chemical within a phase (e.g., water) from a location of relatively high concentration to a place of lower concentration until the chemical is homogeneously distributed throughout the phase. Likewise, diffusive transport will drive a chemical between media (e.g., water and air) until their equilibrium concentrations are reached and thus, the chemical potentials or fugacities are equal in each phase.

Diffusion within a Phase   Diffusional transport within a phase can result from random (thermal) motion of the chemical (molecular diffusion), the random turbulent mixing of the transport medium (turbulent diffusion), or a combination of both. Turbulent diffusion usually dominates the diffusive (but not necessarily the advective) chemical transport in air and water due to the turbulent motions or eddies that are common in nature. In porous media (sediment and soil), the water velocities are typically too low to create eddies, but random mixing still occurs as water tortuously flows around particles. This mechanical diffusion is often called dispersion by hydrologists, and dispersion on larger scales, such as when groundwater detours around large areas of less permeable soil, is called macrodispersion.

Note, however, that the term dispersion often is used by meteorologists and engineers to describe any turbulent diffusion.

Although different physical mechanisms can cause diffusive mixing, they all cause a net transport of a chemical from areas of higher concentration to areas of lower concentration. All diffusive processes are also referred to as *Fickian* transport because they all can be described mathematically by Fick's first law, which states that the flow (or flux) of a chemical (N, g/h) is proportional to its concentration gradient (dC/dx):

$$N = -DA(dC/dx),$$

where D is the diffusivity or the mass transfer coefficient (m^2/h), A is the area through which the chemical is passing (m^2), C is the concentration of the diffusing chemical (g/m^3), and x is the distance being considered (m). The negative sign is simply the convention that the direction of diffusion is from high to low concentration (diffusion is positive when dC/dx is negative). Note that many scientists and texts define diffusion as an area-specific process with units of gram per square meter hour (g/m^2h), and thus the area term (A) is not included in the diffusion equation. This is simply an alternative designation that describes transport as a flux density (g/m^2h) rather than as a flow (gram per hour). In either case, the diffusion equation can be integrated numerically and can even be expressed in three dimensions using vector notation. However, for most environmental situations, we usually have no accurate estimate of D or dx, so we combine the two into a one-dimensional mass transfer coefficient (k_M) with units of velocity (meter per hour). The chemical flux is then the product of this velocity, area, and concentration:

$$N = -k_M AC.$$

Mass transfer coefficients can be estimated from laboratory, mesocosm, and field studies and are widely used in environmental fate models. Mass transfer coefficients can be derived separately for molecular diffusion, turbulent diffusion, and dispersion in porous media, and all three terms can be added to the chemical flux equation. This is usually not necessary because one term often dominates the transport in specific environmental regions. Consider the vertical diffusion of methane gas generated by methanogenic bacteria in deep sediments. Molecular diffusion dominates in the highly compacted and low-porosity deeper sediments. Dispersion becomes important as methane approaches the more porous surface sediments. Following methane gas ebulation from the sediment pore water, turbulent diffusion will dominate transport in a well-mixed water column (i.e., not a stagnant pool or beneath a thermocline where molecular diffusion will dominate). At the water surface, eddies tend to be damped and molecular diffusion may again dominate transport. Under stagnant atmospheric conditions (i.e., a temperature inversion), molecular diffusion will continue to dominate but will yield to more rapid mixing when typical turbulent conditions are reached. The magnitude and variability of the transport rate generally increase as the methane moves vertically through the environment, except when very stagnant conditions are encountered in the water or in air. Modeling the transport of a chemical in air is particularly difficult because of the high spatial and temporal variability of air movement. Note also that advective processes in water or air usually transport chemicals at a faster rate than either molecular or turbulent diffusion.

Diffusion between Phases   The transport of a chemical between phases is sometimes treated as a third category of transport processes or even as a transformation reaction. Interphase or intermedia transport is not a transformation reaction because the chemical is moving only between phases; it is not reacting with anything or changing its chemical structure. Instead, intermedia transport is simply driven by diffusion between two phases. When a chemical reaches an interface such as air–water, particle–water, or (biological) membrane–water, two diffusive regions are created at either side of the interface. The classical description of this process is the Whitman two-film or two-resistance mass transfer theory, where chemicals pass through two stagnant boundary layers by molecular diffusion, while the two bulk phases are assumed to be homogeneously mixed. This allows us to use a first-order function of the concentration gradient in the two phases, where the mass transfer coefficient will depend only on the molecular diffusivity of the chemical in each phase and the thickness of the boundary layers. This is fairly straightforward for transfer at the air–water interface (and often at the membrane–water interface), but not for the particle–water or particle–air interfaces.

Diffusive transport between phases can be described mathematically as the product of the departure from equilibrium and a kinetic term:

$$N = kA(C_1 - C_2 K_{12}),$$

where N is the transport rate (g/h), k is the transport rate coefficient (m/h), A is the interfacial area (m^2), C_1 and C_2 are the concentrations in the two phases, and K_{12} is the equilibrium partition coefficient. At equilibrium K_{12} equals C_1/C_2, so the term describing the departure from equilibrium $(C_1 - C_2 K_{12})$ becomes zero, and thus the net rate of transfer is also zero. The partition coefficients are readily obtained from thermodynamic data and equilibrium partitioning experiments. The transport rate coefficients are usually estimated from the transport rate equation itself by measuring intermedia transport rates (N) under controlled laboratory conditions (temperature, wind, and water velocities) at known values of A, C_1, C_2, and K_{12}. These measurements must then be extrapolated to the field, sometimes with great uncertainty. This uncertainty, along with the knowledge that many interfacial regions have reached or are near equilibrium, has led many to simply assume that equilibrium exists at the interface. Thus, the net transport rate is zero and the phase distribution of a chemical is simply described by its equilibrium partition coefficient.

26.4 EQUILIBRIUM PARTITIONING

When a small amount of a chemical is added to two immiscible phases and then shaken, the phases will eventually separate and the chemical will partition between the two phases according to its solubility in each phase. The concentration ratio at equilibrium is the partition coefficient:

$$C_1/C_2 = K_{12}.$$

In the laboratory, we usually determine K_{12} from the slope of C_1 versus C_2 over a range of concentrations. Partition coefficients can be measured for essentially any two-phase system: air–water, octanol–water, lipid–water, particle–water, and so on. *In situ* partition coefficients also can be measured where site-specific environmental conditions might influence the equilibrium phase distribution.

26.4.1   Air–Water Partitioning

Air–water partition coefficients ($K_{\text{air–water}}$) are essentially Henry's law constants (H):

$$K_{\text{air–water}} = H = C_{\text{air}}/C_{\text{water}},$$

where H can be expressed in dimensionless form (same units for air and water) or in units of pressure divided by concentration (e.g., Pa·m³/mol). The latter is usually written as

$$H' = P_{\text{air}}/C_{\text{water}},$$

where P_{air} is the partial vapor pressure of the chemical. When H is not measured directly, it can be estimated from the ratio of the chemical's vapor pressure and aqueous solubility, although one must be careful about using vapor pressures and solubilities that apply to the same temperature and phase. Chemicals with high Henry's law constants (such as alkanes and many chlorinated solvents) have a tendency to escape from water to air and typically have high vapor pressures, low aqueous solubilities, and low boiling points. Chemicals with low Henry's law constants (such as alcohols, chlorinated phenols, larger polycyclic aromatic hydrocarbons, lindane, atrazine) tend to have high water solubility and/or very low vapor pressure. Note that some chemicals that are considered to be "nonvolatile," such as DDT, are often assumed to have low Henry's law constants. However, DDT also has a very low water solubility yielding a rather high Henry's law constant. Thus, DDT readily partitions into the atmosphere as is now apparent from the global distribution of DDT.

26.4.2   Octanol–Water Partitioning

For many decades, chemists have been measuring the octanol–water partition coefficient (K_{OW}) as a descriptor of hydrophobicity or how much a chemical "hates" to be in water. It is now one of the most important and frequently used physicochemical properties in toxicology and environmental chemistry. In fact, toxicologists often simply use the symbol P, for partition coefficient, as if no other partition coefficient is important. Strong correlations exist between K_{OW} and many biochemical and toxicological properties. Octanol has a similar carbon:oxygen ratio as lipids and the K_{OW} correlates particularly well with lipid–water partition coefficients. This has led many to use K_{OW} as a measure of lipophilicity or how much a chemical "loves" lipids. This is really not the case because most chemicals have an equal affinity for octanol and other lipids (within about a factor of 10), but their affinity for water varies by many orders of magnitude. Thus, it is largely aqueous solubility that determines

K_{OW}, not octanol or lipid solubility. We generally express K_{OW} as log K_{OW} because K_{OW} values range from less than one (alcohols) to over one billion (larger alkanes and alkyl benzenes).

26.4.3 Lipid–Water Partitioning

In most cases, we can assume that the equilibrium distribution and partitioning of organic chemicals in both mammalian and nonmammalian systems is a function of lipid content in the animal and that the lipid–water partition coefficient (K_{LW}) is equal to K_{OW}. Instances where this is not the case include specific binding sites (e.g., kepone in the liver) and nonequilibrium conditions caused by slow elimination rates of higher-level organisms or structured lipid phases that sterically hinder accumulation of very hydrophobic chemicals. For aquatic organisms in constant contact with water, the bioconcentration factor or fish–water partition coefficient (K_{FW})is simply

$$K_{FW} = f_{lipid} K_{OW},$$

where f_{lipid} is the mass fraction of lipid in the fish (g lipid/g fish). Several studies have shown that this relationship works well for many fish and shellfish species, and an aggregate plot of K_{FW} versus K_{OW} for many different fish species yields a slope of 0.048, which is about the average lipid concentration of fish (5%). Again, nonequilibrium conditions will cause deviation from this equation. Such deviations are found at both the top and bottom of the aquatic food chain. Phytoplanktons can have higher apparent lipid–water partition coefficients because their large surface area : volume ratios increase the relative importance of surface sorption. Top predators such as marine mammals also have high apparent lipid–water partition coefficients because of very slow elimination rates. Thus, the deviations occur not because "there is something wrong with the equation," but because the underlying assumption of equilibrium is not appropriate in these cases.

26.4.4 Particle–Water Partitioning

It has been known for several decades that many chemicals preferentially associate with soil and sediment particles rather than the aqueous phase. The particle–water partition coefficient (K_P) describing this phenomenon is

$$K_P = C_S/C_W,$$

where C_S is the concentration of chemical in the soil or sediment (mg/kg dry weight) and C_W is the concentration in water (mg/L). Using this form, K_P has units of liter per kilogram or reciprocal density. Dimensionless partition coefficients are sometimes used where K_P is multiplied by the particle density (in kg/L). It has also been observed, first by pesticide chemists in soil systems and later by environmental engineers and chemists in sewage effluent and sediment systems, that nonionic organic chemicals were primarily associated with the organic carbon phase(s) of particles. A plot of K_P versus the mass fraction of organic carbon in the soil (f_{OC}, g/g) is linear with a near-zero intercept yielding the simple relationship

$$K_P = f_{OC} K_{OC},$$

where K_{OC} is the organic carbon–water partition coefficient (L/kg). Studies with many chemicals and many sediment/soil systems have demonstrated the utility of this equation when the fraction of organic carbon is about 0.5% or greater. At lower organic carbon fractions, interaction with the mineral phase becomes relatively more important (though highly variable), resulting in a small positive intercept of K_P versus f_{OC}. The strongest interaction between organic chemicals and mineral phases appears to be with dry clays. Thus, K_P will likely change substantially as a function of water content in low organic carbon clay soils.

Measurements of K_{OC} have been taken directly from partitioning experiments in sediment– and soil–water systems over a range of environmental conditions in both the laboratory and the field. Not surprisingly, the K_{OC} values for many organic chemicals are highly correlated with their K_{OW} values. Plots of the two partition coefficients for hundreds of chemicals with widely ranging K_{OW} values yield slopes from about 0.3 to 1.0, depending on the classes of compounds and the particular methods included. Most fate modelers continue to use a slope of 0.41, which was reported by the first definitive study on the subject in the early 1980s. Thus, we now have a means of estimating the partitioning of a chemical between a particle and water by using the K_{OW} and f_{OC}:

$$K_P = f_{OC} K_{OC} = f_{OC} 0.41 K_{OW},$$

This relationship is commonly used in environmental fate models to predict aqueous concentrations from sediment measurements by substituting the equilibrium expression for K_P and by rearranging to solve for C_W:

$$K_P = C_S/C_W = f_{OC} 0.41 K_{OW}$$
$$C_W = C_S/f_{OC} 0.41 K_{OW}.$$

This last equation forms the basis for the Environmental Protection Agency's sediment quality criteria that are to be used to assess the potential toxicity of contaminated sediments. The idea is to simply measure C_S and f_{OC}, look up K_{OW} in a table, compute the predicted C_W, and compare this result to established water quality criteria for the protection of aquatic life or human life (e.g., carcinogenicity risk factors). The use of this simple equilibrium partitioning expression for this purpose is currently the subject of much debate among both scientists and policymakers.

26.5 TRANSFORMATION PROCESSES

The potential environmental hazard associated with the use of a chemical is directly related to its persistence in the environment (see Chapter 25), which in turn depends on the rates of chemical transformation reactions. Transformation reactions can be divided into two classes: reversible reactions that involve continuous exchange among chemical states (ionization, complexation) and irreversible reactions that permanently transform a parent chemical into a daughter or reaction product (photolysis, hydrolysis, and many redox reactions). Reversible reactions are usually abiotic, although biological processes can still exert great influence over them (e.g., via production of complexing agents or a change in pH). Irreversible reactions can be abiotic or can be mediated directly by biota, particularly bacteria.

26.5.1 Reversible Reactions

Ionization   Ionization refers to the dissociation of a neutral chemical into charged species. The most common form of neutral toxicant dissociation is acid–base equilibria. The hypothetical monoprotic acid, HA, will dissociate in water to form the conjugate acid–base pair (H^+, A^-) usually written as

$$HA + H_2O = H_3O^+ + A^-.$$

The equilibrium constant for this reaction, the acidity constant (K_a), is defined by the law of mass action and is given by

$$K_a = [H_3O^+][A^-]/[HA].$$

For convenience, we often express equilibrium constants as the negative logarithm, or the pK value. Thus, the relative proportion of the neutral and charged species will be a function of the pK_a and solution pH. When the pH is equal to pK_a, equal concentrations of the neutral and ionized forms will be present. When pH is less than the pK_a, the neutral species will be predominant; when pH is greater than pK_a, the ionized species will be in excess. The exact equilibrium distribution can be calculated from the equilibrium expression above and the law of mass conservation.

The fate of a chemical is often a function of the relative abundance of a particular chemical species as well as the total concentration. For example, the neutral chemical might partition into biological lipids or organic carbon in soil to a greater extent than the ionized form. Many acidic toxicants (pentachlorophenol) exhibit higher toxicities to aquatic organisms at lower pH where the neutral species predominates. However, specific ionic interactions will take place only with the ionized species. A classic example of how pH influences the fate and effects of a toxicant is with hydrogen cyanide. The pK_a of HCN is about 9 and the toxicity of CN^- is much higher than that of HCN for many aquatic organisms. Thus, the discharge of a basic (high pH) industrial effluent containing cyanide would pose a greater hazard to fish than a lower pH effluent (everything else being equal). The effluent could be treated to reduce the pH well below the pK_a according to the reaction

$$CN^- + H^+ = HCN_{(aq)},$$

thus reducing the hazard to the fish. However, HCN has a rather high Henry's law constant and will partition into the atmosphere:

$$HCN_{(aq)} = HCN_{(air)}.$$

This may be fine for the fish, but birds in the area and humans working at the industrial plant will now have a much greater exposure to HCN. Thus, both the fate and toxicity of a chemical can be influenced by simple ionization reactions.

Precipitation and Dissolution A special case of ionization is the dissolution of a neutral solid phase into soluble species. For example, the binary solid metal sulfide, CuS, dissolves in water according to

$$CuS(s) + H^+ = Cu^{2+} + HS^-.$$

The equilibrium constant for this reaction, the solubility product (K_{sp}), is given by

$$K_{sp} = [Cu^{2+}][HS^-]/[H^+].$$

The solubility product for CuS is very low ($K_{sp} = 10^{-19}$ as written) so that the presence of sulfide in water acts to immobilize Cu (and many other metals) and to reduce effective exposure. The formation of metal sulfides is important in anaerobic soil and sediment, in stagnant ponds and basins, and in many industrial and domestic sewage treatment plants and discharges. Co-precipitation of metals also can be a very important removal process in natural waters. In aerobic systems, the precipitation of hydrous oxides of manganese and iron often incorporates other metals as impurities. In anaerobic systems, the precipitation of iron sulfides can include other metals as well. These co-precipitates are usually not thermodynamically stable, but their conversion to stable mineral phases often takes place on geological timescales.

Complexation and Chemical Speciation Natural systems contain many chemicals that undergo ionic or covalent interactions with toxicants to change toxicant speciation, and chemical speciation can have a profound effect on both fate and toxicity. Again, using copper as an example, inorganic ions (Cl⁻, OH⁻) and organic detritus (humic acids, peptides) will react with dissolved Cu^{2+} to form various metal–ligand complexes. Molecular diffusivities of complexed copper will be lower than uncomplexed (hydrated) copper and will generally decrease with the size and number of ligands. The toxicity of free, uncomplexed Cu^{2+} to many aquatic organisms is much higher than Cu^{2+} that is complexed to chelating agents such as ethylenediaminetetraacetic acid (EDTA) or glutathione (GSH). Many transition metal toxicants, such as Cu, Pb, Cd, and Hg, have high binding constants with compounds that contain amine, sulfhydryl, and carboxylic acid groups. These groups are quite common in natural organic matter. Even inorganic complexes of OH⁻ and Cl⁻ reduce Cu^{2+} toxicity. In systems where a mineral phase is controlling Cu^{2+} solubility, the addition of these complexing agents will shift the solubility equilibrium according to Le Chatelier's principle as shown here for CuS and OH⁻, Cl⁻, and GSH:

$$CuS(s) + H^+ = Cu^{2+} + HS^- \quad K_{sp}$$
$$Cu^{2+} + OH^- = CuOH^+ \quad K_{CuOH}$$
$$Cu^{2+} + Cl^- = CuCl^+ \quad K_{CuCl}$$
$$Cu^{2+} + GSH = CuGS^+ + H^+ \quad K_{CuGS}.$$

Each successive complexation reaction "leaches" Cu^{2+} from the solid mineral phase, thereby increasing the total copper in the water but not affecting the concentration of (or exposure to) Cu^{2+}. These equilibria can be combined into one reaction:

$$4\,CuS(s) + 3H^+ + OH^- + Cl^- + GSH = Cu^{2+} + 4HS^- + CuOH^+ + CuCl^+ + CuGS^+,$$

and the overall equilibrium constant derived as shown:

$$K_{overall} = (4)\,K_{sp} \times K_{CuOH} \times K_{CuCl} \times K_{CuGS}$$
$$= [Cu^{2+}][CuOH^+][CuCl^+][CuGS^+][HS^-]^4/[H^+]^3[OH^-][Cl^-][GS^-].$$

A series of simultaneous equations can be derived for these reactions to compute the concentration of individual copper species, and the total concentration of copper, $[Cu]_T$, would be given by

$$[Cu]_T = [Cu^{2+}] + [CuOH^+] + [CuCl^+] + [CuGS^+].$$

Thus, the total copper added to a toxicity test or measured as the exposure (e.g., by atomic absorption spectroscopy) may be much greater than that which is available to an organism to induce toxicological effects.

Literally, hundreds of complex equilibria like this can be combined to model what happens to metals in aqueous systems. Numerous speciation models exist for this application, which include all of the necessary equilibrium constants. Several of these models include surface complexation reactions that take place at the particle–water interface. Unlike the partitioning of hydrophobic organic contaminants into organic carbon, metals actually form ionic and covalent bonds with surface ligands such as sulfhydryl groups on metal sulfides and oxide groups on the hydrous oxides of manganese and iron. Metals also can be biotransformed to more toxic species (e.g., conversion of elemental mercury to methylmercury by anaerobic bacteria) or less toxic species (oxidation of tributyl tin to elemental tin), or can be temporarily immobilized (e.g., via microbial reduction of sulfate to sulfide, which then precipitates as an insoluble metal sulfide mineral).

26.5.2 Irreversible Reactions

The reversible transformation reactions discussed above alter the fate and toxicity of chemicals, but they do not irreversibly change the structure or properties of the chemical. An acid can be neutralized to its conjugate base and vice versa. Copper can precipitate as a metal sulfide, dissolve and form a complex with numerous ligands, and later reprecipitate as a metal sulfide. Irreversible transformation reactions alter the structure and properties of a chemical forever.

Hydrolysis   Hydrolysis is the cleavage of organic molecules by reaction with water with a net displacement of a leaving group (X) with OH^-:

$$RX + H_2O = ROH + HX.$$

Hydrolysis is part of the larger class of chemical reactions called nucleophilic displacement reactions in which a nucleophile (electron-rich species with an unshared pair of electrons) attacks an electrophile (electron deficient) cleaving one covalent bond to form a new one. Hydrolysis is usually associated with surface waters but also takes place in the atmosphere (fogs and clouds), in groundwater, at the particle–water interface of soils and sediments, and in living organisms.

Hydrolysis can proceed through numerous mechanisms via attack by H_2O (neutral hydrolysis) or by acid (H^+) or base (OH^-) catalysis. Acid and base-catalyzed reactions proceed via alternative mechanisms that require less energy than neutral hydrolysis. The combined hydrolysis rate term is a sum of these three constituent reactions and is given by

$$d[RX]/dt = k_{obs}[RX] = k_a[H^+][RX] + k_n[RX] + k_b[OH^-][RX],$$

where $[RX]$ is the concentration of the hydrolyzable chemical; k_{obs} is the macroscopic observed hydrolysis rate constant; and k_a, k_n, and k_b are the rate constants for the acid-catalyzed, neutral, and base-catalyzed hydrolysis. If we assume that the hydrolysis can be approximated by first-order kinetics with respect to RX (which is usually true), the rate term is reduced to

$$k_{obs} = k_a[H^+] + k_n + k_b[OH^-].$$

Neutral hydrolysis is dependent only on water that is present in excess so k_n is a simple pseudo-first-order rate constant (with units t^{-1}). The acid- and base-catalyzed hydrolysis depends on the molar quantities of $[H^+]$ and $[OH^-]$, respectively, so k_a and k_b have units of $M^{-1}t^{-1}$. The observed or apparent hydrolysis half-life at a fixed pH is then given by

$$t_{1/2} = \ln 2/k_{obs}.$$

Compilations of hydrolysis half-lives at pH and temperature ranges encountered in nature can be found in many sources. Reported hydrolysis half-lives for organic compounds at pH 7 and 298 K range at least 13 orders of magnitude. Many esters hydrolyze within hours or days, whereas some organic chemicals will never hydrolyze. For halogenated methanes, which are common groundwater contaminants, half-lives range from about 1 year for CH_3Cl to about 7000 years for CCl_4. The half-lives of halomethanes follow the strength of the carbon–halogen bond with half-lives decreasing in the order F > Cl > Br. Small structural changes can dramatically alter hydrolysis rates. An example is the difference between tetrachloroethane ($Cl_2HC–CHCl_2$) and tetrachloroethene ($Cl_2C=CCl_2$), which have hydrolysis half-lives of about 0.5 and 10^9 years, respectively. In this case, the hydrolysis rate is affected by the C–Cl bond strength and the steric bulk at the site of nucleophilic substitution.

The apparent rate of hydrolysis and the relative abundance of reaction products is often a function of pH because alternative reaction pathways are preferred at different pH. Using halogenated hydrocarbons as an example, base-catalyzed hydrolysis will result in elimination reactions, while neutral hydrolysis will take place via nucleophilic displacement reactions. An example of the pH dependence of hydrolysis is illustrated by the base-catalyzed hydrolysis of the structurally similar insecticides DDT and methoxychlor. Under a common range of natural pH (5–8), the hydrolysis rate of methoxychlor is invariant, while the hydrolysis of DDT is about 15-fold faster at pH 8 compared with pH 5. Only at higher pH (>8) does the hydrolysis rate of methoxychlor increase. In addition, the major product of DDT hydrolysis throughout this pH range is the same 1,1-Dichloro-2,2-bis(p-chlorophenyl) ethylene (DDE), while the methoxychlor hydrolysis product shifts from the alcohol at pH 5–8 (nucleophilic substitution) to the dehydrochlorinated 1,1-bis(p-methoxyphenyl)-2,2-dichloroethylene (DMDE) at pH > 8 (elimination). This illustrates the necessity to conduct detailed mechanistic experiments as a function of pH for hydrolytic reactions.

Photolysis   Photolysis of a chemical can proceed either by direct absorption of light (direct photolysis) or by reaction with another chemical species that has been produced or excited by light (indirect photolysis). In either case, photochemical transformations such as bond cleavage, isomerization, intramolecular rearrangement, and various intermolecular reactions can result. Photolysis can take place wherever sufficient light energy exists, including the atmosphere (in the gas phase and in aerosols and fog/cloud droplets), surface waters (in the dissolved phase or at the particle–water interface), and in the terrestrial environment (on plant and soil/mineral surfaces).

Photolysis dominates the fate of many chemicals in the atmosphere because of the high solar irradiance. Near the earth's surface, chromophores such as nitrogen oxides, carbonyls, and aromatic hydrocarbons play a large role in contaminant fate in urban areas. In the stratosphere, light is absorbed by ozone, oxygen, organohalogens, and hydrocarbons with global environmental implications. The rate of photolysis in surface waters depends on light intensity at the air–water interface, the transmittance through this interface, and the attenuation through the water column. Open ocean waters ("blue water") can transmit blue light to depths of 150 m, while highly eutrophic or turbid waters might absorb all light within 1 cm of the surface.

Oxidation–Reduction Reactions   Although many redox reactions are reversible, they are included here because many of the redox reactions that influence the fate of toxicants are irreversible on the temporal and spatial scales that are important to toxicity.

Oxidation is simply defined as a loss of electrons. Oxidizing agents are electrophiles and thus gain electrons upon reaction. Oxidations can result in the increase in the oxidation state of the chemical as in the oxidation of metals, or oxidation can incorporate oxygen into the molecule. Typical organic chemical oxidative reactions include dealkylation, epoxidation, aromatic ring cleavage, and hydroxylation. The term auto-oxidation or weathering is commonly used to describe the general oxidative degradation of a chemical (or chemical mixture such as petroleum) upon exposure to air. Chemicals can react abiotically in both water and air with oxygen, ozone, peroxides, free radicals, and singlet oxygen. The last two are common intermediate reactants in indirect photolysis. Mineral surfaces are known to catalyze many oxidative reactions. Clays and oxides of silicon, aluminum, iron, and manganese can provide surface active sites that increase rates of oxidation. There are a variety of complex mechanisms associated with this catalysis, so it is difficult to predict the catalytic activity of soils and sediment in nature.

Reduction of a chemical species takes place when an electron donor (reductant) transfers electrons to an electron acceptor (oxidant). Organic chemicals typically act as the oxidant, while abiotic reductants include sulfide minerals, reduced metals or sulfur compounds, and natural organic matter. There are also extracellular biochemical reducing agents such as porphyrins, corrinoids, and metal-containing coenzymes. Most of these reducing agents are present only in anaerobic environments where anaerobic bacteria are themselves busy reducing chemicals. Thus, it is usually very difficult to distinguish biotic and abiotic reductive processes in nature. Well-controlled, sterile laboratory studies are required to measure abiotic rates of reduction. These studies indicate that many abiotic reductive transformations could be important in the environment, including dehalogenation, dealkylation, and the reduction of quinones, nitrosamines, azoaromatics, nitroaromatics, and sulfoxides. Functional groups that are resistant to reduction include aldehydes, ketones, carboxylic acids (and esters), amides, alkenes, and aromatic hydrocarbons.

Biotransformations   As we have seen throughout much of this textbook, vertebrates have developed the capacity to transform many toxicants into other chemicals, sometimes detoxifying the chemical and sometimes activating it. The same or similar biochemical processes that hydrolyze, oxidize, and reduce toxicants in vertebrates also take place in many lower organisms. In particular, bacteria, protozoans, and fungi provide a significant capacity to biotransform toxicants in the environment. Although many vertebrates can metabolize toxicants faster than these lower forms of life, the aggregate capacity of vertebrates to biotransform toxicants (based on total biomass and exposure) is insignificant to the overall fate of a toxicant in the environment. In this section, we use the term *biotransformation* to include all forms of direct biological transformation reactions.

Biotransformations follow a complex series of chemical reactions that are enzymatically mediated and are usually irreversible reactions that are energetically favorable, resulting in a decrease in the Gibbs free energy of the system. Thus, the potential for the biotransformation of a chemical depends on the reduction in free energy that results from reacting the chemical with other chemicals in its environment (e.g., oxygen). As with inorganic catalysts, microbes simply use enzymes to lower the activation energy of the reaction and to increase the rate of the transformation. Each successive chemical reaction further degrades the chemical, eventually mineralizing it to inorganic compounds (CO_2, H_2O, inorganic salts) and continuing the carbon and hydrologic cycles on earth.

Usually, microbial growth is stimulated because the microbes capture the energy released from the biotransformation reaction. As the microbial population expands, overall biotransformation rates increase even though the rate for each individual microbe may be constant or may even decrease. This complicates the modeling and prediction of biotransformation rates in nature. When the toxicant concentration (and potential energy) is small relative to other substrates or when the microbes cannot efficiently capture the energy from the biotransformation, microbial growth is not stimulated but biotransformation often still proceeds inadvertently through co-metabolism.

Biotransformation can be modeled using simple Michaelis–Menten enzyme kinetics, Monod microbial growth kinetics, or more complex numerical models that incorporate various environmental parameters and even the formation of microbial mats or slime, which affects diffusion of the chemical and nutrients to the microbial population. Microbial ecology involves a complex web of interaction among numerous environmental processes and parameters. The viability of microbial populations and the rates of biotransformation depend on many factors such as genetic adaptation, moisture, nutrients, oxygen, pH, and temperature. Although a single factor may limit biotransformation rates at a particular time and location, we cannot generalize about what limits biotransformation rates in the environment. Biotransformation rates often increase with temperature (according to the Arrhenius law) within the optimum range that supports the microbes, but many exceptions exist for certain organisms and chemicals. The availability of oxygen and various nutrients (C, N, P, Fe, Si) often limits microbial growth, but the limiting nutrient often changes with space (e.g., downriver) and time (seasonally and even diurnally).

Long-term exposure of microbial populations to certain toxicants often is necessary for adaptation of enzymatic systems capable of degrading those toxicants. This was the case with the *Exxon Valdez* oil spill in Alaska in 1989. Natural microbial populations in Prince William Sound, Alaska, had developed enzyme systems that oxidize petroleum hydrocarbons because of long-term exposure to natural oil seeps and to hydrocarbons that leached from the pine forests in the area. Growth of these natural microbial populations was nutrient limited during the summer. Thus, the application of nutrient formulations to the rocky beaches of Prince William Sound stimulated microbial growth and helped to degrade the spilled oil.

In terrestrial systems with high nutrient and oxygen content, low moisture and high organic carbon can control biotransformation by limiting microbial growth and the availability of the toxicant to the microbes. For example, the biotransformation rates of certain pesticides have been shown to vary two orders of magnitude in two separate agricultural fields that were both well aerated and nutrient rich, but spanned the common range of moisture and organic carbon content.

26.6 ENVIRONMENTAL FATE MODELS

The discussion above provides a brief qualitative introduction to the transport and fate of chemicals in the environment. The goal of most fate chemists and engineers is to translate this qualitative picture into a conceptual model and ultimately into a quantitative description that can be used to predict or reconstruct the fate of a chemical in the environment (Figure 26.1). This quantitative description usually takes the form of a mass balance model. The idea is to compartmentalize the environment into defined units (control volumes) and to write a mathematical expression for the mass balance within the compartment. As with pharmacokinetic models, transfer between compartments can be included as the complexity of the model increases. There is a great deal of subjectivity to assembling a mass balance model. However, each decision to include or exclude a process or compartment is based on one or more assumptions, most of which can be tested at some level. Over time, the applicability of various assumptions for particular chemicals and environmental conditions becomes known and model standardization becomes possible.

The construction of a mass balance model follows the general outline of this chapter. First, one defines the spatial and temporal scales to be considered and establishes the environmental compartments or control volumes. Second, the source emissions are identified and quantified. Third, the mathematical expressions

$$Vdc/dt = (I_{air} + I_{river} + I_{wastes}) - (O_{air} + O_{sed} + O_{dissolved} + O_{particle} + O_{rx})$$

Figure 26.4   An illustration of constructing a simple chemical mass balance model.

for advective and diffusive transport processes are written. And lastly, chemical transformation processes are quantified. This model-building process is illustrated in Figure 26.4. In this example, we simply equate the change in chemical inventory (total mass in the system) with the difference between chemical inputs and outputs to the system. The inputs could include numerous point and nonpoint sources or could be a single estimate of total chemical load to the system. The outputs include all of the loss mechanisms: transport out of the compartment and irreversible transformation reactions. If steady state can be assumed (i.e., the chemical's concentration in the compartment is not changing over the timescale of the model), the inventory change is zero and we are left with a simple mass balance equation to solve. Unsteady-state conditions would require a numerical solution to the differential equations.

There are many tricks and shortcuts to this process. For example, rather than compiling all of the transformation rate equations (or conducting the actual kinetic experiments yourself), there are many sources of typical chemical half-lives based on pseudo-first-order rate expressions. It is usually prudent to begin with these "best estimates" of half-lives in air, water, soil, and sediment and to perform a sensitivity analysis with the model to determine which processes are most important. One can return to the most important processes to assess whether a more detailed rate expression is necessary. An illustration of this mass balance approach is given in Figure 26.5 for benzo[a]pyrene. This approach allows a first-order evaluation of how chemicals enter the environment, what happens to them in the environment, and what the exposure concentrations will be in various environmental media. Thus, the chemical mass balance provides information relevant to toxicant exposure to both humans and wildlife.

Mass budget for benzo(*a*)pyrene

Figure 26.5   An illustration of the information provided by a chemical mass balance model. The annual mass budget of benzo[*a*]pyrene in Massachusetts Bay is shown.

BIBLIOGRAPHY AND SUGGESTED READING

Hemond, H. F. and E. J. Fechner. *Chemical Fate and Transport in the Environment.* New York: Academic Press, 1994.

Mackay, D. *Multimedia Environmental Models: The Fugacity Approach*, 2nd ed. Boca Raton, FL: Lewis Publishers, 2001.

Mackay, D., W. Y. Shiu, and K. C. Ma. *Physical-Chemical Properties and Environmental Fate and Degradation Handbook.* CRCnetBASE 2000 CR-ROM. Boca Raton, FL: CRC Press LLC, 2000.

Rand, G. M., ed. *Fundamentals of Aquatic Toxicology: Part II Environmental Fate.* Washington, DC: Taylor & Francis, 1995.

Schnoor, J. L. *Environmental Modeling: Fate and Transport of Pollutants in Water, Air, and Soil.* New York: Wiley, 1996.

Schwarzenbach, R. P., P. M. Gschwend, and D. M. Imboden. *Environmental Organic Chemistry*, 2nd ed. New York: Wiley, 2002.

SAMPLE QUESTIONS

1. Describe how the transport and fate of a chemical fits within the risk assessment paradigm.

2. Describe the difference between point source and nonpoint source pollution and give examples of each.

3. Describe how our knowledge of the behavior of nonpolar chemicals can be used to assess the toxicity of contaminated sediments.

4. List two (or three) types of reversible transformation reactions and give an example of each.

5. List two (or three) types of irreversible transformation reactions and give an example of each.

6. Describe how the knowledge of the source and fate of a chemical can be combined to develop a model to predict exposure in an environmental compartment.

Environmental Risk Assessment

DAMIAN SHEA

27.1 INTRODUCTION

Risk assessment is the process of assigning magnitudes and probabilities to adverse effects associated with an event. The development of risk assessment methodology has focused on accidental events (e.g., an airplane crash) and specific environmental stresses to humans (exposure of humans to chemicals), and thus most risk assessment is characterized by discrete events or stresses affecting well-defined end points (e.g., incidence of human death or cancer). This *single stress–single end-point* relationship allows the use of relatively simple statistical and mechanistic models to estimate risk and is widely used in human health risk assessment. However, this simple paradigm has only partial applicability to ecological risk assessment because of the inherent complexity of ecological systems and the exposure to numerous physical, chemical, and biological stresses that have both direct and indirect effects on a diversity of ecological components, processes, and end points. Thus, although the roots of ecological risk assessment can be found in human health risk assessment, the methodology for ecological risk assessment is not well developed and the estimated risks are highly uncertain. Despite these limitations, resource managers and regulators are looking to ecological risk assessment to provide a scientific basis for prioritizing problems that pose the greatest ecological risk and to focus research efforts in areas that will yield the greatest reduction in uncertainty.

To this end, the United States Environmental Protection Agency (EPA) has issued guidelines for planning and conducting ecological risk assessments. Because of the complexity and uncertainty associated with ecological risk assessment, the EPA guidelines provide only a loose framework for organizing and analyzing data, information, assumptions, and uncertainties to evaluate the likelihood of adverse ecological effects. However, the guidelines represent a broad consensus of the present scientific knowledge and experience on ecological risk assessment. This chapter presents a brief overview of the ecological risk assessment process as presently described by the EPA.

Ecological risk assessment can be defined as

> The process that evaluates the likelihood that adverse ecological effects may occur or are occurring as a result of exposure to one or more stressors.

A Textbook of Modern Toxicology, Fourth Edition. Edited by Ernest Hodgson
Copyright © 2010 John Wiley & Sons, Inc.

Estimating the *likelihood* can range from qualitative judgments to quantitative probabilities, though quantitative risk estimates still are rare in ecological risk assessment. The *adverse ecological effects* are changes that are considered undesirable because they alter valued structural or functional characteristics of ecological systems and usually include the type, intensity, and scale of the effect as well as the potential for recovery. The statement that effects *may occur or are occurring* refers to the dual *prospective* and *retrospective* nature of ecological risk assessment. The inclusion of *one or more stressors* is a recognition that ecological risk assessments may address single or multiple chemical, physical, and/or biological stressors. Because risk assessments are conducted to provide input to management decisions, most risk assessments focus on stressors generated or influenced by anthropogenic activity. However, natural phenomena also will induce stress that results in adverse ecological effects and cannot be ignored.

The overall ecological risk assessment process is shown in Figure 27.1 and includes three primary phases: (1) problem formulation, (2) analysis, and (3) risk

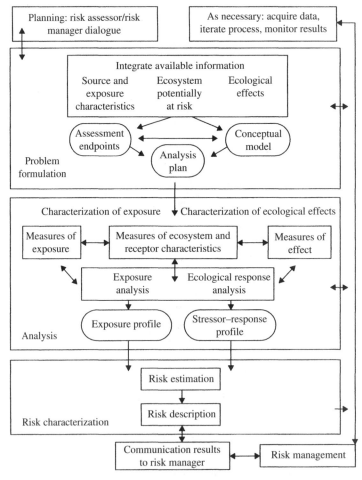

Figure 27.1 The ecological risk assessment framework as set forth by the U.S. Environmental Protection Agency.

characterization. Problem formulation includes the development of a conceptual model of stressor–ecosystem interactions and the identification of risk assessment end points. The analysis phase involves evaluating exposure to stressors and the relationship between stressor characteristics and ecological effects. Risk characterization includes estimating risk through the integration of exposure and stressor–response profiles, describing risk by establishing lines of evidence and by determining ecological effects, and communicating this description to risk managers. While discussions between risk assessors and risk managers are emphasized both at risk assessment initiation (planning) and completion (communicating results), usually, a clear distinction is drawn between risk assessment and risk management. Risk assessment focuses on scientifically evaluating the likelihood of adverse effects, and risk management involves the selection of a course of action in response to an identified risk that is based on many factors (e.g., social, legal, or economic) in addition to the risk assessment results. Monitoring and other data acquisition is often necessary during any phase of the risk assessment process, and the entire process is typically iterative rather than linear. The evaluation of new data or information may require revisiting a part of the process or conducting a new assessment.

27.2 FORMULATING THE PROBLEM

Problem formulation is a process for generating and evaluating preliminary hypotheses about why ecological effects have occurred, or may occur, because of human activities. During problem formulation, management goals are evaluated to help establish objectives for the risk assessment; the ecological problem is defined; and the plan for analyzing data and characterizing risk is developed. The objective of this process is to develop (1) assessment end points that adequately reflect management goals and the ecosystem they represent and (2) conceptual models that describe critical relationships between a stressor and assessment end point or among several stressors and assessment end points. The assessment end points and the conceptual models are then integrated to develop a plan or proposal for risk analysis.

27.2.1 Selecting Assessment End Points

Assessment end points are *explicit expressions of the actual environmental value that is to be protected* and they link the risk assessment to management concerns. Assessment end points include both a valued or key ecological entity and an attribute of that entity that is important to protect and that is potentially at risk. The scientific basis for a risk assessment is enhanced when assessment end points are both ecologically relevant and susceptible to the stressors of concern. Assessment end points that also logically represent societal values and management goals will increase the likelihood that the risk assessment will be understood and used in management decisions.

Ecological Relevance Ecologically relevant end points reflect important attributes of the ecosystem and can be functionally related to other components of the ecosystem; they help sustain the structure, function, and biodiversity of an ecosystem. For example, ecologically relevant end points might contribute to the food base (e.g., primary production), provide habitat, promote regeneration of critical resources (e.g., nutrient cycling), or reflect the structure of the community, ecosystem, or landscape (e.g., species diversity). Ecological relevance becomes most useful when it is possible to identify the potential cascade of adverse effects that could result from a critical initiating effect such as a change in ecosystem function. The selection of assessment end points that address both specific organisms of concern and landscape-level ecosystem processes becomes increasingly important (and more difficult) in landscape-level risk assessments. In these cases, it may be possible to select one or more species and an ecosystem process to represent larger functional community or ecosystem processes. Extrapolations like these must be explicitly described in the conceptual model (see Section 27.2.2).

Susceptibility to Stressors   Ecological resources or entities are considered susceptible if they are sensitive to a human-induced stressor to which they are exposed. *Sensitivity* represents how readily an ecological entity responds to a particular stressor. Measures of sensitivity may include mortality or decreased growth or fecundity resulting from exposure to a toxicant, or behavioral abnormalities such as avoidance of food source areas or nesting sites because of the proximity of stressors such as noise or habitat alteration. Sensitivity is directly related to the mode of action of the stressors. For example, chemical sensitivity is influenced by individual physiology, genetics, and metabolism. Sensitivity also is influenced by individual and community life history characteristics. For example, species with long life cycles and low reproductive rates will be more vulnerable to extinction from increases in mortality than those with short life cycles and high reproductive rates. Species with large home ranges may be more sensitive to habitat fragmentation compared with those species with smaller home ranges within a fragment. Sensitivity may be related to the life stage of an organism when exposed to a stressor. Young animals often are more sensitive to stressors than adults. In addition, events like migration and molting often increase sensitivity because they require significant energy expenditure, which makes these organisms more vulnerable to stressors. Sensitivity also may be increased by the presence of other stressors or natural disturbances.

Exposure is the other key determinant in susceptibility. In ecological terms, exposure can mean co-occurrence, contact, or the absence of contact, depending on the stressor and assessment end point. The characteristics and conditions of exposure will influence how an ecological entity will respond to a stressor and thus will determine what ecological entities might be susceptible. Therefore, one must consider information on the proximity of an ecological entity to the stressor along with the timing (e.g., frequency and duration relative to sensitive life stages) and intensity of exposure. Note that adverse effects may be observed even at very low stressor exposures if a necessary resource is limited during a critical life stage. For example, if fish are unable to find suitable nesting sites during their reproductive phase, risk is significant even when water quality is high and food sources are abundant.

Exposure may take place at one point in space and time, but effects may not arise until another place or time. Both life history characteristics and the circumstances of exposure influence susceptibility in this case. For example, exposure of a population to endocrine-modulating chemicals can affect the sex ratio of the offspring, but the population impacts of this exposure may not become apparent until years later when the cohort of affected animals begins to reproduce. Delayed effects and multiple stressor exposures add complexity to evaluations of susceptibility. For example, although toxicity tests may determine receptor sensitivity to one stressor, the degree of susceptibility may depend on the co-occurrence of another stressor that significantly alters receptor response. Again, conceptual models need to reflect these additional factors.

Defining Assessment End Points   Assessment end points provide a transition between management goals and the specific measures used in an assessment by helping to identify measurable attributes to quantify and model. However, in contrast to management goals, no intrinsic value is assigned to the end point so it does not contain words such as *protect* or *maintain*, and it does not indicate a desirable direction for change. Two aspects are required to define an assessment end point. The first is the valued ecological entity such as a species, a functional group of species, an ecosystem function or characteristic, or a specific valued habitat. The second is the characteristic about the entity of concern that is important to protect and is potentially at risk.

Expert judgment and an understanding of the characteristics and function of an ecosystem are important for translating general goals into usable assessment end points. End points that are too broad and vague (ecological health) cannot be linked to specific measurements. End points that are too narrowly defined (hatching success of bald eagles) may overlook important characteristics of the ecosystem and may fail to include critical variables. Clearly defined assessment end points provide both direction and boundaries for the risk assessment.

Assessment end points directly influence the type, characteristics, and interpretation of data and information used for analysis and the scale and character of the assessment. For example, an assessment end point such as "fecundity of bivalves" defines local population characteristics and requires very different types of data and ecosystem characterization compared to "aquatic community structure and function." When concerns are on a local scale, the assessment end points should not focus on landscape concerns. But if ecosystem processes and landscape patterns are being considered, survival of a single species would provide inadequate representation of this larger scale.

The presence of multiple stressors also influences the selection of assessment end points. When it is possible to select one assessment end point that is sensitive to many of the identified stressors yet responds in different ways to different stressors, it is possible to consider the combined effects of multiple stressors while still discriminating among effects. For example, if the recruitment of a fish population is the assessment end point, it is important to recognize that recruitment may be adversely affected at several life stages, in different habitats, through different ways, by different stressors. The measures of effect, exposure, and ecosystem and receptor characteristics chosen to evaluate recruitment provide a basis for discriminating among different stressors, individual effects, and their combined effect.

Although many potential assessment end points may be identified, practical considerations often drive their selection. For example, assessment end points usually must reflect environmental values that are protected by law or that environmental managers and the general public recognize as a critical resource or an ecological function that would be significantly impaired if the resource were altered. Another example of a practical consideration is the extrapolation across scales of time, space, or level of biological organization. When the attributes of an assessment end point can be measured directly, extrapolation is unnecessary and this uncertainty is avoided. Assessment end points that cannot be linked with measurable attributes are not appropriate for a risk assessment. However, assessment end points that cannot be measured directly but can be represented by surrogate measures that are easily monitored and modeled can still provide a good foundation for the risk assessment.

27.2.2 Developing Conceptual Models

Conceptual models link anthropogenic activities with stressors and evaluate the relationships among exposure pathways, ecological effects, and ecological receptors. The models also may describe natural processes that influence these relationships. Conceptual models include a set of risk hypotheses that describe predicted relationships between stressor, exposure, and assessment end-point response, along with the rationale for their selection. Risk hypotheses are hypotheses in the broad scientific sense; they do not necessarily involve statistical testing of null and alternative hypotheses or any particular analytical approach. Risk hypotheses may predict the effects of a stressor or they may postulate what stressors may have caused observed ecological effects.

Diagrams can be used to illustrate the relationships described by the conceptual model and risk hypotheses. Conceptual model diagrams are useful tools for communicating important pathways and for identifying major sources of uncertainty. These diagrams and risk hypotheses can be used to identify the most important pathways and relationships to consider in the analysis phase. The hypotheses considered most likely to contribute to risk are identified for subsequent evaluation in the risk assessment.

The complexity of the conceptual model depends on the complexity of the problem, the number of stressors and assessment end points being considered, the nature of effects, and the characteristics of the ecosystem. For single stressors and single assessment end points, conceptual models can be relatively simple relationships. In cases where conceptual models describe both the pathways of individual stressors and assessment end points and the interaction of multiple and diverse stressors and assessment end points, several submodels would be required to describe individual pathways. Other models may then be used to explore how these individual pathways interact. An example of a conceptual model for a watershed is shown in Figure 27.2.

27.2.3 Selecting Measures

The last step in the problem formulation phase is the development of an analysis plan or proposal that identifies measures to evaluate each risk hypothesis and that describes the assessment design, data needs, assumptions, extrapolations, and specific methods for conducting the analysis. There are three categories of measures

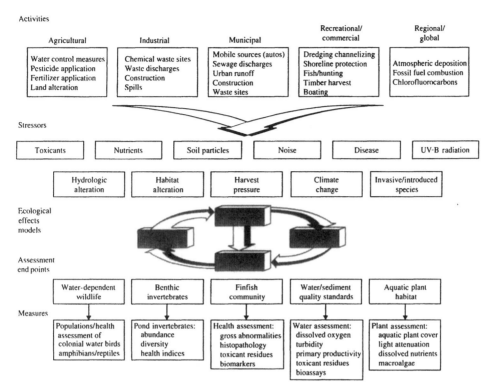

Figure 27.2   An example of a conceptual model for a watershed. Human activities, shown at the top of the diagram, result in various stressors that induce ecological effects. Assessment end points and related measures that are associated with these effects are shown at the bottom of the diagram.

that can be selected: (1) *measures of effect* (also called *measurement end points*) are measures used to evaluate the response of the assessment end point when exposed to a stressor; (2) *measures of exposure* are measures of how exposure may be occurring, including how a stressor moves through the environment and how it may co-occur with the assessment end point; and (3) *measures of ecosystem and receptor characteristics* include ecosystem characteristics that influence the behavior and location of assessment end points, the distribution of a stressor, and the life history characteristics of the assessment end point that may affect exposure or response to the stressor. These diverse measures increase in importance as the complexity of the assessment increases.

An important consideration in the identification of these measures is their response sensitivity and ecosystem relevance. Response sensitivity is usually highest with measures at the lower levels of biological organization, but the ecosystem relevance is highest at the higher levels of biological organization. This dichotomy is illustrated in Figure 27.3. In general, the time required to illicit a response also increases with the level of biological organization. Note that toxicologists focus on measures at lower levels of biological organization, relying upon an extrapolation based on the tenet that effects of toxicants on populations and communities are

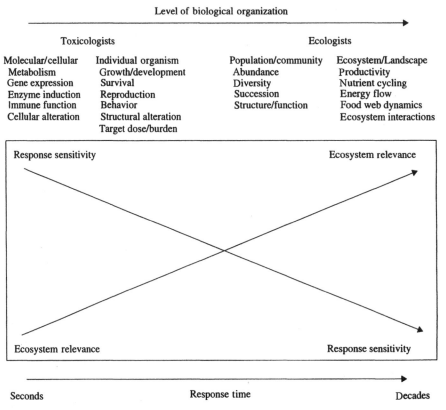

Figure 27.3 The response time and sensitivity of an ecological receptor is a function of the level of biological organization. Higher levels of organization have greater ecosystem relevance. However, as the level of biological organization increases, response time increases, sensitivity decreases, and causal relationships become more uncertain. Ecological risk assessments must balance the need for sensitive, timely, and well-established responses with ecological relevance.

initiated at the molecular/cellular level, and if this insult is not corrected for, or adapted to, then effects on physiological systems and individual organisms will result. For certain toxic modes of action (e.g., reproductive toxicity), this could result in effects at the population and community levels. In contrast, ecologists focus on measures at the population level or higher for obvious reasons of ecological relevance. A combination of measures is often necessary to provide reasonable sensitivity, ecosystem relevance, and causal relationships.

27.3   ANALYZING EXPOSURE AND EFFECTS INFORMATION

The second phase of ecological risk assessment, the analysis phase, includes two principal activities: characterization of exposure and characterization of ecological effects (Figure 27.1).

27.3.1 Characterizing Exposure

In exposure characterization, credible and relevant data are analyzed to describe the source(s) of stressors, the distribution of stressors in the environment, and the contact or co-occurrence of stressors with ecological receptors. An exposure profile is developed that identifies receptors and exposure pathways, describes the intensity and spatial and temporal extent of exposure, describes the impact of variability and uncertainty on exposure estimates, and presents a conclusion about the likelihood that exposure will occur.

A source description identifies where the stressor originates, describes what stressors are generated, and considers other sources of the stressor. Exposure analysis may start with the source when it is known, but some analyses may begin with known exposures and attempt to link them to sources, while other analyses may start with known stressors and attempt to identify sources and to quantify contact or co-occurrence. The source description includes what is known about the intensity, timing, and location of the stressor and whether other constituents emitted by the source influence transport, transformation, or bioavailability of the stressor of interest.

Many stressors have natural counterparts and/or multiple sources that must be considered. For example, many chemicals occur naturally (e.g., most metals), are generally widespread due to multiple sources (e.g., polycyclic aromatic hydrocarbons), or may have significant sources outside the boundaries of the current assessment (e.g., regional atmospheric deposition of polychlorinated biphenyl (PCBs)). Many physical stressors also have natural counterparts such as sedimentation from construction activities versus natural erosion. In addition, human activities may change the magnitude or frequency of natural disturbance cycles such as the frequency and severity of flooding. Source characterization can be particularly important for new biological stressors (e.g., invasive species) since many of the strategies for reducing risks focus on preventing entry in the first place. Once the source is identified, the likelihood of entry may be characterized qualitatively.

Because exposure occurs where receptors co-occur with or contact stressors in the environment, characterizing the spatial and temporal distribution of a stressor is a necessary precursor to estimating exposure. The stressor's spatial and temporal distribution in the environment is described by evaluating the pathways that stressors take from the source as well as the formation and subsequent distribution of secondary stressors. For chemical stressors, the evaluation of pathways usually follows the type of transport and fate modeling described in Chapter 26. Some physical stressors such as sedimentation also can be modeled, but other physical stressors require no modeling because they eliminate entire ecosystems or portions of them, such as when a wetland is filled, a resource is harvested, or an area is flooded.

The movement of biological stressors has been described as diffusion and/or jump–dispersal processes. Diffusion involves a gradual spread from the site of introduction and is a function primarily of reproductive rates and motility. Jump–dispersal involves erratic spreads over periods of time, usually by means of a vector. The gypsy moth and zebra mussel have spread this way: the gypsy moth via egg masses on vehicles and the zebra mussel via boat ballast water. Biological stressors can use both diffusion and jump–dispersal strategies, which makes it difficult to predict dispersal rates. An additional complication is that biological stressors are influenced by their own survival and reproduction.

The creation of secondary stressors can greatly alter risk. Secondary stressors can be formed through biotic or abiotic transformation processes and may be of greater or lesser concern than the primary stressor. Physical disturbances can generate secondary stressors, such as when the removal of riparian vegetation results in increased nutrients, sedimentation, and altered stream flow. For chemicals, the evaluation of secondary stressors usually focuses on metabolites or degradation products. In addition, secondary stressors can be formed through ecosystem processes. For example, nutrient inputs into an estuary can decrease dissolved oxygen concentrations because they increase primary production and subsequent decomposition. A changeover from an aerobic to an anaerobic environment often is accompanied by the production of sulfide via sulfate-reducing bacteria. Sulfide can act as a secondary stressor to oxygen-dependent organisms, but it also can reduce exposure to metals through the precipitation of metal sulfides (see Chapter 26).

The distribution of stressors in the environment can be described using measurements, models, or a combination of the two. If stressors have already been released, direct measurements of environmental media or a combination of modeling and measurement is preferred. However, a modeling approach may be necessary if the assessment is intended to predict future scenarios or if measurements are not possible or practicable.

27.3.2 Characterizing Ecological Effects

In ecological effect characterization, relevant data are analyzed to evaluate stressor–response relationships and/or to provide evidence that exposure to a stressor causes an observed response. The characterization describes the effects that are elicited by a stressor, links these effects with the assessment end points, and evaluates how the effects change with varying stressor levels. The conclusions of the ecological effect characterization are summarized in a stressor–response profile.

Analyzing Ecological Response   Ecological response analysis has three primary components: determining the relationship between stressor exposure and ecological effects, evaluating the plausibility that effects may occur or are occurring as a result of the exposure, and linking measurable ecological effects with the assessment end points.

Evaluating ecological risks requires an understanding of the relationships between stressor exposure and resulting ecological responses. The stressor–response relationships used in a particular assessment depend on the scope and nature of the ecological risk assessment as defined in problem formulation and as reflected in the analysis plan. For example, a point estimate of an effect (such as an median lethal concentration (LC50)) might be compared with point estimates from other stressors. The stressor–response function (e.g., shape of the curve) may be critical for determining the presence or absence of an effect threshold or for evaluating incremental risks, or stressor–response functions may be used as input for ecological effects models. If sufficient data are available, cumulative distribution functions can be constructed using multiple point estimates of effects. Process models that already incorporate empirically derived stressor–response functions can also be used.

However, many stressor–response relationships are very complex and ecological systems frequently show responses to stressors that involve abrupt shifts to new community or system types.

In simple cases, the response will be one variable (e.g., mortality) and quantitative univariate analysis can be used. If the response of interest is composed of many individual variables (e.g., species abundances in an aquatic community), multivariate statistical techniques must be used. Multivariate techniques (e.g., factor and cluster analysis) have a long history of use in ecology but have not yet been extensively applied in risk assessment. Stressor–response relationships can be described using any of the dimensions of exposure (i.e., intensity, time, or space). Intensity is probably the most familiar dimension and is often used for chemicals (e.g., dose, concentration). The duration of exposure can also be used for chemical stressor–response relationships; for example, median acute effect levels are always associated with a time parameter (e.g., 24, 48, and 96 h). Both the time and spatial dimensions of exposure can be important for physical disturbances such as flooding. Single-point estimates and stressor–response curves can be generated for some biological stressors. For pathogens such as bacteria and fungi, inoculum levels may be related to the level of symptoms in a host or to the actual signs of the pathogen. For other biological stressors such as introduced species, developing simple stressor–response relationships may be inappropriate.

Causality is the relationship between cause (one or more stressors) and effect (assessment end-point response to one or more stressors). Without a sound basis for linking cause and effect, uncertainty in the conclusions of an ecological risk assessment will be high. Developing causal relationships is especially important for risk assessments driven by observed adverse ecological effects such as fish kills or long-term declines in a population. Criteria need to be established for evaluating causality. For chemicals, ecotoxicologists have slightly modified Koch's postulates to provide evidence of causality:

1. The injury, dysfunction, or other putative effect of the toxicant must be regularly associated with exposure to the toxicant and with any contributory causal factors.

2. Indicators of exposure to the toxicant must be found in the affected organisms.

3. The toxic effects must be seen when normal organisms or communities are exposed to the toxicant under controlled conditions, and any contributory factors should be manifested in the same way during controlled exposures.

4. The same indicators of exposure and effects must be identified in the controlled exposures as in the field.

While useful as an ideal, this approach may not be practical if resources for experimentation are not available or if an adverse effect may be occurring over such a wide spatial extent that experimentation and correlation may prove difficult or may yield equivocal results. In most cases, extrapolation will be necessary to evaluate causality. The scope of the risk assessment also influences extrapolation through the nature of the assessment end point. Preliminary assessments that evaluate risks to general trophic levels, such as fish and birds, may extrapolate among different genera or families to obtain a range of sensitivity to the stressor. On the other hand, assessments concerned with management strategies for a particular species may employ population models.

Whatever methods are employed to link assessment end points with measures of effect, it is important to apply the methods in a manner consistent with sound ecological and toxicological principles. For example, it is inappropriate to use structure–activity relationships to predict toxicity from chemical structure unless the chemical under consideration has a similar mode of toxic action to the reference chemicals. Similarly, extrapolations from upland avian species to waterfowl may be more credible if factors such as differences in food preferences, physiology, and seasonal behavior (e.g., mating and migration habits) are considered.

Finally, many extrapolation methods are limited by the availability of suitable databases. Although these databases are generally largest for chemical stressors and aquatic species, even in these cases, data do not exist for all taxa or effects. Chemical effects databases for mammals, amphibians, or reptiles are extremely limited, and there is even less information on most biological and physical stressors. Extrapolations and models are only as useful as the data on which they are based and should recognize the great uncertainties associated with extrapolations that lack an adequate empirical or process-based rationale.

Developing a Stressor–Response Profile   The final activity of the ecological response analysis is developing a stressor–response profile to evaluate single species, populations, general trophic levels, communities, ecosystems, or landscapes—whatever is appropriate for the defined assessment end points. For example, if a single species is affected, effects should represent appropriate parameters such as effects on mortality, growth, and reproduction, while at the community level, effects may be summarized in terms of structure or function depending on the assessment end point. At the landscape level, there may be a suite of assessment end points and each should be addressed separately. The stressor–response profile summarizes the nature and intensity of effect(s), the timescale for recovery (where appropriate), causal information linking the stressor with observed effects, and uncertainties associated with the analysis.

27.4   CHARACTERIZING RISK

Risk characterization is the final phase of an ecological risk assessment (Figure 27.1). During risk characterization, risks are estimated and interpreted, and the strengths, limitations, assumptions, and major uncertainties are summarized. Risks are estimated by integrating exposure and stressor–response profiles using a wide range of techniques such as comparisons of point estimates or distributions of exposure and effects data, process models, or empirical approaches such as field observational data. Risks are described by evaluating the evidence supporting or by refuting the risk estimate(s) and interpreting the adverse effects on the assessment end point. Criteria for evaluating adversity include the nature and intensity of effects, spatial and temporal scales, and the potential for recovery. Agreement among different lines of evidence of risk increases confidence in the conclusions of a risk assessment.

27.4.1 Estimating Risk

Risk estimation determines the likelihood of adverse effects to assessment end points by integrating exposure and effects data and by evaluating any associated uncertainties. The process uses the exposure and stressor–response profiles. Risks can be estimated by one or more of the following approaches: (1) estimates based on best professional judgment and expressed as qualitative categories such as low, medium, or high; (2) estimates comparing single-point estimates of exposure and effects such as a simple ratio of exposure concentration to effects concentration (quotient method); (3) estimates incorporating the entire stressor–response relationship often as a nonlinear function of exposure; (4) estimates incorporating variability in exposure and effects estimates providing the capability to predict changes in the magnitude and likelihood of effects at different exposure scenarios; (5) estimates based on process models that rely partially or entirely on theoretical approximations of exposure and effects; and (6) estimates based on empirical approaches, including field observational data. An example of the first approach, using qualitative categorization, is shown in Figure 27.4.

27.4.2 Describing Risk

After risks have been estimated, available information must be integrated and interpreted to form conclusions about risks to the assessment end points. Risk descriptions include an evaluation of the lines of evidence supporting or refuting the risk estimate(s) and an interpretation of the adverse effects on the assessment end point. Confidence in the conclusions of a risk assessment may be increased by using several lines of evidence to interpret and compare risk estimates. These lines of evidence may be derived from different sources or by different techniques relevant to adverse effects on the assessment end points, such as quotient estimates, modeling results, field experiments, or field observations. Some of the factors to consider when evaluating separate lines of evidence are

- the relevance of evidence to the assessment end points,
- the relevance of evidence to the conceptual model,
- the sufficiency and quality of data and experimental designs used in supporting studies,
- the strength of cause/effect relationships, and
- the relative uncertainties of each line of evidence and their direction.

At this point in risk characterization, the changes expected in the assessment end points have been estimated and described. The next step is to interpret whether these changes are considered adverse and meaningful. Meaningful adverse changes are defined by ecological and/or social concerns and, thus, usually depend on the best professional judgment of the risk assessor. Five criteria have been proposed by EPA for evaluating adverse changes in assessment end points:

1. Nature of effects
2. Intensity of effects

Resources at risk

Sources of stressors	Phytoplankton	Zooplankton	Fish eggs and larvae	Soft-bottom benthos	Hard-bottom benthos	Bottom fish	Predatory fish	Plant life	Wetlands and estuaries	Amphibians/reptiles	Mammals	Birds	Water quality	Human health
Pesticide runoff					?				?	?				
Point sources		?						?	?	?		?	?	?
Oil/gas production									?	?				
Atmospheric input		?							?	?	?	?		
Contaminated sediments					?		?	?	?	?		?		?
Diseases/pathogens					?				?	?			?	
Particle loading					?		?							
Carbon loading			?		?									
Nutrient loading														
Dredging			?						?	?				
River flow					?									
Flooding					?						?			
Fishing pressure														
Commerce and boating	?	?	?	?	?	?	?	?	?	?	?	?	?	?
Habitat loss/modification					?					?			?	?
Blooms/invasions		?	?	?	?	?	?	?	?	?	?	?	?	
Global climate change				?	?				?	?	?	?	?	?

Potential influence in resource (position within cell)
High   Moderate   low

Utility as assessment indicator
High   Moderate   low

Empty cells = no interaction
? = high level of uncertainty.

Figure 27.4 An example of a qualitative categorization of an ecological risk for a hypothetical matrix of stressors and resources at risk.

3. Spatial scale
4. Temporal scale
5. Potential for recovery

The extent to which the five criteria are evaluated depends on the scope and complexity of the ecological risk assessment. However, understanding the underlying assumptions and science policy judgments is important even in simple cases. For example, when exceedance of a previously established decision rule such as a benchmark stressor level or water quality criteria is used as evidence of adversity, the reasons why exceedances of the benchmark are considered adverse should be clearly understood.

To distinguish ecological changes that are adverse from those ecological events that are within the normal pattern of ecosystem variability or result in little or no meaningful alteration of biota, it is important to consider the nature and intensity of effects. For example, an assessment end point involving survival, growth, and reproduction of a species must consider whether predicted effects involve survival and reproduction or only growth, or if the survival of the offspring is affected, the relative loss must be considered.

It is important to consider both the ecological and statistical contexts of an effect when evaluating intensity. For example, a statistically significant 1% decrease in fish growth may not be relevant to an assessment end point of fish population viability, and a 10% decline in reproduction may be worse for a population of slowly reproducing marine mammals than for rapidly reproducing planktonic algae.

Natural ecosystem variation can make it very difficult to observe (detect) stressor-related perturbations. For example, natural fluctuations in marine fish populations are often very large and cyclic events (e.g., fish migration) are very important in natural systems. Predicting the effects of anthropogenic stressors against this background of variation can be very difficult. Thus, a lack of statistically significant effects in a field study does not automatically mean that adverse ecological effects are absent. Rather, factors such as statistical power to detect differences, natural variability, and other lines of evidence must be considered in reaching conclusions about risk.

Spatial and temporal scales also need to be considered in assessing the adversity of the effects. The spatial dimension encompasses both the extent and pattern of effect as well as the context of the effect within the landscape. Factors to consider include the absolute area affected, the extent of critical habitats affected compared to a larger area of interest, and the role or use of the affected area within the landscape. Adverse effects to assessment end points vary with the absolute area of the effect. A larger affected area may be (1) subject to a greater number of other stressors, increasing the complications from stressor interactions; (2) more likely to contain sensitive species or habitats; or (3) more susceptible to landscape-level changes because many ecosystems may be altered by the stressors.

Nevertheless, a smaller area of effect is not always associated with lower risk. The function of an area within the landscape may be more important than the absolute area. Destruction of small but unique areas, such as submerged vegetation at the land–water margin, may have important effects on local wildlife populations. Also, in river systems, both riffle and pool areas provide important microhabitats that maintain the structure and function of the total river ecosystem. Stressors acting on some of these microhabitats may present a significant risk to the entire system. Spatial factors also are important for many species because of the linkages between ecological landscapes and population dynamics. Linkages between one or more landscapes can provide refuge for affected populations, and species may require adequate corridors between habitat patches for successful migration.

The temporal scale for ecosystems can vary from seconds (photosynthesis, pro-karyotic reproduction) to centuries (global climate change). Changes within a forest ecosystem can occur gradually over decades or centuries and may be affected by slowly changing external factors such as climate. The timescale of stressor-induced changes operates within the context of multiple natural timescales. In addition, temporal responses for ecosystems may involve intrinsic time lags so that responses from a stressor may be delayed. Thus, it is important to distinguish the long-term impacts of a stressor from the immediately visible effects. For example, visible changes resulting from eutrophication of aquatic systems (turbidity, excessive mac-rophyte growth, population decline) may not become evident for many years after initial increases in nutrient levels.

Considering the temporal scale of adverse effects leads us to a consideration of recovery. Recovery is the rate and extent of return of a population or community to a condition that existed before the introduction of a stressor. Because ecosystems are dynamic and even under natural conditions are constantly changing in response to changes in the physical environment (weather, natural catastrophes, etc.) or other factors, it is unrealistic to expect that a system will remain static at some level or will return to exactly the same state that it was before it was disturbed. Thus, the attributes of a "recovered" system must be carefully defined. Examples might include productivity declines in a eutrophic system, reestablishment of a species at a particular density, species recolonization of a damaged habitat, or the restoration of the health of diseased organisms.

Recovery can be evaluated in spite of the difficulty in predicting events in ecological systems. For example, it is possible to distinguish changes that are usually reversible (e.g., recovery of a stream from sewage effluent discharge), frequently irreversible (e.g., establishment of introduced species), and always irreversible (e.g., species extinction). It is important to consider whether significant structural or functional changes have occurred in a system that might render changes irreversible. For example, physical alterations such as deforestation can change soil structure and seed sources such that forests cannot easily grow again.

Natural disturbance patterns can be very important when evaluating the likelihood of recovery from anthropogenic stressors. Ecosystems that have been subjected to repeated natural disturbances may be more vulnerable to anthropo-genic stressors (e.g., overfishing). Alternatively, if an ecosystem has become adapted to a disturbance pattern, it may be affected when the disturbance is removed (fire-maintained grasslands). The lack of natural analogues makes it difficult to predict recovery from novel anthropogenic stressors such as exposure to synthetic chemicals.

The relative rate of recovery also can be estimated. For example, fish popu-lations in a stream are likely to recover much faster from exposure to a degradable chemical than from habitat alterations resulting from stream channelization. It is critical to use knowledge of factors such as the temporal scales of organisms' life histories, the availability of adequate stock for recruitment, and the interspecific and trophic dynamics of the populations in evaluating the relative rates of recovery. A fisheries stock or forest might recover in several decades, a benthic infaunal community in years, and a planktonic community in weeks to months.

27.5 MANAGING RISK

When risk characterization is complete, a description of the risk assessment is communicated to the risk manager (Figure 27.1) to support a risk management decision. This communication usually is a report and might include a

- Description of risk assessor/risk manager planning results;
- Review of the conceptual model and the assessment end points;
- Discussion of the major data sources and analytical procedures used;
- Review of the stressor–response and exposure profiles;
- Description of risks to the assessment end points, including risk estimates and adversity evaluations;
- Summary of major areas of uncertainty and the approaches used to address them; and
- Discussion of science policy judgments or default assumptions used to bridge information gaps, and the basis for these assumptions.

After the risk assessment is completed, risk managers may consider whether additional follow-up activities are required. Depending on the importance of the assessment, confidence level in the assessment results, and available resources, it may be advisable to conduct another iteration of the risk assessment in order to facilitate a final management decision. Ecological risk assessments are frequently designed in sequential tiers that proceed from simple, relatively inexpensive evaluations to more costly and complex assessments. Initial tiers are based on conservative assumptions, such as maximum exposure and ecological sensitivity. When an early tier cannot sufficiently define risk to support a management decision, a higher assessment tier that may require either additional data or applying more refined analysis techniques to available data may be needed. Higher tiers provide more ecologically realistic assessments while making less conservative assumptions about exposure and effects.

Another option is to proceed with a management decision based on the risk assessment and to develop a monitoring plan to evaluate the results of the decision. For example, if the decision was to mitigate risks through exposure reduction, monitoring could help determine whether the desired reduction in exposure (and effects) was achieved. Monitoring is also critical for determining the extent and nature of any ecological recovery that may be occurring.

Ecological risk assessment is important for environmental decision making because of the high cost of eliminating environmental risks associated with human activities and the necessity of making regulatory decisions in the face of uncertainty. Ecological risk assessment provides only a portion of the information required to make risk management decisions, but this information is critical to scientifically defensible risk management. Thus, ecological risk assessments should provide input to a diverse set of environmental decision-making processes, such as the regulation of hazardous waste sites, industrial chemicals, and pesticides, or the management of watersheds affected by multiple nonchemical and chemical stressors.

BIBLIOGRAPHY AND SUGGESTED READING

Bartell, S. M., R. H. Gardner, and R. V. O'Neill. *Ecological Risk Estimation*. Boca Raton, FL: Lewis Publishers, 1992.

Cardwell, R. D., B. R. Parkhurst, W. Warren-Hicks, et al. Aquatic ecological risk. *Water Environ. Technol.* **5**:47–51, 1993.

Harwell, M. A., W. Cooper, and R. Flaak. Prioritizing ecological and human welfare risks from environmental stresses. *Environ. Manag.* **16**:451–464, 1992.

Kendall, R. J., T. E. Lacher, C. Bunck, et al. An ecological risk assessment of lead shot exposure in non-waterfowl avian species: Upland game birds and raptors. *Environ. Toxicol. Chem.* **15**:4–20, 1996.

National Research Council. A paradigm for ecological risk assessment. In *Issues in Risk Assessment*, pp. 241–272. Washington, DC: National Academy Press, 1993.

National Research Council. *Science and Judgment in Risk Assessment*. Washington, DC: National Academy Press, 1994.

National Research Council. *Understanding Risk: Informing Decisions in a Democratic Society*. Washington, DC: National Academy Press, 1996.

Ruckelshaus, W. D. Science, risk, and public policy. *Science* **221**:1026–1028, 1983.

Solomon, K. R., D. B. Baker, R. P. Richards, et al. Ecological risk assessment of atrazine in North American surface waters. *Environ. Toxicol. Chem.* **15** (1): 31–76, 1996.

Suter, G. W. II. Endpoints for regional ecological risk assessments. *Environ. Manag.* **14**: 19–23, 1990.

Suter, G. W. II. *Ecological Risk Assessment*. Boca Raton, FL: Lewis Publishers, 1993a.

Suter, G. W. II. A critique of ecosystem health concepts and indexes. *Environ. Toxicol. Chem.* **12**:1533–1539, 1993b.

United States Environmental Protection Agency. *Summary report on issues in ecological risk assessment*. Washington, DC: Risk Assessment Forum, United States Environmental Protection Agency. EPA/625/3-91/018, 1991.

United States Environmental Protection Agency. *Framework for ecological risk assessment*. Washington, DC: Risk Assessment Forum, United States Environmental Protection Agency. EPA/630/R-92/001, 1992a.

United States Environmental Protection Agency. *Peer review workshop report on a framework for ecological risk assessment*. Washington, DC: Risk Assessment Forum, United States Environmental Protection Agency. EPA/625/3-91/022, 1992b.

United States Environmental Protection Agency. *A review of ecological assessment case studies from a risk assessment perspective*. Washington, DC: Risk Assessment Forum, United States Environmental Protection Agency. EPA/630/R-92/005, 1993.

United States Environmental Protection Agency. *Ecological risk assessment issue papers*. Washington, DC: Risk Assessment Forum, United States Environmental Protection Agency. EPA/630/R-94/009, 1994.

United States Environmental Protection Agency. *Proposed guidelines for ecological risk assessment*. Washington, DC: Risk Assessment Forum, United States Environmental Protection Agency. EPA/630/R-95/002B, 1996.

SAMPLE QUESTIONS

1. Describe the process of assessing the risk of a stressor to an ecosystem.

2. Describe the difference and relationship between an assessment end point and a measurement end point. Give an example.

3. Describe one (or two) fundamental differences between human health risk assessment and ecological risk assessment.

4. Describe the general relationship among the response time, sensitivity, level of biological organization, and ecological relevance of measurements that can be used to assess the response of an ecological receptor to a stressor.

NEW APPROACHES IN TOXICOLOGY

Perspectives on Informatics in Toxicology

SETH W. KULLMAN, CAROLYN J. MATTINGLY, JOEL N. MEYER, and ANDREW WHITEHEAD

28.1 INTRODUCTION

In this chapter, we discuss the application of bioinformatics and genomics as it applies to toxicology. This interaction, referred to as toxicogenomics, represents a component of toxicology that elucidates how the entire genome is involved in biological responses of organisms exposed to environmental chemical stressors. It combines information from studies of genomic-scale mRNA profiling, cell- or tissue-wide protein profiling (proteomics), alterations in the metabolome (metabolomics), genetic susceptibility, and computational models to understand the roles of gene–environment interactions in disease (Tennant, 2002). The goals of toxicogenomics are twofold: (1) to understand the underlying mechanism of toxicity and (2) to explain relationships between environmental or chemical stress and human disease (Waters and Fostel, 2004; Borgert, 2007). From the molecular perspective, chemical stressors are exogenous signals that influence gene expression, protein synthesis, and the resultant metabolome.

The knowledge that pollutant exposure influences gene expression was established long ago and historically, this relationship has been examined "one gene at a time." This approach has led numerous significant detailed discoveries in toxicology and has been the basis for our initial understanding and insight into the mechanistic action of specific genes and the regulatory pathways through which toxicants influence cell homeostasis. The logical extension of "one gene at a time" has been the development of transgenic and knockout animals. The development of these organisms has been instrumental in establishing the role of single genes in toxicology and disease and continues to be an important component relating gene function to animal health. With the development of new genomics tools and techniques including comparative genomics, gene expression profiling, genome association studies, and bioinformatics, complex interactions between genotype, phenotype, and environmental interaction are being established. Equally important is the development of similar approaches for the analysis of nontranscriptional events occurring in the proteome (e.g., protein phosphorylation and glycosylation) and the analysis of resultant metabolomes and the epigenome (e.g., histone modifications, DNA methylation). These new tools have facilitated new paradigms in biological sciences based upon the development of patterns of alteration termed "systems biology." The use of such technology is rapidly changing the way in which toxicology is conducted, emphasizing global changes at varying levels of biological organization.

The availability of draft genome sequences for human, mouse, rat, and other species extends the ability to study aspects of toxicology as it relates to the whole genome level. Integration of structural, comparative, and functional genomics through robust bioinformatics is facilitating a comprehensive view of the complex interactions between exposure and genetics. Genetic variation within populations and individuals is additionally a critical determinant of differences in response to xenobiotics and disease susceptibilities. The most common form of genetic variation is the single nucleotide polymorphism (SNP), a single-base substitution within the DNA. To date, over 2.8 million SNPs have been discovered and linked to discreet locations in the human genome. SNPs are highly stable and occur approximately once in every 1000–2000 nucleotides (http://www.ncbi.nlm.nih.gov/SNP/). With the development of genetic polymorphism databases, great strides have been made in understanding gene-by-environment interactions, and it is now well substantiated that many major diseases including cardiovascular disease, diabetes, obesity, and cancer result from the interaction between genetic susceptibility and environmental factors.

The reliance of bioinformatics in these studies is paramount. Development in computerized data management has made much of genomics possible, and advancements in bioinformatics continue to broaden genomic approaches. In this mini chapter, we detail a few genomic and bioinformatic platforms important in toxicology. We discuss computational tools and provide an overview of their application in toxicology.

28.2 TRANSCRIPTOMICS

Transcriptome profiling is useful in toxicogenomics as a discovery tool for identifying the biochemical mechanisms underlying compensatory responses to xenobiotics, illuminating potential mechanisms of toxicity, and as a biomarker for drug safety evaluation. Microarrays remain as the primary technology appropriate for transcriptome profiling, though an emerging alternative is to use next-generation massively parallel sequencing (MPS) to directly sequence and count transcripts from experimental samples (RNAseq or digital transcriptomics) (Wang et al., 2009). The new generation of MPS technologies overcomes many technical limitations inherent to the microarray platform. MPS offers a much wider dynamic range of detection (Nagalakshmi et al., 2008), higher throughput, and is more amenable to automation. Importantly, for research in species for which genomic resources are undeveloped, MPS offers species-specific transcript abundance data thereby avoiding biases associated with heterologous hybridization (Bar-Or et al., 2007). Additionally, the information content of RNAseq exceeds mRNA abundance, including alternative splicing (Wang et al., 2008) and small RNA detection.

Though advances in RNAseq are exciting, microarrays remain as the method of choice for most investigators for three main reasons. First, it is a mature technology with platforms available for most model species in toxicology research and for many non-model species. The barrier for entry (a reference transcriptome for probe design) has been significantly lowered in recent years with the advent of highly efficient cDNA library normalization techniques and sequencing technologies (e.g., see Meyer et al. 2009). Second, microarray-generated expression profiles are available for many species and chemicals in searchable databases to serve as a foundation for the pattern matching of expression profiles induced by novel compounds (Hayes et al., 2005; Smith et al., 2008). Finally, few individual investigators have access to the computational and bioinformatic resources necessary to manage and analyze the massive data sets generated by MPS, and the development of normalization and analysis pipelines is in its infancy (though likely not for long).

28.3 ANNOTATION RESOURCES

Because of the high-volume nature of -omics data, nearly all -omics research is dependent on annotation resources for both logistic and statistical reasons. Annotation refers to attaching biological information to -omics entities (a gene, transcript, protein, etc.). This information can be of many kinds, as described below; examples are identification of splice sites in gene structure, description of the function of a protein or its pattern of expression, identification of a gene's transcription factor binding sites or protein motifs, and so on.

28.3.1 Logistics

One of the strengths of -omics approaches in toxicology is the ability to detect responses to a stressor that are completely unexpected, via interrogation of the entire transcriptome, proteome, and so on. A difficulty associated with this benefit is that logistically, the sheer amount of data generated can be challenging to analyze. For example, if a microarray experiment identifies several hundred transcripts that are altered at different times after exposure to a stressor, it is likely that the investigator will be only superficially familiar with most. Performing exhaustive searches of the scientific literature in order to learn about each such transcript would be prohibitive; annotation resources permit researchers to quickly learn a great deal about long lists of -omics features and to organize those lists in a biologically meaningful fashion (e.g., biological function, expression pattern, and chromosomal location).

28.3.2 Statistics

Statistically, the large number of data points measured (e.g., 20,000+ transcripts per sample in a microarray experiment) means that it is rarely feasible to have a high enough n to obtain sufficient power to identify statistically significant changes via a traditional multiple testing adjustment such as a Bonferroni correction. Statistical approaches to this difficulty exist, including clustering algorithms and the adoption of less conservative multiple test corrections and false discovery rate analyses, sometimes in combination with a "fold-change" cut-off that must be arbitrarily chosen. Another is to statistically analyze only a subset of the data points measured, based on a prior hypothesis (Slotkin and Seidler, 2007). An approach that is both much more powerful statistically and based on decades of biological research is to carry out statistical tests on data points after grouping them in a biologically meaningful way, in an annotation-based fashion. Thus, for example, if transcripts are organized according to their biological function and then each group of biologically related transcripts is tested for unexpected enrichment in altered transcript levels, the resultant statistical tests are much more powerful. Analysis of transcriptomic data is more robust to laboratory-to-laboratory and platform-to-platform variability when using annotation-based grouping, than when using a one-gene-at-a-time approach (Bammler et al., 2005).

28.3.3   Types of Annotation

There are a large and growing number of annotation resources. The following list is by no means complete but will provide a sense of the types of annotation available.

Gene Ontology   Gene ontologies (http://www.geneontology.org/; The Gene Ontology Consortium, 2008) are attributes of gene products that are presumably species independent. Three general categories exist: molecular function (e.g., metal ion binding), biological process (e.g., oxidative phosphorylation), and cellular component (e.g., cytoplasmic). Gene products are described within each category in a hierarchical fashion. For example, the hierarchy of molecular functions for glutamate–cysteine ligase (the rate-limiting enzyme in glutathione synthesis) is glutamate–cysteine ligase activity; acid–amino acid ligase activity; ligase activity, forming carbon–nitrogen bonds; ligase activity; and catalytic activity. Many software tools have been developed to analyze -omics data in a GO context (Khatri and Draghici, 2005), including BiNGO (Maere et al., 2005), GOMiner (Zeeberg et al., 2003), and DAVID (Huang et al., 2009).

Kyoto Encyclopedia of Genes and Genomes (KEGG)   The KEGG strives to achieve "a complete computer representation of the cell, the organism, and the biosphere" (http://www.kegg.com/; Kanehisa et al., 2006). KEGG maintains several linked databases, including molecular pathways, gene catalogs, chemical compounds (including endogenous compounds, drugs, and pollutants) that function as ligands in cells, and pathway analysis and manipulation tools.

Interactomes   Interactomes are networks of all known interacting partners in a cell. The partners may be proteins, or DNA sequences, or other biological entities (e.g., calcium, ATP, and glutathione). For example, the aryl hydrocarbon receptor (AHR) and AHR nuclear translocator proteins would be connected by a protein–protein link, and both would connect to genes containing functional dioxin response elements (DREs) via protein–DNA links. It is also possible to use nonphysical links such as genetic interactions (two genes that exhibit synthetic lethality). It is possible to use interactomes that only contain very well-curated and high-confidence links or larger interactomes that include more entities and more links but also more uncertainty. There are a variety of databases that offer interactomes or interaction information, including Biomolecular Interaction Network Database (BIND; http://bond.unleashedinformatics.com/), Database of Interacting Proteins (DIP; http://dip.doe-mbi.ucla.edu/dip/Main.cgi), and species-specific interactomes published in the scientific literature. Many interactome analysis tools exist; one of the most well established is Cytoscape (Cline et al., 2007).

Pathways   Although some of the annotation resources described above incorporate pathway relationships, there are other projects that focus more directly on organizing genes or proteins by signaling, biochemical, or gene regulation pathways. These include Signaling Pathway Database (SPAD; http://www.grt.kyushu-u. ac.jp/spad/), TRANSFAC (http://www.gene-regulation.com/pub/databases.html), Biocarta (http://www.biocarta.com/genes/index.asp), PathDB (Biochemical Pathways; http://www.ncgr.org/pathdb/index.html), Gene MicroArray Pathway Profiler (GenMAPP; http://www.genmapp.org/), and Ingenuity Pathways Analysis (IPA; http://www.ingenuity.com/). Multiple analysis tools that facilitate the detection of pathways in which changes are observed are available.

Gene Expression Profile Similarity   Finally, there are a variety of tools and annotation resources available that allow a researcher to compare a novel -omic data set with previously published data sets. These include Gene Set Enrichment Analysis (GSEA; http://www.broadinstitute.org/gsea/) and the Cancer Genome Anatomy Project (CGAP; http://cgap.nci.nih.gov/).

28.4 GENOME SEQUENCING, RESEQUENCING AND GENOTYPING

Whole genome sequencing has generated reference genomes for many important toxicology models including human, mouse, rat, zebrafish, and Japanese medaka (http://www.ebi.ac.uk/genomes/). *De novo* sequencing and assembly of whole genomes is still not a mature science as sequencing technologies and assembly algorithms are constantly evolving, but much attention is now focusing on approaches for genome resequencing and high-coverage genotyping. Objectives common to many research programs focus on characterizing variation within and among human populations with the goal of identifying genetic polymorphisms that are predictive of or functionally related to disease or xenobiotic susceptibility. However, other species could emerge as powerful toxicogenomic models, for example, those in which tolerance to environmental pollution is genetically based, well defined, and clearly varies among populations such as the killifish *Fundulus heteroclitus* (Van Veld and Nacci, 2008; Williams and Oleksiak, 2008).

Massively parallel technologies have largely replaced Sanger technology for de novo and resequencing of genomes because of dramatic cost per base savings. However, this has so far come at the cost of shorter read length, which poses difficulties for assembly or mapping to a reference genome, thereby prompting the development of new assembly algorithms (such as Newbler designed by 454/ Roche, and ELAND developed by Illumina) coupled with other strategies such as large-insert paired-end sequencing. The massively parallel technologies that are currently available represent a huge advance over Sanger sequencing but are unlikely to yield the coveted "$1000 genome." This will require the advent of fundamentally different "third-generation" technologies such as single-molecule real-time (SMRT) sequencing (Eid et al., 2009) or nanopore sequencing (Branton et al., 2008; Clarke et al., 2009). Nanopore technology may offer very long reads, may require minimal sample preparation, and may directly detect not only the four typical bases but also methylated cytosine, thereby extending applications to epigenomic profiling (see below) (Clarke et al., 2009).

Though large-scale resequencing of genomes holds promise for yielding important insights into regions of genomic variation in humans, these goals are beyond those of most research programs. However, similar omic-scale technologies are relevant for association mapping and comparative population genomic studies that use high-coverage genotyping to identify the genes or genomic markers associated with disease or xenobiotic susceptibility (Jorde et al., 2001). Association studies survey variation in hundreds or thousands of markers distributed throughout the genome in many individuals and seek variants statistically associated with some disease phenotype, such as lung cancer (Amos et al., 2008), for example.

SNPs are the most common type of sequence variation and are often the markers of focus for association studies, but variation in gene copy number can also affect physiology or disease traits (Gonzalez et al., 2005; Perry et al., 2007). The most common platform for analyzing genome-wide patterns of copy number variation is comparative genomic hybridization to microarrays. SNPs are also typically assayed by hybridization to microarrays, though MPS-based approaches are gaining momentum (see, e.g., Baird et al., 2008). These techniques are most appropriate for organisms with sequenced genomes, but they also offer utility for organisms with reference transcriptomes and in some cases for organisms closely related to species that are genome enabled.

28.5 EPIGENOMIC PROFILING

Epigenomics refers to cellular phenomena (such as DNA methylation) that regulate gene expression independent of gene or promoter sequence at a global scale. Methylation patterns appear to vary significantly across genomic regions among plants, vertebrates, and invertebrates, prompting the question of whether common underlying mechanisms are conserved across species (Suzuki and Bird, 2008). Although epigenetic mechanisms are usually considered in the context of developmental biology, some recent evidence suggests that epigenetic phenomena may mediate a link between environmental exposures and disease (Jirtle and Skinner, 2007; Szyf, 2007) and are associated with the etiologies of some cancers (Feinberg and Vogelstein, 1983). Recent evidence is challenging the notion that methylation serves only to repress gene expression, since the inner regions of many active genes appear targeted for methylation in many organisms. Indeed, the "active" copy of the X chromosome in female mammals is more densely methylated than the "inactive" copy (Hellman and Chess, 2007).

The first complete methylome has been mapped for *Arabidopsis thaliana* (Zhang et al., 2006; Zilberman et al., 2007) and the first mammalian methylome is expected soon; the National Institutes of Health (NIH) has initiated a comprehensive program to develop resources and techniques for epigenomics research (http:// nihroadmap.nih.gov/epigenomics/). Several approaches are currently available for global methylome profiling (Beck and Rakyan, 2008). The first viable and currently most common approaches are microarray based, though these are now being challenged by MPS-based approaches. Regardless of the profiling techniques, genomic DNA is typically manipulated either by sodium bisulfite treatment, which converts only unmethlyated cytosines to uracils, restriction digestion by methylation-sensitive enzymes, or by immunoprecipitation of methylated DNA using an antibody specific to methylated cytosine (see Beck and Rakyan, 2008 for review). All of these approaches have been adapted for microarray profiling. Bisulfite conversion techniques offer single-base resolution of the methylation status, but often at the expense of genome coverage. However, immunoprecipitation-based techniques, which were used to map the *Arabidopsis* methylome, offer lower resolution, though single-base resolution may rarely be critical since the methylation status of neighboring regions is often correlated.

Though these techniques are powerful and are commonly applied, hybridization-based approaches are likely soon to be eclipsed by massively parallel bisulfite sequencing, which promises very high-throughput and high-genome coverage at a single-base resolution. This has been accomplished for *Arabidopsis* (Lister et al., 2008), though whole genome coverage for much larger mammalian genomes remains unfeasible and currently requires enrichment of target genomic regions. The reduced sequence complexity of bisulfite-treated samples (only three bases instead of four) provides a challenge for mapping these sequences to the genome, especially for short-read technologies. Zeschnigk et al. (2009) have designed algorithms to map the longer bisulfite-modified and -unmodified reads generated by 454/Roche sequencing to the reference genome in a study of methylation patterns in human blood and sperm.

28.6 COMPUTATIONAL TOXICOLOGY

Computational toxicology integrates mathematical and computer models with molecular biological and chemical approaches. This integration seeks to establish both qualitative and quantitative relationships between toxicant exposure and adverse health outcomes. Computational models in toxicology are applied to aid in the predictive power of toxicity evaluation (Kavlock et al., 2008). This evaluation can occur with the development of novel pharmaceuticals (Johnson and Rodgers, 2006) and/or the evaluation of risk posed by environmental stressors (Kavlock et al., 2008). Computational toxicology seeks to integrate large-scale chemical and biological data sets with predictive capabilities across multiple levels of biological organization. Thus, this process is reliant upon access to chemical and biological repositories that curate detailed chemical and biological information including absorption, distribution, metabolism, excretion (ADME) parameters, toxicity pathways, genomic data, pathway analysis and results of high-throughput screening, (HTS) and high-content screening (HCS). It is the combination of all these complementary types of data that is needed to build realistic models of the potential toxic effects of a compound on an organism (Nigsch et al., 2009).

Much of computational toxicology is driven by identifying structural features of chemicals known to cause defined toxicities (Kruhlak et al., 2007). The basis of this thinking is rooted in the toxicant–target paradigm (Rabinowitz et al., 2008) where an interaction is determined between a small molecule and a molecular target including genetic material, receptors, transport molecules, and enzymes. Based on these observations of interaction, chemicals with similar structural features may be predicted to act through the same biological mode of action. Many molecular models have been developed to assess theses interactions and have been employed for pharmaceutical development and toxicity assessment (Muster et al., 2008). Numerous commercial and free Web-based programs for this type of analysis are available (see Muster et al., 2008 for a partial listing).

The toxicant–target paradigm is currently used in many computational models for predicting toxicity. These models incorporate mathematical equations for defining physical and chemical features governing the behavior of interacting molecules. Experimental data are then incorporated to define putative targets and toxicities. When combined, computational descriptions provide a linkage between biomolecular structure and biological effect (Rabinowitz et al., 2008). Additionally, a chemoinformatic approach may be taken when structural information is limiting. Chemoinformatics is designed to integrate textual information available from data repositories on the biological activity of putative ligands (Kavlock et al., 2008).

There are many different computational models from which toxicity predictions can be established including mathematical or foundation algorithm, and the nature of predictions is either continuous or discrete (Nigsch et al., 2009). The common theme, however, is the relation of the toxicological indication to chemical structure. Predictive structurally focused toxicity tools fall into two categories, defining either qualitative rule-based structure–activity relationships (SARs) or quantitative structure–activity relationships (QSARs). SAR models use expert rules developed by panels of researchers that are applied to a test compound and yield a computer-automated prediction of toxicity. QSAR programs generate algorithms (models) by statistical inference from molecular descriptors and substructural molecular attributes that are correlated with toxicity (Kruhlak et al., 2008). Both methods are adept at identifying similar structural determinants associated with defined toxicities; however, QSAR is a more effective method when experimental toxicity data are lacking. Other models include target prediction, ligand-based models and protein structure-based models (Nigsch et al., 2009).

From the pharmaceutical perspective, application of computational models is envisioned to reduce the attrition rate of compounds during lead candidate drug discovery and development (Muster et al., 2008). This is driven by the cost, time, and feasibility of conducting extensive toxicity evaluation for large chemical combinatorial libraries. QSAR screening provides an *in silico* alternative to experimental approaches and enables the ability to flag undesired activities early in the development phase. A similar approach is being employed by the National Center for Computational Toxicology (NCCT), which seeks to develop computational methods to characterize pollutant exposure, hazard, and risk as component components of chemical screening and prioritization programs (http://www.epa.gov/comptox/comptoxfactsheet.html). The NCCT comprises multiple components necessary to both generate and house data for establishing risk models. The ToxCast program (http://www.epa.gov/ncct/toxcast/) incorporates state-of-the-art high-throughput bioassays and is poised to aid in the prioritization of chemicals for more detailed toxicological evaluations and for animal testing (Dix et al., 2007). Supporting ToxCast are several relational databases including Toxicity Reference Database

(ToxRefDB) (http://www.epa.gov/ncct/toxrefdb/), which compiles and stores reference *in vivo* animal toxicity data including chronic, subchronic, developmental cancer and reproductive end points; the Distributed Structure-Searchable Toxicity (DSSTox) Database Network, a chemical data foundation for improved structure–activity and predictive toxicology capabilities (Williams-DeVane et al., 2009), and the Aggregate Computational Toxicology Resource (ACToR) (http://actor.epa.gov/ actor/faces/ACToRHome.jsp;jsessionid=6E2B41F7618FF9D9276B6FBC4671 59C0), a tool designed to aid in the management of large-scale bioassay and toxicology data (Judson et al., 2009). A similar electronic toxicology knowledge-based computational database has also been established through the Food and Drug Administration (FDA) Center for Drug Evaluation and Research (CDER) (http:// www.fda.gov/Drugs/default.htm). Here the FDA aims to develop QSAR models to facilitate enhanced scientific decision support for the regulatory review process.

28.7 INFORMATICS TOOLS IN TOXICOLOGY

The increasing number of publicly available biological databases reflects their essential role in biological and clinical research as well as the evolving need for managing and evaluating abundant and complex data. Currently, there are over 1000 biologically relevant databases in the public domain (Galperin, 2008), although only a few are dedicated to environmental chemicals and to understanding their mechanisms of action or impact on human health. Three unique resources are described here that highlight different toxicological perspectives and applications, the Comparative Toxicogenomics Database (CTD; http://ctd.mdibl.org), Chemical Effects in Biological Systems (CEBS; http://www.niehs.nih.gov/research/resources/ databases/cebs/index.cfm), and Environment, Drugs and Gene Expression (EDGE; http://edge.oncology.wisc.edu).

CTD is a publicly available resource that promotes understanding about the effects of environmental chemicals on human health (Davis et al., 2008, 2009). Despite the many technological advances incorporated into toxicology research in recent years, the mechanisms of action of most chemicals and their biological impact remain poorly understood. CTD provides manually curated data that describe chemical–gene/protein interactions (e.g., transcription, binding, methylation) and chemical–disease and gene–disease relationships. Currently, CTD contains over 184,600 interactions between over 5000 chemicals and 16,000 genes as well as more than 9300 gene–disease and 5900 chemical–disease relationships. All data are curated using controlled vocabularies that ensure consistency of curation and data retrieval in addition to providing hierarchical searching options. For example, users may search for chemicals or diseases by specific terms (e.g., sodium arsenite or kidney cancer, respectively) or by broad categories (e.g., heavy metals or cancer, respectively). Curated data are also integrated with select external data sets such as those from Gene Ontology and KEGG pathways. The level of integration in CTD allows users to conduct very sophisticated queries easily (e.g., which genes involving membrane transport are affected by arsenic? Which neurological diseases are associated with heavy metal exposure?). Several analysis tools are now offered. Batch Query allows users to retrieve associated data sets for lists of chemicals, genes/proteins or diseases of interest. VennViewer generates a Venn diagram comparing associated data sets for up to three user-defined chemicals, genes, or diseases. MyGeneVenn allows users to compare their gene lists of interest (e.g., from a microarray experiment) to CTD gene sets for user-defined chemicals or diseases. CTD also computes similarity indices to provide lists of comparable genes (GeneComps) or chemicals (ChemComps) based on shared chemical–gene interaction profiles. These comparable data sets provide alternative methods for grouping chemicals or genes based on biological effects or chemical responses, respectively. The novel data sets and deep integration of data in CTD provide opportunities to develop novel hypotheses about chemical–genes/proteins–disease interaction networks.

CEBS is a data repository and tool for primary toxicogenomics data including microarray, proteomics, clinical chemistry, hematology, and histopathology data (Waters et al., 2008). All data in CEBS are highly annotated and include many details about study design, including treatment parameters and study timeline. Query capabilities allow users to search across data domains and studies by a range of criteria such as subject characteristics (e.g., strain, age), subject responses (e.g., histopathology, clinical chemistry results), protocol parameters (e.g., diet, time of sacrifice), and study characteristics (e.g., experimental factors, principal investigator name). Query results can be downloaded as an integrated data matrix for import into statistical or visualization tools for further analysis.

EDGE is a Web-based application for toxicology-related gene expression information that enables storage, sophisticated analysis, and sharing of microarray data among diverse research groups (Vollrath et al., 2009). This combination of features is important because it provides access to data management and analysis capabilities without significant financial investment or training and allows straightforward, customized queries of microarray data, which can be conducted across multiple experiments. Such cross-experimental search capabilities have many valuable applications such as gaining insight into genes that are commonly affected by exposure to a particular compound or, alternatively, are *only* affected under certain conditions (e.g., liver vs. kidney).

BIBLIOGRAPHY AND SUGGESTED READING

Amos, C. I., X. F. Wu, P. Broderick, et al. Genome-wide association scan of tag SNPs identifies a susceptibility locus for lung cancer at 15q25.1. *Nat. Genet.* **40**:616–622, 2008.

Baird, N. A., P. D. Etter, T. S. Atwood, et al. Rapid SNP discovery and genetic mapping using sequenced RAD markers. *PLoS ONE* **3**:e3376, 2008.

Bammler, T., R. P. Beyer, S. Bhattacharya, et al. Standardizing global gene expression analysis between laboratories and across platforms. *Nat. Methods* **2**:351–356, 2005.

Bar-Or, C., H. Czosnek, and H. Koltai. Cross-species microarray hybridizations: A developing tool for studying species diversity. *Trends Genet.* **23**:200–207, 2007.

Beck, S. and V. K. Rakyan. The methylome: Approaches for global DNA methylation profiling. *Trends Genet.* **24**:231–237, 2008.

Borgert, C. J. Predicting interactions from mechanistic information: Can omic data validate theories? *Toxicol. Appl. Pharmacol.* **223**:114–120, 2007.

Branton, D., D. W. Deamer, A. Marziali, et al. The potential and challenges of nanopore sequencing. *Nat. Biotechnol.* **26**:1146–1153, 2008.

Clarke, J., H.-C. Wu, L. Jayasinghe, et al. Continuous base identification for single-molecule nanopore DNA sequencing. *Nature Nanotech.* **4**:265–270, 2009.

Cline, M. S., M. Smoot, E. Cerami, et al. Integration of biological networks and gene expression data using Cytoscape. *Nat. Protoc.* **2**:2366–2382, 2007.

Davis, A. P., C. G. Murphy, M. C. Rosenstein, et al. The Comparative Toxicogenomics Database facilitates identification and understanding of chemical-gene-disease associations: Arsenic as a case study. *BMC Med. Genomics* **1**:48, 2008.

Davis, A. P., C. G. Murphy, C. A. Saraceni-Richards, et al. Comparative Toxicogenomics Database: A knowledgebase and discovery tool for chemical-gene-disease networks. *Nucleic Acids Res.* **37** (Database issue): D786–D792, 2009.

Dix, D. J., K. A. Houck, M. T. Martin, et al. The ToxCast program for prioritizing toxicity testing of environmental chemicals. *Toxicol. Sci.* **95** (1): 5–12, 2007.

Eid, J., A. Fehr, J. Gray, et al. Real-time DNA sequencing from single polymerase molecules. *Science* **323**:133–138, 2009.

Feinberg, A. P. and B. Vogelstein. Hypomethylation distinguishes genes of some human cancers from their normal counterparts. *Nature* **301**:89–92, 1983.

Galperin, M. Y. The molecular biology database collection: 2008. *Nucleic Acids Res.* **36** (Database issue): D2–D4, 2008.

Gonzalez, E., H. Kulkarni, H. Bolivar, et al. The influence of CCL3L1 gene-containing segmental duplications on HIV-1/AIDS susceptibility. *Science* **307**:1434–1440, 2005.

Hayes, K. R., A. L. Vollrath, G. M. Zastrow, et al. EDGE: A centralized resource for the comparison, analysis, and distribution of toxicogenomic information. *Mol. Pharmacol.* **67**:1360–1368, 2005.

Hellman, A. and A. Chess. Gene body-specific methylation on the active X chromosome. *Science* **315**:1141–1143, 2007.

Huang, D. W., B. T. Sherman, and R. A. Lempicki. Systematic and integrative analysis of large gene lists using DAVID bioinformatics resources. *Nat. Protoc.* **4**:44–57, 2009.

Jirtle, R. L. and M. K. Skinner. Environmental epigenomics and disease susceptibility. *Nat. Rev. Genet.* **8**:253–262, 2007.

Jorde, L. B., W. S. Watkins, and M. J. Bamshad. Population genomics: A bridge from evolutionary history to genetic medicine. *Hum. Mol. Genet.* **10**:2199–2207, 2001.

Judson, R., A. Richard, D. J. Dix, et al. The toxicity data landscape for environmental chemicals. *Environ. Health Perspect.* **117** (5): 685–695, 2009.

Johnson, D. E. and A. D. Rodgers. Computational toxicology: Heading toward more relevance in drug discovery and development. *Curr. Opin. Drug Discov. Devel.* **9** (1): 29–37, 2006.

Kanehisa, M., S. Goto, M. Hattori, et al. From genomics to chemical genomics: New developments in KEGG. *Nucleic Acids Res.* **34** (Database issue): D354–D357, 2006.

Kavlock, R. J., G. Ankley, J. Blancato, et al. Computational toxicology—A state of the science mini review. *Toxicol. Sci.* **103** (1): 14–27, 2008.

Khatri, P. and S. Draghici. Ontological analysis of gene expression data: Current tools, limitations, and open problems. *Bioinformatics* **21**:3587–3595, 2005.

Kruhlak N. L., J. F. Contrera, R. D. Benz, and E. J. Matthews. Progress in QSAR toxicity screening of pharmaceutical impurities and other FDA regulated products. *Adv. Drug Deliv. Rev.* **59** (1): 43–55, Jan 10, 2007.

Lister, R., R. C. O'Malley, J. Tonti-Filippini, et al. Highly integrated single-base resolution maps of the epigenome in *Arabidopsis*. *Cell* **133**:523–536, 2008.

Maere, S., K. Heymans, and M. Kuiper (2005) BiNGO: A Cytoscape plugin to assess overrepresentation of gene ontology categories in biological networks. *Bioinformatics* **21**:3448–3449.

Meyer, E., G. V. Aglyamova, S. Wang, et al. Sequencing and *de novo* analysis of a coral larval transcriptome using 454 GSFlx. *BMC Genomics* **10**:219, 2009.

Muster, W., A. Breidenbach, H. Fischer, S. Kirchner, L. Müller, and A. Pähler. Computational toxicology in drug development. *Drug Discov. Today*. **7–8**:303–310, 2008.

Nagalakshmi, U., Z. Wang, K. Waern, et al. The transcriptional landscape of the yeast genome defined by RNA sequencing. *Science* **320**:1344–1349, 2008..

Nigsch, F., N. J. Macaluso, J. B. Mitchell, et al. Computational toxicology: An overview of the sources of data and of modelling methods. *Expert Opin. Drug Metab. Toxicol.* **5** (1): 1–14, 2009.

Perry, G. H., N. J. Dominy, K. G. Claw, et al. Diet and the evolution of human amylase gene copy number variation. *Nat. Genet.* **39**:1256–1260, 2007.

Rabinowitz, J. R., M. R. Goldsmith, S. B. Little, et al. Computational molecular modeling for evaluating the toxicity of environmental chemicals: Prioritizing bioassay requirements. *Environ. Health Perspect.* **116** (5): 573–577, 2008.

Slotkin, T. A. and F. J. Seidler. Comparative developmental neurotoxicity of organophosphates in vivo: Transcriptional responses of pathways for brain cell development, cell signaling, cytotoxicity and neurotransmitter systems. *Brain Res Bull.* **72**:232–274, 2007.

Smith, A. A., A. Vollrath, C. A. Bradfield, et al. Similarity queries for temporal toxicogenomic expression profiles. *PLoS Comput. Biol.* **4**:e1000116, 2008.

Suzuki, M. M. and A. Bird. DNA methylation landscapes: Provocative insights from epigenomics. *Nat. Rev. Genet.* **9**:465–476, 2008.

Szyf, M. The dynamic epigenome and its implications in toxicology. *Toxicol. Sci.* **100**:7–23, 2007.

Tennant R. W. The National Center for Toxicogenomics: using new technologies to inform mechanistic toxicology. *Environ. Health Perspect.* **110** (1):A8–A10, 2002.

The Gene Ontology Consortium. The Gene Ontology project in 2008. *Nucleic Acids Res.* **36** (Database issue): D440–D444, 2008.

Van Veld, P. A. and D. E. Nacci. Toxicity resistance. In *The Toxicology of Fishes*, eds. R. T. Di Giulio and D. E. Hinton, pp. 597–641. Boca Raton, FL: Taylor and Francis, 2008.

Vollrath, A. L., A. A. Smith, M. Craven, et al. EDGE3: A web-based solution for management and analysis of Agilent two color microarray experiments. *BMC Bioinformatics*, **10**:280, 2009.

Wang, E. T., R. Sandberg, S. J. Luo, et al. Alternative isoform regulation in human tissue transcriptomes. *Nature* **456**:470–476, 2008.

Wang, Z., M. Gerstein, and M. Snyder. RNA-Seq: A revolutionary tool for transcriptomics. *Nat. Rev. Genet.* **10**:57–63, 2009.

Waters, M., S. Stasiewicz, B. A. Merrick, et al. CEBS—Chemical Effects in Biological Systems: A public data repository integrating study design and toxicity data with microarray and proteomics data. *Nucleic Acids Res.* **36** (Database issue): D892–D900, 2008.

Waters, M. D. and J. M. Fostel. Toxicogenomics and systems toxicology: Aims and prospects. *Nat. Rev. Genet.* **5**:936–948, 2004.

Williams, L. M. and M. F. Oleksiak. Signatures of selection in natural populations adapted to chronic pollution. *BMC Evol. Biol.* **18**:282, 2008.

Williams-DeVane, C. R., M. A. Wolf, and A. M. Richard. DSSTox chemical-index files for exposure-related experiments in ArrayExpress and Gene Expression Omnibus: Enabling toxico-chemogenomics data linkages. *Bioinformatics* **25** (5): 692–694, 2009.

Zeeberg, B. R., W. Feng, G. Wang, et al. GoMiner: A resource for biological interpretation of genomic and proteomic data. *Genome Biol.* **4**:R28, 2003.

Zeschnigk, M., M. Martin, G. Betzl, et al. Massive parallel bisulfite sequencing of CG-rich DNA fragments reveals that methylation of many X-chromosomal CpG islands in female blood DNA is incomplete. *Hum. Mol. Genet.* **18**:1439–1448, 2009.

Zhang, X. Y., J. Yazaki, A. Sundaresan, et al. Genome-wide high-resolution mapping and functional analysis of DNA methylation in *Arabidopsis. Cell* **126**:1189–1201, 2006.

Zilberman, D., M. Gehring, R. K. Tran, et al. Genome-wide analysis of *Arabidopsis thaliana* DNA methylation uncovers an interdependence between methylation and transcription. *Nat. Genet.* **39**:61–69, 2007.

SAMPLE QUESTION

What are some advantages offered by MPS technologies to the toxicogenomics toolkit? What are some drawbacks?

Future Considerations

ERNEST HODGSON

29.1 INTRODUCTION

Since the publication of the third edition of this textbook, there have been developments that will, in the future, dramatically impact toxicology, particularly in the areas of toxicity testing and in human health risk analysis. It is still true, however, that speculation concerning future developments in toxicology can be made only against an assessment of where the science has come from and its current status. Toxicology, despite its use of many state-of-the-art techniques and explorations of the most fundamental molecular mechanisms of toxic action, is, at its heart, an applied science serving the needs of society. Society is served in a number of important ways, including the protection of human health, the protection of the environment, the development of new, safer, chemicals for clinical and occupational use, and the safe manufacture of consumer products without significant toxic effects. Future studies will range from mechanisms of toxic action to *in vivo* toxicity testing, but the ultimate goal will remain: a meaningful assessment of risk resulting from exposure to chemicals.

Biology, including toxicology, has been largely reductionist throughout its history, studying organs as components of organisms, cells as components of organs, proteins, nucleic acids, and so on, as components of cells, with the ultimate goal of describing function in individual organisms at the molecular level. More recent developments, led by rapid and dramatic advances have caused a shift to a more holistic systems biology approach.

The vast increase in public awareness of the potential of chemicals to cause harmful effects and the propensity of the print and electronic media to fan the flames of controversy in this area make certain the continued need for toxicologists. We need to ask, however, what they will be doing during the next few decades compared to what they have been doing in the immediate past.

Through the 1950s and 1960s, toxicology tended to be a largely descriptive science, relating the results of *in vivo* dosing to a variety of toxic end points, in many cases a little more than the median lethal dose (LD_{50}) or the median lethal

A Textbook of Modern Toxicology, Fourth Edition. Edited by Ernest Hodgson
Copyright © 2010 John Wiley & Sons, Inc.

concentration (LC_{50}). However, ongoing studies of xenobiotic-metabolizing enzymes were attracting more attention, and techniques for chemical analysis of toxicants were starting to undergo a remarkable metamorphosis. The 1970s were most remarkable for developments in metabolism and the beginnings of a boom in mechanism of toxic action studies, whereas the 1980s and 1990s saw the incorporation of the techniques of molecular biology into many aspects of toxicology, but perhaps to the greatest effect in studies of the mechanisms of chemical carcinogenesis and the induction of xenobiotic-metabolizing enzymes.

As a consequence of the large number of chemicals to be tested under a number of federal statutes and the extensive backlog of untested chemicals, it is necessary to develop high-throughput, rapid assays that can handle many chemicals simultaneously. From both cost and animal welfare considerations, it will not be possible to use whole animal studies; assays based on human proteins and cell lines will doubtless be used. Testing systems based on quantitative structure–activity relationships (QSARs), using engineered human cell lines and the techniques of genomics, proteomics, metabolomics (below) as well as bioinformatics (Chapter 28), are currently being developed. Systems biology will emerge as the overarching, holistic approach that will make possible the integration of the large databases of the future.

It should be emphasized that all of these activities will proceed simultaneously; for example, the tremendous increase in specificity and sensitivity of chemical methods has proceeded simultaneously with the introduction of molecular biological techniques into studies of mechanisms of toxic action. It is also true, however, that increased emphasis and interest in any particular area is often preceded by the development of new techniques.

The success of the project to sequence the human genome along with the progress of the International HapMap Project to identify polymorphisms in genes coding for human xenobiotic-metabolizing enzymes and other proteins has led to the ability to define populations and individuals at increased risk from a particular chemical insult. This ability is being extended and put on a more mechanistic basis by advances in the new disciplines of genomics, proteomics, and metabolomics. Although privacy issues remain to be resolved, it appears inevitable that in the future, individual genotyping will play a role in medical practice and in the avoidance of toxicity by susceptible individuals.

The future, both immediate and long term, will provide important information on all aspects of toxic action, and the role of toxicology in public life will mature as the importance of toxicology is perceived by the population in general, first in developed countries and ultimately around the world. The fundamental role of the toxicologist, namely, the acquisition and dissemination of information about all aspects of the deleterious effects of chemicals on living organisms, will not change; however, the manner in which it is carried out is changing and will continue to change. The next several decades will be exciting times for toxicologists, and those in training at this time have much to anticipate. Change can be expected in almost every aspect of both the applied and the fundamental aspects of toxicology: risk communication, risk assessment, hazard and exposure assessment, *in vivo* toxicity, development of selective chemicals, *in vitro* toxicology, and biochemical and molecular toxicology will all change, as will the integration of all of these areas into new paradigms of risk assessment and of the ways in which chemicals affect human health and the environment.

29.2 RISK ASSESSMENT

In the past, risk assessment consisted largely of computer-based models written to start from hazard assessment assays, such as chronic toxicity assays on rodents, encompassed the necessary extrapolations between species and between high and low doses, and then produced a numerical assessment of the risk to human health. Although the hazard assessment tests and the toxic end points are different, an analogous situation exists in environmental risk assessment. A matter of considerable importance, now getting some belated attention, is the integration of human health and environmental risk assessments.

Although many of these risk assessment programs were statistically sophisticated, they frequently did not rise above the level of number crunching, and more often than not, different risk assessment programs, starting with the same experimental values, produced very different numerical assessments of risk to human health or to the environment. Although having been a stated goal of regulators for decades, having risk assessment become more science based has not progressed significantly, although the need to incorporate mechanistic data, including mode of action studies and physiologically based pharmacokinetics, has been realized to some extent. Apart from epidemiology and exposure analysis, human studies have not been utilized to a significant extent, despite the fact that many such studies can now be performed using noninvasive and ethical methods.

The immediate future in risk assessment will focus on the difficult but necessary task of integrating experimental data from all levels into the risk assessment process. A continuing challenge to toxicologists engaged in hazard or risk assessment is that of risk from chemical mixtures. Neither human beings nor ecosystems are exposed to chemicals one at a time, yet logic dictates that the initial assessment of toxicity starts with individual chemicals. The resolution of this problem will require considerable work at all levels, *in vivo* and *in vitro*, into the implications of chemical interactions for the expression to toxicity, particularly chronic toxicity.

29.2.1 Hazard and Exposure Assessment

The enormous cost of multispecies, multidose, lifetime evaluations of chronic effects has already made the task of carrying out hazard assessments of all chemicals in commercial use impossible and, as indicated above, we are in a phase of development of high-throughput techniques that can investigate many chemicals simultaneously. QSARs, which will be an important adjunct to such methods, are not yet predictive enough to indicate which chemicals should be tested further and which chemicals need not be tested. It is likely that for some time, the new techniques will function to establish priorities as to which chemicals need to be investigated by the traditional toxicity testing techniques currently in use (Chapter 20). In exposure assessment, continued development of analytical methods will permit ever more sensitive and selective determinations of toxicants in food and the environment, as well as the effects of chemical mixtures and the potential for interactions that affect the ultimate expression of toxicity. Developments in QSAR, in short-term tests based on the expected mechanism of toxic action and simplification of chronic testing procedures, will all be necessary if the chemicals to which the public and the environment are exposed are to be assessed adequately for their potential to cause harm.

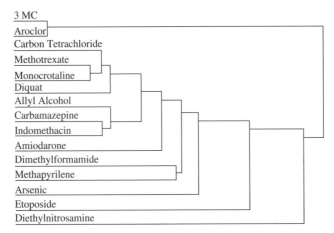

Figure 29.1 Dendrogram showing clustering of hepatotoxicants based on gene expression profiles generated from RNA and derived from microarrays. Adapted from Waring et al., *Toxicol. Lett.* **120**:359–368, 2001.

29.2.2 Toxicogenomics

Toxicogenomics is the initial step in a "systems toxicology" approach starting with genome-scale RNA expression (transcriptomics) followed by other emerging subdisciplines such as protein expression (proteomics), metabolite profiling (metabolomics), bioinformatics, and conventional toxicology/pathology to elucidate gene–environment interactions and can be utilized to identify biomarkers of toxicity.

The sometimes stated distinction that genomics deals with genomes while molecular biology deals with single genes is unrealistic and unnecessary; it is more appropriate to regard genomics as an aspect of molecular biology that deals not only with genomes and gene expression but also with such important aspects as genetic polymorphisms, particularly single nucleotide polymorphisms (SNPs). Techniques, such as microarrays, are now available to examine simultaneously the expression of very large numbers of genes. An example of the use of these techniques is seen in Figure 29.1.

29.2.3 Proteomics

Proteomics deals with the protein complement of organisms, the entire complement being known as the proteome. Thus, while genomics is concerned with gene expression, proteomics examines the products of the expressed genes.

The proteome is defined as the protein complement present in the biological unit (e.g., cell, organ, and organism) and represents that portion of the genome being expressed at that point in time. Proteomics is represented by broad, inclusive techniques to separate, identify, and study the structure of the proteins of the proteome. Separation is usually by two-dimensional polyacrylamide gel electrophoresis and identification by a number of variants of mass spectrometry.

29.2.4 Metabolomics

Metabolomics is the next step in the sequence following genomics and proteomics and is concerned with the profile of small molecules produced by the metabolic processes of an organism. Changes in the profile in response to chemical stress are of importance to both fundamental and applied toxicology.

While genomics has the goal of determining, through analysis of mRNA, which genes are being expressed and proteomics has the goal of determining whether expression of mRNA results in protein synthesis, metabolomics has the goal of determining whether the expressed proteins are metabolically active. Metabolomics is, therefore, the identification and quantification of all of the metabolites in a biological system at some point in time using blood or urine samples. It is important to remember that the metabolites in question are the products of the normal endogenous metabolism of the cell, organ, or organism and are not the metabolic products of toxicants or other xenobiotics, although in the latter case, the techniques of metabolomics can be invaluable.

Given the large number, chemical diversity, and concentration range of the entire metabolome, of necessity, a number of techniques are needed to obtain the complete picture needed. Initially, an unbiased extraction technique must be selected or developed. Since no single extraction technique is likely to extract all metabolites, several techniques are usually employed. Metabolite identification depends on two sensitive techniques: mass spectrometry and nuclear magnetic resonance spectroscopy.

29.2.5 Systems Biology Approach to Risk Assessment

These integrative developments, including the establishment and development of toxicogenomics, proteomics, and metabolomics as important subdisciplines, have led to the integrative subdiscipline of bioinformatics (see Chapter 28) and the overarching discipline of systems biology. The original definition of bioinformatics was the application of information technology to molecular biology. While still the most important aspect, bioinformatics is increasingly applied to other fields of biology, including molecular and other aspects of toxicology. It is characterized by computationally intensive methodology and includes the design of large databases and the development of techniques for their manipulation, including data mining.

Although systems biology has been defined in a number of ways, some involving quite simple approaches to limited problems, in the currently most commonly accepted sense, it is an integrative approach to biological structure and function that will be of increasing importance to biology in general and toxicology in particular. In a large part, biology has been reductionist throughout its history, studying organs as components of organisms, cells as components of organs, and proteins, nucleic acids, and so on, as components of cells and subcellular organelles, with the goal of describing function at the molecular level. Systems biology, on the other hand, is holistic and has the objective of discerning interactions between components of biological systems and integrating these interactions into rigorous mathematical models. Furthermore, the ultimate goal of the proponents of systems biology is to integrate these models at higher and higher levels of organization in order to develop an integrated model of an entire organism.

Clearly, systems biology is in its infancy. In order to attain the ambitious goal of creating a mathematically rigorous model at even the cellular level, it is necessary to have available results from global genomic, proteomic, and metabolomic studies. To extend this to whole organisms will be even more difficult.

Difficulties aside, the ultimate value of having an integrative model that could clarify all of the effects, from the most proximate to the ultimate, of a toxicant on a living organism will provide enormous benefits not only for fundamental studies but also in such applied areas as human health risk assessment.

Movement in this direction had begun with the formation of the National Institute for Environmental Health Sciences' (NIEHS) National Center for Toxicogenomics and Toxicogenomic Research Consortium, and in 2005 when the United States Environmental Protection Agency (EPA) established the National Center for Computational Toxicology, a part of which is the ToxCast program. The ToxCast program has tested several hundred chemicals using high-throughput methodology. The impetus for these developments was maintained by an influential 2007 National Academy study, "Toxicity Testing in the 21st century; a Vision and a Strategy." The ultimate goal is to identify signature effects of chemicals on biological systems and pathways, such as signaling pathways, and to categorize classes of chemicals by their ability to illicit toxicity in specific organs. The number and complexity of these pathways to be assessed have yet to be fully elucidated, but identifying such signatures of toxicity could greatly aid in the early screening of new chemicals for their potential to cause toxicity and could provide insight into the mechanisms responsible.

29.2.6 Endocrine Disruptors

An area that involves a more traditional approach, testing chemicals for activity as endocrine disruptors for regulatory purposes, will see increased activity in the near future. How this should be done has been under consideration by EPA for over a decade. During that time, some new tests have been developed and both new and existing tests have been validated. Those assays under consideration in 2007 (http://www.epa.gov/endo/pubs/assayvalidation/status.htm) are shown in Table 29.1.

As recently as April 2009, EPA released a "Final List of Chemicals for Tier 1 Screening in the EDSP" that consisted of 67 chemicals, primarily pesticide active ingredients. In fact, this is not a "final" list; rather, it is the final decision on those chemicals to be initially tested to inaugurate the program.

29.2.7 Genetically Modified Plants (GMPs)

The importance of a new group of potential toxicants, GMPs and their constituents, has emerged in the past two decades. Potentially a boon to the human race, they have already generated considerable controversy. While these products of applied molecular biology appear to be relatively harmless both to human health and to the environment, they will need to be monitored as they increase in number and complexity.

TABLE 29.1   Endocrine Disruptor Assays under Consideration by the U.S. EPA in 2007

Tier 1: Screening Assays

Amphibian (frog) metamorphosis
Receptor binding *in vitro* assays
   Androgen receptor binding utilizing rat prostate cytosol
   Androgen receptor binding utilizing rat recombinant androgen receptor
   Estrogen receptor binding utilizing rat uterine cytosol
   Estrogen receptor binding utilizing the alpha isoform of human recombinant estrogen
     receptor
Aromatase inhibition
Fish screen assay for estrogenic and androgenic effects
Hershberger test for androgenic and anti-androgenic effects
Pubertal female test for estrogenic and thyroid effects during maturation
Pubertal male test for androgenic, anti-androgenic, and thyroid effects during maturation
Steroidogenesis tests for effects on steroid hormone production
Uterotrophic assay for estrogenic effects in rats
15-day adult intact male test for anti-androgenic and thyroid effects

Tier 2: Screening Assays

Amphibian development
Reproductive and development effects in Japanese quail
Reproductive and development effects in rats
Fish and invertebrate life cycle effects

29.3 RISK MANAGEMENT

Public decisions concerning the use of chemicals will continue to be a blend of science, politics, and law, with the media spotlight continuing to shine on the most contentious aspects: the role of the trained toxicologist to serve as the source of scientifically sound information and as the voice of reason will be even more critical. As the chemist extends our ability to detect smaller and smaller amounts of toxicants in food, air, and water, the concept that science, including toxicology, does not deal in certainty but only in degrees of certitude must be made clear to all. Although this concept is easy for most scientists to grasp, it appears difficult, even arcane, to the general public and almost impossible to the average attorney or politician. Risk will have to be managed in the light of our newfound ability to identify individuals and populations at increased risk and to accommodate new legislation such as, for example, the Food Quality Protection Act.

29.4 RISK COMMUNICATION

Risk communication is essentially education and should take at all levels of formal education and to the general public. This approach is summarized in Chapter 22, specifically Section 22.4. The use of electronic media will continue to be of increasing importance. Since much of the toxicological information to be found on the web is unedited, is not peer reviewed, and is frequently both incorrect and alarmist, it will be essential that the trained toxicologist misses no opportunity to be "the voice of reason, presenting a balanced view of risks and benefits, and outlining alternatives whenever possible."

Risk communication to risk managers, to regulators, and to legislators is a specialized area of risk communication that will become increasingly important. If both overregulation and underregulation are to be avoided, not only must risk assessments be science based, but they must also be realistic and communicated without bias.

29.5 *IN VIVO* TOXICITY

Although developments continue in elucidating the mechanisms of chemical carcinogenicity, much remains to be done with regard to this and other chronic end points, particularly developmental and reproductive toxicity, chronic neurotoxicity, and immunotoxicity. The further utilization of the methods of molecular biology will bring rapid advances in all of these areas and will aid in the identification of biomarkers that arise early in the processes of chemical toxicity and carcinogenesis. *In vivo* studies utilizing multiple rodent strains that are genetically diverse will aid in identifying useful biomarkers as an attempt to mimic the genetic heterogeneous nature of human populations. It will be a challenge to integrate all of this information into useful paradigms for responsible and meaningful risk assessments.

29.6 *IN VITRO* TOXICITY

In vitro studies of toxic mechanisms will depend heavily on developments in molecular biology, and great advances can be expected. Many of the ethical problems associated with carrying out studies on the effects of toxicants on humans will be circumvented at the *in vitro* level by the use of human cell lines, genetically engineered human cells, cloned and expressed human enzymes, receptors, and so on. Mechanistic toxicity studies and the integration of these data into intact organism models will doubtless still require experimental animals. Ultimately, systems biology will be essential if intact organism models, whether of humans or of surrogate animals, are to be achieved. High-throughput technology in genomics, proteomics, and metabolomics will greatly facilitate these studies.

29.7 MOLECULAR AND BIOCHEMICAL TOXICOLOGY

Molecular biology, the study of the biochemical and molecular processes within cells, especially the processes of DNA replication, RNA transcription, and protein translation, has been widely adopted in toxicology. Molecular toxicology, the study of the effects of toxicants on these parameters, relies heavily on the same methods as molecular biology, namely, molecular cloning, polymerase chain reaction (PCR), Northern blotting and real-time PCR, transfection assays, microarray assays, bioinformatics, and systems biology. A useful example of this is to be found in the systems approach to dose–response modeling (Figure 29.2).

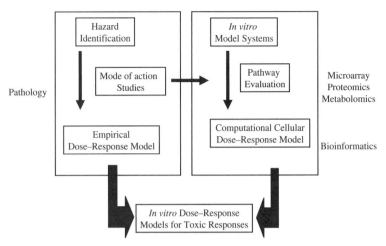

Figure 29.2 Systems approaches for dose–response modeling. Adapted from Anderson et al., *Reprod. Toxicol.* **19**:327–337, 2005.

Smart and Hodgson (2008) is an extensive summary of the current status of molecular and biochemical toxicology. As indicated previously, contributions to all aspects of the mechanistic study of toxic action from the use of biochemical and molecular techniques can be expected. There is no doubt that new techniques will be developed; answers will be found to many questions that did not yield to earlier techniques; and new questions will be raised. The challenge, as always, will be to integrate the results from these studies—studies that will reach new levels of sophistication—into useful and productive approaches to reduce chemical effects on human health and the environment. Again, the techniques of genomics, proteomics, metabolomics, bioinformatics, and ultimately, systems biology, will be essential.

29.8 DEVELOPMENT OF SELECTIVE TOXICANTS

Almost all aspects of contemporary human society depend on the use of numerous chemicals. Except in the unlikely event that society decides to return to a more simplistic and, in fact, more primitive, more unhealthy, and more demanding lifestyle, the challenge is in learning how to live with anthropomorphic chemicals, not in learning how to live without them. In many aspects, such as the production of food and fiber and the maintenance of human health, the development of selective pesticides, drugs, and so on, is still needed. New techniques in molecular biology, in particular the availability of cloned and expressed human enzymes and receptors and new knowledge of human polymorphisms, will make this task easier, as will similar knowledge of target species, including microorganisms causing human disease, and insects and weeds affecting the production of food and fiber.

High-throughput techniques will not only speed up the search in this area, as in other aspects of toxicology, bioinformatics, and systems biology will be necessary, not only for correlating the data from many sources but also for reducing it for practical applications.

29.9   SUMMARY AND CONCLUSIONS

Toxicology is undergoing a paradigm shift, from reductionist to systems biology, and is at the beginning of dramatic changes that will affect not only the screening of large numbers of chemicals for regulatory purposes but also fundamental studies of mechanisms of toxic action. However, some words of caution are appropriate. The reductionist approach is a critical necessity to uncover molecular mechanisms of toxicity of individual toxicants, and traditional screening tests will continue to be used for the foreseeable future. Indeed, the latter must be improved since, despite the potential of the new techniques to screen large numbers of potential toxicants quickly, they are in their infancy. Even when mature, they will undoubtedly suffer from the twin drawbacks of all toxicity tests, false positives and false negatives, the latter of which can have serious and far-reaching consequences. In the meantime, systems-based testing protocols will probably serve primarily to establish priorities for existing traditional protocols.

BIBLIOGRAPHY AND SUGGESTED READING

Anderson, M. E., R. S. Thomas, K. W. Gaido, et al. Dose-response modeling in reproductive toxicology in the systems biology era. *Reprod. Toxicol.* **19**:327–337, 2005.

Deighton, N. Metabolomics. In *Molecular and Biochemical Toxicology*, eds. R. C. Smart and E. Hodgson, pp. 67–79. Hoboken, NJ: Wiley, 2008.

Edwards, S. W. and R. J. Preston. Systems biology and mode of action based risk assessment. *Toxicol. Sci.* **106**:312–318, 2008.

Harrill, A. H., P. K. Ross, D. M. Gatti, et al. Population-based discovery of toxicogenomics biomarkers for hepatotoxicity using a laboratory strain diversity panel. *Toxicol. Sci.* **110**:235–243, 2009.

Merrick, B. A. Proteomics. In *Molecular and Biochemical Toxicology*, eds. R. C. Smart and E. Hodgson, pp. 41–66. Hoboken, NJ: Wiley, 2008.

National Research Council. Toxicity testing in the 21st century: A vision and a strategy. Washington, DC: National Research Council Committee on Toxicity Testing and Assessment of Environmental Agents, National Academy Press, 2007.

Olelsiak, M. F. Toxicogenomics. In *Molecular and Biochemical Toxicology*, eds. R. C. Smart and E. Hodgson, pp. 25–39. Hoboken, NJ: Wiley, 2008.

Plant, N. Can systems toxicology identify common biomarkers of non-genotoxic carcinogenesis? *Toxicology* **254**:164–169, 2008.

Smart, R. C. and E. Hodgson, eds. *Molecular and Biochemical Toxicology*. Hoboken, NJ: John Wiley and Sons, 2008.

Stone, E. A. and D. M. Nielsen. Bioinformatics. In *Molecular and Biochemical Toxicology*, eds. R. C. Smart and E. Hodgson, pp. 81–107. Hoboken, NJ: Wiley, 2008.

Waring, J. F., R. Ciurlionis, R. A. Jolly, et al. Microarray analysis of hepatotoxins in vitro reveals a correlation between gene expression profiles and mechanisms of toxicity. *Toxicol. Lett.* **120**:359–368, 2001.

Waters, M. D., K. Olden, and R. W. Tennant. Toxicogenomic approach for assessing toxicant-related disease. *Mutat. Res.* **544**:415–424, 2003.

Waters, M. D., J. K. Selkirk, and K. Olden. The impact of new technologies on human population studies. *Mutat. Res.* **544**:349–360, 2003.

SAMPLE QUESTIONS

1. Summarize the justification for the development of high-throughput techniques for toxicity testing that do not involve *in vivo* experiments on surrogate (experimental) animals. Which scientific subdisciplines will be involved in these techniques?

2. Give examples of Tier 1 tests being considered for regulation of endocrine-disrupting chemicals.

3. Define systems biology and summarize its potential role in human health risk assessment.

acceptable daily intake (ADI) Amount of exposure determined to be "safe"; usually derived from the lowest No Effect Level in an experimental study, divided by a safety factor such as 100. Also known as the Reference Dose (RfD).

acetylation The addition of an acetyl group from acetyl coenzyme A to a xenobiotic or xenobiotic metabolite by the enzyme N-acetyltransferase. Polymorphisms in this enzyme can be important in the expression of toxicity in humans.

acetylator phenotype Variation in the expression of N-acetyltransferase isoforms in humans gives rise to two subpopulations—fast and slow acetylators. Slow acetylators are more susceptible to the toxic effects of toxicants that are detoxified by acetylation.

acid deposition Wet and dry air pollutants that lower the pH of deposition and subsequently the pH of the environment. Acid rain with a pH of 4 or lower refers to the wet components. Normal rain has a pH of about 5.6. Sulfuric acid from sulfur and nitric acid from nitrogen oxides are the major contributors. In lakes in which the buffering capacity is low, the pH becomes acidic enough to cause fish kills, and the lakes cannot support fish populations. A contributing factor is the fact that acidic conditions concurrently release toxic metals, such as aluminum, into the water.

activation (bioactivation) In toxicology, this term is used to describe metabolic reactions of a xenobiotic in which the product is more toxic than is the substrate. Such reactions are most commonly monooxygenations, the products of which are electrophiles that, if not detoxified by Phase II (conjugation) reactions, may react with nucleophilic groups on cellular macromolecules such as proteins and DNA.

active oxygen Term used to describe various short-lived highly reactive intermediates in the reduction of oxygen. Active oxygen species such as superoxide anion and hydroxyl radical are known or believed to be involved in several toxic actions. Superoxide anion is detoxified by superoxide dismutase.

acute toxicity tests The most common tests for acute toxicity are the LC_{50} and LD_{50} tests, which are designed to measure mortality in response to an acute toxic insult. Other tests for acute toxicity include dermal irritation tests, dermal sensitization tests, eye irritation tests, photoallergy tests, and phototoxicity tests. *See also* eye irritation tests; LC_{50}; and LD_{50}.

acute toxicity Refers to adverse effects on, or mortality of, organisms following soon after a brief exposure to a chemical agent. Either a single exposure or multiple exposures within a short time period may be involved, and an acute

A Textbook of Modern Toxicology, *Fourth Edition*. Edited by Ernest Hodgson
Copyright © 2010 John Wiley & Sons, Inc.

effect is generally regarded as an effect that occurs within the first few days after exposure, usually less than 2 weeks.

adaptation to toxicants   Refers to the ability of an organism to show insensitivity or decreased sensitivity to a chemical that normally causes deleterious effects. The terms resistance and tolerance are closely related and have been used in several different ways. However, a consensus is emerging to use the term *resistance* to mean that situation in which a change in the genetic constitution of a population in response to the stressor chemical enables a greater number of individuals to resist the toxic action than were able to resist it in the previous unexposed population. Thus, an essential feature of resistance is selection and then inheritance by subsequent generations. In microorganisms, this frequently involves mutations and induction of enzymes by the toxicant; in higher organisms, it usually involves selection for genes already present in the population at low frequency. The term *tolerance* is then reserved for situations in which individual organisms acquire the ability to resist the effect of a toxicant, usually as a result of prior exposure.

Ah locus   A gene(s) controlling the trait of responsiveness for induction of enzymes by aromatic hydrocarbons. In addition to aromatic hydrocarbons such as the polycyclics, the chlorinated dibenzo-*p*-dioxins, dibenzofurans, and biphenyls, as well as the brominated biphenyls, are also involved. This trait, originally defined by studies of induction of hepatic aryl hydrocarbon hydroxylase activity following 3-methylcholanthrene treatment, is inherited by simple autosomal dominance in crosses and backcrosses between C57BL/6 (Ah-responsive) and DBA/2 (Ah-nonresponsive) mice.

Ah receptor (AhR)   A protein coded for by a gene of the Ah locus. The initial location of the Ah receptor is believed to be in the cytosol and, after binding to a ligand such as TCDD, is transported to the nucleus. Binding of aromatic hydrocarbons to the Ah receptor of mice is a prerequisite for the induction of many xenobiotic-metabolizing enzymes, as well as for two responses to TCDD: epidermal hyperplasia and thymic atrophy. Ah responsive mice have a high-affinity receptor, whereas the Ah-nonresponsive mice have a low-affinity receptor.

air pollution   In general, the principal air pollutants are carbon monoxide, oxides of nitrogen, oxides of sulfur, hydrocarbons, and particulates. The principal sources are transportation, industrial processes, electric power generation, and the heating of buildings. Hydrocarbons such as benzo(a)pyrene are produced by incomplete combustion and are associated primarily with the automobile. They are usually not present at levels high enough to cause direct toxic effects but are important in the formation of photochemical air pollution, formed as a result of interactions between oxides of nitrogen and hydrocarbons in the presence of ultraviolet light, giving rise to lung irritants such as peroxyacetyl nitrate, acrolein, and formaldehyde. Particulates are a heterogeneous group of particles, often seen as smoke, that are important as carriers of absorbed hydrocarbons and as irritants to the respiratory system.

alkylating agents   These are chemicals that can add alkyl groups to DNA, a reaction that can result either in mispairing of bases or in chromosome breaks. The mechanism of the reaction involves the formation of a reactive carbonium ion that combines with electron-rich bases in DNA. Thus, alkylating agents such as dimethylnitrosomine are frequently carcinogens and/or mutagens.

Ames test   An *in vitro* test for mutagenicity utilizing mutant strains of the bacterium *Salmonella typhimurium*, which is used as a preliminary screen of chemicals for assessing potential carcinogenicity. Several strains are available that cannot grow in the absence of histidine because of metabolic defects in histidine biosynthesis. Mutagens and presumed carcinogens can cause mutations that enable the strains to regain their ability to grow in a histidine-deficient medium. The test can be performed in the presence of the *S*-9 fraction from rat liver to allow the metabolic activation of promutagens. There is a high correlation between bacterial mutagenicity and carcinogenicity of chemicals.

antagonism   In toxicology, antagonism is usually defined as that situation in which the toxicity to two or more compounds administered together is less than that expected from consideration of their toxicities when administered alone. Although this includes lowered toxicity resulting from induction of detoxifying enzymes, this is frequently considered separately because of the time that must elapse between treatment with the inducer and subsequent treatment with the toxicant. Antagonism not involving induction is often at a marginal level of detection and is consequently difficult to explain. Such antagonism may involve competition for receptor sites or nonenzymatic combination of one toxicant with another to reduce the toxic effect. Physiological antagonism, in which two agonists act on the same physiological system but produce opposite effects, may also occur.

antibody   A large protein first expressed on the surface of the B cells of the immune system, followed by a series of events resulting in a clone of plasma cells that secrete the antibody into body fluids. Antibodies bind to the substance (generally a protein) that stimulated their production but may cross-react with related proteins. The natural function is to bind foreign substances such as microbes or microbial products but, because of their specificity, antibodies are used extensively in research and in diagnostic and therapeutic procedures.

antidote   A compound administered in order to reverse the harmful effect(s) of a toxicant. They may be toxic mechanism specific, as in the case of 2-pyridine aldoxime (2-PAM) and organophosphate poisoning, or nonspecific, as in the case of syrup of ipecac, used to induce vomiting and, thereby, elimination of toxicants from the stomach.

behavioral toxicity   Behavior may be defined as an organism's motor or glandular response to changes in its internal or external environment. Such changes may be simple or highly complex, innate or learned, but in any event represent one of the final integrated expressions of nervous system function. Behavioral toxicity is adverse or potentially adverse effects on such expression brought about by exogenous chemicals.

binding, covalent   *See* covalent binding.

bioaccumulation   The accumulation of a chemical either from the medium (usually water) directly or from consumption of food containing the chemical. Biomagnification is often used as a synonym for bioaccumulation, but is more correctly used to describe an increase in concentration of a chemical as it passes from organisms at one tropic level to organisms at higher tropic levels.

bioactivation   *See* activation.

bioassay   This term is used in two distinct ways. The first and most appropriate is the use of a living organism to measure the amount of a toxicant present in a sample or the toxicity of a sample. This is done by comparing the toxic effect of the sample with that of a graded series of concentrations of a known standard. The second and less appropriate meaning is the use of animals to investigate the toxic effects of chemicals as in chronic toxicity tests.

bioinformatics   In the narrow and original meaning, bioinformatics was the application of information technology to molecular biology. While this is still the most important aspect of bioinformatics, it is increasingly applied to other fields of biology, including molecular and other aspects of toxicology. It is characterized by computationally intensive methodology and includes the design of large databases and the development of techniques for their manipulation, including data mining.

burden of proof   Responsibility for determining whether a substance is safe or hazardous; a range of approaches can be seen when comparing laws. For example, for OSHA, regulators show a substance is hazardous before exposure is restricted, with the government conducting the tests; for FDA, manufacturers must show lack of hazard before marketing.

biomagnification   *See* bioaccumulation.

carcinogen   Any chemical or process involving chemicals that induces neoplasms that are not usually observed, the earlier induction of neoplasms than are commonly observed, and/or the induction of more neoplasms than are usually found.

carcinogen, epigenetic   Cancer-causing agents that exert their carcinogenic effect by mechanisms other than genetic, such as by immunosuppression, hormonal imbalance, or cytotoxicity. They may act as cocarcinogens or promoters. Epigenetic carcinogenesis is not as well understood a phenomenon as is genotoxic carcinogenesis.

carcinogen, genotoxic   Cancer-causing agents that exert their carcinogenic effect by a series of events that is initiated by an interaction with DNA, either directly or through an electrophilic metabolite.

carcinogen, proximate   *See* carcinogen, ultimate.

carcinogen, ultimate   Many, if not most, chemical carcinogens are not intrinsically carcinogenic but require metabolic activation to express their carcinogenic potential. The term procarcinogen describes the initial unreactive compound, the term proximate carcinogen describes its more active products, and the term ultimate carcinogen describes the product that is actually responsible for carcinogenesis by its interaction with DNA.

carcinogenesis   This is the process encompassing the conversion of normal cells to neoplastic cells and the further development of these neoplastic cells into a tumor. This process results from the action of specific chemicals, certain viruses, or radiation. Chemical carcinogens have been classified into those that are genotoxic and those that are epigenetic (i.e., not genotoxic).

chronic toxicity   This term is used to describe adverse effects manifested after a long time period of uptake of small quantities of the toxicant in question. The dose is small enough that no acute effects are manifested, and the time period is frequently a significant part of the expected normal lifetime of the organism. The most serious manifestation of chronic toxicity is carcinogenesis, but other types of chronic toxicity are also known (e.g., reproductive effects, behavioral effects).

chronic toxicity tests   Chronic tests are those conducted over a significant part of the life span of the test species or, in some cases, more than one generation. The most important tests are carcinogenicity tests, and the most common test species are rats and mice.

clinical toxicology   Clinical toxicology addresses the diagnosis, treatment, and prevention of chemical poisonings of humans as well as domestic and companion animals, and includes aspects of occupational and emergency medicine, poison control, and public health.

cocarcinogenesis   Cocarcinogenesis is the enhancement of the conversion of normal cells to neoplastic cells. This process is manifested by enhancement of carcinogenesis when the agent is administered either before or together with a carcinogen. Cocarcinogenesis should be distinguished from promotion as, in the latter case, the promoter must be administered after the initiating carcinogen.

comparative toxicology   The study of the variation in the expression of the toxicity of exogenous chemicals toward organisms of different taxonomic groups or of different genetic strains.

compartment   In pharmaco(toxico)kinetics, a compartment is a hypothetical volume of an animal system wherein a chemical acts homogeneously in transport and transformation. These compartments do not correspond to physiological or anatomic areas but are abstract mathematical entities useful for predicting drug or toxicant concentrations. Transport into, out of, or between compartments is described by rate constants, which are used in models of the intact animal.

conjugation reactions   *See* Phase II reactions.

covalent binding   This involves the covalent bond or "shared electron pair" bond. Each covalent bond consists of a pair of electrons shared between two atoms and occupying two stable orbitals, one of each atom. Although this is distinguished from the ionic bond or ionic valence, in fact, chemical bonds may show both covalent and ionic character. In toxicology, the term covalent binding is used less precisely to refer to the binding of toxicants or their reactive metabolites to endogenous molecules (usually macromolecules) to produce stable adducts resistant to rigorous extraction procedures. A covalent bond between ligand and macromolecule is generally assumed. Many forms of chronic toxicity involve covalent binding of the toxicant to DNA or protein molecules within the cell.

cross-resistance, cross-tolerance   These terms describe the situation in which either resistance or tolerance to a particular toxicant (as defined under adaptation to toxicants) is induced by exposure to a different toxicant. This is commonly seen in resistance of insects to insecticides in which selection with one insecticide brings about a broad spectrum of resistance to insecticides of the same or different chemical classes. Such cross-resistance is usually caused by the inheritance of a high level of nonspecific xenobiotic-metabolizing enzymes.

cytotoxicity   Cellular injury or death brought about by chemicals external to the cell. Such chemicals may be soluble mediators produced by the immune system, or they may be chemicals (toxicants) to which the organism has been exposed.

Delaney Amendment   *See* Food, Drug and Cosmetics Act.

detoxication   A metabolic reaction or sequence of reactions that reduces the potential for adverse effect of a xenobiotic. Such sequences normally involve an increase in water solubility that facilitates excretion and/or the reaction of a reactive product with an endogenous substrate (conjugation), thereby not only increasing water solubility but also reducing the possibility of interaction with cellular macromolecules. Not to be confused with detoxification. *See also* detoxification.

detoxification   Treatment by which toxicants are removed from intoxicated patients or a course of treatment during which dependence on alcohol or other drugs of abuse is reduced or eliminated. Not to be confused with detoxication. *See also* detoxication.

distribution   The term distribution refers both to the movement of a toxicant from the portal of entry to the tissue and also to the description of the different concentrations reached in different locations. The first involves the study of transport mechanisms primarily in the blood, and both are subject to mathematical analysis in toxicokinetic studies.

dosage   The amount of a toxicant, drug, or other chemical administered or taken expressed as some function of the organism (e.g., mg/kg body weight/day).

dose   The total amount of a toxicant, drug, or other chemical administered to or taken in by the organism.

dose–response relationship   In toxicology, the quantitative relationship between the amount of a toxicant administered or taken and the incidence or extent of the adverse effect.

dose–response assessment   A step in the risk-assessment process to characterize the relationship between the dose of a chemical administered to a population of test animals and the incidence of a given adverse effect. It involves mathematical modeling techniques to extrapolate from the high dose effects observed in test animals to estimate the effects expected from exposure to the typically low doses that may be encountered by humans.

Draize Test   *See* eye irritation test.

drugs of abuse   Although all drugs may have deleterious effects on humans, drugs of abuse either have no medicinal function or are taken at higher than therapeutic doses. Some drugs of abuse may affect only higher nervous functions (i.e., mood, reaction time, and coordination), but many produce physical dependence and have serious physical effects, with fatal overdose being a common occurrence. The drugs of abuse include central nervous system (CNS) depressants such as ethanol, methaqualone (Quaalude), and secobarbital; CNS stimulants such as cocaine, methamphetamine (speed), caffeine, and nicotine; opioids such as heroin and morphine; hallucinogens such as lysergic acid diethylamide (LSD), phencyclidine (PCP), and tetrahydrocannabinol (THC), the most important active principle of marijuana.

drugs, therapeutic   All therapeutic drugs can be toxic at some dose. The danger to the patient is dependent on the nature of the toxic response, the dose necessary to produce the toxic response, and the relationship between the therapeutic and the toxic dose. Drug toxicity is affected by all of those factors that affect the toxicity of xenobiotics, including (genetic) variation, diet, age, and the presence of other exogenous chemicals. The risk of toxic side effects from a particular drug must be weighed against the expected benefits; the use of a quite dangerous drug with only a narrow tolerance between the therapeutic and toxic doses might be justified if it is the sole treatment for an otherwise fatal disease. For example, cytotoxic agents used in the treatment of cancer are known carcinogens.

ecotoxicology   *See* environmental toxicology.

electron transport system (ETS)   This term is often restricted to the mitochondrial system, although it applies equally well to other systems, including that of microsomes and chloroplasts. The mitochondrial ETS (also termed respiratory chain or cytochrome chain) consists of a series of cytochromes and other electron carriers arranged in the inner mitochondrial membrane. These components transfer the electrons from NADH or $FADH_2$ generated in metabolic oxidations to oxygen, the final electron acceptor, through a series of alternate oxidations and reductions. The energy that these electrons lose during these transfers is used to pump H^+ from the matrix into the intermembrane space, creating an electrochemical proton gradient that drives oxidative phosphorylation. The energy is conserved as adenosine triphosphate (ATP).

electron transport system (ETS) inhibitors   The three major respiratory enzyme complexes of the mitochondrial electron transport system can all be blocked by inhibitors. For example, rotenone inhibits the NADH dehydrogenase complex, antimycin A inhibits the b-c complex, and cyanide and carbon monoxide inhibits the cytochrome oxidase complex. Although oxidative phosphorylation inhibitors prevent phosphorylation while allowing electron transfers to proceed, ETS inhibitors prevent both electron transport and ATP production.

electrophilic   Electrophiles are chemicals that are attracted to and react with electron-rich centers in other molecules in reactions known as electrophilic reactions. Many activation reactions produce electrophilic intermediates such as epoxides, which exert their toxic action by forming covalent bonds with nucleophilic centers in cellular macromolecules such as DNA or proteins.

endocrine disruptors   An endocrine disruptor is an exogenous chemical that interacts with the endocrine system of an organism to produce one or more deleterious effects. These effects may be brought about in a number of ways including serving as ligands for hormone receptors and inhibition or induction of hormone-metabolizing enzymes.

endoplasmic reticulum   The endoplasmic reticulum (ER) is an extensive branching and anastomosing double membrane distributed in the cytoplasm of eukaryotic cells. The ER is of two types: rough ER (RER) contains attached ribosomes on the cytosolic surface and smooth ER (SER) is devoid of ribosomes. Ribosomes are involved in protein biosynthesis, and RER is abundant in cells specialized for protein synthesis. Many xenobiotic-metabolizing enzymes are integral components of both SER and RER, such as the cytochrome P450-dependent monooxygenase system and the flavin-containing monooxygenase, although the specific content is usually higher in SER. When tissue or cells are disrupted by homogenization, the ER is fragmented into many smaller (c. 100 nm diameter) closed vesicles called microsomes, which can be isolated by differential centrifugation.

enterohepatic circulation   This term describes the excretion of a compound into the bile and its subsequent reabsorption from the small intestine and transport back to the liver, where it is available again for biliary excretion. The most important mechanism is conjugation in the liver, followed by excretion into the bile. In the small intestine, the conjugation product is hydrolyzed, either nonenzymatically or by the microflora, and the compound is reabsorbed to become a substrate for conjugation and re-excretion into the bile.

environmental toxicology   This is concerned with the movement of toxicants and their metabolites in the environment and in food chains and the effect of such toxicants on populations of organisms.

epigenetic carcinogen   *See* carcinogen, epigenetic.

exposure assessment   A component of risk assessment. The number of individuals likely to be exposed to a chemical in the environment or in the workplace is assessed, and the intensity, frequency, and duration of human exposure are estimated.

eye irritation test (Draize Test)   Eye irritation tests measure irritancy of compounds applied topically to the eye. These tests are variations of the Draize test, and the experimental animal used is the albino rabbit. The test consists of adding the material to be tested directly into the conjunctival sac of one eye of each of several albino rabbits, the other eye serving as the control. This test is probably the most controversial of all toxicity tests, being criticized primarily on the grounds that it is inhumane. Moreover, because both concentrations and volumes used are high and show high variability, it has been suggested that these tests cannot be extrapolated to humans. However, because visual impairment is a critical toxic end point, tests for ocular toxicity are essential. Attempts to solve the dilemma have taken two forms: to find substitute *in vitro* tests and to modify the Draize test so that it becomes not only more humane, but also more predictive for humans.

Federal Insecticide, Fungicide and Rodenticide Act (FIFRA)   This law is the basic U.S. law under which pesticides and other agricultural chemicals distributed in interstate commerce are registered and regulated. First enacted in 1947, FIFRA placed the regulation of agrochemicals under the control of the U.S. Department of Agriculture. In 1970, this responsibility was transferred to the newly created Environmental Protection Agency (EPA). Subsequently, FIFRA has been revised extensively by the Federal Environmental Pesticide Control Act (FEPCA) of 1972 and by the FIFRA amendments of 1975, 1978, and 1980. Under FIFRA, all new pesticide products used in the United States must be registered with the EPA. This requires the registrant to submit information on the composition, intended use, and efficacy of the product, along with a comprehensive database establishing that the material can be used without causing unreasonable adverse effects on humans or on the environment. The Food Quality Protection Act of 1996 is an amendment to FIFRA.

fetal alcohol syndrome (FAS)   FAS refers to a pattern of defects in children born to alcoholic mothers. Three criteria for FAS are prenatal or postnatal growth retardation; characteristic facial anomalies such as microcephaly, small eye opening, and thinned upper lip; and central nervous system dysfunction, such as mental retardation and developmental delays.

food additives   Chemicals may be added to food as preservatives (either antibacterial or antifungal compounds or antioxidants) to change the physical characteristics, for processing, or to change the taste or odor. Although most food additives are safe and are without chronic toxicity, many were introduced when toxicity testing was relatively unsophisticated and some have been shown subsequently to be toxic. The most important inorganic additives are nitrate and nitrite. Well-known examples of food additives include the antioxidant butylatedhydroxyanisole (BHA), fungistatic agents such as methyl *p*-benzoic acid, the emulsifier propylene glycol, sweeteners such as saccharin and aspartame, and dyes such as tartrazine and Sunset Yellow.

food contaminants (food pollutants)   Food contaminants, as opposed to food additives, are those compounds included inadvertently in foods that are raw, cooked, or processed. They include bacterial toxins such as the exotoxin of *Clostridium botulinum*, mycotoxins such as aflatoxins from *Aspergillus flavus*, plant alkaloids, animal toxins, pesticide residues, residues of animal food additives such as diethylstilbestrol (DES) and antibiotics, and a variety of industrial chemicals such as polychlorinated biphenyls (PCBs) and polybrominated biphenyls (PBBs).

Food, Drug and Cosmetics Act   The Federal Food, Drug and Cosmetic Act is administered by the Food and Drug Administration (FDA). It establishes limits for food additives, sets criteria for drug safety for both human and animal use, and requires proof of both safety and efficacy. This act contains the Delaney Amendment, which states that food additives that cause cancer in humans or animals at any level shall not be considered safe and are, therefore, prohibited. The Delaney Amendment has been modified to permit more flexible use of mechanistic and cost–benefit data. This law also empowers the FDA to establish and modify the "Generally Recognized as Safe" (GRAS) list and to establish Good Laboratory Practice (GLP) rules. As a result of the Food Quality Protection Act (1996), the Delaney amendment no longer applies to chemicals regulated under FIFRA.

forensic toxicology   Forensic toxicology is concerned with the medicolegal aspects of the adverse effects of chemicals on humans and animals. Although primarily devoted to the identification of the cause and circumstances of death and the legal issues arising therefrom, forensic toxicologists also deal with sublethal poisoning cases.

free radicals   Molecules that have unpaired electrons. Free radicals may be produced metabolically from xenobiotics and, because they are extremely reactive, may be involved in interactions with cellular macromolecules, giving rise to adverse effects. Examples include the trichloromethyl radical produced from carbon tetrachloride or the carbene radical produced by oxidation of the acetal carbon of methylenedioxyphenyl synergists.

genomics   The sometimes stated distinction that genomics deals with genomes while molecular biology deals with single genes is unrealistic and unnecessary; it is more appropriate to regard genomics as an aspect of molecular biology that deals not only with genomes and gene expression but also such important aspects as genetic polymorphisms, particularly single nucleotide polymorphisms (SNPs). Techniques, such as microarrays, are now available to examine simultaneously the expression of very large numbers of genes.

genotoxic carcinogen   *See* carcinogen, genotoxic.

genotoxicity   Genotoxicity is an adverse effect on the genetic material (DNA) of living cells that, on the replication of the cells, is expressed as a mutagenic or a carcinogenic event. Genotoxicity results from a reaction with DNA that can be measured either biochemically or in short-term tests with end points that reflect DNA damage.

Good Laboratory Practice (GLS)   In the United States, this is a code of laboratory procedures laid down under federal law and to be followed by laboratories undertaking toxicity tests, the results of which will be used for regulatory or legal purposes.

Generally Regarded as Safe (GRAS) list   *See* Food, Drug and Cosmetics Act.

hazard identification   Considered the first step in risk assessment, hazard identification involves the qualitative determination of whether exposure to a chemical causes an increased incidence of an adverse effect, such as cancer or birth defects, in a population of test animals and an evaluation of the relevance of this information to the potential for causing similar effects in humans.

hepatotoxicity   Hepatotoxicants are those chemicals causing adverse effects on the liver. The liver may be particularly susceptible to chemical injury because of its anatomic relationship to the most important portal of entry, the gastrointestinal (GI) tract, and its high concentration of xenobiotic-metabolizing enzymes. Many of these enzymes, particularly cytochrome P450, metabolize xenobiotics to produce reactive intermediates that can react with endogenous macromolecules such as proteins and DNA to produce adverse effects.

immunotoxicity   This term can be used in either of two ways. The first refers to toxic effects mediated by the immune system, such as dermal sensitivity reactions to compounds like 2,4-dinitrochlorobenzene. The second and currently most acceptable definition refers to toxic effects that impair the functioning of the immune system—for example, the ability of a toxicant to impair resistance to infection.

in vitro tests   Literally, these are tests conducted outside of the body of the organism. In toxicity testing, they would include studies using isolated enzymes, subcellular organelles, or cultured cells. Although technically the term would not include tests involving intact eukaryotes (e.g., the Ames test), it frequently is used by toxicologists to include all short-term tests for mutagenicity that are normally used as indictors of potential carcinogenicity.

***in vivo* tests**   Tests carried out on the intact organism, although the evaluation of the toxic end point almost always requires pathological or biochemical examination of the test organism's tissues. They may be acute, subchronic, or chronic. The best known are the lifetime carcinogenesis tests carried out on rodents.

induction   The process of causing a quantitative increase in an enzyme as a result of de novo protein synthesis following exposure to an inducing agent. This can occur either by a decrease in the degradation rate or an increase in the synthesis rate or both. Increasing the synthesis rate is the most common mechanism for induction by xenobiotics. Coordinate (pleiotypic) induction is the induction of multiple enzymes by a single-inducing agent. For example, phenobarbital can induce isoforms of both cytochrome P450 and glutathione *S*-transferase.

industrial toxicology   A specific area of environmental toxicology dealing with the work environment; it includes risk assessment, establishment of permissible levels of exposure, and worker protection.

inhibition   In its most general sense, inhibition means a restraining or a holding back. In biochemistry and biochemical toxicology, inhibition is a reduction in the rate of an enzymatic reaction, and an inhibitor is any compound causing such a reduction. Inhibition of enzymes important in normal metabolism is a significant mechanism of toxic action of xenobiotics, whereas inhibition of xenobiotic-metabolizing enzymes can have important consequences in the ultimate toxicity of their substrates. Inhibition is sometimes used in toxicology in a more general and rather ill-defined way to refer to the reduction of an overall process of toxicity, as in the inhibition of carcinogenesis by a particular chemical.

initiation   The initial step in the carcinogenic process involving the conversion of a normal cell to a neoplastic cell. Initiation is considered to be a rapid and essentially irreversible change involving the interaction of the ultimate carcinogen with DNA; this change primes the cell for subsequent neoplastic development via the promotion process.

intoxication   In the general sense, this term refers primarily to inebriation with ethyl alcohol, secondarily to excitement or delirium caused by other means, including other chemicals. In the clinical sense, it refers to poisoning or becoming poisoned. In toxicology, it is sometimes used as a synonym for activation—that is, the production of a more toxic metabolite from a less toxic parent compound. This latter use of intoxication is ambiguous and should be abandoned in favor of the aforementioned general meanings, and activation or bioactivation used instead.

isozymes (isoenzymes)   Isozymes, also known as isoforms, are multiple forms of a given enzyme that occur within a single species or even a single cell and that catalyze the same general reaction but are coded for by different genes. Different isozymes may occur at different life stages and/or in different organs and tissues, or they may coexist within the same cell. The first well-characterized isozymes were those of lactic dehydrogenase. Most xenobiotic-metabolizing enzymes exist in multiple isozymes, including cytochrome P450 and glucuronosyltransferase.

LC_{50} (median lethal concentration)   The concentration of a test chemical that, when a population of test organisms is exposed to it, is estimated to be fatal to 50% of the organisms under the stated conditions of the test. Normally used in lieu of the LD_{50} test in aquatic toxicology and inhalation toxicology.

LD_{50} (median lethal dose)   The quantity of a chemical compound that, when applied directly to test organisms, is estimated to be fatal to 50% of those organisms under the stated conditions of the test. The LD_{50} value is the standard for comparison of acute toxicity between toxicants and between species. Because the results of LD_{50} determinations may vary widely, it is important that both biological and physical conditions be narrowly defined (e.g., strain, gender, and age of test organism; time and route of exposure; environmental conditions). The value may be determined graphically from a plot of log dose against mortality expressed in probability units (probits) or, more recently, by using one of several computer programs available.

lethal synthesis   This term is used to describe the process by which a toxicant, similar in structure to an endogenous substrate, is incorporated into the same metabolic pathway as the endogenous substrate, ultimately being transformed into a toxic or lethal product. For example, fluoroacetate simulates acetate in intermediary metabolism, being transformed via the tricarboxylic acid cycle to fluorocitrate, which then inhibits aconitase, resulting in disruption of the TCA cycle and energy metabolism.

lipophilic   The physical property of chemical compounds that causes them to be soluble in nonpolar solvents (e.g., chloroform and benzene) and, generally, relatively insoluble in polar solvents such as water. This property is important toxicologically because lipophilic compounds tend to enter the body easily and to be excretable only when they have been rendered less lipophilic by metabolic action.

maximum tolerated dose (MTD)   The MTD has been defined for testing purposes by the U.S. environmental Protection Agency as the highest dose that causes no more than a 10% weight decrement, as compared to the appropriate control groups, and does not produce mortality, clinical signs of toxicity, or pathologic lesions (other than those that may be related to a neoplastic response) that would be predicted to shorten the animals' natural life span. It is an important concept in chronic toxicity testing; however, the relevance of results produced by such large doses has become a matter of controversy.

mechanism of action   *See* mode of action.

membranes   Membranes of tissues, cells, and cell organelles are all basically similar in structure. They appear to be bimolecular lipid leaflets with proteins embedded in the matrix and also arranged on the outer polar surfaces. This basic plan is present in spite of many variations, and it is important in toxicological studies of uptake of toxicants by passive diffusion and active transport.

metabolomics   Metabolomics is concerned with the profile of small molecules produced by the metabolic processes of an organism. Changes in the profile in response to chemical stress are of importance to both fundamental and applied toxicology.

microarray   Microarrays are based on the principle that any gene being expressed at any point in time is giving rise to a specific, corresponding mRNA. The microarray itself consists of spots of DNA (c. 200μ) bound to a suitable matrix. The mRNAs in the biological sample in question bind to the corresponding DNA and can be visualized by techniques involving dyes. Given the complexity of the data obtained (often thousands of genes are evaluated on a single microarray), special techniques have been developed for array scanning, data extraction, and statistical analysis.

microsomes   Microsomes are small closed vesicles (c.110 nm diameter) that represent membrane fragments formed form the endoplasmic reticulum when cells are disrupted by homogenization. Microsomes are separated from other cell organelles by differential centrifugation. The cell homogenate contains rough microsomes that are studded with ribosomes and are derived from rough endoplasmic reticulum, and smooth microsomes that are devoid of ribosomes and are derived from smooth endoplasmic reticulum. Microsomes are important preparations for studying the many processes carried out by the endoplasmic reticulum, such as protein biosynthesis and xenobiotic metabolism.

mode of action (mode of toxic action)   Terms used to describe the mechanism(s) that enables a toxicant to exert its toxic effect. The term(s) may be narrowly used to describe only those events at the site of action (perhaps better referred to as mechanism of action) or more broadly, to describe the sequence of events from uptake from the environment, through metabolism, distribution, and so on, up to and including events at the site of action.

monooxygenase (mixed-function oxidase)   An enzyme for which the cosubstrates are an organic compound and molecular oxygen. In reactions catalyzed by these enzymes, one atom of a molecule of oxygen is incorporated into the substrate whereas the other atom is reduced to water. Monooxygenases of importance in toxicology include cytochrome P450 and the flavin-containing monooxygenase, both of which initiate the metabolism of lipophilic xenobiotics by the introduction of a reactive polar group into the molecule. Such reactions may represent detoxication or may generate reactive intermediates of importance in toxic action. The term mixed-function oxidase is now considered obsolete and should not be used. The term multifunction oxidase was never widely adopted and also should not be used.

mutagenicity   Mutations are heritable changes produced in the genetic information stored in the DNA of living cells. Chemicals capable of causing such changes are known as mutagens and the process is known as mutagenesis.

mycotoxins   Toxins produced by fungi. Many, such as aflatoxins, are particularly important in toxicology.

nephrotoxicity   A pathologic state that can be induced by chemicals (nephrotoxicants) and in which the normal homeostatic functioning of the kidney is impaired. It is often associated with necrosis of the proximal tubule.

neurotoxicity   This is a general term referring to all toxic effects on the nervous system, including toxic effects measured as behavioral abnormalities. Because the nervous system is complex, both structurally and functionally, and has considerable functional reserve, the study of neurotoxicity is a many-faceted branch of toxicology. It involves electrophysiology, receptor function, pathology, behavior, and other aspects.

No Observed Effect Level (NOEL)   This is the highest dose level of a chemical that, in a given toxicity test, causes no observable effect in the test animals. The NOEL for a given chemical varies with the route and duration of exposure and the nature of the adverse effect (i.e., the indicator of toxicity). The NOEL for the most sensitive test species and the most sensitive indicator of toxicity is usually employed for regulatory purposes. Effects considered are usually adverse effects, and this value may be called the No Observed Adverse Effects level (NOAEL).

nuclear receptors   These are cellular proteins that bind steroid and other hormones as well as some toxicants. As part of a process involving other proteins as well as migration into the nucleus of the cell, nuclear receptors regulate gene expression. These genes may be involved in development, metabolism, or toxicity. The role of nuclear receptors in the induction of xenobiotic-metabolizing enzymes is of particular importance in metabolic interactions of toxicants.

Occupational Safety and Health Administration (OSHA) In the United States, OSHA is the government agency concerned with health and safety in the workplace. Through the administration of the Occupational Safety and Health Act, OSHA sets the standards for worker exposure to specific chemicals, for air concentration values, and for monitoring procedures. OSHA is also concerned with research (through the National Institute for Occupational Safety and Health [NIOSH]), information, education, and training in occupational safety and health.

oncogenes Oncogenes are genes that, when activated in cells, can transform the cells from normal to neoplastic. Sometimes, oncogenes are carried into normal cells by infecting viruses, particularly RNA viruses or retroviruses. In some cases, however, the oncogene is already present in the normal human cell, and it needs only a mutation or other activating event to change it from a harmless and possible essential gene, called *proto-oncogene*, into a cancer-producing gene. More than 30 oncogenes have been identified in humans.

oxidative phosphorylation The conservation of chemical energy extracted from fuel oxidations by the phosphorylation of adenosine diphosphate (ADP) by inorganic phosphate to form adenosine triphosphate (ATP) is accomplished in several ways. The majority of ATP is formed by respiratory chain-linked oxidative phosphorylation associated with the electron transport system in the mitochondrial inner membrane. The oxidations are tightly coupled to phosphorylations through a chemiosmotic mechanism in which H^+ are pumped across the inner mitochondrial membrane. Uncouplers of oxidative phosphorylation serve as H^+ ionophores to dissipate the H^+ gradient, and thus uncouple the phosphorylations from the oxidations.

oxidative stress Damage to cells and cellular constituents and processes by reactive oxygen species generated *in situ*. Oxidative stress may be involved in such toxic interactions as DNA damage, lipid peroxidation, and pulmonary and cardiac toxicity. Because of the transitory nature of most reactive oxygen species, although oxidative stress is often invoked as a mechanism of toxicity, rigorous proof may be lacking. (See also reactive oxygen species.)

partition coefficient This is a measure of the relative lipid solubility of a chemical and is determined by measuring the partitioning of the compound between an organic phase and an aqueous phase (e.g., octanol and water). The partition coefficient is important in studies of the uptake of toxicants because compounds with high coefficients (lipophilic compounds) are usually taken up more readily by organisms and tissues.

pharmacokinetics The study of the quantitative relationship between absorption, distribution, and excretion of chemicals and their metabolites. It involves derivation of rate constants for each of these processes and their integration into mathematical models that can predict the distribution of the chemical throughout the body compartments at any point in time after administration. Pharmacokinetic studies have been carried out most extensively in the case of clinical drugs. When applied specifically to toxicants, the term *toxicokinetics* is often used.

Phase I reactions These reactions introduce a reactive polar group into lipophilic xenobiotics. In most cases, this group becomes the site for conjugation during Phase II reactions. Such reactions include microsomal monooxygenations, cytosolic and mitochondrial oxidations, co-oxidations in the prostaglandin synthetase reaction, reductions, hydrolyses, and epoxide hydrolases. The products of Phase I reactions may be potent electrophiles that can be conjugated and detoxified in Phase II reactions or that may react with nucleophilic groups on cellular constituents, thereby causing toxicity.

Phase II reactions   Reactions involving the conjugation with endogenous substrates of Phase I products and other xenobiotics that contain functional groups such as hydroxyl, amino, carboxyl, epoxide, or halogen. The endogenous metabolites include sugars, amino acids, glutathione, and sulfate. The conjugation products, with rare exceptions, are more polar, less toxic, and more readily excreted than are their parent compounds. There are two general types of conjugations: type I (e.g., glycoside and sulfate formation), in which an activated conjugating agent combines with substrate to yield the conjugated product; and type II (e.g., amino acid conjugation), in which the substrate is activated and then combines with an amino acid to yield a conjugated product.

poison (toxicant)   A poison (toxicant) is any substance that causes a harmful effect when administered to a living organism. Due to a popular connotation that poisons are, by definition, fatal in their effects and that their administration is usually involved with attempted homicide or suicide, most toxicologists prefer the less prejudicial term toxicant. Poison is a quantitative concept. Almost any substance is harmful at some dose and, at the same time, is harmless at a very low dose. There is a range of possible effects, from subtle long-term chronic toxicity to immediate lethality.

pollution   This is contamination of soil, water, food, or the atmosphere by the discharge or admixture of noxious materials. A pollutant is any chemical or substance contaminating the environment and contributing to pollution.

portals of entry   The sites at which xenobiotics enter the body. They include the skin, the gastrointestinal (GI) tract, and the respiratory system.

potentiation   *See* synergism and potentiation.

procarcinogen   *See* carcinogen, ultimate.

promotion   The facilitation of the growth and development of neoplastic cells into a tumor. This process is manifested by enhancement of carcinogenesis when the agent is given after a carcinogen.

proteomics   Deals with the protein complement of organisms, the entire complement being known as the proteome. Thus, while genomics is concerned with gene expression, proteomics examines the products of the expressed genes.

pulmonary toxicity   This term refers to the effects of compounds that exert their toxic effects on the respiratory system, primarily the lungs.

quantitative structure activity relationships (QSAR)   The relationship between the physical and/or chemical properties of chemicals and their ability to cause a particular effect, enter into particular reactions, and so on. The goal of QSAR studies in toxicology is to develop procedures whereby the toxicity of a compound can be predicted from its chemical structure by analogy with the properties of other toxicants of known structure and toxic properties.

reactive intermediates (reactive metabolites)   Chemical compounds, produced during the metabolism of xenobiotics that are more chemically reactive than is the parent compound. Although they are susceptible to detoxication by conjugation reactions, these metabolites, as a consequence of their increased reactivity, have a greater potential for adverse effects than does the parent compound. A well-known example is the metabolism of benzo(a)pyrene to its carcinogenic dihydrodiol epoxide derivative as a result of metabolism by cytochrome P450 and epoxide hydrolase. Reactive intermediates involved in toxic effects include epoxides, quinones, free radicals, reactive oxygen species, and a small number of unstable conjugation products.

reactive oxygen species   Molecular oxygen normally exists in a relatively unreactive triplet state ($3O_2$). However, reactive species such as superoxide anion, hydrogen peroxide, singlet oxygen, and the highly reactive hydroxyl radical are also known. Reactive oxygen species are formed *in vivo*, either during, or as a consequence of, aerobic metabolism. There is a great deal of evidence that these reactive oxygen species are linked to a number of toxic end points, and this phenomenon is known as oxidative stress.

Reference Dose (RfD)   See Acceptable Daily Intake (ADI).

resistance   See adaptation to toxicants.

Resource Conservation and Recovery Act (RCRA)   Administered by the EPA, the RCRA is the most important act governing the disposal of hazardous wastes in the United States; it promulgates standards for identification of hazardous wastes, their transportation, and their disposal. Included in the last are siting and construction criteria for landfills and other disposal facilities as well as the regulation of owners and operators of such facilities.

risk analysis   This term includes risk assessment (below) together with consideration of risk communication and risk management.

risk assessment   The process by which the potential adverse health effects of human exposure to chemicals are characterized; it includes the development of both qualitative and quantitative expression of risk. The process of risk assessment may be divided into four major components: hazard identification, dose–response assessment (high-dose to low-dose extrapolation), exposure assessment, and risk characterization.

risk, toxicologic   The probability that some adverse effect will result from a given exposure to a chemical is known as the risk. It is the estimated frequency of occurrence of an event in a population and may be expressed in absolute terms (e.g., 1 in 1 million) or in terms of relative risk (i.e., the ratio of the risk in question to that in an equivalent unexposed population).

safety factor (uncertainty factor)   A number by which the no observed effect level (NOEL) is divided to derive the reference dose (RfD), the reference concentration (RfC), or minimum risk level (MRL) of a chemical from experimental data. The safety factor is intended to account for the uncertainties inherent in estimating the potential effects of a chemical on humans from results obtained with test species. The safety factor allows for possible differences in sensitivity between the test species and humans, as well as for variations in the sensitivity within the human population. The size of safety factor (e.g., 100–1000) varies with confidence in the database and the nature of the adverse effects. Small safety factors indicate a high degree of confidence in the data, an extensive database, and/or the availability of human data. Large safety factors are indicative of an inadequate and uncertain database and/or the severity of the unexpected toxic effect.

selectivity (selective toxicity)   A characteristic of the relationship between toxic chemicals and living organisms whereby a particular chemical may be highly toxic to one species but relatively innocuous to another. The search for and study of selective toxicants is an important aspect of comparative toxicology because chemicals toxic to target species but innocuous to nontarget species are extremely valuable in agriculture and medicine. The mechanisms involved vary from differential penetration rates through different metabolic pathways to differences in receptor molecules at the site of toxic action.

solvents   In toxicology, this term usually refers to industrial solvents. These belong to many different chemical classes, and a number of these are known to cause problems of toxicity to humans. They include aliphatic hydrocarbons (e.g., hexane), halogenated aliphatic hydrogens (e.g., methylene chloride), aliphatic alcohols (e.g., methanol), glycols and glycol ethers (e.g., propylene and propylene glycol), and aromatic hydrocarbons (e.g., toluene).

subchronic toxicity   Toxicity due to chronic exposure to quantities of a toxicant that do not cause any evident acute toxicity for a time period that is extended but is not so long as to constitute a significant part of the life span or the species in question. In subchronic toxicity tests using mammals, a 30–90 day period is considered appropriate.

synergism and potentiation   The terms synergism and potentiation have been variously used and defined but in any case involve a toxicity that is greater when two compounds are given simultaneously or sequentially than would be expected from a consideration of the toxicities of the compounds given alone. In an attempt to make the use of these terms uniform, it is a suggested that, insofar as toxic effects are concerned, they be used as defined as follows: both involve toxicity greater than would be expected from the toxicities of the compounds administered separately, but in the case of synergism, one compound has little or no intrinsic toxicity administered alone, whereas in the case of potentiation, both compounds have appreciable toxicity when administered alone.

systems biology   Although systems biology has been defined in a number of ways, some involving quite simple approaches to limited problems, in the currently most commonly accepted sense, it is an integrative approach to biological structure and function. In large part, biology has been reductionist throughout its history, studying organs as components of organisms, cells as components of organs, enzymes, nucleic acids, and so on, as components of cells, with the goal of describing function at the molecular level. Systems biology, on the other hand, is holistic and has the objective of discerning interactions between components of biological systems and describing these interactions in rigorous mathematical models. Furthermore, the proponents of systems biology aim to integrate these models at higher and higher levels or organization in order to develop an integrated model of the entire organism.

teratogenesis   This term refers to the production of defects in the reproduction process resulting in either reduced productivity due to fetal or embryonic mortality or the birth of offspring with physical, mental, behavioral, or developmental defects. Compounds causing such defects are known as teratogens.

therapy   Poisoning therapy may be nonspecific or specific. Nonspecific therapy is treatment for poisoning that is not related to the mode of action of the particular toxicant. It is designed to prevent further uptake of the toxicant and to maintain vital signs. Specific therapy, however, is therapy related to the mode of action of the toxicant and not simply to the maintenance of vital signs by treatment of symptoms. Specific therapy may be based on activation and detoxication reactions, on mode of action, or on elimination of the toxicant. In some cases, more than one antidote, with different modes of action, are available for the same toxicant.

threshold dose   The dose of a toxicant below which no adverse effect occurs. The existence of such a threshold is based on the fundamental tenet of toxicology that, for any chemical, there exists a range of doses over which the severity of the observed effect is directly related to the dose, the threshold level representing the lower limit of this dose range. Although practical thresholds are considered to exist for most adverse effects, for regulatory purposes it is assumed that there is no threshold dose for carcinogens.

threshold limit value (TLV)   Upper permissive limit of airborne concentrations of substances. They represent conditions under which it is believed that nearly all workers may be exposed repeatedly, day after day, without adverse effect. Threshold limits are based on the best available information from industrial experience, from experimental human and animal studies, and when possible, from a combination of the three.

threshold limit value–ceiling (TLV-C)   This is the concentration that should not be exceeded even momentarily. For some substances (e.g., irritant gases), only one TLV category, the TLV-C, may be relevant. For other substances, two or three TLV categories may need to be considered.

threshold limit value–short-term exposure limit (TLV-STEL)   This is the maximal concentration to which workers can be exposed for a period up to 15 min. continuously without suffering from (1) irritation, (2) chronic or irreversible tissue change, or (3) narcosis of sufficient degree to increase accident proneness.

threshold limit value–time-weighted average (TLV-TWA)   This is the TWA concentration for a normal 8-hr workday or 40-hr workweek to which nearly all workers may be exposed repeatedly day after day, without adverse effect. Time-weighted averages allow certain permissible excursions above the limit, provided they are compensated by equivalent excursions below the limit during the workday. In some instances, the average concentration is calculated for a workweek rather than for a workday.

tolerance   *See* adaptation to toxicants.

Toxic Substances Control Act (TSCA)   Enacted in 1976, the TSCA provides the EPA with the authority to require testing and to regulate chemicals, both old and new, entering the environment. It was intended to supplement sections of the Clean Air Act, the Clean Water Act, and the Occupational Safety and Health Act that already provide for regulation of chemicals. Manufacturers are required to submit information to allow the EPA to identify and evaluate the potential hazards of a chemical prior to its introduction into commerce. The act also provides for the regulation of production, use, distribution, and disposal of chemicals.

toxicant   *See* poison.

toxicogenomics   Those aspects of genomics of relevance to toxicology (see genomics).

toxicokinetics   *See* pharmacokinetics.

toxicology   Toxicology is defined as that branch of science that deals with poisons (toxicants) and their effects; a poison is defined as any substance that causes a harmful effect when administered, either by accident or design, to a living organism. There are difficulties in bringing a more precise definition to the meaning of poison and in the definition and measurement of toxic effect. The range of deleterious effects is wide and varies with species, gender, developmental stage, and so on, while the effects of toxicants are always dose dependent.

toxin   A *toxin* is a toxicant produced by a living organism. Toxin should never be used as a synonym for toxicant.

transport   In toxicology, this term refers to the mechanisms that bring about movement of toxicants and their metabolites from one site in the organism to another. *Transport* usually involves binding to either blood albumins or blood lipoproteins.

ultimate carcinogen   *See* carcinogen, ultimate.

venom   A venom is a toxin produced by an animal specifically for the poisoning of other species via a mechanism designed to deliver the toxin to its prey. Examples include the venom of bees and wasps, delivered by a sting, and the venom of snakes, delivered by fangs.

water pollution   Water pollution is of concern in both industrialized and nonindustrialized nations. Chemical contamination is more common in industrialized nations, whereas microbial contamination is more important in nonindustrialized areas. Surface water contamination has been the primary cause for concern but, since the discovery of agricultural and industrial chemicals in groundwater, contamination of water from this source is also a problem. Water pollution may arise from runoff of agricultural chemicals, from sewage or from specific industrial sources. Agricultural chemicals found in water include insecticides, herbicides, fungicides, and nematocides; fertilizers, although less of a toxic hazard, contribute to such environmental problems as eutrophication. Other chemicals of concern include low molecular-weight halogenated hydrocarbons such as chloroform, dichloroethane, and carbon tetrachloride; polychlorinated biphenyls (PCBs); chlorophenols; 2,3,7,8-tetrachlorodibenzo-*p*-dioxin (TCDD); phthalate ester plasticizers; detergents; and a number of toxic inorganics.

xenobiotic   A general term used to describe any chemical interacting with an organism that does not occur in the normal metabolic pathways of that organism. The use of this term in lieu of "foreign compound," among others, has gained wide acceptance.

A Textbook of Modern Toxicology, Fourth Edition. Edited by Ernest Hodgson
Copyright © 2010 John Wiley & Sons, Inc.